The Evolution of Hominin Diets

Vertebrate Paleobiology and Paleoanthropology Series

Edited by

Eric Delson
Vertebrate Paleontology, American Museum of Natural History,
New York, NY 10024, USA
delson@amnh.org

Ross D. E. MacPhee
Vertebrate Zoology, American Museum of Natural History,
New York, NY 10024, USA
macphee@amnh.org

Focal topics for volumes in the series will include systematic paleontology of all vertebrates (from agnathans to humans), phylogeny reconstruction, functional morphology, paleolithic archaeology, taphonomy, geochronology, historical biogeography, and biostratigraphy. Other fields (e.g., paleoclimatology, paleoecology, ancient DNA, total organismal community structure) may be considered if the volume theme emphasizes paleobiology (or archaeology). Fields such as modeling of physical processes, genetic methodology, nonvertebrates or neontology are out of our scope.

Volumes in the series may either be monographic treatments (including unpublished but fully revised dissertations) or edited collections, especially those focusing on problem-oriented issues, with multidisciplinary coverage where possible.

Editorial Advisory Board

For other titles published in this series, go to
www.springer.com/series/6978

A volume in the

Max Planck Institute Subseries in Human Evolution

Coordinated by

Jean-Jacques Hublin
Max Planck Institute for Evolutionary Anthropology, Department of Human Evolution, Leipzig, Germany

The Evolution of Hominin Diets

Integrating Approaches to the Study of Palaeolithic Subsistence

Edited by

Jean-Jacques Hublin
Max Planck Institute for Evolutionary Anthropology,
Leipzig, Germany

and

Michael P. Richards
Max Planck Institute for Evolutionary Anthropology,
Leipzig, Germany

Springer

Editors
Jean-Jacques Hublin
Max Planck Institute for Evolutionary Anthropology
Leipzig, Germany

Michael P. Richards
Max Planck Institute for Evolutionary Anthropology
Leipzig, Germany

ISBN 978-1-4020-9698-3 e-ISBN 978-1-4020-9699-0

Library of Congress Control Number: 2009927462

Cover illustrations:
A three-dimensional photosimulation produced from a point cloud of an area of the occlusal surface of a human molar. The area was scanned at 100 times magnification using a Sensofar μ Confocal Imaging Profiler at the Paleoanthropology Laboratory, University of Arkansas at Fayetteville.

Microgravette point from the Gravettian site Grub/Kranawetberg (Austria) representing possible hafted tip of hunting weapon, illustration by: Walpurga Antl-Weiser (Naturhistorisches Museum, Vienna)

Printed on acid-free paper

springer.com

Contents

List of Contributors

Daniel S. Adler
University of Connecticut
Department of Anthropology, U-2176
Storrs, CT 06269, USA
daniel.adler@uconn.edu

Zeresenay Alemseged
Department of Anthropology
California Academy of Sciences
55 Concourse Dr.
San Francisco, CA 94118, USA
zeray@calacademy.org

Kai Yang Ang
Department of Biological Sciences National
University of Singapore
14 Science Drive 4
Singapore 117543
Republic of Singapore

Guy Bar-Oz
University of Haifa
Zinman Institute of Archaeology
Mount Carmel, 31905 Haifa, Israel
guybar@research.haifa.ac.il

René Bobe
Department of Anthropology
The University of Georgia
250 A Baldwin Hall, Jackson St.
Athens, GA 30602, USA
renebobe@uga.edu

Hervé Bocherens
Institute für Geowissenschaften, Biogeologie
Universität Tübingen, Sigwartstrasse 10
D-72076 Tübingen, Germany
herve.bocherens@uni-tuebingen.de

Steven E. Churchill
Department of Evolutionary Anthropology
and Anatomy
Box 90383
Duke University
Durham, NC 27708-0383, USA
churchy@duke.edu

Darna L. Dufour
Department of Anthropology
University of Colorado at Boulder
Boulder, CO 80309, USA
Darna.Dufour@Colorado.edu

Sabine Gaudzinski-Windheuser
Palaeolithic Research Unit
Römisch-Germanisches Zentralmuseum
Schloss Monrepos
56567 Neuwied, Germany
gaudzinski@rgzm.de

John F. Hoffecker
Institute of Arctic and Alpine Research
University of Colorado at Boulder
Boulder, Colorado, USA
John.Hoffecker@colorado.edu

Gottfried Hohmann
Max-Planck-Institute for Evolutionary Anthropology
Deutscher Platz 6
04103 Leipzig, Germany
hohmann@eva.mpg.de

Martin Jones
Department of Archaeology
University of Cambridge, UK
Cambridge CB2 3DZ, UK
mkj12@cam.ac.uk

Sheau Horng King
Department of Biological Sciences
National University of Singapore
14 Science Drive 4
Singapore 117543
Republic of Singapore

Richard G. Klein
Program in Human Biology
Stanford University
Stanford, CA 94305, USA
rklein@stanford.edu

Steven L. Kuhn
Department of Anthropology
P.O. Box 210030
University of Arizona
Tucson, AZ 85721-0030, USA
skuhn@email.arizona.edu

Michel Lenoir
Institut de Préhistoire et Géologie du Quaternaire
PACEA-UMR 5199 du C.N.R.S.
Université Bordeaux 1
33405 Talence, France
m.lenoir@ipgq.u-bordeaux1.fr

William R. Leonard
Department of Anthropology
Northwestern University
1810 Hinman Ave.
Evanston, IL 60208, USA
w-leonard1@northwestern.edu

Staffan Lindeberg
Department of Clinical Sciences
Lund University
SE-221 85 Lund, Sweden
staffan.lindeberg@med.lu.se

Peter W. Lucas
Department of Anthropology
George Washington University
2110 G St NW
Washington DC, USA
pwlucas@gwu.edu

Kathy MacDonald
Faculty of Archeology
Leiden University
P.O. Box 9515
2300 RA Leiden, The Netherlands
k.macdonald@arch.leidenuniv.nl

Natalie D. Munro
University of Connecticut
Department of Anthropology,
Unit 2176 354 Mansfield Road
Storrs, CT, USA
Natalie.Munro@uconn.edu

Laura Niven
Palaeolithic Research Unit
Römisch-Germanisches Zentralmuseum
Schloss Monrepos
56567 Neuwied, Germany;
Max Planck Institute for
Evolutionary Anthropology
Department of Human Evolution
Deutscher Platz 6
04103 Leipzig, Germany
laura.niven@eva.mpg.de

Neeraja Peri
Department of Anthropology
George Washington University
2110 G St NW
Washington DC, USA

Jill A. Rhodes
Department of Anthropology
Bryn Mawr College
101 N. Merion Avenue
Bryn Mawr, PA 19010 USA
jarhodes@brynmawr.edu

Mike P. Richards
Department of Human Evolution
Max Planck Institute for
Evolutionary Anthropology
Deutscher Platz 6
D 04103 Leipzig, Germany
Richards@eva.mpg.de

Marcia L. Robertson
Department of Anthropology
Northwestern University
1810 Hinman Ave.
Evanston, IL 60208, USA

Wil Roebroeks
Faculty of Archeology
Leiden University
P.O. Box 9515
2300 RA Leiden, The Netherlands
w.roebroeks@arch.leidenuniv.nl

Brooke Sadler
Department of Anthropology
George Washington University
2110 G St NW
Washington DC, USA

Margaret J. Schoeninger
Department of Anthropology
University of California at San Diego
La Jolla, CA 92093-0532, USA
mjschoen@ucsd.edu

John J. Shea
Anthropology Department
Stony Brook University
Stony Brook, NY 11794-4364, USA
John.Shea@sunysb.edu

J. Josh Snodgrass
Department of Anthropology
University of Oregon
1218 University of Oregon
Eugene, OR 97403, USA
jjosh@uoregon.edu

Matt Sponheimer
Department of Anthropology
University of Colorado at Boulder
Boulder, CO 80309, USA
msponheimer@yahoo.com

Teresa E. Steele
Department of Human Evolution
Max Planck Institute for Evolutionary Anthropology
Deutscher Platz 6
D 04103 Leipzig, Germany;
Department of Anthropology
University of California – Davis
One Shields Avenue
Davis, CA 95616, USA
testeele@ucdavis.edu

Mary C. Stiner
Department of Anthropology
P.O. Box 210030
University of Arizona
Tucson, AZ 85721-0030, USA
mstiner@email.arizona.edu

Zhongquan Sui
Department of Botany
University of Hong Kong
Pokfulam Road, Hong Kong

Hugh Tiang Wah Tan
Department of Biological Sciences
National University of Singapore
14 Science Drive 4, Singapore 117543
Republic of Singapore

Alexander Verpoorte
Faculty of Archeology
Leiden University
P.O. Box 9515
2300 RA Leiden, The Netherlands
a.verpoorte@arch.leidenuniv.nl

Paola Villa
University of Colorado Museum
UCB 265
Boulder, Colorado 80309-0265, USA
villap@colorado.edu

Preface

Michael P. Richards and Jean-Jacques Hublin

The study of hominin diets, and especially how they have evolved throughout time, has long been a core research area in archaeology and paleoanthropology, but it is also becoming an important research area in other fields such as primatology, nutrition science, and evolutionary medicine. Although this is a fundamental research topic, much of the research continues to be undertaken by specialists and there is, with some notable exceptions (e.g., Stanford and Bunn, 2001; Ungar and Teaford, 2002; Ungar, 2007) relatively little interaction with other researchers in other fields. This is unfortunate, as recently it has appeared that different lines of evidence are causing similar conclusions about the major issues of hominid dietary evolution (i.e., the recognition of the important role of meat eating in brain evolution in early Homo, as well as the subsistence strategies of Neanderthals). However, multidisciplinary or integrated, approaches to the study of hominid diets remain rare. Therefore, we wanted to address this issue through a symposium we organized at the Department of Human Evolution in the Max-Planck Institute for Evolutionary Anthropology in Leipzig from May 17th to 21st 2006.

The symposium had two main goals. The first was to bring together key researchers to provide a current state-of-the-art report of their research area. The second and main goal of the symposium was to bring together researchers who may not normally meet, as they work in different regions, on time periods, and with different analytical tools. With this meeting we aimed to address three main issues of dietary evolution:

(a) Meat eating: when did it start and how did it intensify?
(b) Hunting technologies: What is the first evidence of hunting, and how did it develop over time?
(c) Resource intensification: When did this first occur, and how do the species chosen for intensification differ over time and between regions (i.e., use of marine/aquatic resources, small mammals).

To do this, we invited participants from different fields who all had researched these topics, or some aspects of them. These four general research areas were (1) modern studies (primates, modern humans), (2) faunal and plant studies, (3) archaeology and paleoanthropology, and (4) isotopic studies.

This volume therefore presents research articles by most of these participants that are mainly based on their presentations at the symposium. As can hopefully be seen in the volume, these papers provide important reviews of the current research in these areas, as well as often present new research on dietary evolution.

In the section on modern studies Hohmann provides a review of the diets of non-human primates, including an interesting discussion of the role of food-sharing amongst these primates. Snodgrass, Leonard, and Roberston provide a review of the evidence for the change in brain size of the earliest hominids and how this is linked with diet quality and body composition. Lucas presents an informative look at the mechanics of food processing in the mouth and how this may have changed over time related to differing foods used by different hominins. Finally, Lindeberg presents the evidence for hominin dietary evolution based on the studies of modern human nutrition and health, arguing that many modern diet-related diseases are linked with a mismatch between the diets we should have based on our dietary evolution (lean animal protein and simple carbohydrates) the diets most people have (fatty animal foods and complex carbohydrates).

In the faunal and floral research section, Villa and Lenoir provides a summary of the faunal evidence for diets in Lower and Middle Paleolithic Europe. Comparisons between Middle Paleolithic (or Middle Stone age) and Late Paleolithic (or Late Stone Age) assemblages are proposed for different geographical areas. Hoffecker explores Neanderthals and modern human diets in the Late Pleistocene environments of Eastern Europe. Gaudzinzki-Windhauser and Niven provide a review of the faunal evidence for subsistence in the Middle Paleolithic and Upper Paleolithic of Europe, focusing on reindeer exploitation. Steele and Klein review the available zooarcheological data documenting the "transition" from the Middle to the Late Stone Age in South Africa. They examine how subsistence changes relate to demographic changes and models of modern human origin. Adler and Bar-Oz conduct a comparison between Neanderthal and the first modern human game exploitation in Southern Caucasus and conclude they occupied the same ecological niche with clear consequences regarding mutual exclusion. Munro presents a synthesis of the evidence for changes in dietary adaptations towards smaller game and intensification in South-West Asia in the later Paleolithic, and Stiner and Kuhn move beyond discussion of the faunal evidence directly to discuss the implications of this work and especially division of labor between males and females in Paleolithic Mediterranean Eurasia. Jones provides a useful review of the (limited) evidence for the exploitation of plant foods in the European Paleolithic.

The archaeological and paleoanthropological data also contribute to a better understanding of food acquisition by ancient hominins. Alemseged and Bobe use a paleoenvironmental approach to infer possible dietary adaptation of two hominin genera: *Paranthropus* and *Homo*, particularly in the way they could exploit fallback resource. Shea analyzes the emergence of long range projectile technology likely in Africa at approximately 50–100 Ka and its subsequent spread into Eurasia. Churchill and Rhodes' review of the anatomical features related to the development throwing bring support to the claim that projectile weapons arose in the African later MSA and moved into Europe in the hands of modern humans. However Villa and Lenoir also provide possible evidence for the use of stone-tipped spears by Neanderthals in Western Europe. MacDonald, Roebroeks, and Verpoorte examine the Neanderthal archaeological record and how energetic issues in the use of space may in part explain it.

Finally, in the isotopic studies section Schoeninger provides a review of the small number of studies on the isotopic evidence for the diets of living non-human primates and critically relates it to the isotopic evidence for the diets of early hominins. Sponheimer and Dufour provide a critical assessment of the isotopic evidence for diets of early hominins with a focus on whether there is evidence for increasing dietary breadth in australopithecines and early Homo. Bocherens presents a selection of the isotopic data for European Neanderthals, emphasizing the importance of large herbivores in Neanderthal diets. Finally, Richards provides a summary of the isotopic evidence for Upper Paleolithic humans in Eurasia that also shows the dietary importance of animal foods, including small game such as aquatic foods.

This wide range of papers will hopefully be of interest to researchers who work in dietary evolution to provide a current account of the field, and will provide an introduction into the research that is being undertaken into this topic in fields that are not the reader's own. The research on dietary evolution continues, and hopefully through further interdisciplinary forums such as this we can come closer to knowing how hominin diets changed and evolved over time, as well as learning how to use this knowledge to help us understand the origin and implications of the range of diets of modern humans living in different parts of the world today.

Acknowledgments. We are grateful to the Max Planck Society for funding the workshop, as well as all of the participants for agreeing to take part in this, which we hope they found as interesting and enjoyable as we did. We are especially grateful to Diana Carstens and Silke Streiber as well as to Michelle Hänel, Jörn Scheller, and Annette Weiske for assistance in organizing the workshop. Allison Cleveland and Stefanie Altman played an essential role in the preparation of this volume.

References

Stanford, C.B., Bunn, H.T. (Eds.), 2001. Meat-Eating and Human Evolution. Oxford University Press, Oxford.

Ungar, P. (Ed.), 2007. Evolution of the Human Diet: The Known, the Unknown, and the Unknowable. Oxford University Press, Oxford.

Ungar, P.S., Teaford, M.F. (Eds.), 2002. Human Diet: Its Origin and Evolution. Bergin and Garvey, Westport, CT.

1. The Diets of Non-human Primates: Frugivory, Food Processing, and Food Sharing

Gottfried Hohmann
Max Planck Institute for Evolutionary Anthropology
Deutscher Platz 6
D-04103 Leipzig, Germany
hohmann@eva.mpg.de

Keywords Primates • diet composition • frugivory • provisioning • meat eating

Abstract Most nonhuman primates have a mixed diet that consists of a wide spectrum of plant foods and a relatively small spectrum of animal foods. Patterns of food selection shift in relation to seasonal changes in food availability. Food habits may also vary within and between groups of the same species for other reasons, such as inter-specific competition and local traditions. Primates practice various forms of food processing and by doing so modify the physical structure of food. Food processing is likely to affect food intake rate, passage time, and nutrient absorption. The diets of Hominids (*Pan*, *Gorilla*, and *Pongo*) are dominated by plant foods. In some species, food sharing is habitual. Food may be transferred between mature individuals (e.g., chimpanzee, bonobo) or from mature individuals to immatures (e.g., bonobo). Consumption of animal food is widespread, but may vary considerably between species. Except for bonobos, insectivory is common among Pongids, and except for gorillas and orangutans, consumption of vertebrate meat is habitual. Hunting and meat eating occurs equally in forests and in open habitats. In forest populations, hunting increases at times when plant food is highly abundant, suggesting that hunting activities are triggered by high energy intake from plant foods. Patterns of meat sharing by chimpanzees and bonobos differ: in chimpanzees, the majority of the prey is consumed by adult males and adult females. In bonobos, adult females control access to carcasses and share meat among each other and with immatures. Information from bonobos and chimpanzees suggest that neither food sharing nor provisioning is causally related to hunting and meat eating. Instead, evidence suggests that food sharing and provisioning have developed in the context of frugivory when primates consume fruits that are large and/or difficult to process. Patterns of food sharing by bonobos offer a model for how a change in the diet of adults could have precipitated to immatures.

J.-J. Hublin and M.P. Richards (eds.), *The Evolution of Hominin Diets: Integrating Approaches to the Study of Palaeolithic Subsistence,* 1–14.

Introduction

Tree branches intertwine overhead, arching high above the forest floor. Multiple layers of leaves keep the sun out, making the air on the forest ground remain humid and cool. Not far away, work is in progress. The sound of hammers carries through the green light. Their echoes duplicate the original sounds and create an acoustic ambience reminiscent of carpenters constructing the roof truss of a new building. Only the whistles and voices of the craftsman are missing. Not surprisingly, the sounds come from a group of chimpanzees cracking *Coula* nuts. To access the nutritional seeds, chimpanzees place the nuts on a rock and use clubs or stones to open the hard shell. Watching nut-cracking chimps is like zooming back into a world without human primates. Is this how our ancestors branched out of the line of nonhuman primates? Patterns of dentition of australopithecines have been interpreted as adaptations to nut-cracking (Peters, 1987). Nuts provide high-quality food that preserves well and can bridge lean seasons when other plant foods are scarce. In some species, density of fat is high and converts into metabolic energy to a high degree (Peters and O'Brien, 1981). The technology to open the shells is relatively simple, but tools are not always available at the food patch. Chimpanzees have been seen carrying tools from far away, suggesting that they prepare for the task of nut cracking long before they arrive at the food patch (Boesch and Boesch, 1983). Studies of chimpanzees at Taï have demonstrated that acquisition of the hammer-anvil technique involves both learning and teaching (Boesch and Boesch, 1984), and it is easy to imagine how extractive feeding and tool use could have pushed the development of primates from nonhuman species to ancestral hominins (Peters, 1987). However, while nut cracking by primates is a remarkable pattern of food acquisition, alternative models of human evolution associate the dietary shift from non-human apes to ancestral hominins with the consumption of meat (Dart, 1953; Lee and DeVore, 1968), changes in nutrient density in plant foods (Conklin-Brittain et al., 2002), or foraging style (O'Connell et al., 2002). While the discussion about meat versus plant foods continues (e.g., Bunn, 2007), there is large agreement that the split between nonhuman primates and proto-hominins was marked by a shift in diet composition (Leonard, 2002). Most models associate this development with a series of environmental changes that transformed tropical forest into savannah woodlands (review by Reed and Rector, 2007). It is thought that as a consequence of climatic changes, frugivore primates faced longer periods of reduced food abundance and were forced to use alternative sources such as meat, nuts, or underground storage organs of plants (Tooby and DeVore, 1987; Bunn and Ezzo, 1993). In contrast to models that associate the shift in diet with a particular diet component (e.g., meat), and a specific habitat (dry savannah), it has been proposed that subsistence patterns of early hominins were more flexible and adapted to a variety of environmental conditions and food sources (Ungar et al., 2006).

Models of feeding patterns of early humans use various sources of information; data from nonhuman primates is one source. The construction of conceptual models has a long history in research on human and nonhuman primates, and has turned out to be very useful for testing the causality of links between ecology, behavior of contemporary species, and fossil evidence (Tooby and DeVore, 1987; McGrew, 1992; Milton, 1999; Leonard et al., 2003). The strength of such models depends on the data that are fed into the model. As new information accumulates, the database requires updates.

The goal of this paper is to review recent reports on primate feeding ecology, focusing on three corresponding topics. It begins with a summary of published information on dietary composition and its variation between and within populations. This is followed by a section on how food items are processed prior to ingestion and how food is distributed among group members. Given the prominent role of animal matter in the diet of early humans, the final part focuses on the role of fauna in the diet of primates.

Diet Composition: Variation Between and Within Species

Nonhuman primates exploit a variety of food sources, including reproductive and non-reproductive plant parts, exudates of plants, invertebrates, and the meat of vertebrates. Notable exceptions to the average primate diet are species which have specialized in the consumption of secretions from insects (e.g., *Microcebus*), the ingestion of lichen (e.g., *Pygathix bieti*), needles from conifers (e.g., *Macaca sylvanus*), and species that depend heavily or exclusively on insects and other animal food, such as *Tarsius* and many Galagos (Harcourt and Nash, 1986; Bearder, 1987; Richard et al., 1989; Bennett and Davies, 1994; Mittermeier et al., 1994).

Primate omnivory can be portrayed as a mixture of a large number of plant food species and a small number of animal foods (Milton, 1987). This classification is not unjustified, but does not account for the extensive variability of food habits. Considering the proportions that different plant food items contribute to diet and/or feeding time, one item dominates all others: fruit.

From the species listed in Table 1.1, the majority (N = 90, 73%) have a food repertoire that includes some fruit; in 49 species (40%), fruit contributes 50% or more of the diet. In addition, the diet is often skewed towards a few species of fruit (Table 1.2). Black howler monkeys consume fruit and leaves from various sources, but more than 80% comes from a single species. The diet of the bamboo lemur is dominated by shoots and leaves of bamboo (Wright and Randrimanantena, 1989), and in gelada baboons 90% of the food comes from seeds, leaves, and bulbs of monocotyledons (Dunbar, 1977). In some cases like gelada baboons, the diet composition reflects ecological constraints. In some Malagasy lemurs, variation of the concentration of macronutrients narrows the niches that are exploited by different species (Ganzhorn, 1988), and the same may apply to other sympatric primate communities (Terborgh, 1983). Hanuman langurs, crab-eating macaques, and olive baboons are able to sustain themselves on sources that are nutrient-poor and difficult to digest (Koenig and Borries, 2001), and respond flexibly to environmental changes by switching to novel food sources that are highly nutritious (Richard et al., 1989).

The diets of chimpanzees and bonobos are dominated by ripe fruit and complemented by leaves, flowers, and animal food. Comparative nutritional analyses of samples from Salonga bonobos and Gashaka chimpanzees showed that the food items of both species contain similar amounts of nutrients (Hohmann et al., 2006). Plant food samples from the two sites differed in terms of associations of nutritionally relevant components. In chimpanzee samples, anti-feedants such as tannins and phenols were found to be associated with lipids whereas in the food samples eaten by bonobos, concentration of lipids was independent of tannins and phenols. Another difference was that ambient levels of tannins and phenols were significantly higher in the habitat of chimpanzees (Hohmann et al., 2006). However, food samples from both species contained very similar amounts of these anti-feedants, suggesting that chimpanzees are more selective in choosing food plants than bonobos. The effects of these differences are still unknown, but they are likely to be reflected in species-specific patterns of food selection, food processing, and/or digestion.

The dietary patterns shown in Table 1.1 depict major trends of food selection, but these patterns do not account for extensive intraspecific variability. Some of the variation between populations of the same species is probably based on differences in ecology. Hanuman langurs occupy a variety of different habitats, and dietary composition tends to correlate with the abundance of young leaves, flowers and fruit, while the intake of mature leaves is considered to be independent of forest productivity (Newton, 1992). In rich habitats, common langurs concentrate on high quality food and have relatively large food repertoires (Koenig and Borries, 2001).

Early field studies on woolly spider monkeys, the largest New World primate, suggested that the species adapted to a low quality diet dominated by leaves (Milton, 1984). More recent work revealed that some populations are highly

TABLE 1.1. Size of food repertoire of primate species, proportions of different food components in the total diet, and body mass (kg) (From Rowe, 1996).

Species	Food	Fruits	Leaves	Flowers	Seeds	Animal prey	Body mass Female	Body mass Male
Prosimians								
Arctocebus aureus		14				85		
Arctocebus calabarensis						100	0.31	0.32
Loris tardigradus						99	0.26	0.29
Nycticebus coucang		50				30	1.20	1.21
Perodicticus potto		65				10	1.08	1.02
Euoticus elegantulus		5				20	0.27	0.30
Galago alleni		73				25	0.27	0.23
Galago senegalensis						52	0.19	0.21
Galagoides demidoff		19				70	0.07	0.08
Galagoides zanzibaricus		30				70	0.14	0.15
Otolemur garnettii		50				50	0.72	0.82
Phaner furcifer							0.40	0.44
Lepilemur leucopus							0.58	0.54
Lepilemur mustelinus							0.62	0.59
Lemur catta		70	25	5			2.68	2.71
Eulemur rubriventer	67						2.14	2.27
Hapalemur aureus							1.50	1.66
Hapalemur griseus							0.89	0.94
Hapalemur simus							2.40	2.15
Propithecus diadema		20	26	3	25		7.50	7.50
Propithecus tattersalli		37	39					
Propithecus verreauxi	102						3.48	3.64
Indri indri	62	25	50–75		19		7.14	5.83
Tarsius bancanus						100	0.11	0.12
Tarsius dianae						100	0.11	0.09
Tarsius pumilus						100		
Tarsius spectrum						100	0.11	0.13
Tarsius syrichta						100	0.12	0.13
New World Monkeys								
Callithrix flaviceps	30	14				20		
Callithrix humeralifer	52	83					0.31	0.28
Callithrix kuhlii		67						
Callithrix penicillata		30					0.18	0.23
Callithrix pygmaea	38						0.13	0.13
Saguinus bicolor	21	96					0.43	0.43
Saguinus fuscicollis		96					0.40	0.39
Saguinus geoffroyi		60				30	0.54	0.55
Saguinus imperator		97		4				
Saguinus nigricollis	41						0.48	0.47
Leontopithecus rosalia		78					0.58	0.57
Aotus nigriceps		65–75	5–30			5–20	0.94	0.94
Aotus trivirgatus							0.92	0.95
Callicebus brunneus		47	28	2		15	0.85	0.85
Callicebus moloch		54	28			17	0.86	1.00
Callicebus personatus	91	81				0	1.38	1.27
Callicebus torquatus	57	65	15			16	1.31	1.30
Cebus albifrons	68	53			42		1.81	2.48
Cebus paella	96	66			25		2.39	3.05
Cebus capucinus	119	65	15			20	2.67	3.87
Saimiri boliviensis		18				82	0.80	1.03
Saimiri oerstedii							0.70	0.85
Saimiri sciureus							0.95	0.85
Pithecia albicans	81	34	10	7	46			
Pithecia monachus		55	4	3	38		1.90	2.80
Pithecia pithecia		60		7	30		1.26	1.73
Chiropotes albinasus		54		3	36		2.51	3.02

(continued)

TABLE 1.1. (continued)

Species	Food	Fruits	Leaves	Flowers	Seeds	Animal prey	Body mass	
							Female	Male
Chiropotes satanas	86	30		3	66		2.60	3.10
Cacajao calvus		18		6	67	5	2.88	3.45
Alouatta caraya		24	76				4.61	6.64
Alouatta fusca		2–29	66–77	6–20			4.55	6.23
Alouatta palliata	100	13	70	18			5.35	7.15
Alouatta seniculus	195	42	53	5			5.60	7.20
Ateles belzebuth	33	83	7	1			9.85	8.53
Ateles chamek		80	17	2				
Ateles geoffroyi		78	8	13	11	1	7.46	8.21
Ateles paniscus	171	83	6	6			8.75	7.34
Lagothrix lagotricha	225	68	14	3	7		5.00	6.80
Brachyteles arachnoides		32	51				9.45	12.13
Old World Monkeys								
Macaca fascicularis		64					4.10	6.50
Macaca fuscata	192						13.15	14.50
Macaca mulatta	92						7.65	8.25
Macaca nemestrina	160	74	5	4		12	7.80	10.35
Macaca nigra	120						6.60	10.40
Macaca radiata	39	47–53					4.17	7.13
Macaca sinica		75	23				3.85	6.40
Macaca sylvanus	100						10.60	16.15
Papio hamadryas cynocephalus	180						12.30	24.89
Papio hamadryas papio							13.00	26.00
Mandrillus sphinx	113	92					11.50	26.90
Theropithecus gelada					90		11.70	20.00
Cercocebus galeritus	61	73–78	11–14	1		1–3	5.40	10.20
Lophocebus albigena		59	5	3		11	5.67	7.26
Allenopithecus nigroviridis		81	2				3.70	5.95
Miopithecus talapoin		43					0.78	1.27
Cercopithecus ascanius	104	62	7			25	3.30	4.21
Cercopithecus cephus	47	78	7			15	2.90	4.10
Cercopithecus mitis		55	19			17	4.23	7.35
Cercopithecus neglectus		74	9	3		5	4.46	7.55
Cercopithecus nictitans		90					4.08	6.35
Colobus polykomos							8.30	9.90
Procolobus pennantii		6	72	12			7.00	10.50
Procolobus rufomitratus		25	63	6				
Procolobus verus		3	73	7	14		4.20	4.70
Presbytis comata	64	14	65	7	1		6.67	6.40
Presbytis femoralis		49						
Presbytis hosei	47	30	65	2			5.57	6.20
Presbytis melalophos	55	24	40	12	25		5.80	5.90
Presbytis potenziani		32	55				6.40	6.50
Presbytis rubicunda		19	37	11	30	1	5.70	6.22
Semnopithecus entellus		24	45	20		3	11.20	18.30
Trachypithecus auratus		35	47	7		1		
Trachypithecus cristatus	94	10	90				5.70	6.60
Trachypithecus johnii	114	25	58	9			10.90	12.71
Trachypithecus obscurus		35	58	7			6.60	7.32
Trachypithecus phayrei		24	68				6.95	7.93
Trachypithecus vetulus		28	60	12			6.30	7.55
Pygathrix nemaeus	50						8.20	10.90
Nasalis concolor		25–30	60				7.10	8.75
Nasalis larvatus		17	44	3	2	1	10.00	21.20
Apes								
Hylobates agilis		58	39	3		1	5.70	6.00
Hylobates concolor		21	72	7			5.80	5.60
Hylobates hoolock		65	13	17		5	6.10	6.90

(continued)

TABLE 1.1. (continued)

							Body mass	
Species	Food	Fruits	Leaves	Flowers	Seeds	Animal prey	Female	Male
Hylobates klossii		70	2			25	5.90	5.70
Hylobates lar		50	29	7		13	5.60	6.29
Hylobates muelleri		62	32	4		2		
Hylobates pileatus		71	11	15		2	7.50	9.16
Hylobates syndactylus		31	59	8	3		10.57	13.52
Pongo abelii	200							
Pongo pygmaeus	400	60					37.00	77.50
Gorilla gorilla beringei	145	2	86	2		0	97.70	159.2
Gorilla gorilla gorilla		67	17			3	71.50	169.5
Gorilla gorilla graueri	102						80.00	175.2
Pan troglodytes	250	45–76	12–45	1–18	1–11	0–5	39.50	50.00

TABLE 1.2. Number of main food items in relation to total food repertoire, the proportion of main food items in the total diet, or the time spent feeding on main food items (From Rowe, 1996).

	Total repertoire	Number of main food items	Proportion of diet (%)	Feeding time (%)
Eulemur fulvus	57	10	60	
Eulemur rubriventer	67	10	50	
Indri indri	62	5	50	
Pithecia pithecia		2	49	
Alouatta pigra		1	86	
Cercocebus galeritus		1	61	
Colobus angolensis	46	5	47	
Colobus guereza		1	69	
Trachypithecus pileatus	35	2		30
Nasalis larvatus	55	4	60	

frugivorous, and it seems that the proportion of fruit in the diet of woolly spider monkeys increases whenever the forest offers sufficient resources (Strier, 1991; Talebi et al., 2005).

Data from chimpanzees also indicate that populations in wet forest habitats have a larger food repertoire than those living in dry forest (Hunt and McGrew, 2002; Pruetz, 2006). This difference is considered to reflect variation in the abundance of fruit in relation to climatic factors but may also be related to the quality of food sources. Another force that has been found to affect food choice is the presence of competing species. Where chimpanzees coexist with gorillas, dietary patterns may reflect niche separation (Yamagiwa et al., 1996) or geographic differences in forest vegetation (Ganas et al., 2004). In Western gorillas, differences between populations are most prominent in terms of fruit consumption (Rogers et al., 2004). Flexibility of food choice is not restricted to natural sources, but extends to cultivated plants. Some populations of baboons, macaques, chimpanzees and gorillas have adapted well to exploit human crops (Naughton-Treves et al., 1998), and translocation of wild communities has shown that some species are able to switch to a completely different set of foods (Silver and Marsh, 2003).

Even in areas where climatic conditions are relatively stable, fruit production is a seasonal phenomenon and fruiting intervals are often separated by several years, making the availability of certain species difficult to predict (Leigh and Windsor, 1982). When preferred food sources are scarce, most species turn to foods that are not used at other times of the year. The type of fallback food varies between, and sometimes also within species. Cercopithecines and chimpanzees of Kanyawara Community, Kibale forest, Uganda, both have a diet that is dominated by fruit. At times when ripe fruit are scarce, cercopithecines turn to unripe fruit, leaves, and other food sources while chimpanzees maintain feeding primarily on ripe fruit (Wrangham et al., 1998).

Data from sympatric populations of gorillas and chimpanzees show variation in terms of dynamics of diet composition during lean seasons. In the lowland forest of Lope, Gabon, gorillas focus on leaves of dicotyledons while chimpanzees maintain a high intake of fruit throughout the year (Tutin and Fernandez, 1993). In the montane forest of Kahuzi Biega, D.R. Congo, gorillas stick to a folivore diet while chimpanzees turn from fruit to monocotyledon herbs when fruit becomes scarce (Basabose, 2002).

Comparison of data from two populations of orangutans provides a particularly interesting example of the nutritional consequences of differences in the availability of fallback foods. In both populations, food availability is affected by a community-wide fluctuation of fruiting in which brief periods of high fruit abundance are separated by long periods of low fruit availability (Medway, 1972). At times of food shortage, Borneo orangutans are forced to use cambium of trees, a food that is characterized by very low nutrient concentration and high fiber content, as fallback food. Urine samples collected during the lean season contained high concentrations of keton, suggesting that subjects had lost body mass because of a lack of energy (Knott, 1998). Orangutans from Sumatra experience a similar fluctuation in fruit abundance. However, due to the higher density of *Ficus* trees, fruit availability does not reach the low level recorded at Borneo (Wich et al., 2004). Instead, measurements of social parameters (e.g., group size) and energy constitution (e.g., keton excretion) suggest that individuals from Sumatran population are able to maintain body mass during the lean seasons by using fig trees (Wich et al., 2006).

Digestive Strategies

The variation in diet composition between species reflects to some extent the functional morphology of the digestive tract (Chivers and Hladik, 1980). Omnivore generalists like macaques, baboons, and chimpanzees have a simple digestive tract that facilitates digestion of various types of food. In capuchins and other fruit eaters, small intestines make up a large proportion of the digestive tract, but in most species the volume of large intestines exceeds that of the small intestines. Species that depend largely on a diet of leaves and other non-reproductive plant parts show adaptations to ferment food in the stomach or in the large intestines (Milton and McBee, 1983). Evidence suggests that adaptations in gut morphology constrain flexible responses to changes in diet in terms of digestive kinetics and digestive efficiency. The diet of howler monkeys is dominated by leaves, and nutrient extraction is enhanced by fermentation in the large intestines and slow passage rate of ingesta (Milton et al., 1980). Whenever available, howlers increase intake of fruit, but feeding experiments suggest that this change in diet composition does not shorten gut passage time (Milton, 1981). In this case, the digestive morphology seems to relate to the dominant section of the diet and not to sporadic bursts of frugivory. Woolly spider monkeys show a strong tendency to feed on fruit whenever resources are abundant, but are also able to sustain themselves on a highly folivore diet when fruit is scarce. However, unlike howlers, woolly spider monkeys have a rapid passage rate (Milton, 1984), suggesting that their digestive system is adapted to a diet that is dominated by fruit (Strier, 1991).

In a long-term study on red colobus monkeys, Chapman et al. (2002) found that diet composition changed on a spatial and a temporal scale. While all groups focused on vegetation, the proportion of different food items varied within and across groups.

Although the diets of chimpanzees and orangutans are characterized mainly by ripe fruit, the large intestines remain spacious. Analyzing data on food intake and nutritional quality of foods eaten by chimpanzees at Kanyawara, Uganda, Conklin-Brittain et al. (2006) found that at times when ripe fruit is in short supply, individuals with a particularly high energy demand (e.g., lactating females) would not be able to cover their energy requirements without fermentation. Similarly, the long passage of digesta by orangutans is seen as an adaptation to the high concentration of fermentable fibers in the natural food of this species (Caton et al., 1999).

The examples given above do not necessarily contradict the predicted relationship between diet and digestive system (Chivers and Haldik, 1980). However, the flexibility of diet composition indicates that gut size alone does not allow for inferences about the nature and quality of food items. Temporary changes in diet composition may affect parameters such as intake rate, transit time, gut motility, and digestive efficiency of nutrients and other food components. So far, the digestive physiology of primates remains largely unexplored and offers a challenging field for future studies.

Processing Plant Foods

As do other frugivores, primates deal with items that are often difficult to access because of hardness, protective structures like spines and thorns, or large size (Leighton and Leighton, 1983). Moreover, accessibility of nutrients is not homogenous, but is concentrated in certain parts of fruit (e.g., mesocarp, exocarp, arille). Seeds and other parts of the fruit are usually protected from consumption by secondary compounds (Levey and Cipollini, 1998). Because of this, consumption of fruit requires some kind of processing. Many species of the Old World have cheek pouches. Cheek pouches serve as containers for storing food, which allows individuals to collect food as they move and exposes food items to salivary enzymes that extract part of the nutrients. Storage of food in cheek pouches also promotes retention of seeds (Corlett and Lucas, 1990) and in this way, affects the structure of ingesta, and has important consequences for factors such as digestive kinetics and nutrient absorption (Lambert, 1998). However, inter-species differences in seed handling are not always linked to cheek pouches. Comparing feeding behavior of woolly monkeys and spider monkeys, two frugivore species from the neotropics without cheek pouches, it was found that woolly monkeys, like cheek-pouched monkeys of the Old World, remove most seeds from fruit while spider monkeys ingest fruit together with seeds (Dew, 2005).

Another form of oral food processing occurs in orangutans and the two *Pan* species. When eating ripe fruit, individuals may compress fruit pulp into wedges that are chewed on for extended periods of time (Fig. 1.1). Small parts of the wedge are probably ingested, but the majority is discarded after chewing. This type of food processing is usually seen when fruit crop has reached an advanced stage of ripeness, suggesting that wedging serves to avoid certain components

FIG. 1.1. Male chimpanzee from Ngogo (Uganda) chewing on a wadge of *Ficus mucuso* (photograph by K. Langergraber).

such as non-digestible fibers that are not so prominent at an earlier stage.

In addition to oral food processing, primates have a tendency to process food manually, including removal of spines and thorns, breaking hard shells by smashing fruit on hard surfaces, and washing food items (Boesch and Boesch, 1990; Panger et al., 2002). Panger et al. (2002) compared food processing techniques performed by different populations of wild capuchin monkeys and found that 16 types (80%) of processing involved processing of fruit and four of animal foods. Other taxa well known for investing in manual food processing are orangutans and chimpanzees. Orangutans have exceptional manipulative skills that are habitually used for food preparation. These include the use of leaves as gloves to assist manipulation of spiny fruit, and use of extractive tools to access the hard shelled nuts of *Nesia* fruit (van Schaik et al., 1996).

The catalogue of food processing behaviors by chimpanzees is impressively large and biased towards tool use (McGrew, 1992; Matsuzawa, 1994; Whiten et al., 1999). Long-term studies of nut cracking by chimpanzees at Taï, Ivory Coast, and Bossou, Guinea, indicate that consumers know about material characteristics of both food items and tools (Boesch and Boesch, 1983, 1984; Matsuzawa, 1994). Reports on other types of plant food processing by chimpanzees are rare and anecdotal (Nishida et al., 1983). At Lope, Gabon, chimpanzees were seen to remove the stingy hairs from the skin of *Diospyros* before feeding (Tutin et al., 1996). Perhaps the most detailed analysis of food processing has been made by Corp and Byrne (2002), who investigated techniques of manual food processing of leaves that are physically defended. Comparing manipulation of defended and undefended leaves, the study revealed not only the complexity of manual skills but also the hierarchical organization of different actions.

Dividing Plant Foods

Most fruits eaten by primates are small compared to the body mass of the consumers. Small size promotes an individualistic style of feeding; that is, food items are collected and consumed by the same individual, and competition occurs over food patches rather than food items. However, some species of trees and climbers produce larger fruit that can not be consumed in a single bite. For example, Asian breadfruit trees carry fruit that weigh up to 10kg (Fig. 1.2). The fruit is rich in nutrients but difficult to process. Lion tailed macaques are eager to feed on these fruit but it seems that only males with their long canines are able to open the tough outer skin (own observations). After opening a fruit, males and females alternate in extracting the sweet husk covering each single seed. This type of communal feeding requires social tolerance but does not include food sharing. Division of plant food has been reported from both marmosets and chimpanzees (Feistner and McGrew, 1989) and was thought to be most frequent among females sharing fruit with dependent offspring (McGrew, 1975; Silk, 1978). In addition to provisioning kin with nutrients, the transfer of food from the mother to her infant is an important way to transfer information about what and what not to eat. During the dry season, chimpanzees at Tongo (D.R. Congo) are reported to dig for tubers to obtain the moisture as an alternative source of water (Lanjouw, 2002). Digging for one tuber takes up to 15 minutes, and immatures may dig for their own food or beg for a share from older individuals. Given that all sources of surface water disappear during this period, division of tubers may be under strong selective pressure. In bonobos, division of plant food seems to be more frequent than in other hominoids, resembling meat sharing in many ways (see below). Bonobos share fruit from a number of species; one species, African bread fruit, *Treculia africana* (Fig. 1.3), seems

FIG. 1.2. Liontailed macaque feeding on Asian breadfruit, *Artocarpus heteropyllus* (photograph by G. Hohmann).

Fig. 1.3. Wild fruit of African breadfruit, *Treculia africana*, may weigh up to 30kg and are consumed by bonobos, chimpanzees, gorillas and humans (photograph by G. Hohmann).

to be more important than others (Hohmann and Fruth, 1996). Fruit of *Treculia* are consumed by all African ape species living in forested habitats and is also widely used for production of human food (Akubor and Badifu, 2004). Ripe fruit can weigh up to 30kg; they are rich in protein, unstructured carbohydrates, and lipids. Using nutritional data from samples collected at the site of Lui Kotal, a 10kg *Treculia* provides up to 20,000 calories. Given that consumption of this resource does not require climbing or other forms of locomotion, it is reasonable to assume that the nutritional reward is likely to be unusually high. At Lomako, bonobos consumed, on average, one fruit per week and in the majority of cases, the fruit was shared with other adult and immature individuals (Fruth and Hohmann, 2002). In most cases, females were in control and shared with other adult females and immatures. While access of other adult females was often constrained, females were usually very tolerant towards immatures independent of their kin relationships, and it was obvious that substantial amounts of food were consumed by members of this age class (Fruth and Hohmann, 2002). The mode of food division included tolerated theft, delivery of partly processed food items, and handouts of fresh food (Hohmann and Fruth, 1996).

Faunivory

Animal matter is a common component in the diet of nonhuman primates. Most species select some kind of fauna, some depend entirely on animal food (Table 1.1); missing scores in Table 1.1 are more likely to reflect a lack of observation than a strict vegetarian life style. The data compiled by Rowe (1996) suggest that there are few species that have been studied intensely and long enough without providing any evidence for using sources of animal food (e.g., *Cercopithecus mitis*, *Trachypethecus christatus*).

Looking at the nature of animal foods, insects dominate all other sources (McGrew, 1992). In spite of the importance of insects in the diet of primates, little is known about the nutritional value of insects and evidence from other species is somewhat contradictory. Some studies found that ants and termites yield few calories in relation to volume and are often difficult to digest because of high concentrations of secondary compounds (McNab, 1984). However, insects that are eaten by humans have relatively high levels of crude protein and fat, and they also contain some essential amino acids and therefore remain important supplements in the diet of many populations (Bukkens, 2005). Given the wide use of insects as food source, it is reasonable to assume that nonhuman primates are equally selective in feeding on species that are relatively easy to digest and/or have behavioral or physiological adaptations that enable them to reduce the negative effects of secondary compounds. Most primates browse for insects while moving on the ground or through the canopy. Prey is sometimes processed (with or without tools) in order to remove parts containing toxic compounds before ingestion (Panger et al., 2002). Hominoids differ from many other primates because they focus on social insects such as ants, bees, and termites that occur in large patches, and chimpanzees enhance collection of insects further by using probes and other tools for "fishing." Similar reports from bonobos are missing, raising the question on whether or not the two *Pan* species have access to the same resources. In a recent study at Salonga National Park, McGrew et al. (2007) found that insect species known to be eaten by chimpanzees at other sites are also available to bonobos. Consequently, there must be other reasons for the absence of ants and termites from the diet of bonobos.

Evidence for the consumption of marine animals is rare and limited to some populations of capuchins, macaques, and baboons inhabiting natural or cultivated shoreline habitat (Freese and Oppenheimer, 1981; Napier, 1981 cited in Rowe, 1996). Swamp monkeys, crab eating macaques, and dwarf guenons are well adapted to life in swamp areas, and anecdotal evidence suggests that lakes, rivers, and streams are commonly used as sources for animal foods, but this has been studied for very few species and the significance of aquatic animals in the diet of nonhuman primates remains anecdotal and fragmentary.

The third type of fauna that primates consume is the meat of other vertebrates. Occasional consumption of the meat has been reported from various taxa, but most reports are restricted to few species: capuchins (Perry and Rose, 1994), baboons (Strum, 1981), and chimpanzees (Teleki, 1973). One trait that these species have in common is that they prey on vertebrates that are relatively large in relation to the body size of the predator. Often, acquisition of prey is described as being opportunistic (catching a prey upon encounter) and individualistic (prey taken by one individual) but chimpanzees may also actively search for prey (Boesch, 1994b).

Evidence for cooperative hunting is still rare, and comparison of different sites shows large variation in hunting frequency, strategies for obtaining prey, and temporal fluctuation of meat eating (Stanford, 1996; Watts and Mitani, 2002). While intra-specific variability of hunting by chimpanzees offers a fascinating model on how environment and sociality affect hunting and meat consumption in hominoids, it should be noted that most data are biased to one prey species, the red colobus monkey, and also to hunting by males. Much less is known about acquisition of other vertebrate species known to be eaten by *Pan*, and information on hunting by female chimpanzees is almost nonexistent. The recent report from Fongoli, Senegal, by Pruetz and Bertolani (2007) is a notable exception. The study provides rare evidence for the use of tools in the context of vertebrate hunting. As interesting as the tool use is that the tool-aided prey acquisition was mainly by adult females and immatures.

With this note of caution in mind, one can ask how differences in ecology and social behavior influence hunting frequency, hunting success, meat acquisition and meat sharing. One critical issue that has been raised in many empirical and theoretical studies is the question of whether hunting is a means to procure food when other sources are scarce, or whether hunting activity is restricted to times of high food supply (Stanford, 1996). Evidence from the field is mixed. Observations of chimpanzees at Gombe and Mahale suggest that hunting and meat consumption coincides with temporal shortages of plant foods (Takahata et al., 1984; Stanford, 1996). In contrast, data from Taï and Ngogo show that hunting occurs most frequently, and is most successful, at times when fruit are abundant (Boesch, 1994b; Watts and Mitani, 2002).

Hunting success seems to be triggered by social and ecological variables. At Taï it increases with the number of adult males, but data from other sites indicates that this is not always the case. Another variable that seems to influence the chance of catching prey is forest structure: hunting success is highest in areas with broken canopy where escape routes are limited; travel patterns of hunting parties suggest that chimpanzees are aware of the spatial constraints (Boesch et al., 2002; Watts and Mitani, 2002).

Meat Sharing

The proximate goal of hunting is acquisition of meat. Figures from Gombe chimpanzees suggest that during the dry season, adult individuals consume between 33 and 98 g of meat per day (Stanford, 1996). At Taï, the average daily intake of meat is estimated to be 186 g for adult males and 25 g for adult females (Boesch and Boesch-Achermann, 2000). Although such an amount seems to be small, it may still be an important source for nutrients that otherwise would be difficult to obtain. And still, is the nutritional pay-off sufficient to compensate for the energetic costs of hunting? To answer this question, a number of variables have to go into the equation: time and energy required to search and capture the prey, the probability of obtaining a share, and the energetic reward. Evidence from the field demonstrates that some individuals gain more than others. In chimpanzees, it is usually one of the hunters controlling access of other group members to the food source, but members of a hunting group who have not contributed to make the catch may also obtain meat (Boesch 1994a, b). Figure 1.4 shows that adult males and cycling females are the preferred beneficiaries of meat sharing, while younger individuals receive very little (Boesch and Boesch-Achermann, 2000). Estrus females are more likely to receive meat than non-estrus females (Stanford, 1996) suggesting that the nutritional benefits from meat consumption may offer reproductive advantages. Indeed, McGrew (1996) found that females who obtain meat from males had a higher reproductive success than females who received meat rarely or not at all. The chimpanzee model fits nicely with what one would predict: close relatives (males) cooperate in hunting, cooperation raises the per capita intake of meat, and meat sharing with females increases mating success and reproductive success of males. Bonobos provide an interesting case that shows how prey capture and meat sharing may have evolved in a different way. Combined information from Lomako (Hohmann and Fruth, 1996) and Lui Kotal (G. Hohmann and Fruth 2008) indicates that duikers, solitary living forest antelopes, are the major prey, and that meat consumption always involves meat sharing. Current evidence suggests that most if not all hunts are opportunistic actions by single individuals. The adult sex ratio within bonobo parties is female biased (Hohmann and Fruth, 2002) and therefore, one would assume that females catch prey twice as often as males. Direct observations on prey acquisition are rare and do not allow estimates of hunting success. However, observations from Lomako and Lui Kotal show that it is almost always adult females who control the prey (Hohmann and

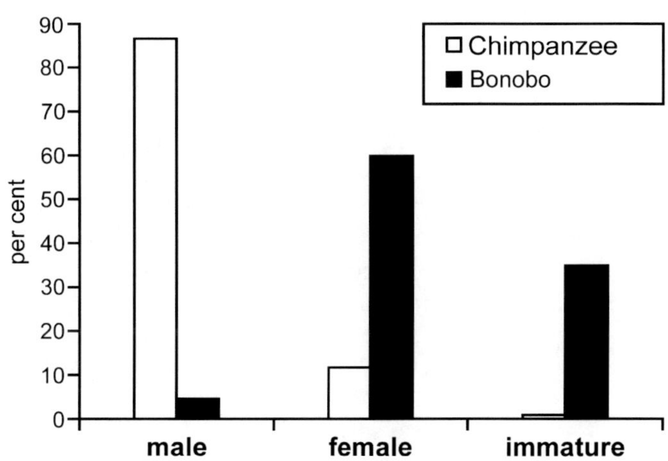

FIG. 1.4. Distribution of meat among chimpanzees and bonobos. Data of chimpanzees are from Taï forest, Ivory Coast, published by Boesch and Boesch-Achermann (2000, Table 8.4, p. 165). Data from bonobos are from Lomako, DRC, published by Fruth and Hohmann, 2002.

Fruth, 1996). Female prey owners share with other adult females and immatures, and both classes receive substantial shares from the prey (Fig. 1.4). Males also make efforts to get some meat but receive nothing or very little. In this regard, the bonobo model does not fit predictions: close relatives (males) do not cooperate in hunting and, although larger than females, are unable to obtain a share of meat. Instead, unrelated females cooperate in monopolizing the catch and divide the source with other females and immatures. Given the energetic demands of pregnancy, lactation, and infant transport on one hand, and the low cost of catching terrestrial prey, the gain from a portion of meat by females will certainly compensate for any costs deriving from catching and processing the food. Meat consumption by immatures seems to be mainly constrained by the toughness of meat and by the physical difficulties to remove meat from bones. Our observations from bonobos at Lomako and Lui Kotal suggest that independent of kinship, females may assist immatures by handing out or offering partly processed pieces of meat, and by fixing the carcass when immatures try to remove meat.

Inferences on the Feeding Behavior of Early Hominins: A Primate Perspective

The diet of any species depends on the types of resources that are available in the habitat at a given time. Analyzing lists of plant food species eaten by hominoids suggest that consumers take food from whatever plant taxa are available (Rodman, 2002). The picture changes immediately when variation in abundance of different plant species is taken into account. Data on diet composition such as those published by Rowe (1996) demonstrate that the time individuals allocate to feed on different food sources is skewed and that preferred foods are rarely the most abundant species. Distinctions between what is eaten and what is not are probably based on the nutritional quality, the density and distribution of resources, and food selection of other species with a similar diet.

In addition to selecting certain species, primates use various techniques of external food processing to modify food items in a way that is likely to enhance digestion. This allows exploiting items that are physically or chemically protected or dissecting items of very large size. Whether or not the food processing abilities of primates exceed those of other species remains to be explored, but the combination of dental features, functional morphology of hands and feet, and cognitive abilities provides a set of tools to improve the quality of what is ingested.

As in other vertebrates, food acquisition is mainly individualistic. However, in some species food sharing is habitual, and the mode of distribution may vary consistently between different species (Fig. 1.5). Data from chimpanzees show that food sharing mainly involves meat and that acquisition of meat does not only benefit the hunter, but also adult females. If food sharing has the potential to promote the reproductive

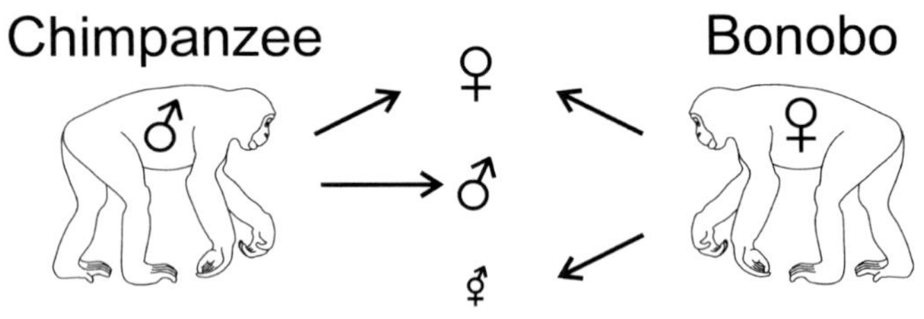

Fig. 1.5. Differences in ownership of divisible food and directions of food sharing in chimpanzees and bonobos. Arrows indicate the major trajectories of food transfer.

success of females, and if males share food preferentially with mating partners, the behavior has an immediate selective advantage in terms of reproductive performance. Bonobos add another component to the primate model: food is controlled by females and distributed to other females *and* to immatures. Food sharing involves large fruit and the meat of vertebrates. Provisioning is not a daily behavior and it is not known how critical it is for the survival of the young. Still, if the tendency to provision offspring varies between individuals – a question that needs to be explored – even modest differences in the nutrition of offspring could influence physical fitness and life history parameters.

Communal feeding, food sharing and provisioning are common practice in other mammals like social carnivores (Pusey and Packer, 1994; East and Hofer, 2002). Nevertheless, for good reasons it is assumed that these behaviors became prime movers during early stages of human evolution (Milton, 1999; O'Connell et al., 2002). Models such as the one proposed by Lovejoy (1981) assume that the behavioral suite of (a) hunting (or scavenging), (b) meat eating, and (c) food sharing emerged when hominins lived in dry habitats where plant food was often in short supply and where vertebrate prey was permanently or temporarily abundant. This implies that meat served as a fallback food during periods in which plant food was not abundant. However, evidence from recent studies of *Pan* differs from these assumptions. First, hunting and meat eating occurs equally in forests and in open habitats. Second, food sharing may have evolved independently of meat eating. Third, at least in some populations, hunting and meat consumption increase at times when plant food is highly abundant, suggesting that hunting activities are triggered by high energy intake from plant foods. The sources that provide high energetic rewards are usually the fruit of forest trees, and not roots and tubers. In sum, evidence from primate studies differs to some extent from predictions made in the models on human evolution. This is not to say that primate studies are irrelevant for modelling human evolution. What it suggests is that the common ancestor of *Pan* and *Homo* engaged already in food sharing, hunting, meat eating, and provisioning, and that this behavioral suite was later modified when populations of forest apes moved into open and dry areas. Communal attacks of other predators and the defense of stolen carcasses may have set the direction for a divergent development that gradually led into independence of subsistence on forest fruit.

Given the fact that hunting and meat sharing of chimpanzees is biased to red colobus, a forest species that does not occur in dry savannah habitat, and given the behavioral diversity of food sharing and provisioning within the genus *Pan*, there is room for refinement of existing models on hunting and meat eating in early hominins. Three topics, amongst others, are likely to provide useful hints: hunting of terrestrial prey, prey acquisition and food sharing by females, and scavenging by contemporary human populations.

Nonhuman primates are not an ideal model for reconstructing the evolution of a tribe that became extinct a couple of million years ago (Wrangham, 1987). Most species have adapted to niches that may not have existed before. However, nonhuman primates are the best model that we have. Moreover, primates offer a view on the behavioral ecology of species that are closely related to us, not only with their genes, but also in their ability to copy and transmit technical and behavioral innovations of others.

Acknowledgments. I want to thank Jean-Jaques Hublin and Mike Richard for inviting me to the symposium on The Evolution of Human Diet: Integrating Approaches to the Study of Palaeolithic Subsistence. Funding for field research on bonobos at Lomako and Lui Kotal came from the Max-Planck-Society, Deutsche Forschungsgemeinschaft, The Leakey Foundation for Anthropological Research, Volkswagen Foundation, and National Geographic Society. Thanks are due to Kevin Langergraber for the photograph shown in Fig. 1.3 and to Knut Finstermeier for help with the other graphics.

References

Akubor, P.I., Badifu, G.I.O., 2004. Chemical composition, functional properties and baking potential of African breadfruit kernel and wheat flour blends. International Journal of Food Science and Technology 39, 223–229.

Basabose, A.K., 2002. Diet composition of chimpanzees inhabiting the montane forest of Kahuzi, Democratic Republic of Congo. American Journal of Primatology 58, 1–21.

Bearder, S.K., 1987. Lorises, bush babies, and tarsiers: Diverse societies in solitary foragers. In: Smuts, B.B., Cheney, D.L., Seyfarth, R.M., Wrangham, R.W., Struhsaker, T.T. (Eds.), Primate Societies. University of Chicago Press, Chicago, IL, pp. 11–24.

Bennett, E.L., Davies, A.G.,1994. The ecology of Asian colobines. In: Davies, A.G., Oates, J.F. (Eds.), Colobine Monkeys: Their Ecology, Behavior, and Evolution. Cambridge University Press, Cambridge, pp. 129–172.

Boesch, C., 1994a. Cooperative hunting in wild chimpanzees. Animal Behavior 48, 653–667.

Boesch, C., 1994b. Chimpanzees-red colobus monkeys: A predator-prey system. Animal Behavior 47, 1135–1148.

Boesch, C., Boesch, H., 1983. Optimization of nutcracking with natural hammers by wild chimpanzees. Behavior 83, 265–286.

Boesch, C., Boesch, H., 1984. Possible causes of sex differences in the use of natural hammers by wild chimpanzees. Journal of Human Evolution 13, 415–440.

Boesch, C., Boesch, H., 1990. Tool use and tool making in wild chimpanzees. Folia Primatologica 54, 86–99.

Boesch, C., Boesch-Achermann, H., 2000. The Chimpanzees of the Taï Forest. Behavioral Ecology and Evolution, Oxford University Press, Oxford.

Boesch, C., Uehara, S., Ihobe, H., 2002. Variation in chimpanzee-red colobus interactions. In: Boesch, C., Hohmann, G., Marchant, L.F. (Eds.), Behavioral Diversity in Chimpanzees and Bonobos. Cambridge University Press, Cambridge, pp. 221–230.

Bukkens, S.G.F., 2005. Insects in the human diet: Nutritional aspects. In: Paoletti, M.G. (Ed.), Ecological Implications of Minilivestock, Science Publishers, Plymouth, pp. 545–577.

Bunn, H.T., 2007. Meat made us human. In: Ungar, P.S. (Ed.), Evolution of the Human Diet: The Known, the Unknown, and the Unknowable. Oxford University Press, New York, pp. 191–211.

Bunn, H.T., Ezzo, J.A., 1993. Hunting and scavenging by Plio-Pleistocene hominids: Nutritional constraints, archaeological patterns, and behavioral implications. Journal of Archaeological Science 20, 365–398.

Caton, J.M., Hume, I.D., Hill, D.M., Harper, P., 1999. Digesta retention in the gastro-intestinal tract of the orangutan (*Pongo pygmaeus*). Primates 40, 551–558.

Chapman, C.A., Chapman, L.J., Gillespie, T.R., 2002. Scale issues in the study of primate foraging: Red colobus of Kibale National Park. American Journal of Physical Anthropology 117, 349–363.

Chivers, D.J., Hladik, C.M., 1980. Morphology of the gastrointestinal tract in primates: Comparisons with other mammals in relation to diet. Journal of Morphology 166, 337–386.

Conklin-Brittain, N.L., Wrangham, R.W., Smith, C.C., 2002. A two-stage model of increased dietary quality in early hominid evolution: The role of fiber. In: Ungar, P.S., Teaford, M.F. (Eds.), Human Diet – Its Origin and Function. Bergin & Garvey, WestPoint, CT, pp. 49–60.

Conklin-Brittain, N.L., Knott, C.D., Wrangham, R.W., 2006. Energy intake by wild chimpanzees and orangutans: Methodological considerations and preliminary comparison. In: Hohmann, G., Robbins, M.M., Boesch, C. (Eds.), Feeding Ecology in Apes and Other Primates. Cambridge University Press, Cambridge & London, pp. 445–471.

Corlett, R.T., Lucas, P.W., 1990. Alternative seed-handling strategies in primates: Seed-spitting by long-tailed macaques (*Macaca fascicularis*). Oecologia 82, 166–171.

Corp, N., Byrne, R.W., 2002. Leaf processing by wild chimpanzees: Physical defended leaves reveal complex manual skills. Ethology 108, 673–696.

Dart, R.A., 1953. The predatory transition from ape to man. International Anthropology of Linguatec Review 1, 201–221.

Dew, J.L., 2005. Foraging, food choice, and food processing by sympatric ripe-fruit-specialists: *Lagothrix lagotricha poeppigii* and *Ateles belzebuth belzebuth*. International Journal of Primatology 26, 1107–1135.

Dunbar, R.I.M., 1977. Feeding ecology of gelada baboons: A preliminary report. In: Clutton-Brock, T.H. (Ed.), Primate Ecology: Studies of Feeding and Ranging Behavior in Lemurs, Monkeys, and Apes. Academic Press, London, pp. 251–275.

East, M.L., Hofer, H., 2002. Conflict and cooperation in a female-dominated society: A reassessment of the "hyper aggressive" image of spotted hyenas. Advances in the Study of Behavior 31, 1–30.

Feistner, A.T.C., McGrew, W.C., 1989. Food-sharing in primates: A critical review. In: Sept, P.K., Sept, S. (Eds.), Perspectives in

Primate Biology. Today and Tomorrow´s Printers and Publishers, New Delhi, pp. 21–36.

Freese, C.H., Oppenheimer, J.R., 1981. The capuchin monkey, genus *Cebus*. In: Coimbra-Filho, A.F., Mittermeier, R.A. (Eds.), Ecology and Behavior of Neotropical Primates Vol 1, Academia Brasileira de Ciences, Rio de Janeiro, pp. 331–390.

Fruth, B., Hohmann, G., 2002. How bonobos handle hunts and harvests: Why share food? In: Boesch, C., Hohmann, G., Marchant, L.F. (Eds.), Behavioral Diversity in Chimpanzees and Bonobos. Cambridge University Press, Cambridge, pp. 231–243.

Ganas, J., Robbins, M.M., Nkuruningi, J.B., Kaplin, B.A., McNeilage, A., 2004. Dietary variability of mountain gorillas in Bwindi Impenetrable National Park, Uganda. International Journal of Primatology 25, 1043–1072.

Ganzhorn, J.U., 1988. Food partitioning among Malagasy primates. Oecologia 75, 436–450.

Harcourt, C.S., Nash, L.T., 1986. Species differences in substrate use and diet between sympatric galagos in two Kenyan coastal forests. Primates 27, 41–52.

Hohmann, G., Fruth, B., 1996. Food sharing and status in unprovisioned bonobos. In: Wiessner, P., Schiefenhoevel, W. (Eds.), Food and the Status Quest: An Interdisciplinary Perspective. Berghan Books, Oxford, pp. 47–67.

Hohmann, G., Fruth, B., 2002. Dynamics in social organization of bonobos (*Pan paniscus*). In: Boesch, C., Hohmann, G., Marchant, L.F. (Eds.), Behavioral Diversity in Chimpanzees and Bonobos. Cambridge University Press, Cambridge, pp. 138–150.

Hohmann, G., Fruth, B., 2008. New records on prey capture and meat eating by bonobos at Lui Kotale, Salonga National Park, Democratic Republic of Congo. Folia primatologica 79, 103–110.

Hohmann, G., Fowler, A., Sommer, V., Ortmann, S., 2006. Frugivory and gregariousness of Salonga bonobos and Gashaka chimpanzees: The abundance and nutritional quality of fruit. In: Hohmann, G., Robbins, M.M., Boesch, C. (Eds.), Feeding Ecology in Apes and Other Primates. Cambridge University Press, Cambridge, pp. 123–159.

Hunt, K.D., McGrew, W.C., 2002. Chimpanzees in the dry habitats of Assirik, Senegal and Semliki Wildlife Reserve, Uganda. In: Boesch, C., Hohmann, G., Marchant, L.F. (Eds.), Behavioral Diversity in Chimpanzees and Bonobos. Cambridge University Press, Cambridge, pp. 35–51.

Knott, C.D., 1998. Changes in orangutan diet, caloric intake and ketons in response to fluctuating fruit availability. International Journal of Primatology 19, 1061–1079.

Koenig, A., Borries, C., 2001. Socioecology of Hanuman langurs: The story of their success. Evolutionary Anthropology 10, 122–137.

Lambert, J.E., 1998. Primate digestion: Interactions among anatomy, physiology, and feeding ecology. Evolutionary Anthropology 7, 8–20.

Lanjouw, A., 2002. Behavioral adaptations to water scarcity in Tongo chimpanzees. In: Boesch, C., Hohmann, G., Marchant, L.F. (Eds.), Behavioral Diversity in Chimpanzees and Bonobos. Cambridge University Press, Cambridge, pp. 52–60.

Lee, R.B., DeVore, I. (Eds.) 1968. Man the Hunter. Aldine-Atherton, Chicago, IL.

Leigh, E.G., Windsor, D.M., 1982. Forest production and regulation of primary consumers on Barro Colorado Island. In: Rand, A.S., Windsor, D.M., Leigh, E.G. (Eds.), The Ecology of a Tropical Forest: Seasonal Rhythms and Long-term Changes. Smithsonian Institution Press, Washington, DC, pp. 111–122.

Leighton, M., Leighton, D.R., 1983. Vertebrate responses to fruiting seasonality within a Bornean rain forest. Special Publication of the British Ecological Society 2(1983), 181–196.

Leonard, W.R., 2002. Food for thought. Dietary change was a driving force in human evolution. American Journal of Science 287, 106–115.

Leonard, W.R., Robertson, M.L., Snodgrass, J.J., Kuzawa, C.W., 2003. Metabolic correlates of hominid brain evolution. Comparative Biochemistry Physiology Part A 136, 5–15.

Levey, D.J., Cipollini, M.L., 1998. A glycoalkaloid in ripe fruit deters consumption by cedar waxwings. The Auk 115, 359–367.

Lovejoy, C.O., 1981. The origin of man. Science 211, 341–350.

Matsuzawa, T., 1994. Field experiments on use of stone tools in the wild. In: Wrangham, R.W., McGrew, W.C., de Waal, F.B.M., Heltne, P.G. (Eds.), Chimpanzee Cultures, Harvard University Press, Cambridge, MA, pp. 351–370.

McGrew, W.C., 1975. Patterns of plant food sharing by wild chimpanzees. In: Kondo, S., Kawai, M., Ehara, A. (Eds.), Contemporary Primatology. Karger, Basel, pp. 304–309.

McGrew, W.C., 1992. Chimpanzee Material Culture: Implications for Human Evolution. Cambridge University Press, Cambridge.

McGrew, W.C., 1996. Dominance status, food sharing, and reproductive success in chimpanzees. In: Wiessner, P., Schiefenhoevel, W. (Eds.), Food and the Status Quest: An Interdisciplinary Perspective. Berghan Books, Oxford, pp. 39–45.

McGrew, W.C., Marchant, L.F., Beuerlein, M.M., Vrancken, D., Fruth, B., Hohmann, G., 2007. Prospects for Bonobo Insectivory: Lui Kotal, Democratic Republic of Congo. International Journal of Primatolology 28, 1237–1252.

McNab, B.K., 1984. Physiological convergence amongst ant-eating and termite-eating mammals. Journal of Zoology London 203, 485–510.

Medway, L., 1972. Phenology of a tropical rainforest in Malaya. Biological Journal of Linnean Society 4, 117–146.

Milton, K., 1981. Food choice and digestive strategies of two sympatric primate species. The American Naturalist 117, 476–495.

Milton, K., 1984. Habitat, diet, and activity patterns of free-ranging woolly spider monkeys (*Brachyteles arachnoides* E. Geoffreoy 1806). International Journal of Primatology 5, 491–514.

Milton, K., 1987. Primate diets and gut morphology: Implications for hominid evolution. In: Harris, M., Ross, E.B. (Eds.), Food and Evolution: Toward a Theory of Human Food Habits. Temple University Press, Philadelphia, PA, pp. 93–115.

Milton, K., 1999. A hypothesis to explain the role of meat-eating in human evolution. Evolutionary Anthropology 8, 11–21.

Milton, K., McBee, R.H., 1983. Rates of fermentative digestion in the howler monkey (*Alouatta palliata gray*). Comparative Biochemistry and Physiology 74, 29–31.

Milton, K., van Soest, P.J., Robertson, J.B., 1980. Digestive efficiencies of wild howler monkeys. Physiological and Biochemical Zoology 53, 402–409.

Mittermeier, R.A., Tattersall, I., Konstant, B., Meyers, D.M., Mast, R.B., 1994. Lemurs of Madagaskar. Conservation International, Washington DC.

Napier, P.H., 1981. Catalogue of Primates in the British Museum (Natural History) and Elsewhere in the British Isles, Part 2: Family *Cercopithecidae*, Subfamily *Cercopithecinae*. British Museum (Natural History), London.

Naughton-Treves, L., Treves, A., Chapman, C., Wrangham, R., 1998. Temporal patterns of crop-raiding by primates: Linking food availability in croplands and adjacent forest. Journal of Applied Ecology 35, 596–606.

Newton, P.N., 1992. Feeding and ranging patterns of forest Hanuman langurs (*Presbythis entellus*). International Journal of Primatology. 13, 245–285.

Nishida, T., Wrangham, R.W., Goodall, J., Uehara, S., 1983. Local differences in plant-feeding habits of chimpanzees between Mahale Mountains and Gombe National Park, Tanzania. Journal of Human Evolution. 12, 467–480.

O'Connell, J., Hawkes, K., Blurton-Jones, N., 2002. Meat-eating, grandmothering, and the evolution of early humnan diets. In: Ungar, P.S., Teaford, M.F. (Eds.), Human Diet – Its Origin and Function. Bergin & Garvey, WestPoint, pp. 49–60.

Panger, M.A., Perry, S., Rose, L., Gros-Louis, J., Vogel, E., Mackinnon, K.C., Baker, M., 2002. Cross-site differences in foraging behavior of white-faced Capuchins (*Cebus capuchinus*). American Journal of Physical Anthropology 119, 52–66.

Perry, S., Rose, L., 1994. Begging and transfer of Coati meat by white-faced capuchine monkeys, *Cebus capuchinus*. Primates 35, 409–415.

Peters, C.R., 1987. Nut-like oil seeds: food for monkeys, chimpanzees, humans and probably ape-men. American Journal of Physical Anthropology 73, 333–363.

Peters, C.R., O'Brien, E.M., 1981. The early hominid plant-food niche: Insights from an analysis of plant exploitation by *Homo*, *Pan*, and *Papio* in eastern and southern Africa. Current Anthropology 22, 127–140.

Pruetz, J.D., 2006. Feeding ecology of savanna chimpanzees (*Pan troglodytes verus*) at Fongoli, Senegal. In: Hohmann, G., Robbins, M.M., Boesch, C. (Eds.), Feeding Ecology in Apes and Other Primates. Cambridge University Press, Cambridge and London, pp. 161–182.

Pruetz J.D., Bertolani, P, 2007. Savanna chimpanzees, *Pan troglodytes verus*, hunt with tools. Current Biology 5, 412–417.

Pusey, A.E., Packer, C., 1994. Non-offspring nursing in social carnivores – minimizing costs. Behavioral Ecology 5, 362–374.

Reed, K.E., Rector, A.L., 2007. African Pliocene paleoecology: Hominin habitats, resources, and diets. In: Ungar, P.S. (Ed.), Evolution of the Human Diet: The Known, the Unknown, and the Unknowable. Oxford University Press, New York, pp. 262–288.

Richard, A.F., Goldstein, S.J., Dewar, R.E., 1989. Weed macaques: The evolutionary implications of macaque feeding ecology. International Journal of Primatology 10, 569–591.

Rodman, P.S., 2002. Plants of the apes: Is there a hominoid model for the origins of the hominoid diet? In: Ungar, P.S., Teaford, M.F. (Eds), Human Diet: Its Origin and Function. Bergin & Garvey, WestPoint, CT, pp. 77–109.

Rogers, M.E., Abernethy, K., Bermejo, M., Cipolletta, C., Doran, D., McFarland, K., Nishihara, T., Tutin, C., 2004. Western gorilla diet: A synthesis from six sites. American Journal of Primatology 64, 173–192.

Rowe, N., 1996. The Pictorial Guide to the Living Primates. Pogonias Press, East Hampton and New York.

Silk, J.B., 1978. Patterns of food sharing among mother and infant chimpanzees at Gombe National Park, Tanzania. Folia Primatologica 29, 129–142.

Silver, S.C., Marsh, L.K., 2003. Dietary flexibility, behavioral plasticity, and survival in fragments: Lessons from translocated howlers. In: Marsh, L.K. (Ed.), Primates in Fragments: Ecology in Conservation. Kluwer Academic Press/Plenum Press, New York, pp. 251–265.

Stanford, C.B., 1996. The hunting ecology of wild chimpanzees: Implications for the evolutionary ecology of pliocene hominids. American Anthropologist 98, 96–113.

Strier, K.B., 1991. Diet in one group of woolly spider monkeys, or Muriquis (*Brachyteles arachnoides*). American Journal of Primatology 23, 113–126.

Strum, S.C., 1981. Processes and products of change: Baboon predatory behavior at Gilgil, Kenya. In: Harding, R.S.O., Teleki, G. (Eds.), Omnivorous Primates: Gathering and Hunting in Human Evolution. Columbia University, New York, pp. 255–302.

Takahata, Y., Hasegawa, T., Nishida, T., 1984. Chimpanzee predation in the Mahale Mountains from August 1979 to May 1982. International Journal of Primatology 5, 213–233.

Talebi, M., Bastos, A., Lee, P.C., 2005. Diet of southern muriquis in continuous Brazilian Atlantic forest. International Journal of Primatology 26, 1175–1187.

Teleki, G., 1973. The Predatory Behavior of Wild Chimpanzees. Bucknell University Press, Brunsville, MN.

Terborgh, J., 1983. Five New World Primates: A Study in Comparative Ecology. Princeton, Princeton University Press, Princeton, NJ.

Tooby, J., DeVore, I., 1987. The reconstruction of hominid behavioral evolution through strategic modelling. In: Kinzey, W. (Ed.), The Evolution of Human Behavior: Primate Models. State University of New York Press, Albany, NY, pp. 183–237.

Tutin, C.E.G., Fernandez, M., 1993. Relationships between minimum temperature and fruit production in some tropical forest trees in Gabon. Journal of Tropical Ecology 9, 241–248.

Tutin, C.E.G., Parnell, R.J., White, F., 1996. Protecting seeds from primates: Examples from *Diospyros spp.* in the Lope Reserve, Gabon. Journal of Tropical Ecology 12, 371–384.

Ungar, P.S., Grine, F.W., Teaford, M.F., 2006. Diet of early homo: Evidence and a new model of adaptive versatility. Annual Review of Anthropology 35, 209–228.

van Schaik, C.P., Fox, E.A., Sitompul, A.F., 1996. Manufacture and use of tools in wild Sumatran orangutans: Implications for human evolution. Naturwissenschaften 83, 186–188.

Watts, D.P., Mitani, J.C., 2002. Hunting behavior of chimpanzees at Ngogo Kibale National Park, Uganda. International Journal of Primatology 23, 1–28.

Whiten, A., Goodall, J., McGrew, W.C., Nishida, T., Reynolds, V., Sugiyama, Y., Tutin, C.E.G., Wrangham, R.W., Boesch, C., 1999. Cultures in chimpanzees. Nature 399, 682–685.

Wich, S.A., Utami-Atmoko, S.S., Mitra Seti, T., 2004. Life history of wild Sumatran orangutans (*Pongo abelii*). Journal of Human Evolution 47, 385–398.

Wich, S.A., Geurts, M.L., Mitra Setia, T., Utami-Atmoko, S.S., 2006. Influence of fruit availability on Sumatran orangutan sociality and reproduction. In: Hohmann, G., Robbins, M.M., Boesch, C. (Eds.), Feeding Ecology in Apes and Other Primates. Cambridge University Press, Cambridge and London, pp. 337–358.

Wrangham, R.W., 1987. The significance of African apes for reconstructing human social evolution. In: Kinzey, W.G. (Ed.), The Evolution of Human Behavior. Primate Models. SUNY Press, Albany, NY, pp. 51–71.

Wrangham, R.W., Conklin-Brittain, N.L., Hunt, K.D., 1998. Dietary response of chimpanzees and cercopithecines to seasonal variation in fruit abundance. I. Antifeedants. International Journal of Primatology 19, 949–970.

Wright, P.C., Randrimanantena, M., 1989. Comparative ecology of three sympatric bamboo lemurs in Madagaskar. American Journal of Physical Anthropology 78, 327.

Yamagiwa, J., Maruhashi, T., Yumoto, T., Mwanza, N., 1996. Dietary and ranging overlap in sympatric gorillas and chimpanzees in Kahuzi-Biega National Park, Zaire. In: McGrew, W.C., Marchant, L.F., Nishida, T. (Eds.), Great Ape Societies. Cambridge University Press, Cambridge, pp. 82–89.

2. The Energetics of Encephalization in Early Hominids

J. Josh Snodgrass
Department of Anthropology
University of Oregon
1218 University of Oregon
Eugene, OR 97403 USA
jjosh@uoregon.edu

William R. Leonard
Department of Anthropology
Northwestern University
1810 Hinman Ave.
Evanston, IL 60208 USA
w-leonard1@northwestern.edu

and

Marcia L. Robertson
Department of Anthropology
Northwestern University
1810 Hinman Ave.
Evanston, IL 60208 USA

Keywords Bioenergetics • ecology • hominid evolution • dietary quality • body composition • scaling

Abstract Bioenergetics, the study of the use and transfer of energy, can provide important insights into the ecology and evolution of early hominids. Despite a relatively large brain with high metabolic demands, contemporary humans and other primates have resting metabolic rates (RMRs) that are similar to those of other mammals. As a result, a comparatively large proportion of their resting energy budget is spent on brain metabolism among humans (~20–25%) and other primates (~8–10%) compared to other mammals (~3–5%). To understand this shift in energy budget, Aiello and Wheeler's Expensive Tissue Hypothesis (ETH) posits a metabolic trade-off – a reduction in gut size with brain size increase – to explain this phenomenon. Here, we explore the interrelationships between brain size, body size, diet, and body composition using comparative data for humans, non-human primates, and other mammals. Among living primates, the relative proportion of energy allocated to brain metabolism is positively correlated with dietary quality. Contemporary humans fall at the positive end of this relationship, having both a high quality diet and a large brain. Thus, high costs associated with the large human brain are supported, in part, by energy-rich diets. Although contemporary humans display relatively small guts, primates as a group have gut sizes that are similar to non-primate mammals. In contrast, humans and other primates have significantly less skeletal muscle for their size compared to other mammals. These comparative analyses suggest that alterations in diet quality and body composition were necessary conditions for overcoming the constraints on encephalization. Fossil evidence indicates that brain expansion with the emergence of *Homo erectus* at about 1.8 million years ago was likely associated with important changes in diet, body composition, and body size.

Introduction

Bioenergetics, the study of the use and transfer of energy, can provide important insights into the ecology and evolution of early hominids. Energy dynamics represent a central interface between an organism and its environment; how

J.-J. Hublin and M.P. Richards (eds.), *The Evolution of Hominin Diets: Integrating Approaches to the Study of Palaeolithic Subsistence*, 15–29.

energy is extracted from limited environmental resources and allocated to various somatic functions has consequences in terms of survival and reproduction (Leonard and Ulijaszek, 2002; McNab, 2002; Leonard et al., 2007). Thus, energy provides a useful currency for measuring fitness. Energy dynamics also shape aspects of an organism's life history, given that energy used for functions related to maintenance (e.g., resting metabolic rate [RMR], physical activity, and thermoregulation) cannot be used for production, such as the metabolic costs associated with growth and reproduction.

Energetic studies offer a window into hominid brain evolution, as an increase in the size of this metabolically expensive organ requires a shift in energy allocation – either an absolute increase in energy intake or a reduction in the portion of energy allotted to other components of energy expenditure. Consequently, encephalization may affect an organism's life history pattern and shape variables such as the timing of weaning, age at maturity, and reproductive scheduling (Bogin, 1999, 2002). Non-human primates, including hominids, are distinct from most other mammals in having relatively large brains for their body size, a pattern noted by numerous authors (e.g., Martin, 1990). Modern humans have extended this trend and, with brains averaging approximately 1,300 g, are outside the range of other living primates (Jerison, 1973; Leonard and Robertson, 1992).

The metabolic cost of brain tissue is approximately 240 kcal/kg/day and as such is considerably higher than certain tissues such as skeletal muscle (13 kcal/kg/day, at rest), similar to other organs such as the liver (200 kcal/kg/day), and lower than others such as the heart (440 kcal/kg/day) (Holliday, 1986; Elia, 1992). Given that humans and other primates (including great apes) have RMRs similar to other mammals (Leonard and Robertson, 1992, 1994; Aiello and Wheeler, 1995; Snodgrass et al., 2007) despite their relatively large brains, a comparatively large proportion of the resting energy budget is expended on brain metabolism in living humans (20–25%) and non-human primates (8–10%) compared to other mammals (3–5%) (Leonard and Robertson, 1994; Aiello and Wheeler, 1995).

While many studies of primate brain evolution have concentrated on identifying the causal selective factors associated with encephalization in non-human primates and hominids, other studies have taken a different approach and considered the factors associated with the ability to grow and maintain large brains in these taxa (e.g., Leonard and Robertson, 1994; Aiello and Wheeler, 1995; Leonard et al., 2003). These latter studies have concentrated on elucidating the ways that non-human primates and hominids, in particular, overcame the energetic constraints on encephalization. Following a similar approach, in the present chapter we use comparative data on living mammals (including humans and other primates) coupled with information on fossil hominids to consider the energetics of brain evolution as related to diet, body composition, and body size. We use these comparative data to test several hypotheses. First, we hypothesize that among non-human primates dietary quality (i.e., the energy and nutrient density of the diet) will be inversely related to body mass, as predicted by the Jarman-Bell relationship (Bell, 1971; Jarman, 1974). Second, we predict among non-human primates relative brain size and relative diet quality will be positively associated (i.e., species with relatively large brains will consume relatively high quality diets). Third, we hypothesize that non-human primates will have smaller gut sizes than non-primate mammals, as predicted by the Expensive Tissue Hypothesis (Aiello and Wheeler, 1995). Finally, we predict that non-human primates will have less total skeletal muscle mass compared to non-primate mammals.

Materials and Methods

We compiled data from published sources on RMR, diet, and body size for living humans and non-human primates (Table 2.1). Only adult animals were included, and for each species we calculated a single unweighted combined-sex average for each variable. Additionally, we compiled brain size and body mass estimations for fossil hominid species (Table 2.2).

Resting metabolic rate and body mass (kg) data were obtained for 41 primate species, including humans (Table 2.1). All RMR values are expressed as kilocalories per day (kcal/day) and were converted from other units if necessary. RMR, which is the amount of energy used by an inactive animal under thermoneutral conditions, is only one component of the total energy expenditure (TEE) of an animal and thus does not provide a complete picture of energy dynamics. Unfortunately, only minimal data on other energetic parameters (e.g., physical activity, thermoregulation, and the thermic effect of food) are presently available for free-living non-human primates; this severely limits the ability to perform comparative analyses.

We used the dietary quality (DQ) index of Sailer et al. (1985) to estimate the energy and nutrient density of the diet for a variety of primate species. The DQ index is a weighted average of the proportions of plant structural parts (*s*; leaves, stems, and bark), reproductive parts (*r*; fruits, flowers, and nectar), and animal matter (*a*; vertebrates and invertebrates) and is calculated as:

$$DQ = s + 2r + 3.5a \tag{2.1}$$

DQ ranges from a minimum of 100 (100% foliage) to a maximum of 350 (100% animal material). DQ values were available for 32 species of non-human primate (Table 2.1). The diversity of human diets, past and present, prevents calculation of an all-inclusive DQ. However, to get a general picture of human diet, we used the average DQ of five contemporary forager groups based on data compiled by Leonard and Robertson (1994).

Body composition data considered in the present study include measures of the mass of the gastrointestinal (GI) tract, skeletal

TABLE 2.1. Metabolic rate, body mass, brain mass, and diet quality (DQ) in primates (From Bauchot and Stefan, 1969; Jerison, 1973; Stephan et al., 1981; Richard, 1985; Sailer et al., 1985; McNab and Wright, 1987; Leonard and Robertson, 1994; Thompson et al., 1994; Kappeler, 1996; Rowe, 1996).

	Metabolic data		Brain data		Dietary data
Species	RMR (kcal/day)	Body mass (kg)	Brain mass (g)	Body mass (kg)	Diet quality
Alouatta palliata	231.9	4.670	51	6.400	136
Aotus trivirgatus	52.4	1.020	16	0.850	177.5
Arctocebus calabarensis	15.2	0.206	7.2	0.323	327.5
Callithrix geoffroyi	27.0	0.225	7.6	0.280	235
Callithrix jacchus	22.8	0.356	7.6	0.280	235
Cebuella pygmaea	10.1	0.105	4.5	0.140	249.5
Cercopithecus mitis	407.7	8.500	76	6.500	201.5
Cercocebus torquatus	196.2	4.000	104	7.900	234
Cheirogaleus medius	22.7	0.300	3.1	0.177	
Colobus guereza	357.9	10.450	73	7.000	126
Erythrocebus patas	186.9	3.000	118	8.000	
Eulemur fulvus	42.0	2.397	25.2	2.397	129
Euoticus elegantulus	25.1	0.260	7.2	0.274	230
Galago moholi	13.9	0.155			
Galago senegalensis	18.1	0.215	4.8	0.186	278
Galagoides demidoff	6.3	0.058	3.4	0.081	305
Homo sapiens	1,400.0	53.500	1,295	53.500	263
Hylobates lar	123.4	1.900	102	6.000	181
Lemur catta	45.1	2.678	25.6	2.678	166
Leontopithecus rosalia	51.1	0.718			
Lepilemur ruficaudatus	27.6	0.682	7.6	0.682	149
Loris tardigradus	14.8	0.284	6.6	0.322	327.5
Macaca fascicularis	400.9	7.100	74	5.500	200
Macaca fuscata	485.4	9.580	84	5.900	223
Macaca mulatta	231.9	5.380	110	8.000	159
Microcebus murinus	4.9	0.054	1.8	0.054	
Nycticebus coucang	32.4	1.380	12.5	0.800	
Otolemur crassicaudatus	47.6	0.950	10.3	0.850	195
Otolemur garnettii	47.8	1.028			275
Pan troglodytes	581.9	18.300	420	46.000	178
Papio anubis	342.9	9.500	205	26.000	207
Papio cynacephalus	668.9	14.300	195	19.000	184
Papio papio	297.3	6.230	190	18.000	
Papio ursinus	589.3	16.620	190	18.000	189.5
Perodicticus potto	41.3	1.000	14	1.150	190
Pongo pygmaeus	569.1	16.200	370	55.000	172.5
Propithecus verreauxi	86.8	3.080	26.7	3.480	200
Saguinus geoffroyi	50.5	0.500	10	0.380	263
Saimiri sciureus	68.8	0.850	22	0.680	323
Tarsius syrichta	8.9	0.113			350
Varecia variegata	69.9	3.512	34.2	3.512	

TABLE 2.2. Geological ages (millions of years ago), brain size (cm^3), reconstructed male and female body mass (kg), and postcanine tooth size (surface area; mm^2) for selected fossil hominids (From McHenry and Coffing (2000), except for *Homo erectus*. Early *H. erectus* brain size is the average of African specimens as presented in McHenry (1994b), Indonesian specimens from Antón and Swisher (2001) and Georgian specimens from Gabunia et al. (2000, 2001). Data for late *H. erectus* are from McHenry (1994a)).

			Body mass		
Species	Geological age (million years)	Brain size (cm^3)	Male (kg)	Female (kg)	Postcanine tooth size (mm^2)
A. afarensis	3.9–3.0	438	45	29	460
A. africanus	3.0–2.4	452	41	30	516
A. boisei	2.3–1.4	521	49	34	756
A. robustus	1.9–1.4	530	40	32	588
H. habilis (sensu strictu)	1.9–1.6	612	37	32	478
H. erectus (early)	1.8–1.5	863	66	54	377
H. erectus (late)	0.5–0.3	980	60	55	390
H. sapiens	0.4–0.0	1,350	58	49	334

muscle, and body fat. To examine the relationship between GI mass (g) and body mass (kg) in primates and other mammals, we compiled published data (Pitts and Bullard, 1968; Tipton and Cook, 1969; Chivers and Hladik, 1980); typesetting errors in the Chivers and Hladik (1980) paper were corrected (D.J. Chivers, personal communication 2000). Total GI mass represents the combined mass of the stomach, small intestine, cecum, and colon. Data were available for 23 species of primates (including humans) and 56 species of non-primate mammals.

Information on skeletal muscle mass (g) and body mass (kg) was compiled from published sources (Wang et al., 2001; Muchlinski et al., 2003, in preparation). Data were available for 22 species of primates (including humans) and 56 species of non-primate mammals. We examined the relationship of muscularity to locomotor behavior by classifying each species (excluding humans and bats) as arboreal or terrestrial according to primary locomotor habit; while this dichotomy is overly simplistic, it is used to get a general picture of habitat use.

Allometric relationships were determined using ordinary least squares regressions of $\log_{10}$-transformed data. Pearson's correlations were used to assess the relationship between DQ and body mass, as well as between relative DQ and relative brain mass. Human data were excluded from the calculation of correlations and regression parameters, unless indicated. Differences between primate and non-primate mammalian regression parameters were assessed using Student's t-tests. One-Way ANOVA (Scheffe's post-hoc test) was used to assess differences between terrestrial and arboreal primates and non-primate mammals. All analyses were performed using SPSS 12.0 (Chicago, IL).

Results

The scaling relationship of RMR and body mass among primates (including humans) is $RMR = 54.7 Mass^{0.81}$ ($r^2 = 0.94$). This is similar to the relationship seen across mammals (i.e., the Kleiber scaling relationship): $RMR = 70 Mass^{0.75}$. Humans fall almost exactly on the primate regression line (standardized residual = 0.08).

Dietary quality shows a significant inverse correlation with body mass among primates, with humans excluded (n = 32) ($P < 0.001$; $r^2 = 0.46$). We used a DQ value for humans of 263, which is an average of five human foraging populations (!Kung [235.5], Ache [263.0], Hiwi [287.0], Ituri Pygmies [252.5], and Inuit [343.4]). The human DQ value was substantially higher than expected for body size, falling outside the 95% confidence intervals for a regression of DQ versus body mass for all primates (humans included). Despite considerable dietary differences between contemporary forager groups, including differences in percent of energy derived from animal material (e.g., 33% in !Kung vs. 96% in the Inuit), the diets of all five groups fall substantially above the primate regression line.

We considered the relationship between relative brain size and relative DQ among living primate species, including humans (n = 31). There is a strong positive association between the amount of energy allocated to the brain and the caloric and nutrient density of the diet ($P < 0.001$; $r^2 = 0.41$). Humans fall outside the 95% confidence interval for a regression of relative brain size versus relative DQ; humans are extreme outliers for both relative DQ and relative brain size.

The scaling coefficient between gastrointestinal tract mass and body mass is comparable between the primate (humans excluded) and mammalian samples (0.99 ± 0.05 vs. 0.98 ± 0.02; n.s.) (Fig. 2.1). The primate regression has a slightly higher y-intercept than that of non-primate mammals, although this relationship is not significantly different (y-intercept = 1.66 ± 0.04 vs. 1.63 ± 0.02; n.s.). Humans are outside the 95% confidence intervals from a regression of primates and other mammals.

When the scaling relationship of skeletal muscle mass versus body mass is compared between primates and other mammals, primates have significantly lower muscle masses for their body size. Primates have a significantly lower y-intercept than non-primate mammals (2.53 ± 0.02 vs. 2.65 ± 0.01; $P < 0.001$), although the scaling coefficients are significantly different (1.05 ± 0.02 in primates and 0.99 ± 0.01 in non-primate mammals; $P < 0.05$). Mean z-scores are significantly lower in primates ($z = -0.71 \pm 0.22$ in primates vs. 0.27 ± 0.11 in non-primate mammals; $P < 0.001$). The differences in muscularity between arboreal (n = 23) and terrestrial (n = 45) species are evident from the z-scores from the skeletal muscle mass versus body weight regression for all species. For all mammals (including primates), arboreal species are significantly less muscular than terrestrial species ($z = -0.87 \pm 0.23$ vs. 0.44 ± 0.08; $P < 0.001$). Terrestrial mammals are the most well-muscled group ($z = 0.53 \pm 0.09$), having a significantly greater residual score than arboreal mammals ($z = -0.70 \pm 0.45$; $P < 0.001$) and arboreal primates ($z = -0.98 \pm 0.25$; $P < 0.001$). Terrestrial primates ($z = -0.02 \pm 0.18$) have significantly higher z-scores than arboreal primates ($P < 0.05$). Terrestrial mammals are not significantly different than terrestrial primates ($P = 0.31$). Humans fall slightly below (standardized residual = −0.65), although there are substantial differences between males and females (Fig. 2.2).

Discussion

Diet Quality

The relationship of resting metabolism and body mass in mammals is negatively allometric, as RMR scales to the three-quarters power of body mass (Kleiber, 1961). The energetic consequence of this scaling relationship is that small mammals have low total energy needs but high mass-specific energy demands. Conversely, large mammals have high total

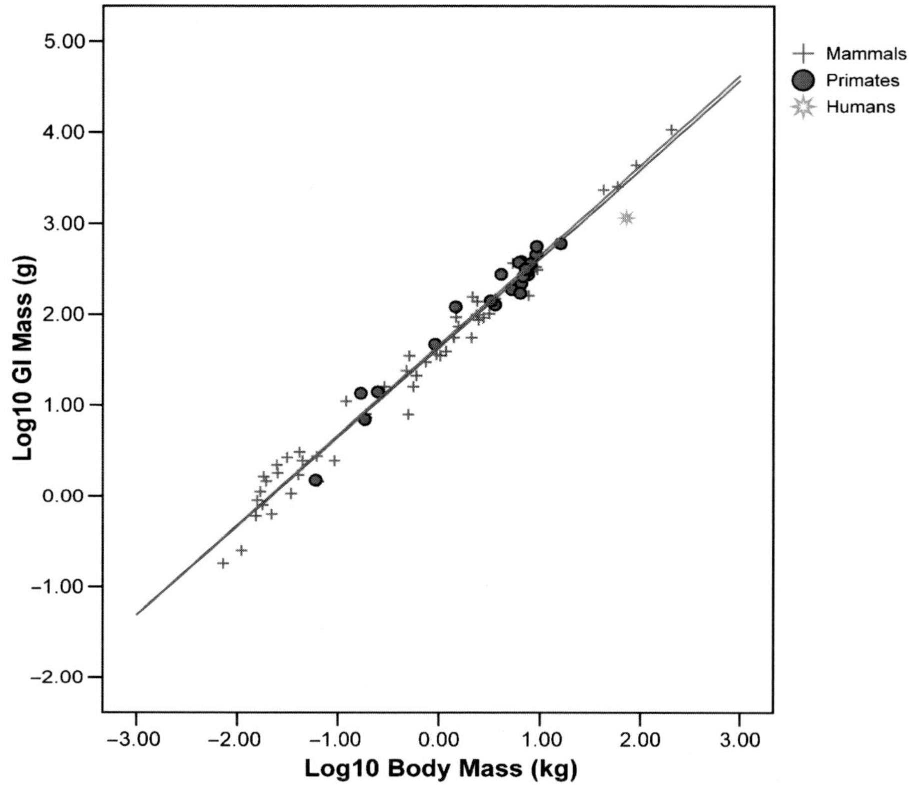

FIG. 2.1. Log-log plot of total gastrointestinal mass (g) vs. body mass (kg) in primates and non-primate mammals. For additional information on sample, see Snodgrass et al., in preparation.

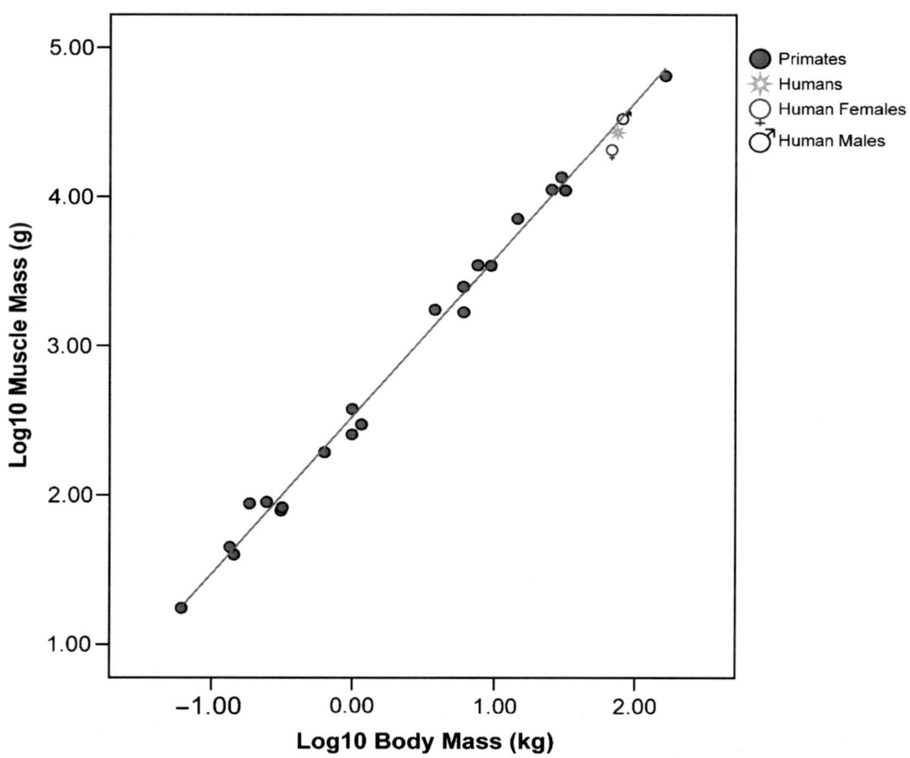

FIG. 2.2. Log-log plot of muscle mass (g) vs. body mass (kg) for primates. Humans fall below the primate regression line (standardized residual = −0.65), indicating they are undermuscled compared to other primates. For additional information on the sample, see Muchlinski et al., in preparation.

energy needs but low mass-specific costs. These metabolic patterns constrain diet and foraging strategies (Bell, 1971; Jarman, 1974; Gaulin, 1979; Leonard and Robertson, 1994; see also Kay, 1984). Small-bodied mammals must consume foods with high caloric and nutritive values (e.g., insects, saps, and gums), which tend to be distributed in patches, while large-bodied mammals typically exploit low quality food items that are nutrient and energy poor and hard to digest (e.g., leaves and bark) but tend to be ubiquitous in the environment.

Results from the present study support the hypothesis that diet quality is inversely related to body mass among primates, and are consistent with results from several earlier studies (Sailer et al., 1985; Leonard and Robertson, 1994). Large-bodied primates (e.g., *Pongo pygmaeus*) generally consume fairly low quality diets that often include leaves, fruit, bark, and shoots, and limited quantities of animal foods, while small primates (e.g., strepsirhines and certain haplorhines), consume higher quality diets that include a considerable amount of animal prey (especially invertebrates) and high quality plant foods.

In contrast to the pattern seen in non-human primates and other mammals, contemporary humans have a diet that is significantly higher in quality than expected for body size. The high quality diet results from the inclusion of energy-dense vegetable foods (e.g., nuts and fruits) and, more importantly, the consumption of large quantities of animal (especially vertebrate) foods. Although contemporary humans have an enormous dietary diversity (even if only foragers are considered), all five foraging populations considered here lie well above the primate regression. Despite variation in the amount and type of foods they eat, most contemporary human foraging populations consume over 50% of their calories from animal sources (Cordain et al., 2000; Kaplan et al., 2000); however, the contribution of hunted foods is influenced by latitude in contemporary foraging populations (Marlowe, 2005). In contrast, fewer than 15% of forager groups obtain more than half their diet from plant foods. The high quality plant and animal foods attractive to human foragers are, in general, more patchily distributed and require more skills to acquire (i.e., through extraction or hunting) than the collected foods that comprise nearly the entire great ape diet (Kaplan et al., 2000). Thus, technology (i.e., tools) and transmission of learned skills and information are particularly important for successful acquisition of these dietary resources.

The inclusion of substantial quantities of animal foods in the human diet contrasts markedly with most other primates who largely rely on plant foods; certain small-bodied species, however, consume large quantities of invertebrates (e.g., insects). Great apes obtain nearly all their calories from plant foods and even the most carnivorous species, the common chimpanzee (*Pan troglodytes*), consumes only 2–13% of its calories from vertebrate foods (Stanford, 1996; Milton, 2003). Field studies indicate that meat is a highly desirable food item for many primate species; modest consumption reflects the limited ability of chimpanzees and other primates to obtain large and consistent quantities of vertebrate foods because of high acquisition costs (Milton, 1999).

In order to test our second hypothesis, which predicts that species with relatively large brains will consume higher quality diets, we examined the relationship between deviations from relative brain size and relative DQ among primates. Consistent with this hypothesis and with our earlier results using a smaller dataset (Leonard and Robertson, 1994) and findings from a recent study (Fish and Lockwood, 2003), we documented a strong positive association between energy allocated to the brain and the caloric and nutrient density of the diet. Therefore, primate species with relatively large brains rely on energy-dense diets to support the high metabolic costs of the brain.

Humans represent an extreme example of this relationship, having the largest brains in the sample and the highest relative DQ. The consumption of an energy-dense and nutrient-rich diet partially offsets the large, metabolically expensive brain, as has been suggested in other studies (Leonard and Robertson, 1994; Aiello and Wheeler, 1995). These empirical findings support Milton's (2003) hypothesis that increased consumption of meat and energy-dense plant foods (e.g., fruit) was necessary for humans to overcome the metabolic constraints on brain expansion. These findings do not imply that dietary change was the impetus for brain expansion among hominids; instead, consumption of a high quality diet was likely a prerequisite for the evolution of a large, energetically expensive brain in hominids. The consumption of nutritionally dense animal foods would have been especially important during early ontogeny, when infants and young children have extremely high metabolic demands from their relatively large energy-expensive brains, yet possess immature digestive morphology and physiology (Kuzawa, 1998; Leonard et al., 2003).

Contemporary humans consume a high quality diet, but to understand the energetics of human brain evolution we must consider the timing of dietary change among earlier hominids. Various lines of evidence (e.g., comparative primate studies, stable isotopes, dental microwear, etc.) suggest that australopithecines consumed a varied and opportunistic diet that was largely composed of plant foods, such as fruits, seeds, and leaves, and included an assortment of C_4 foods (e.g., grasses, sedges, and termites) (Teaford and Ungar, 2000; Sponheimer et al., 2005). Important dietary differences almost certainly existed between species, with certain later australopithecines (e.g., *Australopithecus africanus*) apparently expanding their dietary flexibility and breadth, and robust australopithecines (e.g., *A. robustus*) likely specializing on hard-object feeding. The consumption by australopithecines of limited quantities of animal foods (including invertebrates) is suggested by analogies with living primates (especially *P. troglodytes*), and supported by stable isotope studies and association with putative bone tools likely used for termite extraction (e.g., Backwell and d'Errico, 2001).

Most authorities interpret paleontological and archaeological evidence as suggesting modest dietary change in earliest *Homo* (i.e., *H. habilis*); this species likely incorporated more animal foods in its diet, although the relative amounts obtained through hunting compared to scavenging are debated (Blumenschine, 1987; Harris and Capaldo, 1993; Plummer, 2004). Evidence for dietary change in this species can be seen in the reduced masticatory functional complex (e.g., posterior tooth size); dental reduction in *H. habilis* reversed successive increases in cheek tooth size among the australopithecines (McHenry and Coffing, 2000). Technological advancements, such as the development of Oldowan Industry tools, would likely have allowed easier processing of vertebrate carcasses and increased access to meat, as well as energy and nutrient rich marrow and brains (Semaw et al., 2003; Plummer, 2004).

Multiple lines of evidence suggest a significant dietary shift with the evolution of *H. erectus*; this appears as part of an adaptive shift in this species, which included changes in brain and body size, limb proportions, and various aspects of behavior (Wood and Collard, 1999; Aiello and Key, 2002). However, recent work by Antón (2008) suggests that this foraging shift may have taken place earlier (i.e., with *H. habilis*), and that a smaller dietary shift occurred with the "transition" to *H. erectus*. Fossil evidence suggests that the period beginning approximately 2 million years ago (Ma), with the evolution of *H. habilis* and *H. erectus*, saw the first sizeable increases in brain volume in hominids (Table 2.2). While earlier hominid species showed brain sizes averaging 530 cm^3 or less, brain size increased in *H. habilis* (*sensu strictu*; averaging approximately 610 cm^3) and early *Homo erectus* (averaging approximately 860 cm^3). Although brain size in *H. erectus* is smaller than that of modern humans, it is outside the range seen in non-human primates. However, body size increased too and the enlarged brain size in *H. erectus* may not have represented a grade shift. Further, recent fossil finds attributed by many researchers to *H. erectus*, such as from Ileret (Spoor et al., 2007) and Dmanisi (Lordkipanidze et al., 2007), document fairly small body size in at least some members of the species; this has complicated the picture of this species and has raised questions about the extent of geographic variation and the degree of sexual dimorphism (e.g., Antón et al., 2007).

The adaptive shift seen in *H. erectus*, including dietary change, may have been precipitated by environmental changes in eastern and southern Africa. The first appearance of *H. erectus* at 1.8 Ma (in East Africa; Antón, 2003) is coincident with a punctuated event within the context of a long-term global-scale environmental shift that began in the late Pliocene; this environmental change was characterized by stair step increases in aridity in eastern Africa (deMenocal, 1995; Bobe and Behrensmeyer, 2002; deMenocal, 2004; Wynn, 2004). This climatic shift appears to have heightened climatic variability and led to an overall increase in ecosystem heterogeneity. As a result, this period saw a decrease in forested area and an expansion of open woodlands and grasslands (Hopley et al., 2007). Given this climatic change, the type and distribution of food available to hominids would likely have radically shifted during this period (Behrensmeyer et al., 1997; Plummer, 2004). According to modern savanna ecosystem estimates, primary productivity in early Pleistocene Africa was substantially lower than in the Pliocene, thus limiting the edible plant foods available to hominids (see Leonard and Robertson, 1997, 2000). However, secondary (herbivore) and tertiary (carnivore) trophic-level foods likely increased in abundance; this ecological shift would have increased the overall mammalian biomass-especially of ungulates and other large mammals-available to hominids with the technological and cognitive abilities necessary to exploit this resource (Leonard et al., 2003). In fact, behavioral flexibility within the context of environmental variability and ecosystem heterogeneity may have served as an important selective factor in hominid encephalization (Potts, 1998).

Evidence from archaeological sites has been interpreted by several authorities to suggest a dietary shift in *H. erectus* – specifically, the incorporation of more hunted foods in the diet. *H. erectus* probably occupied a higher predatory position than earlier hominids, given the evidence for early access to mammalian carcasses through hunting and confrontational scavenging (Plummer, 2004). Increasingly sophisticated stone tools (i.e., the Acheulean Industry), which emerged around 1.6–1.4 Ma, almost certainly improved the ability of hominids to process animal and plant materials (Asfaw et al., 1992). Also evident is an increased behavioral complexity that appears to have included food sharing, changes in land-use patterns, and the emergence of a rudimentary hunting and gathering lifeway (Harris and Capaldo, 1993; Rogers et al., 1994).

Dietary change in *H. erectus* has also been inferred from morphological evidence. The reduced size of the posterior teeth and gracility of certain aspects of craniofacial and mandibular morphology are consistent with a diet with less fiber, fewer hard food items, and an overall reduction in the emphasis on mastication (McHenry and Coffing, 2000).

An alternative strategy for increasing dietary quality in early *Homo* was proposed by Wrangham et al. (1999; Wrangham and Conklin-Brittain, 2003; Wrangham, 2007) and focuses on the use of cooking to improve the nutritional density of certain foods. They note that the cooking of savanna tubers and other plant foods would have served to both soften them and increase their energy and nutrient content. In their raw form, the starch in roots and tubers is not absorbed in the small intestine and instead is passed through the body as non-digestible carbohydrate (Tagliabue et al., 1995; Englyst and Englyst, 2005). However, when heated, the starch granules swell and are disrupted from the cell walls, making the starch more accessible to digestive breakdown and increasing the carbohydrate energy available for biological purposes (García-Alonso and Goñi, 2000). Although cooking is clearly an important innovation in hominid evolution that served to increase dietary digestibility and quality, there is very limited evidence for the controlled use of fire by hominids prior to 1.5 Ma (Brain and Sillen, 1988; Bellomo, 1994; Pennisi, 1999). The more widely held view is that the use of fire and cooking did not occur until considerably later in human evolution, probably closer to 200–250,000 years ago (Straus, 1989; Weiner et al., 1998), although possibly as early as 400,000 years ago (Preece et al., 2006). In addition, nutritional analyses of wild tubers used by modern foragers (e.g., Schoeninger et al., 2001) suggest that the energy content of these resources is markedly lower than that of animal foods, even after cooking (Cordain et al., 2001). This, however, does not preclude the possibility that tubers and other plant underground storage organs (USOs) were an important food resource for *H. erectus* and other hominid species, especially as a fallback food (Hatley and Kappelman, 1980; Wrangham et al., 1999; Laden and Wrangham, 2005).

In addition to requiring an energy-dense diet, the human brain has additional demands for essential fatty acids (e.g., long-chain polyunsaturated fatty acids, such as arachidonic acid [AA] and docosahexanoic acid [DHA]) that are critical for optimal neural development and function (Fernstrom and Fernstrom, 2003). As reviewed by Cordain et al. (2001), evolutionary increases in mammalian brain size are apparently constrained by the limited dietary availability in plants of certain fatty acids (i.e., linoleic acid and α-linolenic acid) that are necessary for conversion to AA and DHA. Certain carnivorous species, however, circumvent constraints on endogenous synthesis by directly ingesting AA and DHA in prey species. Limitations in the availability of AA and DHA could have been a barrier to encephalization in australopithecines if they consumed only limited quantities of vertebrate foods. However, early members of the genus *Homo* would have markedly increased their consumption of AA and DHA by direct consumption of these fatty acids in the tissues (e.g., brain, muscle, fat, and liver) of terrestrial mammals (Cordain et al., 2001). Brain tissue is a particularly rich source of both AA and DHA, while liver and muscle are good sources of AA and moderate sources of DHA (Cordain et al., 2001). This scenario is more likely than that proposed by Cunnane and colleagues (e.g., Cunnane and Crawford, 2003), who argue that a shore-based diet (e.g., fish and shellfish) provided the critical nutrients and energy for hominid brain expansion. Given the near complete absence of these foods in early hominid diets and the relatively low energy density of freshwater fish compared to other plant and animal sources, this hypothesis is extremely unlikely (Klein, 1999; Cordain et al., 2001).

Body Composition

Most energetic models use body mass as a single variable without taking into account its constituent components, yet there are dramatic differences among mammals in body composition, even in closely related species. Muscle mass, for example, varies from 24–61% of total body mass in mammals, with slow-moving arboreal mammals (e.g., sloths) occupying the low end and terrestrial carnivores (e.g., felids) occupying the high end (Grand, 1977; Calder, 1984; Muchlinski et al., 2003, in preparation). These differences in body composition contribute to variation in energy demands because of marked differences in mass-specific metabolic rates across organs and tissues. Thus, reductions in organ or tissue mass could theoretically decrease the body's overall energy costs and compensate for the high metabolic demands of the brain. This perspective forms the basis of the Expensive Tissue Hypothesis, which posits that the increased metabolic requirement of an enlarged brain among hominids is offset by a concomitant reduction in gut size since both are metabolically "expensive" tissues (Aiello and Wheeler, 1995; Aiello, 1997; Aiello et al., 2001).

Among mammals, body mass is the prime determinant of the mass of most internal organs. The heart, lungs, kidneys, liver, and spleen all scale with a coefficient nearly identical to or slightly below 1.0, and all regressions have extremely high correlation coefficients (Stahl, 1965). Other tissues (e.g., brain, gut, skeletal muscle, and adipose tissue), seem to be less constrained by body size and vary according to other functional demands (Calder, 1984; Schmidt-Nielsen, 1984; Muchlinski et al., 2003; Wells, 2006). In order to assess whether variation in body composition among primates contributes to the energetics of brain expansion, we compared data on gut mass and skeletal muscle mass in primates with other mammals.

Gut size and proportions are influenced by dietary factors in primates and other mammals (Chivers and Hladik, 1980; Martin et al., 1985; Sussman, 1987). Carnivorous species generally have guts that are dominated by the small intestine, folivores have an enlarged stomach or cecum and colon, and frugivores are morphologically intermediate between carnivores and folivores. Non-human primates have a fairly generalized digestive morphology, which reflects their omnivorous dietary habits; however, certain species (e.g., colobines) have morphological adaptations indicative of a more specialized diet. Most studies to date (e.g., Hladik et al., 1999) have used surface area measures rather than mass to assess gut size (but see Aiello and Wheeler, 1995); however, mass is a more appropriate measure for assessing the energetic implications of interspecific variation in body composition. It should be noted that certain mammals show considerable plasticity in gut dimensions in response to captive diets or seasonal shifts in diet, although this is not true for all species studied (Chivers and Hladik, 1980; Martin et al., 1985). While only limited research has focused on this issue in primates, Milton (2003) notes that humans and great apes display limited gut plasticity and that genetic factors are likely responsible for the divergent gut dimensions in these groups.

Results from the present study indicate that non-human primates have similar sized guts as other mammals. These results do not support our hypothesis that non-human primates have smaller guts than other mammals, and are at odds with the results of Aiello and Wheeler (1995). Although Aiello and Wheeler used a similar primate dataset as the present study, the mammalian sample in that study was largely based on a small number of domesticated species, especially ungulates (based on data from Brody, 1945); consequently, that study likely overestimates "average" mammalian gut size (Snodgrass et al., 1999). In contrast to the Brody (1945) dataset, the sample in the present study included a large number of mammalian species (Pitts and Bullard, 1968; Chivers and Hladik, 1980).

The present study documented a total gut mass in humans significantly smaller than expected for body size, a result similar to that of other studies and reflective of the high quality diet of humans compared to other large-bodied primates (Martin et al., 1985; Aiello and Wheeler, 1995). As suggested by Aiello and Wheeler (1995), the energy cost savings of the reduced GI tract likely play a central role in lowering overall energy costs in humans and help to balance the metabolic costs of an enlarged brain. Studies that have examined gut proportions based on surface area have documented significant differences between humans and other primates, including great apes (e.g., Milton, 1987). The human gut is dominated by the small intestine while the colon is relatively small; in contrast, great apes have relatively modest small intestines and considerably larger colons. These disparities in size and proportions reflect the human adaptation for consumption of a low volume of energy-dense and easily digestible foods, while great apes are adapted for consumption of a fairly low quality diet with greater quantities of difficult to digest plant material. Studies that compared the surface area of gut segments place humans closest to carnivores or to mixed carnivore-frugivores (Martin et al., 1985; Sussman, 1987).

As dietary quality increased during human evolution, the gut likely responded by becoming smaller in overall size and shifting in its proportions in order to maximize extraction from energy and nutrient rich foods. The improved ability of members of the genus *Homo* to process foods extra-orally (i.e., using tools) may also have contributed to the reduction of gut (and tooth) size (Milton and Demment, 1988). It seems unlikely that the small human gut is the result of *direct* selection to decrease metabolic costs and offset the elevated demands of increased brain size, but instead this metabolic balancing was likely an epiphenomenon of increased dietary quality (e.g., animal foods) selecting for smaller gut size (Aiello and Wheeler, 1995; Snodgrass et al., 1999).

The functional dimensions of variation in skeletal muscle mass among primates and other mammals are poorly understood, although studies by Grand (1977, 1978) documented associations between muscularity (i.e., the proportion of total body weight represented by skeletal muscle) and locomotor habits. In general, terrestrial species are more muscular than arboreal species (Grand, 1978). Terrestrial mammals tend to emphasize quick acceleration, attainment of more rapid speeds, and long-distance travel; thus, increased muscularity is likely an adaptation to enhance locomotor performance in order to improve food acquisition capabilities and predator avoidance. Arboreal mammals utilize a strategy that emphasizes locomotor flexibility, passive mechanisms, and reduced activity levels; thus, low muscularity reduces energy costs through decreased RMR and by minimizing the relatively high metabolic costs associated with arboreal movement (Grand, 1978; Elton et al., 1998). Given the arboreal heritage of primates, we hypothesized that primates would have lower levels of skeletal muscle mass compared to non-primate mammals.

Our results indicate that non-human primates are "undermuscled" compared to other mammals, having significantly lower levels of skeletal muscle mass for a given body mass; these results are consistent with our hypothesis. The relatively low levels of skeletal muscle mass may be related to the arboreal heritage of the primate order as, among mammals, arboreal species tend to have lower levels of muscularity (Grand, 1978). Humans fall slightly below the primate regression line, although there are large differences by sex with human females less muscular than males. Relatively low muscularity in humans may reflect a locomotor adaptation to reduce RMR and the costs associated with physical activity. The metabolic costs of skeletal muscle are relatively low at rest (13 kcal/kg/day), and thus small decreases in muscularity are unlikely to substantially lower RMR. However, during physical activity muscle metabolism can increase 100-fold (McArdle et al., 2001). An alternative explanation is that low muscularity, especially among human females, results from increased levels of adipose tissue compared to men (25% vs. 13% on average in non-Western populations; Wells, 2006). As noted by Aiello and Wells (2002), the extent of adiposity in humans (females and males) may partially explain the "location" of humans in interspecific studies of RMR and body mass scaling; greater quantities of adipose tissue, with its low metabolic rate, has the effect of lowering relative metabolic rate.

The extent of human fatness has important implications for the energetics of encephalization. Human adults, including non-Western populations, are fatter than most free-living primates and tropically living mammals (Pond, 1998; Wells, 2006). Energy storage is the primary function of white adipose tissue in humans and other terrestrial mammals; in contrast, aquatic mammals (e.g., cetaceans) apparently store adipose tissue at least in part as an adaptation to cold stress (Kuzawa, 1998; Pond, 1998). The human ability to readily store energy in the form of adipose tissue is a nutritional adaptation to buffer against long-term (e.g., seasonal or periodic) decreases in energy availability. Energy buffering is especially critical during infancy, when the metabolic costs for physical growth and brain metabolism are extremely high (Kuzawa, 1998; Leonard et al., 2003). These metabolic demands are reflected in our unique developmental pattern of fat deposition: human infants are born with high levels of adiposity and continue to gain fat during the first 6 months of postnatal life (Dewey et al., 1993; Kuzawa, 1998; Wells, 2006). In addition, human sex differences in adiposity are shaped by differences in reproductive strategies, given that the enormous energetic costs of pregnancy and lactation are borne largely by females (FAO/WHO/UNU, 1985; Tracer, 1991; Valeggia and Ellison, 2001).

What is most remarkable about human adiposity is our extreme fatness at birth and during early life. At birth, human infants are approximately 15% body fat (Kuzawa, 1998). Compared to the few mammalian species for which published data exist, humans are fatter than domesticated species (e.g., pigs [1.3%]), wild species (e.g., baboons [3%]), and even pinnipeds (e.g., harp seals [10.4%]) (Kuzawa, 1998). Unlike other mammals, humans begin depositing fat prenatally and then continue to accumulate fat during the first 6–9 months of postnatal life (Dewey et al., 1993; Kuzawa, 1998) (see Table 2.3). At its peak in infancy, fat represents on average 25% to over 30% of total body weight (Kuzawa, 1998; Butte et al., 2000).

This unique developmental pattern of fat deposition in humans likely reflects an adaptation to preserve cerebral metabolism in the face of the high metabolic demands of the relatively large brain; these energy demands are obligate and cannot be downregulated in times of energy scarcity (Kuzawa, 1998). The brain relies on glucose as its primary energy source, yet humans have a limited capacity to store glucose. During times of reduced energy intake (e.g., starvation), the primary cerebral energy source is shifted to ketone bodies, which are produced through the mobilization of adipose stores, as well as glucose derived from endogenous production through hepatic gluconeogenesis (Fernstrom and Fernstrom, 2003). The brain is enormously costly in the developing infant, accounting for over 50% of RMR (Holliday, 1986) (see Table 2.3). It is not simply the relative size of the infant brain that explains the high metabolic costs of the brain early in life, but also the rate of energy utilization. Recent studies demonstrate that developmental synaptic overproduction and subsequent pruning results in cerebral metabolic costs that are higher in sub-adults compared to adults – twofold higher glucose utilization uptake rates at 4 years old – and that these rates remain relatively elevated until much later in childhood than previously recognized (Chugani, 1998).

The growing brain is particularly vulnerable to disruptions in energy supply during the nutritional transitions that occur at birth (before the consumption of adequate quantities of breastmilk) and weaning (with complete cessation of the consumption of breastmilk); adipose tissues provide a relatively long-term buffer against limitations in energy intake (Kuzawa, 1998; Pond, 1998; Wells, 2006). The functional link between brain development and body fat is supported by an association between brain size and body fat at birth among mammals. Those species with relatively large adult brain size have larger fat stores at birth; this buffers them against energy disruptions that occur prior to the establishment of adequate energy intake from suckling (Kuzawa, 1998; Leonard et al., 2003).

The other period of heightened vulnerability to nutritional disruption (weaning) also shapes the developmental pattern of fat deposition. In healthy full-term infants fed solely on breast milk, growth typically begins to falter at approximately 6 months of age; supplemental foods (liquids and solids) are typically introduced by this age in virtually all human populations (Whitehead and Paul, 2000; Sellen, 2001; Foote and Marriott, 2003; Kennedy, 2005). Supplemented breastfeeding then continues in most non-Western populations until weaning at 2–3 years of age. This abbreviated weaning schedule departs dramatically from the presumed ancestral condition of protracted lactation in the great apes (e.g., 5 years in *P. troglodytes*). Weaning must occur around this time in humans because the energy and nutrient demands of the relatively large infant brain cannot be met through supplemented breastfeeding (Kuzawa, 1998; Sellen, 2001; Kennedy, 2005). This life history strategy, however, is risky because of the immaturity of the immune and digestive systems and the overall dependence of the child. Consequently, in most traditional human populations, morbidity and mortality rates are high at this age due to the interaction of poor nutritional/dietary quality and increased infectious disease exposure from food and water. Adipose stores accumulated in early infancy provide a critical buffer against the energy disruptions that occur with nutritional "transition" and disease (e.g., diarrhea).

The successful shift to weaning at an earlier age requires weanlings to consume an energy-dense, easily digestible diet to sustain the high metabolic costs of the large brain. High quality foods, such as meat and other animal foods, would have been especially valuable for providing adequate calories and nutrients to the growing child with immature dentition and digestive system (Bogin, 1999; Milton, 1999; Kennedy,

TABLE 2.3. Body mass (kg), brain mass (g), percent body fat (%), resting metabolic rate (RMR; kcal/day), and percent of RMR allocated to brain metabolism (Brain MR; %) for humans of various ages (From Holliday (1986), except body fat data for children (≤18 months) (Dewey et al., 1993)).

Age	Body mass (kg)	Brain mass (g)	Body fat (%)	RMR (kcal/day)	Brain MR (%)
Newborn	3.5	475	16	161	87
3 months	5.5	650	22	300	64
18 months	11.0	1,045	25	590	53
5 years	19.0	1,235	15	830	44
10 years	31.0	1,350	15	1,160	34
Adult male	70.0	1,400	11	1,800	23
Adult female	50.0	1,360	20	1,480	27

2005). Consumption of large quantities of animal foods by young children would have entailed dependence on others for acquisition and preparation (Bogin, 1999; Aiello and Key, 2002). A life history shift that slows growth rates during childhood and the juvenile period also would have helped lower energy costs and allowed for enhanced learning, which would have been particularly important for acquiring hunting and extractive foraging skills (Bogin, 1999; Aiello and Wells, 2002; Kennedy, 2005).

Body Size

Given the limitations inherent in the fossil record, we may never know conclusively when the distinct pattern of body composition emerged in human evolution. However, a considerable amount of information on body size in early hominids can be reconstructed from fossil specimens and provides useful information on the energetics of encephalization. Body size estimates derived from post-cranial fossils suggest that all australopithecine species for which we have adequate information (i.e., *A. afarensis*, *A. africanus*, *A. robustus*, and *A. boisei*) were relatively small-bodied (Table 2.2). Species body weights for males are estimated as 40–49 kg and statures as 130–150 cm. Considerable sexual dimorphism was apparent in all australopithecines; females of each species were on average 29–34 kg and not taller than 125 cm. *Homo habilis* (*sensu strictu*) was no larger in body size than the australopithecines, with male and female body weights of 37 and 32 kg, respectively; stature is reconstructed as 131 cm in males and 100 cm in females. Presently, there is no consensus on whether any postcranial fossils can be definitively assigned to *H. rudolfensis* (Wood and Collard, 1999; McHenry and Coffing, 2000). Given the lack of post-cranial material to serve as a basis for body weight reconstructions, we excluded this species from our study. However, various post-cranial elements from Koobi Fora at ~1.9 Ma, which indicate a fairly large body size (see McHenry and Coffing, 2000), were found in the proximity of cranial remains belonging to *H. rudolfensis* and likely belong to that species. With the appearance of *H. erectus*, body size dramatically increased and reached a weight and height comparable to modern humans; however, as noted above, some recent fossils of apparently small-bodied *H. erectus* (Lordkipanidze et al., 2007; Spoor et al., 2007) have complicated the picture of body size in this species. Based on presently available evidence, body size increase appears to have been most pronounced in *H. erectus* females; if confirmed, one of the remarkable changes in this species is the reduction in sexual dimorphism to within the range of modern humans (Leonard and Robertson, 1997; McHenry and Coffing, 2000; Aiello and Key, 2002). While some evidence suggests a shift in body proportions with the appearance of *H. habilis* (Haeusler and McHenry, 2004), a major shift in body proportions to a linear body form with relatively long legs, was clearly in place in early representatives of *H. erectus* (Ruff and Walker, 1993; Lordkipanidze et al., 2007). This shift likely reflects an adaptation to maximize heat dissipation in the hot and arid environment of eastern and southern Africa (Ruff, 1993). This body size increase and shift to longer legs would have had implications for increased locomotor efficiency, and served to decrease the costs of movement between food sources (Leonard and Robertson, 1997; Steudel-Numbers, 2006).

The larger body size of *H. erectus* would have greatly increased both maintenance (resting) and total energy demands of this species (Leonard and Robertson, 1997). Larger body size coupled with a high quality diet likely would have forced *H. erectus* to expand home ranges, further increasing total energy costs (Leonard and Robertson, 2000). Greater home ranges in this species may help explain why *H. erectus* was the first hominid species to disperse out of Africa (Leonard and Robertson, 2000; Antón et al., 2001, 2002). However, the costs in terms of increased energy needs were likely steep; Leonard and Robertson (1997) estimate a TEE 80–85% greater than that seen in the australopithecines. As indicated by the evolutionary success of this species in its temporal and geographic distributions, greater energy costs were clearly offset by the ability to obtain adequate dietary resources (Leonard and Robertson, 1997, 2000; Aiello and Key, 2002; Aiello and Wells, 2002).

The increase in body mass in *H. erectus*, and in particular the disproportionate increase among females, has implications for the energetics of encephalization (Leonard and Robertson, 1997; Aiello and Key, 2002; Aiello and Wells, 2002). Body size in females is a critical energetic variable because they bear virtually all the costs of reproduction, including providing energy necessary for fetal and early postnatal brain growth and maintenance and fat deposition in the offspring. An absolute increase in metabolism in females affects the ability to transfer energy in reproduction, since greater metabolic turnover allows increasing amounts of energy to be channeled to the offspring (Martin, 1996; Aiello and Key, 2002). Greater female body size also allows for the delivery of a larger brained child, since several important pelvic dimensions are closely associated with stature (Ellison, 2001). Pelvic adaptations to bipedality in humans, however, place important constraints on intra-uterine growth (Rosenberg, 1992). The reduction of gestation lengths and the evolution of secondarily altricial newborns may have emerged with early members of the genus *Homo* as a mechanism for circumventing the pelvic constraints on intra-uterine growth. This change would have extended the very rapid rates of brain growth that are characteristic of fetal development into the first year of an infant's postnatal life (Martin, 1983). Such an important life history shift would have markedly increased maternal energetic demands during lactation. As a consequence, it appears that the evolution of secondary altriciality would have necessitated sufficiently large maternal body sizes that are evident in the hominid lineage only after the emergence of early *Homo*.

Conclusions

In summary, the relative proportion of energy allocated to brain metabolism is positively correlated with diet quality among living primates. Thus, primate species with large brains rely on relatively energy-dense diets to support their high cerebral costs. Australopithecines and other early hominids with brain sizes similar to non-human primates probably increased dietary breadth compared with closely related hominoids but do not appear to have significantly increased diet quality. Thus, dietary factors may have constrained encephalization in the earliest hominids. The remarkable expansion of the brain that began with early *Homo* likely required the following: (1) a shift to a higher quality diet, with a substantial quantity of animal foods; (2) an increase in body size, particularly among females, which allowed greater transfer of energy to the offspring for brain metabolism and fat deposition; and (3) increased levels of body fat early in life to act as an energy buffer for brain metabolism. Important changes in body composition also appear to have resulted from these changes. A reduction in overall gut size and a change in gut proportions was likely a consequence of the shift to a more energetically dense and easily digestible diet. In addition, decreased muscularity was likely a byproduct of increased body fatness. These reductions in gut size and muscle mass would have decreased the metabolic costs associated with somatic maintenance and partially offset increases in cerebral metabolism.

Acknowledgments. We thank L.C. Aiello, S.C. Antón, C.W. Kuzawa, M.N. Muchlinski, and C.J. Terranova for discussions of the project. We thank D.J. Chivers for providing assistance with interpreting gastrointestinal data and for providing corrections for typesetting errors in the original paper.

References

Aiello, L.C., 1997. Brains and guts in human evolution: the expensive tissue hypothesis. Brazilian Journal of Genetics 20, 141–148.

Aiello, L.C., Key, C., 2002. Energetic consequences of being a *Homo erectus* female. American Journal of Human Biology 14, 551–565.

Aiello, L.C., Wells, J.C.K., 2002. Energetics and the evolution of the genus *Homo*. Annual Review of Anthropology 31, 323–338.

Aiello, L.C., Wheeler, P., 1995. The expensive-tissue hypothesis: the brain and the digestive system in human and primate evolution. Current Anthropology 36, 199–221.

Aiello, L.C., Bates, N., Joffe, T., 2001. In defense of the expensive tissue hypothesis. In: Falk, D., Gibson, K. (Eds.), Evolutionary Anatomy of the Primate Cerebral Cortex. Cambridge University Press, Cambridge, pp. 57–78.

Antón, S.C., 2003. A natural history of *Homo erectus*. Yearbook of Physical Anthropology 46, 126–170.

Antón, S.C., 2008. Framing the question: diet and evolution in early *Homo*. In: Vinyard, C., Ravosa, M.J., Wall, C. (Eds.), Primate Craniofacial Function and Biology. Springer, New York.

Antón, S.C., Swisher, C.C., 2001. Evolution of cranial capacity in Asian *Homo erectus*. In: Indriati, E. (Ed.), A Scientific Life: Papers in Honor of Dr. T. Jacob. Bigraf, Yogyakarta, pp. 25–39.

Antón, S.C., Aziz, F., Zaim, Y., 2001. Plio-Pleistocene Homo: patterns and determinants of dispersal. In: Tobias, P.V., Raath, M.A., Moggi-Cecchi, J., Doyle, G.A. (Eds.), Humanity from African Naissance to Coming Millenia. Florence University Press, Florence, pp. 91–102.

Antón, S.C., Leonard, W.R., Robertson, M.L., 2002. An ecomorphological model of the initial hominid dispersal from Africa. Journal of Human Evolution 43, 773–785.

Antón, S.C., Spoor, F., Fellmann, C.D., Swisher, C.C., 2007. Defining *Homo erectus*: size considered. In: Henke, W., Tattersall, I. (Eds.), Handbook of Paleoanthropology, Vol. 3. Springer, Berlin, pp. 1655–1693.

Asfaw, B., Beyene, Y., Suwa, G., Walter, R.C., White, T.D., WoldeGabriel, G., Yemane, T., 1992. The earliest Acheulean from Konso-Gardula. Nature 360, 732–735.

Backwell, L.R., d'Errico, F., 2001. Evidence of termite foraging by Swartkrans early hominids. Proceedings of the National Academy of Sciences of the USA 98, 1358–1363.

Bauchot, R., Stefan, H., 1969. Encephalisation et niveau evolutif ches les simiens. Mammalia 33, 225–275.

Behrensmeyer, K., Todd, N.E., Potts, R., McBrinn, G.E., 1997. Late Pliocene faunal turnover in the Turkana basin, Kenya and Ethiopia. Science 278, 1589–1594.

Bell, R.H., 1971. A grazing ecosystem in the Serengeti. Scientific American 225, 86–93.

Bellomo, R.V., 1994. Methods of determining early hominid behavioral activities associated with the controlled use of fire at FxJj 20 Main, Koobi Fora. Journal of Human Evolution 27, 173–195.

Blumenschine, R.J., 1987. Characteristics of an early hominid scavenging niche. Current Anthropology 28, 383–407.

Bobe, R., Behrensmeyer, A.K., 2002. Faunal change, environmental variability and late Pliocene hominin evolution. Journal of Human Evolution 42, 475–497.

Bogin, B., 1999. Patterns of Human Growth, 2nd ed. Cambridge University Press, Cambridge.

Bogin, B., 2002. The evolution of human growth. In: Cameron, N. (Ed.), Human Growth and Development. Academic Press, Amsterdam, pp. 295–320.

Brain, C.K., Sillen, A., 1988. Evidence from the Swartkrans cave for the earliest use of fire. Nature 336, 464–466.

Brody, S., 1945. Bioenergetics and Growth: With Special Reference to the Efficiency Complex in Domestic Animals. Reinhold, New York.

Butte, N.F., Hopkinson, J.M., Wong, W.W., Smith, E.O., Ellis, K.J., 2000. Body composition during the first 2 years of life: an updated reference. Pediatric Research 47, 578–585.

Calder, W.A., 1984. Size, Function, and Life History. Harvard University Press, Cambridge, MA.

Chivers, D.J., Hladik, C.M., 1980. Morphology of the gastrointestinal tract in primates: comparisons with other mammals in relation to diet. Journal of Morphology 166, 337–386.

Chugani, H.T., 1998. A critical period of brain development: studies of cerebral glucose utilization with PET. Preventative Medicine 27, 184–197.

Cordain, L., Brand Miller, J., Eaton, S.B., Mann, N., Holt, S.H.A., Speth, J.D., 2000. Plant-animal subsistence ratios and macronutrient energy estimations in worldwide hunter-gatherer diets. American Journal of Clinical Nutrition 71, 682–692.

Cordain, L., Watkins, B.A., Mann, N.J., 2001. Fatty acid composition and energy density of foods available to African hominids: evolutionary implications for human brain development. World Review of Nutrition and Dietetics 90, 144–161.

Cunnane, S.C., Crawford, M.A., 2003. Survival of the fattest. Comparative Biochemistry and Physiology 136A, 17–36.

deMenocal, P.B., 1995. Plio-Pleistocene African climate. Science 270, 53–59.

deMenocal, P.B., 2004. African climate change and faunal evolution during the Pliocene-Pleistocene. Earth and Planetary Science Letters 220, 3–24.

Dewey, K.G., Heinig, M.J., Nommsen, L.A., Peerson, J.M., Lonnderdal, B., 1993. Breast-fed infants are leaner than formula-fed infants at 1 y of age: the Darling study. American Journal of Clinical Nutrition 52, 140–145.

Elia, M., 1992. Organ and tissue contribution to metabolic rate. In: Kinney, J.M., Tucker, H.N. (Eds.), Energy Metabolism: Tissue Determinants and Cellular Corollaries. Raven Press, New York, pp. 61–79.

Ellison, P.T., 2001. On Fertile Ground: A Natural History of Human Reproduction. Harvard University Press, Cambridge.

Elton, S., Foley, R., Ulijaszek, S.J., 1998. Habitual energy expenditure of human climbing and clambering. Annals of Human Biology 25, 523–531.

Englyst, K.N., Englyst, H.N., 2005. Carbohydrate bioavailability. British Journal of Nutrition 94, 1–11.

FAO/WHO/UNU., 1985. Energy and Protein Requirements. WHO Technical Report Series #724. World Health Organization, Geneva.

Fernstrom, J.D., Fernstom, M.H., 2003. Nutrition and the brain. In: Gibney, M., MacDonald, I., Roche, H.M. (Eds.), Nutrition and Metabolism. Blackwell, Ames, IA, pp. 145–167.

Fish, J.L., Lockwood, C.A., 2003. Dietary constraints on encephalization in primates. American Journal of Physical Anthropology 120, 171–181.

Foote, K.D., Marriott, L.D., 2003. Weaning of infants. Archives of Disease in Childhood 88, 488–492.

Gabunia, L., Vekua, A., Lordkipanidze, D., Swisher, C.C., Ferring, R., Justus, A., Nioradze, M., Tvalchrelidze, M., Antón, S.C., Bosinski, G., Joris, O., de Lumley, M.-A., Majsuradze, G., Mouskhelishivili, A., 2000. Earliest Pleistocene cranial remains from Dmanisi, Republic of Georgia: taxonomy, geological setting, and age. Science 288, 1019–1025.

Gabunia, L., Antón, S.C., Lordkipanidze, D., Vekua, A., Justus, A., Swisher, C.C., 2001. Dmanisi and dispersal. Evolutionary Anthropology 10, 158–170.

García-Alonso, A., Goñi, I., 2000. Effect of processing on potato starch: in vitro availability and glycemic index. Nahrung 44, 19–22.

Gaulin, S.J.C., 1979. A Jarman/Bell model of primate feeding niches. Journal of Human Evolution 7, 1–20.

Grand, T.I., 1977. Body weight: its relation to tissue composition, segment distribution, and motor function. American Journal of Physical Anthropology 47, 211–239.

Grand, T.I., 1978. Adaptations of tissues and limb segments to facilitate moving and feeding in arboreal folivores. In: Montgomery, G.G. (Ed.), The Ecology of Arboreal Folivores. Smithsonian, Washington, DC, pp. 231–241.

Haeusler, M., McHenry, H.M., 2004. Body proportions of *Homo habilis* reviewed. Journal of Human Evolution 46, 433–465.

Harris, J.W.K., Capaldo, S., 1993. The earliest stone tools: their implications for an understanding of the activities and behavior of late Pleistocene hominids. In: Berthelet, A., Chavaillon, J. (Eds.), The Use of Tools by Human and Nonhuman Primates. Oxford Science Publications, Oxford, pp. 196–220.

Hatley, T., Kappelman, J., 1980. Bears, pigs, and Plio-Pleistocene hominids: a case for the exploitation of below-ground good resources. Human Ecology 8, 371–387.

Hladik, C.M., Chivers, D.J., Pasquet, P., 1999. On diet and gut size in non-human primates and humans: is there a relationship to brain size? Current Anthropology 40, 695–697.

Holliday, M.A., 1986. Body composition and energy needs during growth. In: Falkner, F., Tanner, J.M. (Eds.), Human Growth: A Comprehensive Treatise, 2nd ed., Vol. 2. Plenum Press, New York, pp. 101–117.

Hopley, P.J., Weedon, G.P., Marshall, J.D., Herries, A.I.R., Latham, A.G., Kuykendall, K.L., 2007. High- and low-latitude orbital forcing of early hominin habitats in South Africa. Earth and Planetary Science Letters 256, 419–432.

Jarman, P.J., 1974. The social organization of antelope in relation to their ecology. Behaviour 58, 215–267.

Jerison, H.J., 1973. The Evolution of the Brain and Intelligence. Academic Press, New York.

Kaplan, H., Hill, K., Lancaster, J., Hurtado, A.M., 2000. A theory of human life history evolution: diet, intelligence, and longevity. Evolutionary Anthropology 9, 156–185.

Kappeler, P.M., 1996. Causes and consequences of life-history variation among strepsirhine primates. The American Naturalist 148, 868–891.

Kay, R.F., 1984. On the use of anatomical features to infer foraging behavior in extinct primates. In: Rodman, P.S., Cant, J.G.H. (Eds.), Adaptations for Foraging in Nonhuman Primates. Columbia University Press, New York, pp. 21–53.

Kennedy, G.E., 2005. From the ape's dilemma to the weanling's dilemma: early weaning and its evolutionary context. Journal of Human Evolution 48, 123–145.

Kleiber, M., 1961. The Fire of Life. Wiley, New York.

Klein, R.G., 1999. The Human Career. University of Chicago Press, Chicago, IL.

Kuzawa, C., 1998. Adipose tissue in human infancy and childhood: an evolutionary perspective. Yearbook of Physical Anthropology 41, 177–209.

Laden, G., Wrangham, R.W., 2005. The rise of the hominids as an adaptive shift in fallback foods: plant underground storage organs (USOs) and australopith origins. Journal of Human Evolution 49, 482–498.

Leonard, W.R., Robertson, M.L., 1992. Nutritional requirements and human evolution: a bioenergetics model. American Journal of Human Biology 4, 179–195.

Leonard, W.R., Robertson, M.L., 1994. Evolutionary perspectives on human nutrition: the influence of brain and body size on diet and metabolism. American Journal of Human Biology 6, 77–88.

Leonard, W.R., Robertson, M.L., 1997. Comparative primate energetics and hominid evolution. American Journal of Physical Anthropology 102, 265–281.

Leonard, W.R., Robertson, M.L., 2000. Ecological correlates of home range variation in primates: implications for hominid evolution. In: Boinski, S., Garber, P. (Eds.), On the Move: How and Why Animals Travel in Groups. University of Chicago Press, Chicago, IL, pp. 628–648.

Leonard, W.R., Ulijaszek, S.J., 2002. Energetics and evolution: an emerging research domain. American Journal of Human Biology. 14, 547–550.

Leonard, W.R., Robertson, M.L., Snodgrass, J.J., Kuzawa, C.W., 2003. Metabolic correlates of hominid brain expansion. Comparative Biochemistry and Physiology. 136A, 5–15.

Leonard, W.R., Robertson, M.L., Snodgrass, J.J., 2007. Energetic models of human nutritional evolution. In: Ungar, P.S. (Ed.), Evolution of the Human Diet: The Known, the Unknown, and the Unknowable. Oxford University Press, Oxford, pp. 344–359.

Lordkipanidze, D., Jashashvili, T., Vekua, A., Ponce de León, M.S., Zollikofer, C.P.E., Rightmire, G.P., Pontzer, H., Ferring, R., Oms, O., Tappen, M., Bukhsianidze, M., Agusti, J., Kahlke, R., Kiladze, G., Martinez-Navarro, B., Mouskhelishvili, A., Nioradz, M., Rook, L., 2007. Postcranial evidence from early *Homo* from Dmanisi, Georgia. Nature 449, 305–310.

Marlowe, F.W., 2005. Hunter-gatherers and human evolution. Evolutionary Anthropology 14, 54–67.

Martin, R.D., 1983. Human Brain Evolution in an Ecological Context. American Museum of Natural History, New York.

Martin, R.D., 1990. Primate Origins and Evolution: A Phylogenetic Reconstruction. Chapman & Hall, London.

Martin, R.D., 1996. Scaling of the mammalian brain: the maternal energy hypothesis. News in Physiological Sciences 11, 149–156.

Martin, R.D., Chivers, D.J., MacLarnon, A.M., Hladik, C.M., 1985. Gastrointestinal allometry in primates and other mammals. In: Jungers, W.L. (Ed.), Size and Scaling in Primate Biology. Plenum Press, New York, pp. 61–89.

McArdle, W.D., Katch, F.I., Katch, V.L., 2001. Exercise Physiology: Energy, Nutrition, and Human Performance. Lippincott, Williams, & Wilkins, Philadelphia, PA.

McHenry, H.M., 1994a. Tempo and mode in human evolution. Proceedings of the National Academy of Sciences of the USA 91, 6780–6786.

McHenry, H.M., 1994b. Behavioral ecological implications of early hominid body size. Journal of Human Evolution 27, 77–87.

McHenry, H.M., Coffing, K., 2000. *Australopithecus* to *Homo*: transformations in body and mind. Annual Review of Anthropology 29, 125–146.

McNab, B.K., 2002. The Physiological Ecology of Vertebrates: A View from Energetics. Cornell University Press, Ithaca, NY.

McNab, B.K., Wright, P.C., 1987. Temperature regulation and oxygen consumption in the Philippine tarsier *Tarsius syrichta*. Physiological Zoology 60, 596–600.

Milton, K., 1987. Primate diets and gut morphology: implications for hominid evolution. In: Harris, M., Ross, E.B. (Eds.), Food and Evolution: Toward a Theory of Human Food Habits. Temple University Press, Philadelphia, PA, pp. 93–115.

Milton, K., 1999. A hypothesis to explain the role of meat-eating in human evolution. Evolutionary Anthropology 8, 11–21.

Milton, K., 2003. The critical role played by animal source foods in human (*Homo*) evolution. Journal of Nutrition 133, 3886S–3892S.

Milton, K., Demment, M., 1988. Digestive and passage kinetics of chimpanzees fed high- and low-fiber diets and comparison with human data. Journal of Nutrition 118, 1–7.

Muchlinski, M.N., Snodgrass, J.J., Terranova, C.J., 2003. Scaling of muscle mass in primates. American Journal of Physical Anthropology (Suppl.) 36, 155.

Muchlinski, M.N., Snodgrass, J.J., Terranova, C.J., in preparation. Muscle mass scaling in primates and other mammals. Journal of Mammalian Evolution.

Pennisi, E., 1999. Did cooked tubers spur the evolution of big brains? Science 283, 2004–2005.

Pitts, G.C., Bullard, T.R., 1968. Some interspecific aspects of body composition in mammals. In: National Research Council (Ed.), Body Composition in Animals and Man. National Academy of Sciences Press, Washington, DC, pp. 45–79.

Plummer, T., 2004. Flaked stones and old bones: biological and cultural evolution at the dawn of technology. Yearbook of Physical Anthropology. 47, 118–164.

Pond, C.M., 1998. The Fats of Life. Cambridge University Press, Cambridge.

Potts, R., 1998. Environmental hypotheses of hominin evolution. Yearbook of Physical Anthropology 41, 93–136.

Preece, R.C., Gowlett, J.A.J., Parfitt, S.A., Bridgland, D.R., Lewis, S.G., 2006. Humans in the Hoxnian: habitat, context and fire use at Beeches Pit, West Stow, Suffolk, UK. Journal of Quaternary Science 21, 485–496.

Richard, A.F., 1985. Primates in Nature. WH Freeman, New York.

Rogers, M.J., Harris, J.W.K., Feibel, C.S.S., 1994. Changing patterns of land use by Plio-Pleistocene hominids in the Lake Turkana basin. Journal of Human Evolution 27, 139–158.

Rosenberg, K.R., 1992. The evolution of modern human childbirth. Yearbook of Physical Anthropology 35, 89–124.

Rowe, N., 1996. The Pictorial Guide to Living Primates. Pogonias Press, New York.

Ruff, C.B., 1993. Climatic adaptation and hominid evolution: the thermoregulatory imperative. Evolutionary Anthropology. 2, 53–60.

Ruff, C.B., Walker, A., 1993. Body size and shape. In: Walker, A., Leakey, R. (Eds.), The Nariokotome *Homo erectus* Skeleton. Harvard University Press, Cambridge, MA, pp. 234–265.

Sailer, L.D., Gaulin, S.J.C., Boster, J.S., Kurland, J.A., 1985. Measuring the relationship between dietary quality and body size in primates. Primates 26, 14–27.

Schmidt-Nielsen, K., 1984. Scaling: Why is Animal Size so Important? Cambridge University Press, Cambridge.

Schoeninger, M.J., Bunn, H.T., Murray, S.S., Marlett, J.A., 2001. Nutritional composition of some wild plant foods and honey used by Hadza foragers of Tanzania. Journal of Food Composition and Analysis 14, 3–13.

Sellen, D.W., 2001. Comparison of infant feeding patterns reported for nonindustrial populations with current recommendations. Journal of Nutrition 131, 2707–2715.

Semaw, S., Rogers, M.J, Quade, J., Renne, P.R., Butler, R.F., Dominguez-Rodrigo, M., Stout, D., Hart, W.S., Pickering, T., Simpson, S.W, 2003. 2.6-million-year-old stone tools and associated bones from OGS-6 and OGS-7, Gona, Afar, Ethiopia. Journal of Human Evolution 45, 169–177.

Snodgrass, J.J., Leonard, W.R., Robertson, M.L., 1999. Interspecific variation in body composition and its influence on metabolic variation in primates and other mammals. American Journal of Physical Anthropology (Suppl.) 28, 255.

Snodgrass, J.J., Leonard, W.R., Robertson, M.L., 2007. Primate bioenergetics. In: Ravosa, M.J., Dagosto, M. (Eds.), Primate Origins: Adaptations and Evolution. Springer, New York, pp. 703–737.

Snodgrass, J.J., Leonard, W.R., Robertson, M.L., in preparation. Interspecific variation in body composition in primates and other mammals: implications for human evolution. Journal of Human Evolution.

Sponheimer, M., Lee-Thorp, J., de Ruiter, D., Codron, D., Codron, J., Baugh, A.T., Thackeray, F., 2005. Hominins, sedges, and termites: new carbon isotope data from Sterkfontein valley and Kruger National Park. Journal of Human Evolution 48, 301–312.

Spoor, F., Leakey, M.G., Gathogo, P.N., Brown, F.H., Antón, S.C., McDougall, I., Kiarie, C., Manthi, F.K., Leakey, L.N., 2007. Implications of new early Homo fossils from Ileret, east of Lake Turkana, Kenya. Nature 448, 688–691.

Stahl, W.R., 1965. Organ weights in primates and other mammals. Science 150, 1039–1042.

Stanford, C.B., 1996. The hunting ecology of wild chimpanzees: implications for the evolutionary ecology of Pliocene hominids. American Anthropologist 98, 96–113.

Stephan, H., Frahm, H., Baron, G., 1981. New and revised data on volumes of brain structures in insectivores and primates. Folia Primatologica 35, 1–29.

Steudel-Numbers, K.L., 2006. Energetics in *Homo erectus* and other early hominins: the consequences of increased lower-limb length. Journal of Human Evolution 51, 445–453.

Straus, L.G., 1989. On early hominid use of fire. Current Anthropology 30, 488–491.

Sussman, R.W., 1987. Species-specific dietary patterns in primates and human dietary adaptations. In: Kinzey, W. (Ed.), Evolution of Human Behavior: Primate Models. SUNY Press, Albany, NY, pp. 151–179.

Tagliabue, A., Raben, A., Heijnen, M.L., Duerenberg, P., Pasquali, E., Astrup, A., 1995. The effect of raw potato starch on energy expenditure and substrate oxidation. American Journal of Clinical Nutrition 61, 1070–1075.

Teaford, M.F., Ungar, P.S., 2000. Diet and the evolution of the earliest human ancestors. Proceedings of the National Academy of Sciences of the USA 97, 13506–13511.

Thompson, S.D., Power, M.L., Rutledge, C.E., Kleiman, D.G., 1994. Energy metabolism and thermoregulation in the golden lion tamarin (*Leontopithecus rosalia*). Folia Primatologica 63, 131–143.

Tipton, I.H., Cook, M.J., 1969. Weight of Total Gastrointestinal Tract and Its Subfractions. Oak Ridge National Laboratories, Oak Ridge, TN.

Tracer, D.P., 1991. Fertility-related changes in maternal body composition among the Au of Papua New Guinea. American Journal of Physical Anthropology 85, 393–405.

Valeggia, C.R., Ellison, P.T., 2001. Lactation, energetics and postpartum fecundity. In: Ellison, P.T. (Ed.), Reproductive Ecology and Human Evolution. Aldine, New York, pp. 85–105.

Wang, Z., Moonseong, H., Lee, R.C., Kotler, D.P., Withers, R.T., Heymsfield, S.B., 2001. Muscularity in human adults: proportion of adipose tissue-free body mass in skeletal muscle. American Journal of Human Biology 13, 612–619.

Weiner, S., Qunqu, X., Goldberg, P., Liu, J., Bar-Yosef, O., 1998. Evidence for the use of fire at Zhoukoudian, China. Science 281, 251–253.

Wells, J.C.K., 2006. The evolution of human fatness and susceptibility to obesity: an ethological approach. Biological Reviews 81, 183–205.

Whitehead, R.G., Paul, A.A., 2000. Long-term adequacy of exclusive breast-feeding: how scientific research has led to revised opinions. Proceedings of the Nutrition Society 59, 17–23.

Wood, B., Collard, M., 1999. The human genus. Science 284, 65–71.

Wrangham, R.W., 2007. The cooking enigma. In: Ungar, P.S. (Ed.), Evolution of the Human Diet: The Known, the Unknown, and the Unknowable. Oxford University Press, Oxford, pp. 308–323.

Wrangham, R.W., Conklin-Brittain, N.L., 2003. The biological significance of cooking in human evolution. Comparative Biochemistry and Physiology A 136, 35–46.

Wrangham, R.W., Jones, J.H., Laden, G., Pilbeam, D., Conklin-Brittain, N.L., 1999. The raw and the stolen: cooking and the ecology of human origins. Current Anthropology 40, 567–594.

Wynn, J.G., 2004. Influence of Plio-Pleistocene aridification on human evolution: evidence from paleosols from the Turkana Basin, Kenya. American Journal of Physical Anthropology 123, 106–118.

3. Meals Versus Snacks and the Human Dentition and Diet During the Paleolithic

Peter William Lucas
Department of Anthropology
George Washington University
2110 G St. NW, Washington DC, USA
e-mail: pwlucas@gwu.edu

Zhongquan Sui
Department of Botany
University of Hong Kong
Pokfulam Road
Hong Kong

Kai Yang Ang
Department of Biological Sciences
National University of Singapore
14 Science Drive 4, Singapore 117543
Republic of Singapore

Hugh Tiang Wah Tan
Department of Biological Sciences
National University of Singapore
14 Science Drive 4, Singapore 117543
Republic of Singapore

Sheau Horng King
Department of Biological Sciences
National University of Singapore
14 Science Drive 4, Singapore 117543
Republic of Singapore

Brooke Sadler
Department of Anthropology
George Washington University
2110 G St. NW, Washington DC, USA

and

Neeraja Peri
Department of Anthropology
George Washington University
2110 G St. NW, Washington DC, USA

J.-J. Hublin and M.P. Richards (eds.), *The Evolution of Hominin Diets: Integrating Approaches to the Study of Palaeolithic Subsistence*, 31–41.

Keywords Diet • evolution • digestive system • chemical reactors • cooking

Abstract An adaptive model for the changes in postcanine tooth size that have been observed during human evolution is presented. The argument is based on integrating the oral processing of food (largely mastication) into the kinetics of the mammalian digestive system. It is concluded that the influence of the mouth on the rate of digestion is probably underestimated. Several pieces of recent research support this, including the fact that the oral exposure of at least some food nutrients to taste receptors is important in the rate of uptake of those nutrients later on. These "cephalic phase responses", long known in outline in physiology, thus have a very specific digestive function. The rate of oral processing must match that of the rest of the digestive system with a critical anatomical variable being postcanine tooth size. The chapter reviews digestive function in a broad sense and then applies its logic to selection on tooth size during the Palaeolithic. It is suggested that postcanine tooth size is regulated evolutionarily to provide a rate of digestive throughput appropriate to diet. Cooking and food processing techniques that developed during the Palaeolithic must have reduced the period of oral processing. Coupled with the advent of meals in the life of early humans, a reduction in tooth size would have reduced a potential "avalanche" of food being sent down from the mouth.

Introduction

Diet is critical to a proper understanding of human evolution. This is partly because the acquisition and processing of food is always a primary focus in ecology, but also because, compared to other large primates at least, the diet of modern humans is quite extraordinary. Modern humans acquire and process their food with the help of tools and tend to eat discretely spaced meals of cooked meat, grains, and root vegetables. In contrast, other large primates consume snacks of raw fruits and leaves almost continually, gathering these with just their hands and mouth. This is an overgeneralization of course – there is overlap – but even considered in detail, the contrast seems sharp. So when, why, and where did such a drastic dietary "transition" take place and how can evidence for it be obtained?

Wrangham et al. (1999) have suggested that a potential explanation for many of the anatomical trends seen in the genus *Homo*, particularly rapid brain expansion and decline in tooth size, could be explained by the adoption of cooking practices at least 1.5 million years ago. Evidence of an early use of fire is fragmentary (Gowlett et al., 1981; Brain and Sillen, 1988; Rowlett, 2000) and is not linked to cooking per se. However, Wrangham and Conklin-Brittain (2003) argue that the latter probably constitutes an ancient practice and that it has reduced the need for extensive chewing for, among other things, softening plant fiber.

One aspect of tooth form definitely associated with the rate of size reduction of food particles in the mouth is tooth size (Lucas, 2004). Reduction in tooth size goes back very early in human evolution (Calcagno and Gibson, 1991), though reduction has continued until recent times (Brace et al., 1987). The general explanation for dental reduction still seems to be that tooth sizes will decrease when selective pressure to maintain them is lost. This chapter is intended to offer alternative suggestions for changes in tooth size that are adaptive. To substantiate these suggestions, we draw on analyses of the mammalian dentition in relation to diet presented by Lucas (2004), coupled with more recent evidence. However, the limitations of any analysis of dental adaptation must immediately be recognized: due to the mechanical nature of the chewing process, an analysis of it can only indicate adaptations towards the physical properties of solid foods. Such studies need, therefore, to be combined with other lines of dietary evidence in order to obtain an accurate picture. Before beginning though, the oft-expressed doubts of anthropologists about whether an analysis of the dentitions of (other) mammals would have much relevance for the examination of dental trends in recent human evolution need to be addressed. Do not the tools used in food preparation, coupled with the possible antiquity of the use of heat to transform food textures (Wrangham et al., 1999), shield the teeth from environmental influences? After all, tools and teeth are obvious mechanical parallels (Holloway, 1967). So is the decline in tooth size seen over 2 million years of human evolution (i.e., the approximate period of evolution of the genus *Homo*) not just a product of disuse, as first suggested many years ago by Brace (1963, 1964)? These general issues will be discussed first because resistance to the value of understanding how teeth work appears to penetrate quite far in mammalian biology. It is hoped that some of the following generality may show teeth quite in a different light than that in which anthropologists usually view them.

Broad Overview of the Digestive System

The Role of the Teeth in Digestion

Not many people research oral function, particularly when this topic is compared to studies of locomotion. For the most part, this can be attributed to the rather slim evolutionary role that seems currently to be allotted to the dentition and the mouth by dentists, mammalian physiologists, and ecologists alike. However, this rather derisory attitude is rather strange because (i) as taught in schools, the mouth is the first part of the digestive system, and (ii) mastication is probably the only unique thing about mammals (Lucas, 2004). Notwithstanding, researchers in many biological disciplines, including nutritionists and some food scientists, apparently do not seem to agree. The following paragraphs document this briefly, trying to indicate why it is so and attempting to change the current way of thinking.

Chemical Digestion

A powerful strain of physiological theory treats the digestive process as though it were a series of industrial-style chemical reactors (Penry and Jumars, 1986, 1987; Alexander, 1991, 1993a, b; 1994; Jumars, 2000a, b). Figure 3.1 shows a diagram of the gut pictured as a series of reaction chambers in the order: stomach → small intestine → large intestine. Each digestive compartment acts in sequence on the food as it passes along. The rate of energy intake matches the rate of expenditure in mammals (Hammond and Diamond, 1997), something achieved by a variety of homeostatic mechanisms (Schwartz et al., 2001; Woods, 2005), and so the gut must be built up to provide necessary delivery rate. In particular, each compartment has to play its part at that rate or it will impede the body's provisioning. This is simply because the rate at which the slowest chamber operates gives the overall rate. There are two types of digestive compartments:

(i) *Continuously-stirred tanks*, which are generally rather wide chambers where particles are temporarily trapped while being constantly agitated, and
(ii) *Plug-flow reactors*, which are narrow tubes within which (ideally anyway) the food experiences something close to unidirectional and laminar flow

The stomach and large intestine have been considered to be tanks. In contrast, at least until Jumars (2000b), the small intestine has been assumed to be an example of a plug-flow reactor. The digestive characteristics of these two reactor types are very different. As explained clearly in Alexander (1994), continuously-stirred tanks tend to have low digestive efficiencies because enzymes entering them are constantly being diluted as they are stirred almost immediately into the fluid mix. In contrast, plug-flow reactors keep enzymic concentrations higher for longer simply because such mixing is absent. So if stirred tanks are poor digestive arrangements, what are they there for? Very likely, they are for dealing with microorganisms. The human stomach, with its very low pH, is intended to kill these microbes to prevent infection. In contrast, the high pH of the first compartment of the stomach of a ruminant or colobine monkey, and also the large intestine in many mammals (including humans), allows microorganisms to be incubated in order to enlist their help in the digestion of some plant compounds.

Simulations based on the above concepts have proved capable of imitating digestion in mammals quite accurately (Penry and Jumars, 1986, 1987; Jumars, 2000a, b), and can predict the optimal size of each chamber in mammals with specified diets (Alexander, 1991, 1993a, b, 1994). What these models lack, though (in any depth anyway because Alexander's papers do include discussion of its role), is a mouth. This omission would not be serious if the mouth was simply an ingestion device. Its role then as a transport center, as it is in most non-mammalian vertebrates, would have no more value in these chemical digestion models than the esophagus has because no significant reactions take place there. This apparent indifference to the digestive activities of the mouth of mammals is paralleled in work with artificial (polymer) models of the mammalian gut (Minekus et al., 1995, 1999). As pictured in Dominy et al. (2004), these also feature ingestion directly into (replicas of) the stomach. The models replicate in vivo digestion from the stomach onwards very accurately.

So what is missed by ignoring the mouth? "Not much" would seem the echo from physiologists who study digestion by in vivo techniques. For example, Karasov and Diamond (1995) studied digestion in mammals and reptiles of similar

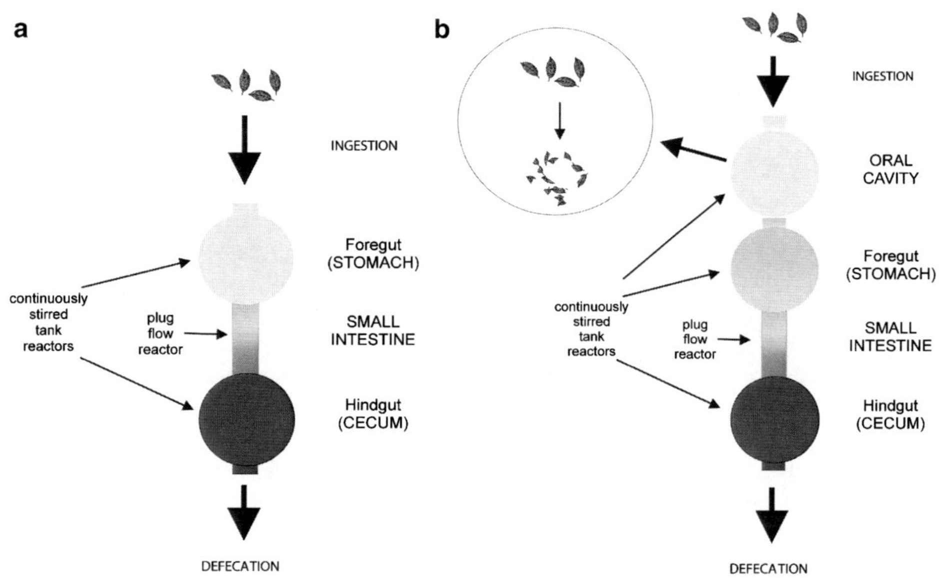

FIG. 3.1. The gut pictured as though it were a sequence of chemical reactors of two types, explained in the text (after Alexander, 1994). The gut shown in (**a**) lacks a mouth, while (**b**) suggests that a mouth could be included by considering it a continuously-stirred tank.

body sizes, concluding that the vastly increased surface area of the mammalian small intestine provides the key to understanding why the mammalian gut delivers energy to the body so much more rapidly than that of reptiles. Yet taking intestinal surface as some evolutionary impediment to digestion in reptiles seems contentious. Obviously, the small intestine has to be up to par just like the rest of the gut; its absorptive capacity must be "enough but not too much" (Diamond, 1991). But are reptiles really restricted from evolving more intestinal surface should this be required? Surely this is a quantitative difference in intestinal physiology appropriate to digestive rate and not the qualitative difference in oral performance provided by the unique and complex set of anatomical modifications that allow mastication in mammals (Lucas, 2004).

The mouth really gets no mention in any of the above work. Yet we believe that there are several aspects of mammalian oral physiology that warrant its inclusion in digestive models. Firstly, there is the physical effect of mastication. Most mammals masticate, and in plant-eating mammals at least, this generally reduces modal food particle sizes by an order of magnitude, reducing from single or multiple particle dimensions measurable somewhere in the centimeter range to the low-millimeter range by the time of swallowing (Lucas, 2004). An order of magnitude of size reduction like this must increase the rate of digestion later on in the gut simply because the reaction rates of liquid enzymes are proportional to the surface areas of the solid food particles on which they act. Since particle size is usually measured as a linear dimension in studies of food comminution in the mouth (Lucas and Luke, 1984), surface area would be expected to increase as the square of particle size. We conclude that if food surfaces are expanded, sometimes by ~100-fold, then the digestive effects of the physical side of mastication cannot easily be dismissed. As a caveat, it should be pointed out that particle size reduction is not the only aim of mastication (Lucas, 2004). The breaking open of plant cells and other chemically-sealed packages like seeds is just as important, because without this, such food may pass through the digestive tract intact. However, having an "all-or-nothing" effect on digestion, this release from chemically-protective packets cannot form part of an argument for extended mastication.

The mouth also contributes to digestion in a chemical way. Usually, such a suggestion would conjure up images of the effect of enzymes in the mouth, particularly the multiple copies of the salivary amylases that constitute the only enzymatic component of the saliva of adult humans. However, here we refer instead to what are usually called "cephalic phase responses." It has long been known that the gut has its own hormonal (neurocrine) regulatory system. Isolated epithelial cells (e.g. those that release gastrin, cholecystokinin, and secretin) monitor gut contents and send (hormonal) signals to chambers further down the gut, preparing them for the food soon to reach them. These signals increase digestive efficiency by preparing the gut for specific food content. What we now know is that some taste receptors are not just detectors that send signals to cortical areas (e.g. the orbitofrontal cortex) of the brain where conscious decisions about food acceptability can be made. They also seem to function analogously to the hormonal cells in providing nutrient-specific information to the gut. The mechanisms of this signaling are different to those lower down the gut, but the general result is the same. They provide a "chemosense" very soon after food ingestion that alerts the body to key food ingredients and gets the body ready for them. The example given below, and which seems most appropriate to a conference's theme (carnivory), is fat detection in the mouth, but sugars could also have been picked (Abdallah et al., 1997).

The brain is largely composed of fat, with an apparent requirement for certain fatty acids (Carlson and Kingston, 2007). The general expense of fuelling it as it enlarged during the last 1–2 million years may have had profound consequences for human body form (Aiello and Wheeler, 1995). As the most energy-dense of foods, fats make sense as key nutrients for members of the genus *Homo*. Hunting activities are now often viewed less as a quest for protein than as a hunt for fats (Speth, 1983; Stanford and Bunn, 2001). While fats themselves are probably not detected chemically in the mouth, the free fatty acids that accompany these foods seem to be (Gilbertson, 1998). This detection mechanism can effectively signal the gut for an input of fat because fatty foods often contain free fatty acids. Via experiments that bypass the mouth, exposure of fatty foods in the mouth has been proven to have a very large effect on the rate of fat absorption in the gut (Mattes, 2001; Tittelbach and Mattes, 2001). In fact, experiments on fatty foods that claim to show the digestive importance of physical aspects of mastication (such as those by Levine and Silva, 1980; Pera et al., 2002) need to be run both with and without oral exposure to foods in order to justify their conclusions.

The final thing to consider in this section, since Alexander (1991, 1993a, b, 1994) has shown the power of such modelling for predicting gut form, is how to fit the mouth onto simulations of the mammalian gut as an effective "front-end" process (Fig. 3.1b). The problem in doing so is twofold. One aspect is the as-yet uncertain concentration effect of what are termed in physiology "cephalic phase responses" on digestive rate. Such cephalic phase responses are pre-absorptive, digestive responses to cues that the special senses pick up about foods, and which help to optimize digestion that will follow after food is swallowed (Nederkoorn et al., 2000). A dog that salivates when it observes food falls into this category (Pavlov, 1927). Recent research in food sciences, though, suggests responses that are far more nutrient-specific than this. The second aspect is the timing of mouth emptying (Alexander, 1994). The mouth does not appear to empty at any particular particle size (among many papers that support this: Yurkstas, 1965; Lucas and Luke, 1986; Hutchings and Lillford, 1988; Lillford, 1991; however, Mishellany et al., 2006 disagree), the "best" models suggest either that there are two separate size and lubrication thresholds that need satisfying before swallowing is initiated (Hutchings and Lillford, 1988), or that some combination of these is in effect via the sensing of bolus formation (Prinz and Lucas, 1997). However, particle size reduction in the mouth should definitely increase the rate at which the stomach empties (as recently-published evidence: Pera et al., 2002) because the pyloric sphincter definitely filters food particle sizes (Meyer, 1980), and the swallowing of large particles of plant foods has been shown to delay stomach emptying significantly (Vincent et al., 1995).

If the mammalian mouth is included in digestive simulations, then it seems clear it should be considered a continuously-stirred tank. The particle mixing that this description assumes is attested to by the fact that food particle size distributions produced by mastication in humans are usually unimodal, even after hundreds of chews (Olthoff et al., 1984; Baragar et al., 1996; but again Mishellany et al., 2006 dissent). Since not all particles get broken in each chew (Lucas and Luke, 1983; van der Glas et al., 1987), this implies that particles are mixed together as they are reduced in size. This tank-like arrangement is probably dictated by comminution – a "batch" style of processing is typical of comminution devices (Rose and Sullivan, 1961; Lowrison, 1974) – and also by the need to wet the surfaces of particles and fragments so as to stick together to form a bolus for efficient transport down the esophagus. The latter factor is insignificant lower down the gut where sufficient fluid is always available to flush out gut chambers; that is, until the distal part of the colon.

Tooth Size

Its Role in Digestion

Having defended the importance of the mouth in mammals to digestion and thus to the rate of energy acquisition, it is now necessary to establish that the size of the teeth within the mouth would have some critical effect on the rate of food processed within it. This is entirely relevant to hominin evolution as a whole where one of the strongest trends has been variation in tooth size (McHenry, 2002). The later (robust) australopithecines tended towards a dentition with large posterior but small anterior teeth. In contrast, early members of the genus *Homo* had relatively larger anterior but much smaller posterior teeth. Since then, there have been periods of marked dental reduction during the evolution of our genus (Brace et al., 1987; Calcagno and Gibson, 1991), making a strong contrast to robust australopithecines where postcanine teeth increased in size. Do these trends correspond to diet? Unfortunately, there are still academic barriers that have to be crossed before deciding this.

The evidence that postcanine tooth size itself affects the rate of food particle size reduction in the mouth is very strong. Half a century of research on humans, mainly stemming from Yurkstas and Manly (1949), Manly and Braley (1950), Yurkstas (1965), and summarized by English et al. (2002), supports this. Very small changes in tooth size (such as those represented by the presence or absence of a single cusp) can make a measurable difference in rate (Luke and Lucas, 1985).

Scaling of Tooth Size

Thus, from the previous section, it would seem reasonable to try to establish a link between tooth size and energy flow. The best way to do this would be to establish the link between tooth size and body mass, since most analyses of energy are tied to it. The rate at which mammals acquire food to satisfy their basal metabolism is well known to be linked to the three-quarter power of their body mass (West et al., 1999, 2001, 2002). The maximum sustainable metabolic rate of humans is about 2–2.5 times the basal rate (without subjects undergoing special training: Westerterp, 2001). This seems broadly true of other larger mammals too, although in some small ones, the multiple can apparently be higher (Hammond and Diamond, 1997). Following West et al., and many other researchers, it seems fair to stick to well-documented trends in basal metabolism rather than to try to analyze active needs.

The chapter has now reached the point that Pilbeam and Gould (1974) reached quite some years ago: what body-mass exponent is expected for scaling patterns of postcanine tooth size? Both Pilbeam and Gould (1974) and Gould (1975) felt that this should be a three-quarters power exponent, because that is how basal metabolic rate scales with body mass (Kleiber, 1961). However, Kay (1975, 1978) and Fortelius (1985) stressed that the rate of energy delivery involves not just the surface areas cleaved by the teeth, but also the frequency of tooth use. This frequency is basically a function of the chewing rate (the timing of a chewing cycle) and frequency of feeding. Dealing with the former first: larger mammals chew more slowly than smaller ones, with the body-mass exponent being between –0.13 (Druzinsky, 1993) and –0.25 (Fortelius, 1985). If the mandible behaves like a swinging limb, then it will have a Froude number – a dimensionless number relating inertia to gravitation that is used heavily in scaling relationships both in engineering and biology (Alexander, 1989). The Froude number has the general form $F = v^2/lg$, where v is a velocity, l is some characteristic length and g is the gravitational constant. Essentially, the dimensionless index reflects the fact that the effect of gravity is to introduce a similarity into structures that move under its influence, e.g. pendulums, such that they have a natural frequency independent of their size. Whether its successful use in biology reflects gravitation (Alexander, 2003), anciently-established neuronal controls (Fischer and Witte, 2007), or both, the gaits of humans and closely-related primates can certainly be compared easily using Froude numbers (Minetti, 2001). Ross and Eckhardt (2006) have recently found that the chewing frequency of the jaw depends on the square root of jaw length, which is a pendular relationship. This result would have chewing frequency scale with a ~–0.13 body-mass exponent (Druzinsky, 1993). However, if inertia does not matter, then chewing frequencies will likely follow most body rhythms that are controlled by central pattern generators (Lund and Lamarre, 1974; Fischer and Witte, 2007) and be related to the –0.25 power of body mass (Peters, 1983). If either hypothesis is true, then this decreases the expected body-mass exponent for postcanine tooth size to below the three-quarters power.

Modelling the process more finely, Fortelius (1985) predicted a body-mass exponent of 0.67 for tooth size, the key assumption being that any given solid food would fail at a given stress. Stress is equivalent to the force applied per unit area, and provided food particles have some standardized shape, then the surface area-to-volume scaling provides the rule. It is, however, rare for solids in which strain energy can be transferred through their constituent parts for the failure stress to be independent of particle size. Instead, such solids get stronger as their particle sizes get smaller (Kendall, 1978). Since particle size reduction is a key element of the chewing process, it makes sense not to assume a constant failure stress. Suspending this assumption gives exponents ranging from 0.50 to 0.67, depending on the way in which a food behaves under load (Lucas, 2004). If the structural components of a food are fully connected, where connectivity refers to the ease of strain energy transmission when the food is loaded, then the lower exponent applies. However, if this is almost completely absent, then the 0.67 exponent predicted by Fortelius (1985) pertains.

A Short Investigation into Chewing Frequencies

When significant sexual dimorphism is present, as in many hominins, there is a problem with the use of body size in understanding allometric trends (Lucas, 1980; Fortelius, 1985). With substantial size differences between the sexes (in primates, males are usually larger, whereas in bats, for example, it can be the reverse – Pickford, 1986), each sex has to be treated separately. This is because the size distributions of the teeth and body mass of adults of such species are bimodal. In the extreme, the pooled means of large male and small females describe almost no individuals in a species. The point is this: when body mass is accounted for, male primates have been found to have relatively smaller postcanine teeth than females (Harvey et al., 1978). This depends totally on the male/female body weight ratio, in that the relatively larger the male, the relatively smaller its dentition becomes (Lucas, 1980; Fortelius, 1985; Lucas et al., 1986a; Cochard, 1987), excepting the permanent canines (Lucas et al., 1986a). The same pattern is found in ungulates (Fortelius (1985) and Carranza et al. (2004) even suggest that the relatively smaller tooth sizes of male red deer are part of an overall design towards the planned senescence of males. Does this all not show then that tooth size is "free" to vary independent of energy flow? Such a conclusion would contradict the above. However, perhaps in a sexually dimorphic species, although larger males have relatively smaller teeth than females, males have a relatively more rapid chewing rate as compensation.

We have started to look at this in primates and have videotaped individuals of nine primate species either at the National Zoo (Washington, DC) or the Singapore Zoological Gardens as they fed. The videotapes were then digitally converted to an MPEG format and the chewing rates measured by watching lip and apparent jaw movements on-screen using a stopwatch, slowing the recordings if necessary to increase the accuracy of timing. The food being eaten was noted, but was not taken into account in the preliminary findings. As shown in Table 3.1, these show no clear relationship between the degree of sexual dimorphism and chewing rate. Smaller females would be predicted to have higher chewing frequencies than larger males. However, this does not always appear to be true because there are contrasting patterns within great apes, for example. There is a case for considering a more accurate method of measurement of jaw movements, but, more likely the sample size in Table 3.1 needs to be vastly increased to expose an overall pattern.

While it is generally true that food texture has little influence on chewing frequencies, they are not obviously fixed by either anatomy or early development of the neural central pattern generators that set bodily rhythms. In an interesting study, Blanchard (2005) provides evidence that in female sheep, chewing frequencies can be elevated during lactation. The prime cause for this, reasons Blanchard, is that lactating sheep have no free time to extend their feeding and need to try to minimize the time spent chewing.

Tooth Size in Human Evolution

We hypothesize that, in sexually-dimorphic species, discrepancies between the sexes in tooth size is compensated for by males chewing relatively faster. Further, we will suppose the arguments presented in this paper so far to apply not just to primates, but to populations of early *Homo* 2 million years ago. These are that (i) the mouth is part of the gut, (ii) the rate at which it delivers comminuted food particles to the gut is critically matched to metabolic needs (just as the small intestine has been shown to be – Diamond, 1991), and (iii) that as in other plant-eating mammals, early *Homo* was time-tight in its feeding habits. Put this all together and postcanine tooth size is clearly going to be one of the critical elements regulating the rate at which the mouth delivers its product to the stomach.

Lucas et al. (1985) made an evolutionary argument for how changes in postcanine tooth size might adapt to patterns of food intake (Table 3.2). Food factors that promote an increase in the sizes of these teeth (towards megadontia) include the frequent presence in the diet of small, chemically sealed particles, ingested in small amounts, fashioned in sheet or rod form, which do not stick together and which are abrasive. Any combination of these factors would tend to slow the rate of food processing in the mouth, either immediately upon ingestion or after the teeth get worn. In contrast, large

Table 3.1. Chewing frequencies in primate species.

Species	Sex	Zoo	Mandibular length[a] (mm)	Body weight[b] (kg)	Average chewing frequency/s^{-1} (No. of chewing cycles)
Pan troglodytes	Male	SING	134	42.7	1.932 (77)
	Female		115	33.7	2.208 (76)
Gorilla gorilla	Male	DC	167	170.4	1.468 (105)
	Female		140	71.5	0.983 (9)
Pongo pygmaeus	Male	DC	151	78.2	1.45 (29)
	Female		133	35.7	1.23 (15)
Hylobates syndactylus	Male	SING	82	11.9	2.42 (418)
	Female		83	10.7	2.148 (94)
Hylobates lar	Male	SING	66	5.9	2.688 (42)
	Female		63	5.34	2.969 (98)
Papio hamadryas	Male	SING	127	19.0	1.991 (124)
	Female			10.7	2.142 (120)
Macaca nigra	Male	DC	95	9.89	2.665 (188)
	Female		74	5.47	2.082 (20)
Nasalis larvatus	Male	SING	83	20.4	2.006 (84)
	Female		71	9.82	2.287 (347)
Lemur catta	Male	SING		2.21	4.776 (104)
	Female			2.21	4.119 (92)

[a] Data from Lucas et al. (1986a).
[b] Data from Smith and Jungers (1997).
SING = Singapore Zoological Gardens; DC = National Zoo, Washington, DC.

Table 3.2. Adaptation of tooth size in relation to dietary properties (From Lucas et al., 1985).

	Microdontia	Megadontia
Ingested food particle size	Large	Small
Ingested batch volume	Large	Small
Particle shape	Thick blocks	Thin sheets
Stickiness	High	Low
Abrasiveness	Low	High

mouthfuls (in terms of overall volume) of large, relatively isodiametric, nonabrasive food particles may be neutral to selection because they are processed so rapidly. Now to this we add the early introduction of cooking, as supposed by Wrangham et al. (1999). Cooking may have two effects: physically, it could affect food texture while chemically, it could accelerate digestion. As yet, there is little published on the chemical effects, but there are clear signs of its textural influence. Lucas (2004) shows this for various foods, but Sui et al. (2006) demonstrates this in detail for a starch-based food, showing that judgments of the cooked state may have a textural basis, i.e., the cooked state could have a physical "signature" that can be perceived.

The effect of cooking food even for a few minutes is evident in the following pilot study carried out on a single subject. Masticating a diet based on 1.5 kg of raw cut carrot (which, as a root vegetable, is relevant to suggested early human diets – Wrangham et al., 1999), when each piece was presented in particle sizes greater than 1 cm in any dimension, took the subject 10,930 chews in all and 318 min of ingestion-mastication-swallowing time. Boiling the carrots for 15 min, but changing nothing else, reduced the number of chews on the same food quantity to 2,558 and 63 min. This represents an enormous time and energy savings. Even though carrot in these quantities does not form an energetically-sustainable diet, these figures are similar to sporadic data given in the literature. The number of chews made per day by modern humans in developed societies has been estimated as ~3,000 (Waters, 1980). This is way below the 16,000+ chews per day recorded for koalas by Logan and Sanson (2002), an accurate estimate made in the wild on animals with unworn teeth. In individuals with unworn teeth, the chew number can reach 44,000.

Trends in Tooth Size: Megadontia Versus Microdontia

Tooth Size Gradients

Predictions about tooth size reduction are complicated by the coordinated growth of tooth germs, which produce morphological gradients of shape and size (Butler, 1939; Sofaer, 1973, 1977; Osborn, 1978). These gradients result from a strip of epithelial tissue, the dental lamina, from which teeth develop, connecting and coordinating the growth of a set of tooth germs (Lumsden, 1979). The number and size of family members is dictated by inhibitory zones around germs that prevent the fusion of neighboring teeth, a feature that applies to all dentitions, even those of non-mammals (Osborn, 1971). Severing the lamina between molars perturbs the gradient, removing some developmental controls (Kavanagh et al., 2007). Evidence from human dental development shows two natural breaks in the lamina, one on either side of the deciduous canine (Öoe, 1957). There are three tooth families or sets thus formed:

the incisors, the canines, and the postcanines, each of which has an internal gradient of size and shape. Their boundaries are clearly indicated in the erupted dentition by visible "breaks" in form and the order in which they form affords a basis for homology (Osborn, 1978). Thus, there is no developmental basis for assuming that the form of the canine might influence the incisors or postcanines. If, say, a lateral incisor actually resembles the canine more than the central incisor, then it might actually be a member of the canine family (Osborn, 1978). The apparent disparity between premolar and molar form is explained by their being from different generations. There are two generations in the mammalian dentition: deciduous and permanent. The deciduous and permanent molars appear to form a single series from the first "deciduous" generation, while the premolars are second generation (Osborn, 1973). The difference in appearance between the last premolar and first permanent molar is due then to difference in generation, with intra-familial gradients within a single generation appearing smooth (Osborn, 1978).

The size relations of the molars are remarkably variable in primates, with (in terms of crown area), M1 < M2 < M3 in cercopithecoids, with M1 < M2 > M3 being more true of apes, and M1 > M2 > M3 in modern humans. Lucas et al. (1986b) showed that most of the variation in the molar gradient of primates could be captured by the M1/M3 ratio (Fig. 3.2) (The M1/M3 ratio can obviously be inverted if this should be felt more appropriate). As currently configured, a low value for the ratio indicates a large molar row while a high ratio indicates a small tooth row. A single value for this ratio, rather than normalizing M2 and M3 to the size of M1 and plotting these against each other, as Kavanagh et al. (2007) and Polly (2007) do, gives the possibility of a regressing a continuous dietary variable against a succinct encapsulation of dental size variation. Figure 3.2 shows a plot of the variation in mandibular M1/M3 as a function of the diet of catarrhine primates. The more leaves and seeds (these are grouped as "small objects") present in the diet, and the more terrestrial the primates are, the smaller the M1/M3 ratio is, i.e., the larger and longer is the molar row. Values for Plio-pleistocene hominins clearly lie in the terrestrial range. This is consistent with a diet for robust australopithecines that includes grasses. It might be gleaned from the graph that it favors early Homo having a diet richer in fruit, but it actually simply states that some other dietary influence is in play that is reducing molar size.

One possibility is the advent of cooking, particularly of underground storage organs. The huge difference in chewing time required by a herbivore like a koala compared to modern humans suggests what cooking may have done for early human populations: it could convert a continually snacking vegetarian animal into one that could limit to food processing to distinct meals. In such circumstances, the ingestion of large quantities of cooked underground storage organs – a diet variously envisaged by Wrangham et al. (1999), Laden and Wrangham (2005) and Yeakel et al. (2007) – might select for a reduction in tooth size (towards microdontia) just to slow the rate of oral processing down. This is the view of molar gradients pictured by Polly (2007).

Of course, humans could have started eating meals when they started to eat meat because, among modern mammals, meals are really the privilege of carnivores. A key dietary factor in humans is the size of the mouthful of food. The bigger the mouthful, the faster it can be processed (Lucas and Luke, 1984). The size of the mouthful probably increased with the advent of the meal, but the size of the average ingested food particle

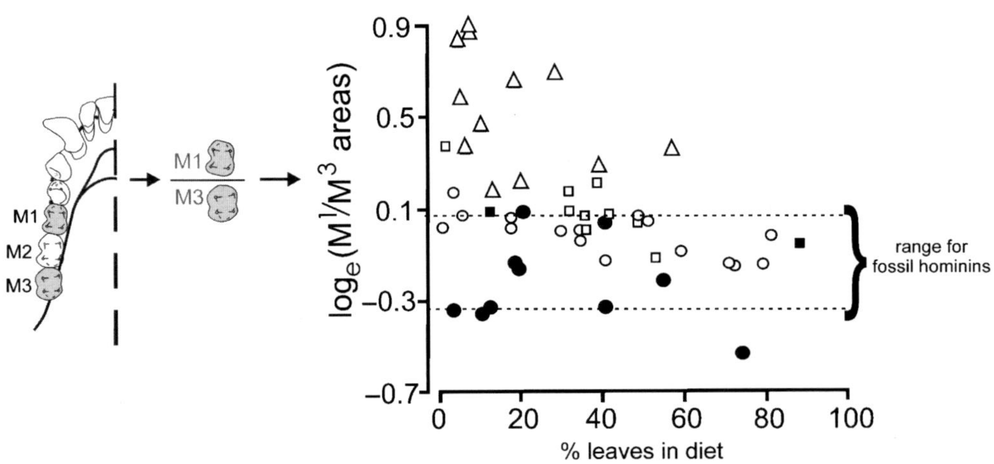

FIG. 3.2. **Left**, the construction of the mandibular M1/M3 ratio as the ratio of the areas of the smallest rectangular box that will fit around each tooth. **Right**, the M1/M3 ratios for maxillary molars in anthropoid primate species (triangles = cebids; circles = cercopithecoids; squares – hominoids) plotted against the percentage of leaves plus seeds reported in their diets (after Lucas et al., 1986b). The lower the value of this ratio, the larger their molar tooth row. Species shown with open symbols are mainly terrestrial, while those with closed symbols are predominantly arboreal. The range for fossil hominins mainly overlaps terrestrial forms.

probably decreased with the advent of tool use. Add this to the potential effect on the mechanical properties of food produced by cooking, and the case for reducing postcanine tooth size for digestive reasons seems to become clear. From the above, cooking may have accelerated the rate at which the mouth provides food to the rest of the gut. If the rest of the gut did not enlarge in some way to accommodate this, then the intestines would fill faster than they could empty (where "emptying" is meant to include the vital function of nutrient delivery to the body). Yet, such gut enlargement would not be needed. Perhaps then, a diminution of postcanine tooth size is nothing more than would be logically expected as a method of lowering this potential food avalanche by decreasing the rate of oral processing. The rate of tooth size decline would reflect the different extra-oral processing techniques that humans developed, including technology as well as cooking refinements. In this context, tooth size is still adaptive: using any part of the body for 3,000 times a day is unlikely to escape selection pressures.

Conclusions

Such a simple analysis might be naïve. Is a reduction in tooth size really that critical in stopping a potential food avalanche to the gut? Might it not be avoided, for example, by a decline in ghrelin release from the stomach (Inui et al., 2004) that alerts an individual to stop feeding? Much depends on the energy density of what is being ingested and the rate of feedback, i.e., on the scale of the "avalanche." However, the analysis might have considerable ramifications. After all, ghrelin and other hormonal controls do not stop obesity in modern humans, so there must be other factors in play. There is much debate in the nutritional literature about the physiological control of feeding, with an underlying assumption about the relationship between nutrient content and food choice. It is often thought that mammals forage on the basis of nutritional need. However, this is easy to question. Wild primates usually have to forage almost ceaselessly, often snacking away on small amounts of differing foods simply because they cannot find any single thing in sufficient abundance. It is not clear how, with such a mixed diet, they can gauge nutrients in their food other than by immediate judgments based on what their special senses tell them (Dominy et al., 2001) and by their individual and group experiences (as appropriate) in this regard. In contrast, a human eats a large amount of a particular food, then stops. In the subsequent period between meals, neural and hormonal feedback from their guts allows association between something like the energy value of food and the item that they have eaten. Perhaps some of the ways in which nutritionists think might have been affected by this, but there appears to be no general evidence that humans can sense the energy value of foods. What is needed now is a joint effort to try to merge nutritional analysis with sensory ecology in primatological studies to establish the principal factors behind foraging and feeding behavior.

References

Abdallah, L., Chabert, M, Louis-Sylvestre, J., 1997. Cephalic phase responses to sweet taste. American Journal of Clinical Nutrition 65, 737–743.

Aiello, L.C., Wheeler P., 1995. The expensive-tissue hypothesis. Current Anthropology 36, 199–221.

Alexander, R.M., 1989. Optimization and gaits in the locomotion of vertebrates. Physiological Reviews 69, 1199–1227.

Alexander, R.M., 1991. Optimization of gut structure and diet for higher vertebrate herbivores. Philosophical Transactions of the Royal Society London Series B 333, 249–255.

Alexander, R.M., 1993a. The energetics of coprophagy: a theoretical analysis. Journal of Zoology 230, 629–637.

Alexander, R.M., 1993b. The relative merits of foregut and hindgut fermentation. Journal of Zoology 231, 391–401.

Alexander, R.M., 1994. Optimum gut structure for specified diets. In: Chivers, D.J., Langer, P. (Eds.) The Digestive System in Mammals. Cambridge University Press. Cambridge, pp. 54–62.

Alexander, R.M., 2003. Principles of Animal Locomotion. Princeton University Press, Princeton.

Baragar, F.A., van der Glas, H.W., van der Bilt, A., 1996. An analytic probability density for particle-size in human mastication. Journal of Theoretical Biology 181, 169–178.

Blanchard, P., 2005. On lactation and rumination in bighorn ewes (*Ovis canadensis*). Journal of Zoology 265, 107–112.

Brace, C.L., 1963. Structural reduction in evolution. American Naturalist 97, 39–49.

Brace, C.L., 1964. The probable mutation effect. American Naturalist 98, 453–455.

Brace, C.L., Rosenberg, K., Hunt, K.D., 1987. Gradual change in human tooth size in the late Pleistocene and post-Pleistocene. Evolution 41, 705–720.

Brain, C.K., Sillen, A., 1988. Evidence from the Swartkrans cave for the earliest use of fire. Nature 336, 464–465.

Butler, P. M., 1939. Studies of the mammalian dentition. Differentiation of the postcanine dentition. Proceedings of the Zoological Society London B 109, 1–36.

Calcagno, J.M., Gibson, K.R., 1991. Selective compromise: evolutionary trends and mechanisms in hominid tooth size. In: Kelley, M.A., Larsen, C.S. (Eds.) Advances in Dental Anthropology. Wiley, New York, pp. 59–76.

Carlson, B.A., Kingston, J.D., 2007. Docohexaenoic acid, the aquatic diet, and hominin encephalization: difficulties in establishing evolutionary links. American Journal of Human Biology 19, 132–141.

Carranza, J., Alarcos, S., Sánchez-Prieto, C.B., Valencia, J., Mateos, C., 2004. Disposable-soma senescence mediated by sexual selection in an ungulate. Nature 432, 215–218.

Cochard, L.R., 1987. Postcanine tooth size in female primates. American Journal of Physical Anthropology 74, 47–54.

Diamond, J., 1991. Evolutionary design of intestinal nutrient absorption: enough but not too much. News in Physiological Sciences 6, 92–96.

Dominy, N.J., Lucas, P.W., Osorio, D., Yamashita, N., 2001. The sensory ecology of primate food perception. Evolutionary Anthropology 10, 171–186.

Dominy, N.J., Davoust, E., Minekus, M., 2004. Adaptive function of soil consumption: an *in vitro* study modeling the human stomach and small intestine. Journal of Experimental Biology 207, 3019–3024.

Druzinsky, R.E., 1993. The time allometry of mammalian chewing movements: chewing frequency scales with body mass in mammals. Journal of Theoretical Biology 160, 427–440.

English, J.D., Buschang, P.H., Throckmorton, G.S., 2002. Does malocclusion affect masticatory performance? The Angle Orthodontist 71, 21–27.

Fischer, M.S., Witte, H., 2007. Legs evolved only at the end! Philosophical Transactions of the Royal Society London series A265, 185–198.

Fortelius, M., 1985. Ungulate cheek teeth: developmental, functional and evolutionary interrelationships. Acta Zoologica Fennica 180, 1–76.

Gilbertson, T.A., 1998. Gustatory mechanisms for the detection of fat. Current Opinions in Neurobiology 8, 447–452.

Gould, S.J., 1975. On the scaling of tooth size in mammals. American Zoologist 15, 351–362.

Gowlett, J.A.J., Harris, J.W.K., Walton, D., Wood, B.A., 1981. Early archaeological sites, hominid remains and traces of fire from Chesowanja, Kenya. Nature 294, 125–129.

Hammond, K.A., Diamond, J., 1997. Maximal sustained energy budgets in humans and animals. Nature 386, 457–462.

Harvey, P.H., Kavanagh, M., Clutton-Brock, T.H., 1978. Sexual dimorphism of primate teeth. Journal of Zoology 186, 475–485.

Holloway, R.L., 1967. Tools and teeth: some speculations regarding canine reduction. American Anthropologist 69, 63–67.

Hutchings, J.B., Lillford, P.J., 1988. The perception of food texture – the philosophy of the breakdown path. Journal of Texture Studies 19, 103–115.

Inui, A., Asakawa, A., Bowyers, C.Y., Mantovani, G., Laviano, A., Meguid, M.M., Fujimiya, M., 2004. Ghrelin, appetite, and gastric motility: the emerging role of the stomach as an endocrine organ. Journal of the Federation of American Societies for Experimental Biology 18, 439–456.

Jumars, P.A., 2000a. Animal guts as ideal chemical reactors: maximizing absorption rates. American Naturalist 155, 527–543.

Jumars, P.A., 2000b. Animal guts as nonideal chemical reactors: partial mixing and axial variation in absorption kinetics. American Naturalist 155, 544–555.

Karasov, W.H., Diamond, J.M., 1995. Digestive adaptations for fuelling the cost of endothermy. Science 228, 202–204.

Kavanagh, K.D., Evans, A.R., Jernvall, J., 2007. Predicting evolutionary patterns of mammalian teeth from development. Nature 449, 427–432.

Kay, R.F., 1975. Allometry and early hominids. Science 189, 63.

Kay, R.F., 1978. Molar structure and diet in extant Cercopithecidae. In: Butler, P.M., Joysey, K.A. (Eds.) Development, Function and Evolution of Teeth. Academic Press, New York, pp. 309–340.

Kendall, K., 1978. The impossibility of comminuting small particles by compression. Nature 272, 710–711.

Kleiber, M., 1961. The Fire of Life: An Introduction to Animal Energetics. Wiley, New York.

Laden, G., Wrangham, R.W., 2005. The rise of the hominids as an adaptive shift in fallback foods: plant underground storage organs (USOs) and australopith origins. Journal of Human Evolution 49, 482–498.

Levine, A.S., Silva, S.E., 1980. Absorption of whole peanuts, peanut oil and peanut butter. New England Journal of Medicine 303, 917–918.

Lillford, P.J., 1991. Texture and acceptability of human foods. In: Vincent, J.F.V., Lillford, P.J. (Eds.) Feeding and the Texture of Food. Cambridge University Press, Cambridge, pp. 93–121.

Logoan, M., Sanson, G.D., 2002. The effect of tooth wear on the activity patterns of free-ranging koalas (*Phascolarctos cinereus*, Goldfuss). Journal of Zoology 50, 281–292.

Lowrison, G.C., 1974. Crushing and Grinding. Butterworths, London.

Lucas, P.W., 1980. Adaptation and Form of the Mammalian Dentition with Special Reference to Primates and the Evolution of Man. Ph.D. thesis. University of London, London.

Lucas, P.W., 2004. Dental Functional Morphology: How Teeth Work. Cambridge University Press, Cambridge.

Lucas, P.W., Luke, D.A., 1983. Methods for analysing the breakdown of food during human mastication. Archives of Oral Biology 28, 813–819.

Lucas, P.W., Luke, D.A., 1984. Optimal mouthful for food comminution in human mastication. Archives of Oral Biology 29, 205–210.

Lucas, P.W., Luke, D.A., 1986. Is food particle size a criterion for the initiation of swallowing? Journal of Oral Rehabilitation 13, 127–136.

Lucas, P.W., Corlett, R.T., Luke, D.A., 1985. Plio-pleistocene hominids: an approach combining masticatory and ecological analysis. Journal of Human Evolution 14, 187–202.

Lucas, P.W., Corlett, R.T., Luke, D.A., 1986a. Sexual dimorphism of teeth in anthropoid primates. Human Evolution 1, 23–39.

Lucas, P.W., Corlett, R.T., Luke, D.A., 1986b. Postcanine tooth size and diet in anthropoids. Zeitschrift für Morphologie und Anthropologie 76, 253–276.

Luke, D.A., Lucas, P.W., 1985. Chewing efficiency in relation to occlusal and other variations in the dentition. British Dental Journal 159, 401–403.

Lumsden, A.G.S., 1979. Pattern formation in the molar dentition of the mouse. Journale de Biologie Buccale 7, 77–103.

Lund, J.P., Lamarre, Y., 1974. Activity of neurons in the lower precentral cortex during voluntary and rhythmical jaw movements in the monkey. Experimental Brain Research 19, 282–299.

Manly, R.S., Braley, L.C., 1950. Masticatory performance and efficiency. Journal of Dental Research 29, 448–462.

Mattes, R.D., 2001. Oral exposure to butter, but not fat replacers elevates post-prandial triaglycerol concentration in humans. Journal of Nutrition 131, 1491–1496.

McHenry, H.M., 2002. Introduction to the fossil record of human ancestry. In: Hartwig, W.C. (Ed.) The Primate Fossil Record. Cambridge University Press, Cambridge, pp. 401–406.

Meyer, J.H., 1980. Gastric emptying of ordinary food: effect of antrum on particle size. American Journal of Physiology 239, G133–135.

Minekus, M., Marteau, P., Havenaar, R., Huis in't Held, J.H.J., 1995. A multicompartmental dynamic computer-controlled model simulating the stomach and small intestine. Alternatives to Laboratory Animals 23, 197–200.

Minekus, M., Smeets-Peters, M., Bernalier, A, Marol, S., Havenaar, R., Marteau, P., Alric, M., Fonty, G., Huis in't Held, J.H.J., 1999. A computer-controlled system to simulate conditions of the large intestine with peristaltic mixing, absorption of fermentation products and a high-density microflora. Applied Microbiology and Biotechnology 53, 108–114.

Minetti, A.E., 2001. Biomechanics: walking on other planets. Nature 409, 467–469.

Mishellany, A., Woda, A., Labas, R., Peyron, M.A., 2006. The challenge of mastication: preparing a bolus suitable for deglutition. Dysphagia 21, 87–94.

Nederkoorn, C., Smulders, F.T., Jansen, A., 2000. Cephalic phase responses, craving and food intake in normal subjects. Appetite 35, 45–55.

Öoe, T., 1957. On the early development of the human dental lamina. Folia Anatomica Japonica 30, 197–211.

Olthoff, L.W., van der Bilt, A., Bosman, F., Kleizen, H.H., 1984. Distribution of particle sizes in food comminuted by human mastication. Archives of Oral Biology 29, 899–903.

Osborn, J.W., 1971. The ontogeny of tooth succession in *Lacerta vvipara* Jacquin (1787). Proceedings of the Royal Society London series B 179, 261–289.

Osborn, J.W., 1973. The evolution of dentitions. American Scientist 61, 548–559.

Osborn, J.W., 1978. Morphogenetic gradients: fields versus clones. In: Butler, P.M., Joysey, K.A. (Eds.) Development, Function and Evolution of Teeth. Academic Press, New York, pp. 171–213.

Pavlov, I.P., 1927. Conditioned Reflexes: An Investigation of the Physiological Activity of the Cerebral Cortex. Anrep, G.V. (Trans. and Ed.). Oxford University Press, London.

Penry, D.L., Jumars, P.A., 1986. Chemical reactor analysis and optimal digestion theory. Bioscience 36, 310–315.

Penry, D.L., Jumars, P.A., 1987. Modeling animal guts as chemical reactors. American Naturalist 129, 69–96.

Pera, P., Bucca, C., Borro, P., Bernocoo, C., De Lillo, A., Carossa, S., 2002. Influence of mastication on gastric emptying. Journal of Dental Research 81, 179–181.

Peters, R.H., 1983. The Ecological Implications of Body Size. Cambridge University Press, Cambridge.

Pickford, M., 1986. On the origins of body size dimorphism in primates. In: Pickford, M., Chiarelli, B. (Eds.) Sexual Dimorphism in living and Fossil Primates. Il Sediciesimo, Firenze, pp. 77–91.

Pilbeam, D.R., Gould, S.J., 1974. Size and scaling in human evolution. Science 186, 892–901.

Prinz, J.F., Lucas, P.W., 1997. An optimization model for mastication and swallowing in mammals. Philosophical Transactions of the Royal Society London series B 264, 1715–1721.

Polly, P.D., 2007. Development with a bite. Nature 449, 413–415.

Rose, H.E., Sullivan, R.M.E., 1961. Vibration Mills and Vibration Milling. Constable Press, London.

Ross, C.F., Eckhardt, A., 2006. Scaling of temporal variables in mammalian mastication. Comparative Biochemistry and Physiology 143A, S87.

Rowlett, R.M., 2000. Fire control by *Homo erectus* in East Africa and Asia. Acta Anthropologica Sinica 19, 198–208.

Schwartz, M.W., Woods, S.C., Porte Jr., D., Seeley, R.L., Baskin, D.G., 2001. Central nervous system control of food intake. Nature 404, 661–671.

Smith, R.J., Jungers, W.L., 1997. Body mass in comparative primatology. Journal of Human Evolution 32, 523–559.

Sofaer, J.A., 1973. A model relating developmental interaction and differential evolutionary reduction of tooth size. Evolution 27, 427–434.

Sofaer, J.A., 1977. Co-ordinated growth of successively initiated tooth germs in the mouse. Archives of Oral Biology 22, 71–72.

Speth, J.D., 1983. Bison Kills and Bone Counts. University of Chicago Press, Chicago.

Stanford, C., Bunn, H.T. (Eds.), 2001. Meat-Eating and Human Evolution. Oxford University Press, New York.

Sui, Z.Q., Lucas, P.W., Corke, H., 2006. Optimal cooking time of noodles related to their notch-sensitivity. Journal of Texture Studies 37, 428–441.

Tittelbach, T.J., Mattes, R.D., 2001. Oral stimulation influences postprandial triaglycerol concentrations in humans: nutrient specificity. Journal of the American College of Nutrition 20, 485–493.

van der Glas, H.W., van der Bilt, A., Olthoff, L.W., Bosman, F., 1987. Measurement of selection chances and breakage functions during chewing in man. Journal of Dental Research 66, 1547–1550.

Vincent, R., Roberts, A., Frier, M., Perkins, A.C., MacDonald, I.A., Spiller, R.C., 1995. Effect of bran particle size on gastric emptying and small bowel transit in humans: a scintigraphic study. Gut 37, 216–219.

Waters, N.E., 1980. Some mechanical and physical properties of teeth. In: Vincent, J.F.V., Currey, J.D. (Eds.) The Mechanical Properties of Biological Materials. Cambridge University Press, Cambridge, pp. 99–135.

West, G.B., Brown, J.H., Enquist, B.J., 1999. The fourth dimension of life: fractal geometry and allometric scaling of organisms. Science 284, 167–169.

West, G.B., Brown, J.H., Enquist, B.J., 2001. A general model for ontogenetic growth. Nature 413, 628–631.

West, G.B., Woodruff, W.H., Brown, J.H., 2002. Allometric scaling of metabolism from molecules and mitochondria to cells and mammals. Proceedings of the National Academy of Sciences 99, 2473–2478.

Westerterp, K., 2001. Limits to maximum sustainable human metabolic rate. Journal of Experimental Biology 204, 3183–3187.

Woods, S.C., 2005. Signals that influence food intake and body weight. Physiology and Behavior 86, 709–716.

Wrangham, R.W., Conklin-Brittain, N.L., 2003. Cooking as a biological trait. Comparative Biochemistry and Physiology series A 136, 35–46.

Wrangham, R.W., Jones, J.H., Laden, G., Pilbeam, D., Conklin-Brittain, N.L., 1999. The raw and the stolen: cooking and the ecology of human origins. Current Anthropology 40, 567–594.

Yeakel, J.D., Bennett, N.C., Koch, P.L., Dominy, N.J., 2007. The isotopic ecology of African mole-rats informs hypotheses on the evolution of human diet. Proceedings of the Royal Society series B 274, 1723–1730.

Yurkstas, A.A., 1965. The masticatory act: a review. Journal of Prosthetic Dentistry 15, 248–260.

Yurkstas, A., Manly, R.S., 1949. Measurement of occlusal contact area effective in mastication. American Journal of Orthodontics 35, 185–195.

4. Modern Human Physiology with Respect to Evolutionary Adaptations that Relate to Diet in the Past

Staffan Lindeberg
Department of Clinical Sciences
Lund University
SE-221 85
Lund, Sweden
staffan.lindeberg@med.lu.se

Keywords Human physiology • evolutionary medicine • nutrition

Abstract This paper reviews evidence from human physiology as to which foods may have been typically consumed by the hominin ancestral lineage up to the advent of anatomically modern humans. Considerable evidence suggests that many common diseases can be prevented by hunter-gatherer diets. Apparently, human nutritional metabolism is not perfectly fine-tuned for recently introduced staple foods, such as cereals, dairy products, added salt, and refined fats and sugar. It is much more uncertain if human physiology can provide direct evidence of which animal and plant foods were regularly consumed during human evolution, and in what proportions. The requirements of ascorbic acid can easily be met by organ meats from large animals, as well as by plant foods. Vitamin B_{12} is absent in plant foods and must be supplied from meat, fish, shellfish, or insects, but the required amounts are apparently small.

Since iodized salt and dairy products were not available before the advent of agriculture, only those ancestors with highly regular access to fish or shellfish would be expected to have reached the currently recommended intake of iodine. However, there is insufficient data to suggest that humans, by way of natural selection, would have become completely dependent on marine food sources. Therefore, it is highly possible that human requirements for iodine are currently increased by some dietary factors. These theoretically include goitrogens in certain roots, vegetables, beans, and seeds. The notion that humans are strictly dependent on marine foods to meet requirements of long-chain omega-3 fatty acids still awaits solid evidence.

Shifting the focus from general human characteristics to ethnic differences, persistent lactase activity in adulthood is obviously not the only characteristic to have emerged under nutritional selection pressure. Other examples are a relative resistance against diseases of affluence in northern Europeans and a relatively low prevalence of gluten intolerance in populations with a long history of wheat consumption.

In conclusion, humans are well adapted for lean meat, fish, insects and highly diverse plant foods without being clearly dependent on any particular proportions of plants versus meat.

J.-J. Hublin and M.P. Richards (eds.), *The Evolution of Hominin Diets: Integrating Approaches to the Study of Palaeolithic Subsistence*, 43–57.

Introduction

Pre-agricultural diets are increasingly acknowledged as templates for healthy diets, partly because contemporary populations with "ancient" dietary habits have shown very low age-adjusted rates of cardiovascular and other nutrition-related diseases compared to people in developed countries (Eaton and Konner, 1985; Lindeberg et al., 2003; Cordain et al., 2005). The !Kung San, hunter-gatherers of Botswana, were thus unaware of spontaneous sudden death in the 1960s (Truswell and Hansen, 1976), and other studies have found highly beneficial risk factor levels in such populations until they have become westernized (Trowell and Burkitt, 1981). We were unable to find any evidence of stroke, coronary heart disease, diabetes, hypertension, or overweight people among traditional horticulturalists of Kitava, Trobriand Islands, Papua New Guinea, despite a considerable number of elderly people (Lindeberg and Lundh, 1993; Lindeberg, 1994; Lindeberg et al., 1997). The bulk of evidence thus strongly suggests that major western diseases can largely be prevented by lifestyles resembling those to which humans were adapted over the long period of hominin and human evolution. In a recent clinical trial, we found marked improvement of glucose tolerance, independent of weight loss, among patients advised to consume a Paleolithic-like diet for 12 weeks, as compared to a traditional healthy, Mediterranean-style diet (Lindeberg et al., 2007). In an uncontrolled Australian trial, a sharp improvement in body weight, blood sugar, triglycerides, and blood pressure was noted in urbanized Aborigines with type-2 diabetes when they returned to a hunter-gatherer lifestyle for 7 weeks, and dramatically increased their protein intake from meat, fish, and shellfish (O'Dea, 1984). Physical activity was also increased in this study. The healthy metabolic and cardiovascular risk profiles of various hunter-gatherer and horticultural groups is likely in part related to fairly high activity levels that result from subsistence participation (Cordain et al., 1998).

Traditional dietary interventions with low-fat, whole-grain foods, enriched with fruit and vegetables have largely failed to prevent cardiovascular disease or cancer (Hooper et al., 2001; Beresford et al., 2006; Howard et al., 2006). Studies specifically testing cereal fiber (Burr et al., 1989), omega-3 fatty acids (Hooper et al., 2004), and minerals and vitamins (Bjelakovic et al., 2004; Miller et al., 2005) have also been disappointing. An exception to these failures is the Lyon Diet Heart Study among French patients with a first myocardial infarction, where a Mediterranean-style diet (emphasizing more bread, more root vegetables and green vegetables; more fish, less beef; lamb and pork replaced with poultry; no day without fruit; and butter and cream replaced with margarine high in alpha-linolenic acid) dramatically reduced the number of new cardiovascular events (de Lorgeril et al., 1999). Patients of the control group received no dietary advice from the investigators, and were only advised to follow a prudent diet by their attending physician.

For the average Westerner, more than 70% of the energy intake (E%) is provided by foods that were practically unavailable in the Paleolithic (or earlier), i.e. cereals, legumes, dairy products, refined sugar, fats and oils (Cordain et al., 2005). In addition, sodium intake is markedly higher today (Eaton and Konner, 1985).

Thus, humans have dramatically shifted their subsistence base during the Neolithic, starting only recently in an evolutionary perspective. This shift is too swift and recent for the human gene pool to have fully adapted to the effects from these new staple foods on human physiology. This is especially true for negative effects, which translate into Western diseases that mainly affect the individual at post-reproductive age. Instead, this dramatic shift in subsistence base implies some physiological pre-adaptation (e.g. salivary amylase) coupled with mainly cultural adaptations (e.g. cooking, baking, thrashing, and industrial preparation). Evidence for still ongoing genetic adaptation comes from studies reporting significant human variation in genes coding for nutrition-related physiology, sometimes showing substantial differences between ethnic groups, families or individuals (Corthesy-Theulaz et al., 2005; Voight et al., 2006). These findings strongly suggest ongoing positive selection, and also pave the way for functional foods "to fit your genes" (Gorman, 2006). Even though not fully adapted, humans still obviously thrive on a Neolithic diet as measured in reproductive terms. However, reproductive success does not automatically mean that the Neolithic diet is good for the human health.

For example, adults who are able to digest lactose are not automatically protected against health problems with milk in later life. Hence, the fact that the majority of elderly carriers of the gene for persistent lactase activity (Caucasians) have advanced atherosclerosis is not at odds with the proposed role of milk in this disease (Lindeberg et al., 2003; Lindeberg, 2005).

While such evolutionary reasoning is plausible and straightforward, it is much more difficult to reverse the argument to see if human physiology can tell us something about genetic adaptations to food habits in the past. This paper reviews such evidence from the medical literature with the primary goal of increasing our knowledge of which foods were typically consumed during the hominin ancestral lineage, up to fully modern humans around 200,000 years BP. Genetic adaptation to the various habitats occupied by fully modern humans after this date will not be covered in detail.

Materials and Methods

Between 1985 and May 2006, more than 200 scientific journals in medicine, nutrition, biology, and anthropology were systematically screened for relevant papers. Computer-based searches and studies of reference lists in journals and books provided a vast number of additional papers. Most areas of physiology were covered, but the majority of papers were related to metabolism or circulation. The main underlying questions were the following: (1) Do the lower (requirements) and upper (toxicity) limits for essential nutrients point in any particular direction as to foods typically consumed in the ancestral human lineage? and (2) Among the foods that were available to hominins and primates, which are apparently beneficial or hazardous for humans?

Results and Discussion

Available evidence suggests that humans are omnivores who can maintain good health and high reproductive success on highly variable proportions of animal and plant foods, as shown by such extremes as meat-eating Inuits consuming mainly protein and fat (Schaefer, 1981) and populations of the Western Pacific with very high intakes of carbohydrates from roots and fruit (Sinnett and Whyte, 1973; Lindeberg and Lundh, 1993). The dentition and gastrointestinal morphology of humans apparently reflect a long ancestral lineage of primates consuming plant foods such as fruits, leaves, flowers and bark (Milton, 2003; Ungar, 2004). A possible increase in the dependence on meat and/or (cooked) underground plant storage organs (corms, bulbs, tubers, roots) during the last

2 million years is consistent with changes in the digestive system, including reduced tooth size and changed length proportions of small versus large intestine (Wrangham et al., 1999; Richards, 2002; Milton, 2003). It has been proposed that the anatomy of the human teeth and jaw reflects past diets of tough texture, such as seeds; but an equally tenable category of tough foods is underground storage organs (Jolly, 1970; Wrangham et al., 1999; Jolly 2001; Lucas et al., 2006). Furthermore, the seeds that may have been consumed were highly diverse dicot seeds from a large variety of plants, not grass seeds from a small number of species, as is the case today (Jönsson et al., 2005).

Published studies in the area of human physiology do not provide solid evidence for which dietary habits humans are best adapted, partly because we are only beginning to understand the complex matters of molecular biology and food-related disorders. In addition, selection pressures may have been low when the relevant foods and nutrients were consumed in sub-optimal amounts. Apparently, species-specific physiological traits in our lineage may often mirror food patterns of ancestors many million years earlier.

Vitamins

Our early primate ancestors lost the ability to synthesize vitamin C (Nishikimi et al., 1994), but this "inborn error of metabolism" was apparently no problem during hominid evolution, since the intake of vitamin C from plant foods is expected to have continuously exceeded requirements, making such biosynthesis redundant. The obligate requirement of vitamin C does not necessarily suggest that early humans regularly consumed plant foods, since organ meats from large animals can easily provide the required amounts (Table 4.1). Sometimes shellfish may also provide enough vitamin C to prevent scurvy.

Vitamin B_{12}, another indispensable (essential) nutrient, is absent from plant foods and must be supplied from meat, fish, shellfish or insects, but the required amounts are apparently small (Table 4.2). Vegans have a very high risk of vitamin B_{12} deficiency and generally need supplements (Stabler and Allen, 2004). Although plant foods may sometimes contain trace amounts of vitamin B_{12} analogues as a result of fermentation, most of these seem to be inactive analogues of the vitamin (Herbert, 1988). Even lacto-ovo vegetarians are at substantially increased risk of deficiency despite the fact that milk products provide some vitamin B_{12} (Hokin and Butler, 1999; Koebnick et al., 2004). From the available knowledge it would seem that early humans must have had access to animal foods in order to avoid vitamin B_{12} deficiency. However, in case of irregular intake there is a high degree of retention of the vitamin in body stores, suggesting that it does not necessarily need to be supplied every week. Another relevant aspect is absorption, which may vary not only between foods, but also between individuals. In the average healthy adult with normal stomach function, approximately 50% of vitamin B_{12} is thought to be absorbed (Writers of Nordic Nutrition Recommendations, 2004). There seems to be a decline with increasing age in western societies, and mild deficiency is common among elderly subjects, but this is not considered to be a consequence of normal biological aging (Carmel, 1997). Non-western populations have not been studied in this regard.

TABLE 4.1. Content of ascorbic acid, expressed as nutrient density, in selected Swedish raw food items (mean SD) (From www.slv.se).

	Ascorbic acid (mg/MJ)
Game muscle	0
Fish	1 (2)
Shellfish	8 (13)
Organ meats	29 (23)
Root vegetables (cultivated)	17 (25)
Fruits (cultivated)	34 (45)
Recommended intake	≥8
Suggested lowest intake to prevent scurvy	≥4

TABLE 4.2. Content of vitamin B_{12}, expressed as nutrient density (µg/MJ), in selected Swedish food items (mean SD in parentheses) (From www.slv.se).

	Vitamin B_{12}
Game muscle, raw	9 (5)
Fish, raw	10 (13)
Shellfish, raw	14 (18)
Organ meats, raw	52 (61)
Vegetables, fruits, legumes, cereals	0
Dairy products	1.5 (0.9)
Recommended intake	≥0.2

Minerals

Contemporary vegetarians are at risk of additional deficiencies, such as iron, zinc, and calcium deficiency (Hunt, 2003), but this is largely due to a high intake of phytate from cereals and legumes, which were apparently not staple foods in the Paleolithic (Hallberg et al., 1989). Cereal grains, pulses, and nuts are rich in phytic acid, the major storage form of phosphorus, which strongly binds to minerals like calcium, iron, zinc, and magnesium to form insoluble salts, phytates. There is overwhelming evidence that whole-grain cereals and legumes decrease the absorption of such minerals through this mechanism. There is no evidence of long-term (months) adaptation to the effects of phytates in humans (Brune et al., 1989).

Phytates are most certainly an important contributing cause of iron deficiency in developing countries, and probably in the western world as well (Hallberg et al., 1989). In contrast to humans, rats have a high capacity to degrade phytates in their intestines, which is one of several adjustment mechanisms to a high consumption of seeds (Iqbal et al., 1994). Nevertheless, dietary phytate impairs mineral absorption in rats as well as in humans (Fairweather and Wright, 1990).

During periods when early humans regularly consumed plant food, vitamin C intake would have been high, and this is known to enhance iron absorption. Meat and fish contain heme iron, which is absorbed to a greater extent than iron from vegetables (Hallberg et al., 2000). In addition, there is some factor in meat that increases iron absorption (Hurrell et al., 2006). Iron deficiency in Western children is effectively prevented by the early introduction of meat (Engelmann et al., 1998; Requejo et al., 1999; Yeung and Zlotkin, 2000). Irrespective of the proportion of plants versus animal foods, iron status is expected to have been excellent in preagricultural humans. Although nuts (hazelnuts, almonds, etc.) may provide enough phytic acid to affect mineral absorption negatively (Hazell and Johnson, 1987), the net effect on iron status of a balanced Paleolithic diet, when compared with an average western diet, is expected to be beneficial.

With regard to calcium deficiency the picture is less clear. The major consequences of calcium deficiency are rickets in children and osteoporosis in adults. However, recent evidence suggests that the influence of calcium intake on bone remodelling has been overestimated (Nordin et al., 1995; Lanou et al., 2005; Jackson et al., 2006; Winzenberg et al., 2006). Mellanby found, back in the 1930s, that young dogs got rickets when they were fed oats, which led to the discovery of phytate as the responsible agent (Mellanby, 1950). The possible absence of rickets in preagricultural skeletons, its apparent increase during medieval urbanization, and its epidemic explosion during industrialism can hardly be explained only in terms of decreasing exposure to sunlight and of decreased length of breast-feeding. An additional possible cause is a trend of increasing inhibition of calcium absorption by phytate from cereals, since cereal intake apparently increased in amount during the Middle Ages, and since old methods of reducing the phytate content by dampening and heat treatment may have been lost during the emergence of large-scale cereal processing (Lindeberg, 1998). A possible role of dietary phytate in osteoporosis has been suggested but not thoroughly evaluated (Miller, 1989; Lau and Woo, 1998). Another suggested mechanism through which ancient humans bones may have been made robust is optimal metabolic acid-base status, owing to a net intake of base-producing foods (Sebastian et al., 2002).

Iodine

One of the most intriguing essential nutrients in terms of evolutionary adaptations to past diets is iodine, a trace element which is required for thyroid hormone synthesis. The thyroid hormones are necessary for growth and development, particularly for the brain and for metabolism. Severe iodine deficiency in infancy causes cretinism, which is considered to be the most common preventable cause of mental retardation worldwide (Hetzel, 1994). In addition to mental retardation, cretinism is characterized by dwarfism and physical malformations, including skull deformities and increased lordosis of the lumbar spine. Dobson has suggested that some of the Upper Paleolithic Venus figurines from mountainous parts of Europe and Asia may represent cretins among iodine deficient terrestrial hunter-gatherers (Dobson, 1998).

Today, milder forms of iodine deficiency with goiter (enlargement of the thyroid gland), but without mental retardation, are common in many regions of the western world (Delange et al., 2000; Rasmussen et al., 2002). The risk is greatest in areas where the soil has been depleted of iodine (mountainous regions or areas that are often flooded), but it is also prevalent in other parts of the world, including many European countries. Of 2,855 Belgian children aged 6–12 years, 5.7% had goiter in 1998 (the frequency varied between 3.9% and 7.7% in the ten provinces of the country). Three years earlier, prior to national campaigns for iodine supplements, the percentage was 11% (Delange et al., 2000). In most European countries, iodine deficiencies were common as late as 1992. The exceptions included Sweden, Norway, Finland, Austria, and Switzerland, where iodine supplementation was introduced early, typically through table salt.

Without iodine enrichment of foods, few people reach the recommended iodine intake: 150 µg/day for adults and higher amounts during pregnancy and breast feeding. Vegans run an increased risk of iodine deficiency (Lightowler and Davies, 1998). As people in developed countries consume less seafood, iodized salt, and dairy products, they are expected to increase their risk of iodine deficiency. The recommended iodine density to be used for planning diets for groups of people is set at 18 µg/MJ. In order to prevent goiter, a daily dose of 50–75 µg, or 1 µg/kg of body weight, is considered sufficient for most people.

Shellfish and fish, especially seafood, and thyroid glands from various animals are exceptionally good sources of iodine (Table 4.3). In contrast, nuts, meat (wild or domestic), organ meats (other than thyroids), fruits, and many roots and leafy vegetables are poor sources. Although preagricultural hunter-gatherers living off-shore sometimes may have had access to

TABLE 4.3. Iodine content in selected Swedish food items (From www.slv.se).

		Iodine content, median (range)	
	n	(µg/100 g)	(µg/MJ)
Fish, raw	25	25 (7–85)	48 (16–222)
Shellfish, raw	3	180 (60–700)	409 (188–1,912)
Meat, raw	22	1 (0–6)	1.5 (0–13)
Thyroid gland	1	50,000	
Other organ meats	18	3 (0–8)	5 (0–19)
Vegetables, raw	26	1 (0–6)	9 (0–132)
Root vegetables, raw	7	1 (0.3–3)	4 (1–21)
Fruits/berries, raw	25	0.4 (0–2)	2 (0–6)
Nuts, raw	5	0.3 (0.05–0.5)	0.2 (0.02–0.2)
Legumes, raw	10	1 (0.2–45)	1 (0.2–36)
Cereals	26	2 (0.3–5)	1 (0.2–3)
Dairy products	7	8 (6–45)	30 (15–38)
Human milk	1	63	211
Table salt	1	5,000	–
Sea salt	1	2,000	–
Recommended intake	–	(150 µg/day)	18
Minimum intake*	–	(50–75 µg/day)	7

*Estimated lowest intake to prevent goiter when daily energy intake is 10 MJ (70 µg/day).

high-iodine vegetables, a fish intake of less than 20% of total food intake (by weight) may not have been enough to meet the above recommendation for dietary iodine. Accordingly, at first glance it would seem that preagricultural humans were dependent on regular access to shellfish or a very high intake of fish, preferably marine fish. Alternatively, they would need to be aware of the necessity that all family members regularly consume small amounts of thyroid glands from terrestrial mammals. In addition, a sufficient intake during pregnancy and long-term breastfeeding would be crucial for the developing infant brain. On the whole, this would seem to support the notion that humans were strictly dependent on marine food sources (Cunnane and Crawford, 2003). However, exploitation of the marine environment is first documented in the archeologic record during the Middle Paleolithic period (approximately 110,000 years BP), long after the emergence of fully modern humans, and stable isotope data suggest that inland aquatic foods were not utilized by hominins living in Europe until the mid-Upper Paleolithic period (approximately 28,000–20,000 years BP) (Richards et al., 2001).

In light of the uncertainty that hominins evolved at fresh- and saltwater shorelines and became physiologically dependent on seafood (Gräslund, 2005), it is important to consider the role of goitrogenic substances in plant foods, substances which increase the requirement of iodine and may cause enlargement of the thyroid gland despite an adequate intake of iodine (Gaitan, 1990). One group of such substances are flavonoids in beans, which are able to suppress the synthesis of thyroid hormones by inhibiting the enzyme thyroxine peroxidase (Doerge and Sheehan, 2002). Another kind of goitrogens are cyanogenic glycosides, which are produced by more than 2,500 plant species. Their production and conversion to cyanide constitutes a well studied defense mechanism against herbivory (Gleadow and Woodrow, 2002). Cyanide can be acutely toxic when consumed in high amounts, but more relevant to this context is when it is ingested regularly in low concentrations. Cyanide is converted to thiocyanate, a potentially goitrogenic substance, which can increase the need for dietary iodine and may cause goiter, even when iodine intake is sufficient (Gaitan, 1990; Doerge and Sheehan, 2002).

These are well documented consequences of certain staple foods in the developing world. In particular, many staple foods that were domesticated during the Neolithic period and later (millet, maize, soy, cassava, sweet potatoes, lima beans, turnips, cabbage, cauliflower, rapeseed, mustard, onion, garlic, bamboo shoots, and palm tree fruit) contain a variety of such goitrogens (Gaitan, 1990; Doerge and Sheehan, 2002). Kopp has suggested that a high carbohydrate intake, for which humans would not be physiologically prepared, increases iodine requirements (Kopp, 2004). His argument is largely based on the notion that thyroid hormone levels are higher after consumption of high-carbohydrate diets.

In conclusion, the reasons why large parts of the European population seem to need supplemental iodine are complex. Further studies are needed. There is insufficient data to suggest that humans, by way of natural selection, would have become totally dependent on marine food sources. Therefore, it is highly possible that human requirements for iodine are currently increased by some dietary factors. Further support for this line of thinking is the observation that iodine-depleted soils often did not coincide historically with endemic goiter (Henschen, 1966).

Protein

Dietary proteins provide energy and amino acids for endogenous protein synthesis. Some amino acids are interchangeable, while others are essential. A third group is thought to be conditionally indispensable, i.e. they must be supplied under certain physiological or pathological conditions. There has been much debate over the roles of different food types with regard to amino acid composition (Young and Pellett, 1994). A commonly held but largely unsubstantiated notion has been that plant proteins would be inferior to meat proteins because they lack specific amino acids. However, the bulk of evidence suggests that a mixture of root vegetables, leaves, fruits, and nuts is adequate with regard to essential amino acids (Young and Pellett, 1994). Wheat and other cereal grains are low in lysine and threonine (Cordain, 1999). Other single plant foods may lack specific amino acids, but a varied vegetarian diet poses no obvious risk of amino acid deficiency.

A contrasting idea with regard to protein quality has been that "animal proteins" are actually unhealthy because of their structure, and because protein digestion is known to be incomplete. By and large, this concept is based on studies of casein, the dominant protein in milk (Hawrylewicz et al., 1991; Kritchevsky, 1995; Anthony et al., 1998; Harrison et al., 1998; Potter et al., 1998; Elliott et al., 1999; Thorsdottir et al., 2000; Wagner et al., 2000; Wasmuth and Kolb, 2000; Wilson et al., 2000; Damasceno et al., 2001; Hakkak et al., 2001; Lavigne et al., 2001; McLachlan, 2001; Adams et al., 2002; Arjmandi et al., 2002; Jenkins et al., 2002; Maddox et al., 2002; Ascencio et al., 2004). Proteins are long chains of amino acids, and their partial degradation may result in shorter chains, peptides, entering the system where they are apparently capable of causing various kinds of inflammation. This may be relevant to the finding in many animal experiments that casein has been shown to enhance pathologic processes such as atherosclerosis, insulin resistance, and lipotoxicity (intracellular lipid overload) (Kritchevsky, 1995; Anthony et al., 1998; Wagner et al., 2000; Wilson et al., 2000; Damasceno et al., 2001; Lavigne et al., 2001; Adams et al., 2002; Ascencio et al., 2004). When discussing these findings it has been common not to call casein by its proper name but rather to talk about "animal protein." This misleading maneuver has obviously supported the widespread misunderstanding that there is evidence of negative health effects of meat proteins.

Total protein intake is another matter of debate as part of a long-term discussion on optimal energy percentages of protein, fat, and carbohydrate. National advisory bodies have for many years recommended that 10–15 E% be provided by protein, and that daily intake should not be less than 0.7 g protein per kg of body weight. The current recommendation for Scandinavian adults is 10–20 E% protein. However, the debate among experts has been rather intense, and both these (upper and lower) limits of protein intake are being questioned by an increasing number of research groups, whose proposed lower limit of protein intake now varies between 0.8 and 1.0 g/kg/day (which is higher than 10 E% for most adults), while the proposed upper limit varies between 20 and 35 E%, or even higher (Kerstetter et al., 2000; Bravata et al., 2003; Appel et al., 2005; Adam-Perrot et al., 2006). A typical intake among elderly Scandinavians is 15 E%, or 1.0 g/kg/day (Writers of Nordic Nutrition Recommendations, 2004).

The suggestion that dietary protein should not exceed 20 E% is not supported by the fact that many twentieth century hunter-gatherers, including the Eskimos, have shown excellent health on very high protein diets (Schaefer, 1981; Dyerberg, 1989). The estimated typical protein intake among 229 populations of twentieth century hunter-gatherers was 19–35 E% (Cordain et al., 2000a). Humans cannot tolerate a sudden increase in protein intake above 250 g/day (individual range 200–300 g/day), because of a limited capacity of the liver to metabolize amino acids (Hamberg et al., 1992). Higher intakes may lead to nausea, diarrhea, and finally death, in a condition which the early white American explorers called "rabbit starvation" (Cordain et al., 2000a). This name refers to their problem of surviving the winter on meat from rabbits and other small game, which have very small fat depots. On a chronically high protein intake the liver can up-regulate its enzymatic activity, but the upper limit has not been established (Hamberg et al., 1992; Nuttall et al., 2006).

A diet based exclusively on (domestic) fruit would rarely meet the lower level of protein requirement (0.75 g/kg/day), while one based on other vegetables (including roots) on average would pose the risk of exceeding the recommended upper limit (Table 4.4). In addition to the content of protein in foods, the capacity of intestinal enzymes to digest the consumed proteins into amino acids may influence the ability to meet requirements. Edible plants contain protease inhibitors, substances that inhibit protein digestion by interfering with intestinal proteases, and particularly high concentrations are found in beans and seeds, and to some extent in roots (Armour et al., 1998; Cordain, 1999; Vanderjagt et al., 2000). Heat treatment partly inactivates some of these substances, which is why cooking may sometimes have been crucial for protein balance for those early humans who relied on plant foods.

TABLE 4.4. Protein content in selected Swedish food items (From www.slv.se).

	Protein content, mean (SD) (g/100 g)	Energy (%)
Meat, raw	20 (25)	59 (19)
Fish, raw	20 (12)	67 (23)
Shellfish, raw	16 (4)	78 (13)
Vegetables, raw	3 (2)	32 (14)
Root vegetables, raw	2 (1)	19 (7)
Nuts, raw	17 (8)	12 (5)
Fruits, raw	0.8 (0.4)	7 (4)
Cereal grains	10 (5)	13 (7)
Dairy products	5 (7)	24 (10)
Legumes, raw	22 (6)	31 (3)
Recommended intake	≥55 g/day	10–15(–20)

An adaptation to very high protein diets, due to long periods of dependence on large game meat during the Paleolithic, has been suggested (Brand Miller and Colagiuri, 1994). The proposed negative selection pressure is low blood sugar due to lack of carbohydrates in the diet. Humans, like other animals, need glucose for the mammary glands, the brain, the red blood cells and, in particular, the developing fetus. However, the capability of humans to synthesize glucose is limited. Dietary protein stimulates release of insulin which may lower blood sugar too much. Therefore, efficient redirection of glucose from muscle to brain and reproductive tissues may have become essential during periods of low carbohydrate intake. Such redirection is typically caused by resistance to the glucometabolic effects of insulin or, for short, insulin resistance. If future studies show that humans are more insulin resistant than other primates, the hypothesis lends support to the notion that meat intake was very high during much of hominid evolution.

Carbohydrate

There is no convincing evidence that a carbohydrate intake above 60 E%, irrespective of food choice, is a health risk to humans. Although avoiding carbohydrates would seem beneficial for patients with diabetes because of a lower increase of blood sugar, many controversies remain, and other dietary factors may be equally or more important (Kennedy et al., 2005). We found diabetes to be non-existent and insulin sensitivity to be excellent among the traditional horticulturalists of Kitava, Trobriand Islands, Papua New Guinea, where the estimated intake of carbohydrate is 69 E%, which is actually higher than the recommended intake of 50–60 E% (Lindeberg et al., 1999, 2001). Marked restriction of carbohydrates means, for the vast majority of Westerners, avoidance of cereal grains and potatoes, and it is often impossible to know if carbohydrates or other constituents of these foods explain the findings in published studies.

Those carbohydrate-rich foods which cause a higher and more persistent increase of blood sugar than others, independent of the total amount of carbohydrates, are said to have a high glycemic index (GI). The product of GI and total carbohydrate content is often termed glycemic load. High-glycemic-load foods like potatoes, breads, and breakfast cereals have been suggested to increase the risk of obesity and type 2 diabetes, but solid evidence is as yet unavailable (Brand Miller et al., 2002). The glycemic load of wild roots and vegetables is typically low (Thorburn et al., 1987).

Fruits differ from other edible plants in that they contain appreciable amounts of fructose, a monosaccharide, which typically constitutes 20–40% of available carbohydrates in wild fruits (Ko et al., 1998; Milton, 1999; Dzhangaliev et al., 2003) and 10–30% in cultivated fruits (National Food Administration, 1986). A (very) high intake of fructose contributes to obesity, insulin resistance and type 2 diabetes in rats and mice, but the effects on humans have not been fully established (Elliott et al., 2002; Vasankari and Vasankari, 2006). Unlike rodents, pre-hominin primates likely consumed considerable quantities of fruit for 50 million years or more, from the time of early primates until the emergence of *Australopithecus* 4 million years ago. An enhanced ability to handle fructose may have been conserved among *Australopithecus* and *Homo*. Therefore, findings in rodents should be interpreted with caution. A daily fructose intake below 60 g, corresponding to 4–5 kg of pineapples, appears safe (Vasankari and Vasankari, 2006). Approximately two thirds of dietary fructose in the US population is provided by non-natural foods and additives, mainly sucrose (a disaccharide built up of fructose and glucose) and high fructose corn syrup (Park and Yetley, 1993). In conclusion, the concerns about dietary fructose are not strong enough to suggest that humans are not designed to eat large amounts of fruit.

Starch, the predominant carbohydrate in modern Western diets, is only present in small amounts in wild edible plants with one category of exception: underground storage organs (roots, corms, bulbs, tubers). It has been suggested that these became important staple foods for *Australopithecus* and early *Homo*, who were able to use tools for procurement and fire for cooking, the latter in order to increase starch digestibility and detoxify phytochemicals (Wrangham et al., 1999; Lucas et al., 2006). In addition, humans have a particularly high activity of salivary amylase, an enzyme for starch digestion which most animals mainly have in the form of pancreatic amylase (Samuelson et al., 1996; Perry et al., 2007). Apparently, human populations with a recent history of high starch consumption have, compared with other human populations, a slightly higher copy number of the gene coding for salivary amylase, a number which explains about 35% of the variation of salivary amylase concentration (Perry et al., 2007). Although this suggests some degree of ongoing positive selection, the fact that "low-starch" human populations have an almost threefold higher copy number than chimpanzees indicates significant adaptation to high-starch root vegetables among hominins, rather than post-agricultural selection among humans.

Fat

Dietary fat provides phospholipids and cholesterol for cell membranes, omega-6 and omega-3 fatty acids for various physiological functions, and concentrated fuel for energy metabolism. Omega-6 and omega-3 fatty acids must be provided, but the requirements are very low, and clinical symptoms of deficiency are only seen in special circumstances, such as chronic disease or prolonged parenteral nutrition. The notion that humans require larger amounts of long-chain omega-3 fatty acids than other mammals still awaits solid evidence (Hooper et al., 2004; Writers of Nordic Nutrition Recommendations, 2004). It is also uncertain if humans, as has been suggested, have a lower capacity than rodents and herbivores to convert the predominant omega-3 fatty acid in plants, alpha-linolenic acid, into the long chain omega-3 fatty acid docosahexenoic acid (DHA) (Writers of Nordic Nutrition Recommendations, 2004). Available knowledge of fatty acid requirements, including requirements for brain development, does not give strong hints about the intake of animal versus plant foods during human evolution, nor about the intake of foods like fish, shellfish, meat, insects, eggs, nuts, roots, leaves, or fruits.

At the other end of the spectrum are those fatty acids which are considered unhealthy, in this context most notably the saturated fatty acids which should not exceed 10 E%, according to nutritional recommendations. The average intake in Western populations is 15–17 E%. Saturated fat raises serum cholesterol and is thereby thought to cause atherosclerosis, the main disorder underlying cardiovascular disease. However, many other factors are involved in this disorder, which affects the vast majority of Westerners including those with a relatively low intake of saturated fat (Ravnskov, 1998). In Kitava, Trobriand Islands, saturated fat intake was approximately 17 E%, mainly from coconut, but cardiovascular disease was virtually nonexistent (Lindeberg, 1994; Lindeberg and Vessby, 1995). Estimated intake of saturated fat across 229 worldwide hunter-gatherers in the twentieth century varied between 4 and 18 E%, with a mean of 11 E% (Cordain, 2006). Even in plant-dominated hunter-gatherer diets (more than 50 E% from plant foods) the average was 11 E%, slightly above recommendations, and surprisingly close to the 15 E% of typical hunter-gatherers relying on meat. Available evidence does not support the notion that humans, irrespective of other dietary factors, are not well adapted to consume diets with 10–15 E% saturates.

Human Variation

One of the most striking findings in the literature was the similarity in physiology across various human populations, and often across the animal kingdom, suggesting a high degree of conservation (Garrow and James, 2000; Berg et al., 2002; Kasper et al., 2005; Shils and Shike, 2006). Apparently, the important dietary shifts that shaped the human genome occured before the appearance of fully modern humans 150,000–200,000 years BP, and indeed long before the emergence of ethnic differences in physionomy. This notion is not in conflict with the fact that there are ethnic differences, it only states that most of the evolution of human characteristics occured during the ancestral lineage up to fully modern humans. Requirements and toxicity levels of dietary nutrients are essentially similar for different populations, and so are the principal reaction patterns to a Western lifestyle.

After the "transition" from hunting and gathering to more westernized dietary habits, a vast number of human populations, irrespective of ethnicity, have shown the same disease patterns with stroke, coronary heart disease, atherosclerosis, diabetes, insulin resistance, and increased levels of blood pressure, serum cholesterol, and blood sugar (Tejada et al., 1968; Trowell and Burkitt, 1981). Migrant studies have found that the whole population is negatively affected (although to a varying degree, see below under "Ethnicity") with increased blood pressure and blood sugar, compared with non-migrants who persist in a traditional lifestyle (Poulter et al., 1988; Cruickshank et al., 1999). In countries with a long history of westernization, the vast majority of elderly people qualify for medical treatment of high blood pressure and cholesterol, and most people above 60 years of age have, by present definitions, a "high" risk of cardiovascular disease (Lindeberg, 2005). Although atherosclerosis is a typical consequence of aging today, early human autopsy studies and animal experiments suggest that it is not an inevitable process (Lindeberg et al., 2003). An international autopsy study in the 1960s comparing middle-aged men in four countries revealed marked differences depending on degree of urbanization (Tejada et al., 1968).

Cereals and Atherosclerosis

Nothing seems to indicate that these reaction patterns are unique for the human species. Many other mammals, including Old and New World monkeys, dogs, cats, and rats become overweight and insulin resistant when they are fed a western diet (Mordes and Rossini, 1981; Feldhahn et al., 1999; Roth et al., 2001; Diez et al., 2002). Free-ranging mammals in their natural habitat are apparently not affected by advanced coronary atherosclerosis, and atherosclerosis promotion and regression in animal experiments is highly dependent on dietary manipulation (Lindeberg et al., 2003). Fiennes noted that mammals and birds (mice, rats, passerines, ostriches, etc.) who are natural seed eaters are relatively resistant to atherosclerosis when they are fed grain-based, atherogenic food, while non-seed eaters (primates, pigs, guinea pigs, parrots, etc.) are more susceptible to the disease (Fiennes, 1965). Therefore, he postulated in 1965 that cereals could contribute to atherosclerosis among *Homo sapiens* and other non-seed-eating species. Compared with unrefined Paleolithic staples (vegetables, fruits, root vegetables, meat, and fish), cereals have several potentially negative effects (Cordain, 1999). Cereals can contribute to deficiencies in most of the nutrients that have been suggested to prevent atherosclerosis (Slavin et al., 1999; Truswell, 2002). Cereals lack vitamin C and vitamin B_{12} and, compared with a mixed hunter-gatherer diet, cereals also contribute toward a lack of zinc, selenium, flavonoids, carotenoids, folic acid, vitamin B_6 (low bioavailability), magnesium (through chelate bonding with phytic acid), potassium, biotin, vitamin D, and taurine. The percentage of omega-6 fatty acids is high, and the percentage of omega-3 fatty acids is low. The documented effects of cereal proteins on autoimmunity (Cordain, 1999; Cordain et al., 2000b) should also be considered in light of the immune system's potential role in atherosclerosis (Hansson, 2001).

Another plausible atherogenic factor is plant lectins, which are generally considered a herbivore defense mechanism. Lectins in seeds (including cereal grains and beans) increase the permeability of the intestinal mucous membrane (Liener, 1986; Pusztai, 1993), and can enter the blood stream when supplied via food (Pusztai et al., 1989; Wang et al., 1998). The peanut lectin contributes to peanut oil's atherosclerosis-promoting effect in rabbits (Kritchevsky et al., 1998). Wheat lectin has many untoward effects which may be relevant to atherosclerosis (Cuatrecasas, 1973; Cuatrecasas and Tell, 1973; Hedo et al., 1981; Livingston and Purvis, 1981; Shechter and Sela, 1981; Purrello et al., 1983; Shechter, 1983; Davis and Glagov, 1986; Kagami et al., 1991; Barbeau et al., 1997; Cyrus et al., 2001; Ohmori et al., 2001; Vacaresse et al., 2001; Wang et al., 2001; Yevdokimova and Yefimov, 2001). Gliadin, a major lectin-like protein in wheat, is toxic to cultured cells, where it causes intracellular lipid accumulation, which may also be relevant (Dolfini et al., 2003).

If cereals have undesired health effects, animal species that are genetically adapted for seed eating can be expected to differ from non-seed eaters in terms of digestion and/or metabolism. Several of the characteristics of rats and mice should be analyzed in this respect. The fact that the intestinal phytase activity of rats is 30 times higher than that of humans (Iqbal et al., 1994) can be seen as an indication that cereals rich in phytic acid can impair the bioavailability of dietary minerals (Cheryan, 1980). Among the other examples of the special traits of rats and/or mice as compared with humans include the fact that high-density lipoprotein is the dominant lipoprotein instead of low-density lipoprotein (Maldonado et al., 2002); the fact that circulating cholesterol is more efficiently converted to bile acids (Chiang et al., 2001); the fact that peroxisome proliferator activated receptor α is more resistant to activation (Laudet and Gronemeyer, 2002); the fact that the genetic expression of resistin, a factor causing insulin resistance, is higher (Nagaev and Smith, 2001); the fact that the release of glucagon-like peptide-1 after consuming protein or fat is higher (Aziz and Anderson, 2002); and the fact that the regulation of liver X receptor α in macrophages is quite different (Laffitte et al., 2001). In addition, a peculiar feature of glucose metabolism has been noted in seed-eating birds compared to meat-eating birds. If the pancreas is removed while retaining the alpha-cells, seed-eating birds will develop a temporary form of insulin-dependant diabetes, in contrast to the type of permanent diabetes that such a procedure causes in non-seed eating birds (Lothrop et al., 1986).

Ethnicity

Although nutritional physiology on the whole is identical across human populations, some ethnic differences are notable, the most conspicuous being persistent lactase activity in adulthood, which is present in 0–100% worldwide, largely depending on time since introduction of dairy farming (Simoons et al., 1977). Lactose, the predominant carbohydrate in milk, is a disaccharide made from galactose and glucose. In order for infants and children to feed on their mother's milk, the enzyme lactase, which is found in the mucous membrane of the small intestines, splits lactose into these two monosaccharides. Without the enzyme, lactose stays unabsorbed in the gut where it is fermented by bacteria and may cause diarrhea and abdominal bloating and discomfort, however many individuals have no symptoms. Apart from decreased absorption of lactose, the consequences in terms of reduced absorption of nutrients (other than lactose) are considered to be small (Debongnie et al., 1979). These results of lactose intolerance are expected to have little impact on reproductive fitness, and do not obviously explain the increase of persistent lactase activity in less than 10,000 years. In contrast, milk may have prevented starvation in some harsh environments, although this does not seem to explain the rapid spread of the trait in Upper Paleolithic Europeans. Possibly, avoidance of low blood sugar is also relevant, as discussed above. A commonly held notion is that milk prevented vitamin D deficiency in northern European populations, although this is at variance with recent findings of limited, if any, beneficial effect of dairy products on bone mass (Weinsier and Krumdieck, 2000; Kanis et al., 2005; Lanou et al., 2005; Winzenberg et al., 2006). Molecular phylogeny suggests that the selection pressure in favor of lactase persistence is one of the strongest yet seen for any gene in the human genome (Bersaglieri et al., 2004). Much of this rapid spread still remains unexplained.

Another trait that may have emerged under nutritional selection pressure in the last few millenia is a relative resistance against the damaging effects of a western lifestyle on glucose metabolism. Type 2 diabetes, the ultimate consequence of deranged glucose metabolism (glucose intolerance), varies dramatically in its prevalence across the urbanized world (King and Rewers, 1993). Although much of this variation relates to different degrees of westernization, there is widespread acceptance in medical science that all populations are not equally vulnerable to the modern lifestyle (Gower, 1999; Abate and Chandalia, 2001; Gunton et al., 2001; Kalhan et al., 2001; Araneta et al., 2002; Zhu et al., 2005). Some of these differences are likely a consequence of ongoing selection in alleles related to diet.

Allen and Cheer considered these differences in an evolutionary perspective (Allen and Cheer, 1996). They dismissed famine as a positive selection factor, as had been suggested by Neel (1962), since Europeans, with well-documented long periods of famine, actually have a relatively low prevalence of diabetes, compared to other westernized populations (King and Rewers, 1993). Instead, they focused on the strong negative correlation worldwide between the prevalence of diabetes and persistent lactase activity in adulthood. After exclusion of African populations, who have largely maintained a traditional

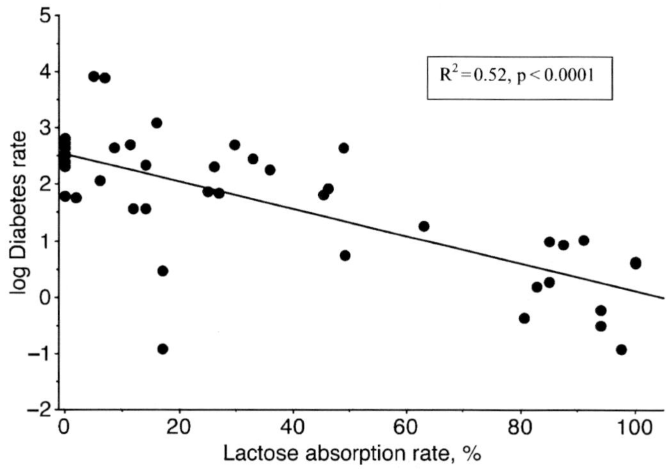

FIG. 4.1. Worldwide prevalence (log-transformed) of diabetes versus prevalence of adult lactase persistence. African populations were excluded because of limited transition to modern lifestyles. Diabetes prevalence was log-transformed in order to obtain normal distribution. On the y axis, values of −1, 0, 1, 2, 3 and 4 correspond, respectively, to a diabetes prevalence of 0.4%, 1.0%, 2.7%, 7.4%, 20.1% and 54.6%. The highest prevalence was seen in Pima and Papago Indians, North America (Modified from Allen and Cheer, 1996).

lifestyle, the prevalence of lactase persistence explains 51% of the international variation in the prevalence of diabetes (Fig. 4.1). Allen and Cheer postulate that when the early cattle herders could take advantage of Europe's barren ecological niche, thanks to their ability to digest lactose as adults, it was at the expense of diabetes, due to the strong insulin response to milk (Gannon et al., 1986; Ercan et al., 1993). A selective pressure would then work against insulin resistance, which is known to precede the development of type 2 diabetes. The hypothesis, which is supported by animal studies of milk as a cause of insulin resistance (Palmquist et al., 1992; Hugi et al., 1998; Lavigne et al., 2001), is one of the few serious attempts to explain why Europeans seem slightly less vulnerable to the perils of a Western lifestyle.

Diabetes and its major inherent dangers – heart disease and stroke – may not be the only consequences of insulin resistance. Increasing evidence suggests that a number of endocrine pathways are disturbed which may cause polycystic ovary syndrome (a common cause of infertility in urbanized females), myopia, cancers of breast, colon, and prostate, and acne vulgaris, as well as the secular trends for increased stature and earlier puberty (Cordain et al., 2002a, b, 2003). Because of the apparent ethnic differences in susceptibility, the consequences of insulin resistance seem especially harmful for non-Europeans. In Singapore, an astonishing 82% of the ethnically Chinese population has myopia (Wu et al., 2001) and in a random sample of Indian women in Britain, the prevalence of polycystic ovary syndrome was 52% (110/212) (Rodin et al., 1998). Both these populations have a high prevalence of insulin resistance. Likewise, the median menarcheal age among 107 girls adopted from India by families in Sweden was 11.6 years, which was significantly lower than for Swedish girls or for girls in India (Proos et al., 1991). Five girls had menarche before the age of 9 years, the earliest at 7.3 years.

Conclusions

Humans are apparently not well adapted for staple foods that were introduced less than 10,000 years ago, including cereal grains, milk, salt, and refined fat and sugar. Although such foods can save lives in the short term, they are highly suspected of causing serious age-related Western diseases, single-handedly or in combination. Human physiology seems well designed for a mixture of lean meat, fish, shellfish, insects, and a large variety of plant foods, including carbohydrate-rich fruits and root vegetables. Although some meat/fish/shellfish must be consumed, available evidence suggests that the proportion of meat versus plant foods has varied considerably during hominin evolution.

References

Abate, N., Chandalia, M., 2001. Ethnicity and type 2 diabetes: focus on Asian Indians. Journal of Diabetes and Its Complications 15(6), 320–327.

Adam-Perrot, A., Clifton, P., Brouns, F., 2006. Low-carbohydrate diets: nutritional and physiological aspects. Obesity Reviews 7(1), 49–58.

Adams, M.R., Golden, D.L., Anthony, M.S., Register, T.C., Williams, J.K., 2002. The inhibitory effect of soy protein isolate on atherosclerosis in mice does not require the presence of LDL receptors or alteration of plasma lipoproteins. Journal of Nutrition 132(1), 43–49.

Allen, J.S., Cheer, S.M., 1996. The non-thrifty genotype. Current Anthropology 37, 831–842.

Anthony, M.S., Clarkson, T.B., Williams, J.K., 1998. Effects of soy isoflavones on atherosclerosis: potential mechanisms. American Journal of Clinical Nutrition 68(6 Suppl), 1390S–1393S.

Appel, L.J., Sacks, F.M., Carey, V.J., Obarzanek, E., Swain, J.F., Miller 3rd, E.R., Conlin, P.R., Erlinger, T.P., Rosner, B.A., Laranjo, N.M., Charleston, J., McCarron, P., Bishop, L.M., 2005. Effects of protein, monounsaturated fat, and carbohydrate intake on blood pressure and serum lipids: results of the OmniHeart randomized trial. Journal of the American Medical Association 294(19), 2455–2464.

Araneta, M.R., Wingard, D.L., Barrett-Connor, E., 2002. Type 2 diabetes and metabolic syndrome in Filipina-American women: a high-risk nonobese population. Diabetes Care 25(3), 494–499.

Arjmandi, B.H., Khalil, D.A., Hollis, B.W., 2002. Soy protein: its effects on intestinal calcium transport, serum vitamin D, and insulin-like growth factor-I in ovariectomized rats. Calcified Tissue International 70(6), 483–487.

Armour, J.C., Perea, R.L.C., Buchan, W.C., Grant, G., 1998. Protease inhibitors and lectins in soya beans and effects of aqueous heat-treatment. Journal of the Science of Food and Agriculture 78, 225–231.

Ascencio, C., Torres, N., Isoard-Acosta, F., Gomez-Perez, F.J., Hernandez-Pando, R., Tovar, A.R., 2004. Soy protein affects serum insulin and hepatic SREBP-1 mRNA and reduces fatty liver in rats. Journal of Nutrition 134(3), 522–529.

Aziz, A., Anderson, G.H., 2002. Exendin-4, a GLP-1 receptor agonist, modulates the effect of macronutrients on food intake by rats. Journal of Nutrition 132(5), 990–995.

Barbeau, W.E., Novascone, M.A., Elgert, K.D., 1997. Is celiac disease due to molecular mimicry between gliadin peptide-HLA class II molecule-T cell interactions and those of some unidentified superantigen? Molecular Immunology 34(7), 535–541.

Beresford, S.A., Johnson, K.C., Ritenbaugh, C., Lasser, N.L., Snetselaar, L.G., Black, H.R., Anderson, G.L., Assaf, A.R., Bassford, T., Bowen, D., Brunner, R.L., Brzyski, R.G., Caan, B., Chlebowski, R.T., Gass, M., Harrigan, R.C., Hays, J., Heber, D., Heiss, G., Hendrix, S.L., Howard, B.V., Hsia, J., Hubbell, F.A., Jackson, R.D., Kotchen, J.M., Kuller, L.H., LaCroix, A.Z., Lane, D.S., Langer, R.D., Lewis, C.E., Manson, J.E., Margolis, K.L., Mossavar-Rahmani, Y., Ockene, J.K., Parker, L.M., Perri, M.G., Phillips, L., Prentice, R.L., Robbins, J., Rossouw, J.E., Sarto, G.E., Stefanick, M.L., Van Horn, L., Vitolins, M.Z., Wactawski-Wende, J., Wallace, R.B., Whitlock, E., 2006. Low-fat dietary pattern and risk of colorectal cancer: the women's health initiative randomized controlled dietary modification trial. Journal of the American Medical Association 295(6), 643–654.

Berg, J.M., Tymoczko, J.L., Stryer, L., Stryer, L., 2002. Biochemistry. W.H. Freeman, New York.

Bersaglieri, T., Sabeti, P.C., Patterson, N., Vanderploeg, T., Schaffner, S.F.,. Drake, J.A, Rhodes, M., Reich, D.E., Hirschhorn, J.N., 2004. Genetic signatures of strong recent positive selection at the lactase gene. American Journal of Human Genetics 74(6), 1111–1120.

Bjelakovic, G., Nikolova, D., Simonetti, R.G., Gluud, C., 2004. Antioxidant supplements for prevention of gastrointestinal cancers: a systematic review and meta-analysis. Lancet 364(9441), 1219–1228.

Brand Miller, J.C., Colagiuri, S., 1994. The carnivore connection: dietary carbohydrate in the evolution of NIDDM. Diabetologia 37. 1280–1286.

Brand Miller, J.C., Holt, S.H., Pawlak, D.B., McMillan, J., 2002. Glycemic index and obesity. American Journal of Clinical Nutrition 76(1), 281S–285S.

Bravata, D.M., Sanders, L., Huang, J., Krumholz, H.M., Olkin, I., Gardner, C.D., Bravata, D.M., 2003. Efficacy and safety of low-carbohydrate diets: a systematic review. Journal of the American Medical Association 289(14), 1837–1850.

Brune, M., Rossander, L., Hallberg, L., 1989. Iron absorption: no intestinal adaptation to a high-phytate diet. American Journal of Clinical Nutrition 49(3), 542–545.

Burr, M.L., Fehily, A.M., Gilbert, J.F., Rogers, S., Holliday, R.M., Sweetnam, P.M., Elwood, P.C., Deadman, N.M., 1989. Effects of changes in fat, fish, and fibre intakes on death and myocardial reinfarction: diet and reinfarction trial (DART). Lancet 2(8666), 757–761.

Carmel, R., 1997. Cobalamin, the stomach, and aging. American Journal of Clinical Nutrition 66(4), 750–759.

Cheryan, M., 1980. Phytic acid interactions in food systems. Critical Reviews in Food Science and Nutrition 13(4), 297–335.

Chiang, J.Y., Kimmel, R., Stroup, D., 2001. Regulation of cholesterol 7alpha-hydroxylase gene (CYP7A1) transcription by the liver orphan receptor (LXRalpha). Gene 262(1–2), 257–265.

Cordain, L., 1999. Cereal grains: humanity's double-edged sword. World Review of Nutrition and Dietetics 84, 19–73.

Cordain, L., 2006. Saturated fat consumption in ancestral human diets: implications for contemporary intakes. In: Meskin, M.S., Bidlack, W.R., Randolph, R.K. (Eds.), Phytochemicals: Nutrient-Gene Interactions. Taylor & Francis, London, pp. 115–126.

Cordain, L., Gotshall, R.W., Eaton, S.B., Eaton, S.B. 3rd, 1998. Physical activity, energy expenditure and fitness: an evolutionary perspective. International Journal of Sports Medicine 19(5), 328–335.

Cordain, L., Miller, J.B., Eaton, S.B., Mann, N., Holt, S.H., Speth, J.D., 2000a. Plant-animal subsistence ratios and macronutrient energy estimations in worldwide hunter-gatherer diets. American Journal of Clinical Nutrition 71(3), 682–692.

Cordain, L., Toohey, L., Smith, M.J., Hickey, M.S., 2000b. Modulation of immune function by dietary lectins in rheumatoid arthritis. British Journal of Nutrition 83(3), 207–217.

Cordain, L., Eaton, S.B., Brand Millers, J., Lindebergs, S., Jensen, C., 2002a. An evolutionary analysis of the aetiology and pathogenesis of juvenile-onset myopia. Acta Ophthalmologica Scandinavica 80(2), 125–135.

Cordain, L., Lindeberg, S., Hurtado, M., Hill, K., Eaton, S.B., Brand Miller, J., 2002b. Acne vulgaris: a disease of civilization. Archives of Dermatology 138(12), 1584–1590.

Cordain, L., Eades, M.R., Eades, M.D., 2003. Hyperinsulinemic diseases of civilization: more than just Syndrome X. Comparative Biochemistry and Physiology – Part A: Molecular and Integrative Physiology 136(1), 95–112.

Cordain, L., Eaton, S.B., Sebastian, A., Mann, N., Lindeberg, S., Watkins, B.A., O'Keefe, J.H., Brand-Miller, J., 2005. Origins and evolution of the Western diet: health implications for the 21st century. American Journal of Clinical Nutrition 81(2), 341–354.

Corthesy-Theulaz, I., den Dunnen, J.T., Ferre, P., Geurts, J.M., Muller, M., van Belzen, N., van Ommen, B., 2005. Nutrigenomics: the impact of biomics technology on nutrition research. Annals of Nutrition and Metabolism 49(6), 355–365.

Cruickshank, J.K., Mbanya, J.C., Wilks, R., Balkau, B., McFarlane-Anderson, N., Forrester, T., 1999. Sick genes, sick individuals or sick populations with chronic disease? An international example from the emergence of diabetes and high blood pressure in African-origin populations. e-Lancet <www.thelancet.com/era/LLAN.ERA.1057>.

Cuatrecasas, P., 1973. Interaction of wheat germ agglutinin and concanavalin A with isolated fat cells. Biochemistry 12(7), 1312–1323.

Cuatrecasas, P., Tell, G.P., 1973. Insulin-like activity of concanavalin A and wheat germ agglutinin – direct interactions with insulin receptors. Proceedings of the National Academy of Sciences of the United States of America 70(2), 485–489.

Cunnane, S.C., Crawford, M.A., 2003. Survival of the fattest: fat babies were the key to evolution of the large human brain. Comparative Biochemistry and Physiology Part A: Molecular and Integrative Physiology 136(1), 17–26.

Cyrus, T., Tang, L.X., Rokach, J., FitzGerald, G.A., Pratico, D., 2001. Lipid peroxidation and platelet activation in murine atherosclerosis. Circulation 104(16), 1940–1945.

Damasceno, N.R., Gidlund, M.A., Goto, H., Dias, C.T., Okawabata, F.S., Abdalla, D.S., 2001. Casein and soy protein isolate in experimental atherosclerosis: influence on hyperlipidemia and lipoprotein oxidation. Annals of Nutrition and Metabolism 45(1), 38–46.

Davis, H.R., Glagov, S., 1986. Lectin binding to distinguish cell types in fixed atherosclerotic arteries. Atherosclerosis 61(3), 193–203.

de Lorgeril, M., Salen, P., Martin, J.L., Monjaud, I., Delaye, J., Mamelle, N., 1999. Mediterranean diet, traditional risk factors, and the rate of cardiovascular complications after myocardial infarction: final report of the Lyon Diet Heart Study. Circulation 99(6), 779–785.

Debongnie, J.C., Newcomer, A.D., McGill, D.B., Phillips, S.F., 1979. Absorption of nutrients in lactase deficiency. Digestive Diseases and Sciences 24(3), 225–231.

Delange, F., Van Onderbergen, A., Shabana, W., Vandemeulebroucke, E., Vertongen, F., Gnat, D., Dramaix, M., 2000. Silent iodine prophylaxis in Western Europe only partly corrects iodine deficiency; the case of Belgium. European Journal of Endocrinology 143(2), 189–196.

Diez, M., Nguyen, P., Jeusette, I., Devois, C., Istasse, L., Biourge, V., 2002. Weight loss in obese dogs: evaluation of a high-protein, low-carbohydrate diet. Journal of Nutrition 132(6 Suppl 2), 1685S–1687S.

Dobson, J.E., 1998. The iodine factor in health and evolution. Geographical Review 88(1), 1–28.

Doerge, D.R., Sheehan, D.M., 2002. Goitrogenic and estrogenic activity of soy isoflavones. Environmental Health Perspectives 110(Suppl 3), 349–353.

Dolfini, E., Elli, L., Ferrero, S., Braidotti, P., Roncoroni, L., Dasdia, T., Falini, M.L., Forlani, F., Bardella, M.T., 2003. Bread wheat gliadin cytotoxicity: a new three-dimensional cell model. Scandinavian Journal of Clinical and Laboratory Investigation 63(2), 135–141.

Dyerberg, J., 1989. Coronary heart disease in Greenland Inuit: a paradox. Implications for western diet patterns. Arctic Medical Research 48(2), 47–54.

Dzhangaliev, A.D., Salova, T.N., Turekhanova, P.M., 2003. The wild fruit and nut plants of Kazakhstan. Horticultural Reviews 29, 305–371.

Eaton, S., Konner, M., 1985. Paleolithic nutrition. A consideration of its nature and current implications. New England Journal of Medicine 312, 283–289.

Elliott, R.B., Harris, D.P., Hill, J.P., Bibby, N.J., Wasmuth, H.E., 1999. Type I (insulin-dependent) diabetes mellitus and cow milk: casein variant consumption. Diabetologia 42(3), 292–296.

Elliott, S.S., Keim, N.L, Stern, J.S., Teff, K., Havel, P.J., 2002. Fructose, weight gain, and the insulin resistance syndrome. American Journal of Clinical Nutrition 76(5), 911–922.

Engelmann, M.D., Sandstrom, B., Michaelsen, K.F., 1998. Meat intake and iron status in late infancy: an intervention study. Journal of Pediatric Gastroenterology and Nutrition 26(1), 26–33.

Ercan, N., Nuttall, F.Q., Gannon, M.C., Redmon, J.B., Sheridan, K.J., 1993. Effects of glucose, galactose, and lactose ingestion on the plasma glucose and insulin response in persons with non-insulin-dependent diabetes mellitus. Metabolism 42(12), 1560–1567.

Fairweather, T.S., Wright, A.J., 1990. The effects of sugar-beet fibre and wheat bran on iron and zinc absorption in rats. British Journal of Nutrition 64(2), 547–552.

Feldhahn, J.R., Rand, J.S., Martin, G., 1999. Insulin sensitivity in normal and diabetic cats. Journal of Feline Medicine and Surgery 1(2), 107–115.

Fiennes, R.N., 1965. Atherosclerosis in wild animals. In: Roberts, J., Straus, R., (Eds.), Comparative Atherosclerosis. The Morphology of Spontaneous and Induced Atherosclerotic Lesions in Animals and Its Relation to Human Disease. Harper & Row, New York, pp. 113–126.

Gaitan, E., 1990. Goitrogens in food and water. Annual Review of Nutrition 10, 21–39.

Gannon, M.C., Nuttall, F.Q., Krezowski, P.A., Billington, C.J., Parker, S., 1986. The serum insulin and plasma glucose responses to milk and fruit products in type 2 (non-insulin-dependent) diabetic patients. Diabetologia 29(11), 784–791.

Garrow, J.S., James, W.P.T., (Eds.), 2000. Human Nutrition and Dietetics. Churchill Livingstone, Edinburgh.

Gleadow, R.M., Woodrow, I.E., 2002. Constraints on effectiveness of cyanogenic glycosides in herbivore defense. Journal of Chemical Ecology 28(7), 1301–1313.

Gorman, C., 2006. Does my diet fit my genes? The new science of nutrigenomics has some answers. It explains why fat and caffeine are worse for some than others. Time 167(24), 69–70.

Gower, B.A., 1999. Syndrome X in children: Influence of ethnicity and visceral fat. American Journal of Human Biology 11(2), 249–257.

Gräslund, B., 2005. Early humans and their world. Routledge, London.

Gunton, J.E., Hitchman, R., McElduff, A., 2001. Effects of ethnicity on glucose tolerance, insulin resistance and beta cell function in 223 women with an abnormal glucose challenge test during pregnancy. Australian and New Zealand Journal of Obstetrics and Gynaecology 41(2), 182–186.

Hakkak, R., Korourian, S., Ronis, M.J., Johnston, J.M., Badger, T.M., 2001. Dietary whey protein protects against azoxymethane-induced colon tumors in male rats. Cancer Epidemiology Biomarkers and Prevention 10(5), 555–558.

Hallberg, L., Brune, M., Rossander, L., 1989. Iron absorption in man: ascorbic acid and dose-dependent inhibition by phytate. American Journal of Clinical Nutrition 49(1), 140–144.

Hallberg, L., Sandström, B., Ralph, A., Arthur, J., 2000. Diseases of the heart and circulation: the role of dietary factors in aetiology and management. In: Garrow, J.S., James, W.P.T. (Eds.), Human Nutrition and Dietetics. Churchill Livingstone, Edinburg, pp. 177–209.

Hamberg, O., Nielsen, K., Vilstrup, H., 1992. Effects of an increase in protein intake on hepatic efficacy for urea synthesis in healthy subjects and in patients with cirrhosis. Journal of Hepatology 14(2–3), 237–243.

Hansson, G.K., 2001. Immune mechanisms in atherosclerosis. Arteriosclerosis, Thrombosis and Vascular Biology 21(12), 1876–1890.

Harrison, E., Adjei, A., Ameho, C., Yamamoto, S., Kono, S., 1998. The effect of soybean protein on bone loss in a rat model of postmenopausal osteoporosis. Journal of Nutritional Science and Vitaminology (Tokyo) 44(2), 257–268.

Hawrylewicz, E.J., Huang, H.H., Blair, W.H., 1991. Dietary soybean isolate and methionine supplementation affect mammary tumor progression in rats. Journal of Nutrition 121(10), 1693–1698.

Hazell, T., Johnson, I.T., 1987. In vitro estimation of iron availability from a range of plant foods: influence of phytate, ascorbate and citrate. British Journal of Nutrition 57(2), 223–233.

Hedo, J.A., Harrison, L.C., Roth, J., 1981. Binding of insulin receptors to lectins: evidence for common carbohydrate determinants on several membrane receptors. Biochemistry 20(12), 3385–3393.

Henschen, F., 1966. The history and geography of diseases. Delacorte, New York.

Herbert, V., 1988. Vitamin B-12: plant sources, requirements, and assay. American Journal of Clinical Nutrition 48(3 Suppl), 852–858.

Hetzel, B.S., 1994. Iodine deficiency and fetal brain damage. New England Journal of Medicine 331(26), 1770–1771.

Hokin, B.D., Butler, T., 1999. Cyanocobalamin (vitamin B-12) status in Seventh-day Adventist ministers in Australia. American Journal of Clinical Nutrition 70(3 Suppl), 576S–578S.

Hooper, L., Summerbell, C.D., Higgins, J.P., Thompson, R.L., Capps, N.E., Smith, G.D., Riemersma, R.A., Ebrahim, S., 2001. Dietary fat intake and prevention of cardiovascular disease: systematic review. British Medical Journal 322(7289), 757–763.

Hooper, L., Thompson, R.L., Harrison, R.A., Summerbell, C.D., Moore, H., Worthington, H.V., Durrington, P.N., Ness, A.R., Capps, N.E., Davey Smith, G., Riemersma, R.A., Ebrahim, S.B., 2004. Omega 3 fatty acids for prevention and treatment of cardiovascular disease. Cochrane Database of Systematic Reviews (4), CD003177.

Howard, B.V., Van Horn, L., Hsia, J., Manson, J.E., Stefanick, M.L., Wassertheil-Smoller, S., Kuller, L.H., LaCroix, A.Z., Langer, R.D., Lasser, N.L., Lewis, C.E., Limacher, M.C., Margolis, K.L., Mysiw, W.J., Ockene, J.K., Parker, L.M., Perri, M.G., Phillips, L., Prentice, R.L., Robbins, J., Rossouw, J.E., Sarto, G.E., Schatz, I.J., Snetselaar, L.G., Stevens, V.J., Tinker, L.F., Trevisan, M., Vitolins, M.Z., Anderson, G.L., Assaf, A.R., Bassford, T., Beresford, S.A., Black, H.R., Brunner, R.L., Brzyski, R.G., Caan, B., Chlebowski, R.T., Gass, M., Granek, I., Greenland, P., Hays, J., Heber, D., Heiss, G., Hendrix, S.L., Hubbell, F.A., Johnson, K.C., Kotchen, J.M., 2006. Low-fat dietary pattern and risk of cardiovascular disease: the women's health initiative randomized controlled dietary modification trial. Journal of the American Medical Association 295(6), 655–666.

Hugi, D., Tappy, L., Sauerwein, R.M., Bruckmaier, R.M., Blum, J.W., 1998. Insulin-dependent glucose utilization in intensively milk-fed veal calves is modulated by supplementing lactose in an age-depenedent manner. Journal of Nutrition 128, 1023–1030.

Hunt, J.R., 2003. Bioavailability of iron, zinc, and other trace minerals from vegetarian diets. American Journal of Clinical Nutrition 78(3 Suppl), 633S–639S.

Hurrell, R.F., Reddy, M.B., Juillerat, M., Cook, J.D., 2006. Meat protein fractions enhance nonheme iron absorption in humans. Journal of Nutrition 136(11), 2808–2812.

Iqbal, T.H., Lewis, K.O., Cooper, B.T., 1994. Phytase activity in the human and rat small intestine. Gut 35(9), 1233–1236.

Jackson, R.D., LaCroix, A.Z., Gasss, M., Wallace, R.B., Robbins, J., Lewis, C.E., Bassford, T., Beresford, S.A., Black, H.R., Blanchette, P., Bonds, D.E., Brunner, R.L.,. Brzyski, R.G, Caan, B., Cauley, J.A., Chlebowski, R.T., Cummings, S.R., Granek, I., Hays, J., Heiss, G., Hendrix, S.L., Howard, B.V., Hsia, J., Hubbell, F.A., Johnson, K.C., Judd, H., Kotchen, J.M., Kuller, L.H., Langer, R.D., Lasser, N.L., Limacher, M.C., Ludlam, S., Manson, J.E., Margolis, K.L., McGowan, J., Ockene, J.K., O'Sullivan, M.J., Phillips, L., Prentice, R.L., Sarto, G.E., Stefanick, M.L., Van Horn, L., Wactawski-Wende, J., Whitlock, E., Anderson, G.L., Assaf, A.R., Barad, D., 2006. Calcium plus vitamin D supplementation and the risk of fractures. New England Journal of Medicine 354(7), 669–683.

Jenkins, D.J., Kendall, C.W., Jackson, C.J., Connelly, P.W., Parker, T., Faulkner, D., Vidgen, E., Cunnane, S.C., Leiter, L.A., Josse, R.G., 2002. Effects of high- and low-isoflavone soyfoods on blood lipids, oxidized LDL, homocysteine, and blood pressure in hyperlipidemic men and women. American Journal of Clinical Nutrition 76(2), 365–372.

Jolly, C.J., 1970. A new model of hominid differentiation based on a baboon analogy. Man 5, 5–26.

Jolly, C.J., 2001. A proper study for mankind: Analogies from the Papionin monkeys and their implications for human evolution. American Journal of Physical Anthropology Supplement 33, 177–204.

Jönsson, T., Olsson, S., Ahrén, B., Bøg-Hansen, T.C., Dole, A., Lindeberg, S., 2005. Agrarian diet and sieases of affluence – Do evolutionary novel lectins cause leptin resistance? BMC Endocrine Disorders 5(10), doi:10.1186/1472-6823-5-10.

Kagami, H., Uryu, K., Okamoto, K., Sakai, H., Kaneda, T., Sakanaka, M., 1991. Differential lectin binding on walls of thoraco-cervical blood vessels and lymphatics in rats. Okajimas Folia Anatomica Japonica 68(2–3), 161–170.

Kalhan, R., Puthawala, K., Agarwal, S., Amini, S.B., Kalhan, S.C., 2001. Altered lipid profile, leptin, insulin, and anthropometry in offspring of South Asian immigrants in the United States. Metabolism 50(10), 1197–1202.

Kanis, J.A., Johansson, H., Oden, A., De Laet, C., Johnell, O., Eisman, J.A., Mc Closkey, E., Mellstrom, D., Pols, H., Reeve, J., Silman, A., Tenenhouse, A., 2005. A meta-analysis of milk intake and fracture risk: low utility for case finding. Osteoporosis International 16(7), 799–804.

Kasper, D.L., Braunwald, E., Fauci, A.S., Hauser, S.L., Longo, D.L., Jameson, J.L. (Eds.), 2005. Harrison's Principles of Internal Medicine. McGraw-Hill, New York.

Kennedy, R.L., Chokkalingam, K., Farshchi, H.R., 2005. Nutrition in patients with Type 2 diabetes: are low-carbohydrate diets effective, safe or desirable? Diabetic Medicine 22(7), 821–832.

Kerstetter, J.E., Svastisalee, C.M., Caseria, D.M., Mitnick, M.E., Insogna, K.L., 2000. A threshold for low-protein-diet-induced elevations in parathyroid hormone. American Journal of Clinical Nutrition 72(1), 168–173.

King, H., Rewers, M., 1993. Global estimates for prevalence of diabetes mellitus and impaired glucose tolerance in adults. WHO Ad Hoc Diabetes Reporting Group. Diabetes Care 16(1), 157–177.

Ko, I.W.P., Corlett, R.T., Xu, R.-J., 1998. Sugar composition of wild fruits in Hong Kong, China. Journal of Tropical Ecology 14, 381–387.

Koebnick, C., Hoffmann, I., Dagnelie, P.C., Heins, U.A., Wickramasinghe, S.N., Ratnayaka, I.D., Gruendel, S., Lindemans, J., Leitzmann, C., 2004. Long-term ovo-lacto vegetarian diet impairs vitamin B-12 status in pregnant women. Journal of Nutrition 134(12), 3319–3326.

Kopp, W., 2004. Nutrition, evolution and thyroid hormone levels – a link to iodine deficiency disorders? Medical Hypotheses 62(6), 871–875.

Kritchevsky, D., 1995. Dietary protein, cholesterol and atherosclerosis: a review of the early history. Journal of Nutrition 125(3 Suppl), 589S–593S.

Kritchevsky, D., Tepper, S.A., Klurfeld, D.M., 1998. Lectin may contribute to the atherogenicity of peanut oil. Lipids 33(8), 821–823.

Laffitte, B.A., Joseph, S.B., Walczak, R., Pei, L., Wilpitz, D.C., Collins, J.L., Tontonoz, P., 2001. Autoregulation of the human liver X receptor alpha promoter. Molecular Cell Biology 21(22), 7558–7568.

Lanou, A.J., Berkow, S.E., Barnard, N.D., 2005. Calcium, dairy products, and bone health in children and young adults: a reevaluation of the evidence. Pediatrics 115(3), 736–743.

Lau, E.M., Woo, J., 1998. Nutrition and osteoporosis. Current Opinion in Rheumatology 10(4), 368–372.

Laudet, V., Gronemeyer, H., 2002. The Nuclear Receptor Facts Book. Academic Press, New York.

Lavigne, C., Tremblay, F., Asselin, G., Jacques, H., Marette, A., 2001. Prevention of skeletal muscle insulin resistance by dietary cod protein in high fat-fed rats. American Journal of Physiology Endocrinology and Metabolism 281(1), E62–E71.

Liener, I.E., 1986. Nutritional significance of lectins in the diet. In: Liener, I.E., Sharon, N., Goldstein, I.J. (Eds.), The Lectins: Properties, Functions and Applications in Biology and Medicine. Academic Press, New York, pp. 527–552.

Lightowler, H.J., Davies, G.J., 1998. Iodine intake and iodine deficiency in vegans as assessed by the duplicate-portion technique and urinary iodine excretion. British Journal of Nutrition 80(6), 529–535.

Lindeberg, S., 1994. Apparent absence of cerebrocardiovascular disease in Melanesians. Risk factors and nutritional considerations – the Kitava Study. M.D.Ph.D. Dissertation, Lund University.

Lindeberg, S., 1998. [Cereals, salt and rickets in historical perspective]. Svensk medicinhistorisk tidskrift 2(1), 79–89.

Lindeberg, S., 2005. Who wants to be normal? European Heart Journal 26(24), 2605–2606.

Lindeberg, S., Lundh, B., 1993. Apparent absence of stroke and ischaemic heart disease in a traditional Melanesian island: a clinical study in Kitava. Journal of Internal Medicine 233(3), 269–275.

Lindeberg, S., Vessby, B., 1995. Fatty acid composition of cholesterol esters and serum tocopherols in Melanesians apparently free from cardiovascular disease – the Kitava study. Nutrition, Metabolism and Cardiovascular Diseases 5, 45–53.

Lindeberg, S., Berntorp, E., Nilsson-Ehle, P., Terent, A., Vessby, B., 1997. Age relations of cardiovascular risk factors in a traditional Melanesian society: the Kitava Study. American Journal of Clinical Nutrition 66(4), 845–852.

Lindeberg, S., Eliasson, M., Lindahl, B., Ahrén, B., 1999. Low serum insulin in traditional Pacific Islanders-the Kitava Study. Metabolism 48(10), 1216–1219.

Lindeberg, S., Soderberg, S., Ahren, B., Olsson, T., 2001. Large differences in serum leptin levels between nonwesternized and westernized populations: the Kitava study. Journal of Internal Medicine 249(6), 553–558.

Lindeberg, S., Cordain, L., Eaton, S.B., 2003. Biological and clinical potential of a palaeolithic diet. Journal of Nutritional and Environmental Medicine 13(3), 1–12.

Lindeberg, S., Jonsson, T., Granfeldt, Y., Borgstrand, E., Soffman, J., Sjostrom, K., Ahren, B., 2007. A Palaeolithic diet improves glucose tolerance more than a Mediterranean-like diet in individuals with ischaemic heart disease. Diabetologia 50, 1795–807.

Livingston, J.N., Purvis, B.J., 1981. The effects of wheat germ agglutinin on the adipocyte insulin receptor. Biochimica et Biophysica Acta 678(2), 194–201.

Lothrop, C., Harrison, G.J., Schultz, D., Utteridge, T., 1986. Miscellaneous diseases. In: Harrison, G.J., Harrison, L.R. (Eds.), Clinical Avian Medicine and Surgery. Saunders, Philadelphia, pp. 525–536.

Lucas, P.W., Ang, K.Y., Sui, Z., Agrawal, K.R., Prinz, J.F., Dominy, N.J., 2006. A brief review of the recent evolution of the human mouth in physiological and nutritional contexts. Physiology and Behavior 89(1), 36–38.

Maddox, D.A., Alavi, F.K., Silbernick, E.M., Zawada, E.T., 2002. Protective effects of a soy diet in preventing obesity-linked renal disease. Kidney International 61(1), 96–104.

Maldonado, E.N., Casanave, E.B., Aveldano, M.I., 2002. Major plasma lipids and fatty acids in four HDL mammals. Comparative

Biochemistry and Physiology Part A: Molecular and Integrative Physiology 132(2), 297–303.
McLachlan, C.N., 2001. Beta-casein A1, ischaemic heart disease mortality, and other illnesses. Medical Hypotheses 56(2), 262–272.
Mellanby, E., 1950. A story of nutrition research. Williams & Wilkins, Baltimore.
Miller, D.D., 1989. Calcium in the diet: food sources, recommended intakes, and nutritional bioavailability. Advances in Food and Nutrition Research 33, 103–156.
Miller, E.R. 3rd, Pastor-Barriuso, R., Dalal, D., Riemersma, R.A., Appel, L.J., Guallar, E., 2005. Meta-analysis: high-dosage vitamin E supplementation may increase all-cause mortality. Annals of Internal Medicine 142(1), 37–46.
Milton, K., 1999. Nutritional characteristics of wild primate foods: do the diets of our closest living relatives have lessons for us? Nutrition 15(6), 488–498.
Milton, K., 2003. The critical role played by animal source foods in human (Homo) evolution. Journal of Nutrition 133(11 Suppl 2), 3886S–3892S.
Mordes, J.P., Rossini, A.A., 1981. Animal models of diabetes. American Journal of Medicine 70(2), 353–360.
Nagaev, I., Smith, U., 2001. Insulin resistance and type 2 diabetes are not related to resistin expression in human fat cells or skeletal muscle. Biochemical and Biophysical Research Communications 285(2), 561–564.
National Food Administration., 1986. Food composition tables. National Food Administration, Uppsala, Sweden.
Neel, J.V., 1962. Diabetes mellitus: a thrifty genotype rendered detrimental by "progress"? American Journal of Human Genetics 14, 353–362.
Nishikimi, M., Fukuyama, R., Minoshima, S., Shimizu, N., Yagi, K., 1994. Cloning and chromosomal mapping of the human nonfunctional gene for L-gulono-gamma-lactone oxidase, the enzyme for L-ascorbic acid biosynthesis missing in man. Journal of Biological Chemistry 269(18), 13685–13688.
Nordin, B.E.C., Need, A.G., Morris, H.A., Horowitz, M., Chatterton, B.E., Sedgwick, A.W., 1995. Bad habits and bad bones. In: Burckhardt, P., Heaney, R.P. (Eds.), Nutritional Aspects Of Osteoporosis '94. Ares-Serono Symposia 7, Rome, pp. 1–25.
Nuttall, F.Q., Schweim, K., Hoover, H., Gannon, M.C., 2006. Metabolic effect of a LoBAG30 diet in men with type 2 diabetes. American Journal of Physiology Endocrinology and Metabolism 291(4), E786–791.
O'Dea, K., 1984. Marked improvement in carbohydrate and lipid metabolism in diabetic Australian aborigines after temporary reversion to traditional lifestyle. Diabetes 33(6), 596–603.
Ohmori, T., Yatomi, Y., Wu, Y., Osada, M., Satoh, K., Ozaki, Y., 2001. Wheat germ agglutinin-induced platelet activation via platelet endothelial cell adhesion molecule-1: involvement of rapid phospholipase C gamma 2 activation by Src family kinases. Biochemistry 40(43), 12992–13001.
Palmquist, D.L., Doppenberg, J., Roehrig, K.L., Kinsey, D.J., 1992. Glucose and insulin metabolism in ruminating and veal calves fed high and low fat diets. Domestic Animal Endocrinology 9, 233–241.
Park, Y.K., Yetley, E.A., 1993. Intakes and food sources of fructose in the United States. American Journal of Clinical Nutrition 58 (5 Suppl), 737S–747S.
Perry, G.H., Dominy, N.J., Claw, K.G., Lee, A.S., Fiegler, H., Redon, R., Werner, J., Villanea, F.A., Mountain, J.L., Misra, R., Carter, N.P., Lee, C., Stone, A.C., 2007. Diet and the evolution of human amylase gene copy number variation. Nature Genetics 39(10), 1256–1260.
Potter, S.M., Baum, J.A., Teng, H., Stillman, R.J., Shay, N.F., Erdman, Jr., J.W., 1998. Soy protein and isoflavones: their effects on blood lipids and bone density in postmenopausal women. American Journal of Clinical Nutrition 68(6 Suppl), 1375S–1379S.
Poulter, N.R., Khaw, K.T., Sever, P.S., 1988. Higher blood pressures of urban migrants from an African low-blood pressure population are not due to selective migration. American Journal of Hypertension 1(3 Pt 3), 143S–145S.
Proos, L.A., Hofvander, Y., Tuvemo, T., 1991. Menarcheal age and growth pattern of Indian girls adopted in Sweden. I. Menarcheal age. Acta Paediatrica Scandinavica 80(8–9), 852–858.
Purrello, F., Burnham, D.B., Goldfine, I.D., 1983. Insulin receptor antiserum and plant lectins mimic the direct effects of insulin on nuclear envelope phosphorylation. Science 221(4609), 462–464.
Pusztai, A., 1993. Dietary lectins are metabolic signals for the gut and modulate immune and hormone functions. European Journal of Clinical Nutrition 47(10), 691–699.
Pusztai, A., Greer, F., Grant, G., 1989. Specific uptake of dietary lectins into the systemic circulation of rats. Biochemical Society Transactions 17, 481–482.
Rasmussen, L.B., Ovesen, L., Bulow, I., Jorgensen, T., Knudsen, N., Laurberg, P., Perrild, H., 2002. Relations between various measures of iodine intake and thyroid volume, thyroid nodularity, and serum thyroglobulin. American Journal of Clinical Nutrition 76(5), 1069–1076.
Ravnskov, U., 1998. The questionable role of saturated and polyunsaturated fatty acids in cardiovascular disease. Journal of Clinical Epidemiology 51(6), 443–460.
Requejo, A.M., Navia, B., Ortega, R.M., Lopez-Sobaler, A.M., Quintas, E., Gaspar, M.J., Osorio, O., 1999. The age at which meat is first included in the diet affects the incidence of iron deficiency and ferropenic anaemia in a group of pre-school children from Madrid. International Journal for Vitamin and Nutrition Research 69(2), 127–131.
Richards, M.P., 2002. A brief review of the archaeological evidence for Palaeolithic and Neolithic subsistence. European Journal of Clinical Nutrition 56(12), 1270–1279.
Richards, M.P., Pettitt, P.B., Stiner, M.C., Trinkaus, E., 2001. Stable isotope evidence for increasing dietary breadth in the European mid-Upper Paleolithic. Proceedings of the National Academy of Sciences of the United States of America 98(11), 6528–6532.
Rodin, D.A., Bano, G., Bland, J.M., Taylor, K., Nussey, S.S., 1998. Polycystic ovaries and associated metabolic abnormalities in Indian subcontinent Asian women. Clinical Endocrinology (Oxford) 49(1), 91–99.
Roth, G.S., Ingram, D.K., Lane, M.A., 2001. Caloric restriction in primates and relevance to humans. Annals of the New York Academy of Sciences 928, 305–315.
Samuelson, L.C., Phillips, R.S., Swanberg, L.J., 1996. Amylase gene structures in primates: retroposon insertions and promoter evolution. Molecular Biology and Evolution 13(6), 767–779.
Schaefer, O., 1981. Eskimos. In: Trowell, H.C., Burkitt, D.P. (Eds.), Western Diseases: Their Emergence and Prevention. Edward Arnold, London, pp. 113–128.
Sebastian, A., Frassetto, L.A., Sellmeyer, D.E., Merriam, R.L., Morris, Jr., R.C., 2002. Estimation of the net acid load of the

diet of ancestral preagricultural Homo sapiens and their hominid ancestors. American Journal of Clinical Nutrition 76(6), 1308–1316.

Shechter, Y., 1983. Bound lectins that mimic insulin produce persistent insulin-like activities. Endocrinology 113(6), 1921–1926.

Shechter, Y., Sela, B.A., 1981. Insulin-like effects of wax bean agglutinin in rat adipocytes. Biochemical and Biophysical Research Communications 98(2), 367–373.

Shils, M.E., Shike, M., 2006. Modern nutrition in health and disease. Lippincott Williams & Wilkins, Philadelphia.

Simmoons, F.J., 1981. Celiac disease as a geographic problem. In: Walcher, N., Kretchmer, N. (Eds.), Food, Nutrition and Evolution. Masson, New York, pp. 179–199.

Simoons, F.J., Johnson, J.D., Kretchmer, N., 1977. Perspective on milk-drinking and malabsorption of lactose. Pediatrics 59(1), 98–108.

Sinnett, P.F., Whyte, H.M., 1973. Epidemiological studies in a highland population of New Guinea: environment, culture, and health status. Human Ecology 1(3), 245–277.

Slavin, J.L., Martini, M.C., Jacobs, Jr., D.R., Marquart, L., 1999. Plausible mechanisms for the protectiveness of whole grains. American Journal of Clinical Nutrition 70(3 Suppl), 459S–463S.

Stabler, S.P., Allen, R.H., 2004. Vitamin B12 deficiency as a worldwide problem. Annual Review of Nutrition 24, 299–326.

Tejada, C., Strong, J.P., Montenegro, M.R., Restrepo, C., Solberg, L.A., 1968. Distribution of coronary and aortic atherosclerosis by geographic location, race, and sex. Laboratory Investigations 18(5), 509–526.

Thorburn, A.W., Brand, J.C., Truswell, A.S., 1987. Slowly digested and absorbed carbohydrate in traditional bushfoods: a protective factor against diabetes? American Journal of Clinical Nutrition 45(1), 98–106.

Thorsdottir, I., Birgisdottir, B.E., Johannsdottir, I.M., Harris, D.P., Hill, J., Steingrimsdottir, L., Thorsson, A.V., 2000. Different beta-casein fractions in Icelandic versus Scandinavian cow's milk may influence diabetogenicity of cow's milk in infancy and explain low incidence of insulin-dependent diabetes mellitus in Iceland. Pediatrics 106(4), 719–724.

Trowell, H.C., Burkitt, D.P. (Eds.), 1981. Western Diseases: Their Emergence and Prevention. Harvard University Press, Cambridge.

Truswell, A.S., 2002. Cereal grains and coronary heart disease. European Journal of Clinical Nutrition 56(1), 1–14.

Truswell, A.S., Hansen, J.D.L., 1976. Medical research among the !Kung. In: Lee, R.B., DeVore, I. (Eds.), Kalahari Hunter-gatherers. Harvard University Press, Cambridge, pp. 166–195.

Ungar, P., 2004. Dental topography and diets of Australopithecus afarensis and early Homo. Journal of Human Evolution 46(5), 605–622.

Vacaresse, N., Vieira, O., Robbesyn, F., Jurgens, G., Salvayre, R., Negre-Salvayre, A., 2001. Phenolic antioxidants trolox and caffeic acid modulate the oxidized LDL- induced EGF-receptor activation. British Journal of Pharmacology 132(8), 1777–1788.

Vanderjagt, D.J., Freiberger, C., Vu, H.T., Mounkaila, G., Glew, R.S., Glew, R.H., 2000. The trypsin inhibitor content of 61 wild edible plant foods of Niger. Plant Foods for Human Nutrition 55(4), 335–346.

Vasankari, T.J., Vasankari, T.M., 2006. Effect of dietary fructose on lipid metabolism, body weight and glucose tolerance in humans. Scandinavian Journal of Food and Nutrition 50(2), 55–63.

Voight, B.F., Kudaravalli, S., Wen, X., Pritchard, J.K., 2006. A map of recent positive selection in the human genome. PLoS Biology 4(3), e72.

Wagner, J.D., Zhang, L., Greaves, K.A., Shadoan, M.K., Schwenke, D.C., 2000. Soy protein reduces the arterial low-density lipoprotein (LDL) concentration and delivery of LDL cholesterol to the arteries of diabetic and nondiabetic male cynomolgus monkeys. Metabolism 49(9), 1188–1196.

Wang, Q., Yu, L.G., Campbell, B.J., Milton, J.D., Rhodes, J.M., 1998. Identification of intact peanut lectin in peripheral venous blood. Lancet 352(9143), 1831–1832.

Wang, X.Y., Bergdahl, K., Heijbel, A., Liljebris, C., Bleasdale, J.E., 2001. Analysis of in vitro interactions of protein tyrosine phosphatase 1B with insulin receptors. Molecular and Cellular Endocrinology 173(1–2), 109–120.

Wasmuth, H.E., Kolb, H., 2000. Cow's milk and immune-mediated diabetes. Proceedings of the Nutrition Society 59(4), 573–579.

Weinsier, R.L., Krumdieck, C.L., 2000. Dairy foods and bone health: examination of the evidence. American Journal of Clinical Nutrition 72(3), 681–689.

Wilson, T.A., Nicolosi, R.J., Marchello, M.J., Kritchevsky, D., 2000. Consumption of ground bison does not increase early atherosclerosis development in hypercholesterolemic hamsters. Nutrition Research 20(5), 707–719.

Winzenberg, T., Shaw, K., Fryer, J., Jones, G., 2006. Effects of calcium supplementation on bone density in healthy children: meta-analysis of randomised controlled trials. British Medical Journal 333(7572), 775–780.

Wrangham, R.W., Holland Jones, J., Laden, G., Pilbeam, D., Conklin-Brittain, N.L., 1999. The raw and the stolen. Cooking and the ecology of human origins. Current Anthropology 40(5), 567–594.

Writers of Nordic Nutrition Recommendations, 2004. Nordic Nutrition Recommendations. Nordic Council of Ministers, Copenhagen.

Wu, H.M., Seet, B., Yap, E.P., Saw, S.M., Lim, T.H., Chia, K.S., 2001. Does education explain ethnic differences in myopia prevalence? A population-based study of young adult males in Singapore. Optometry and Vision Science 78(4), 234–239.

Yeung, G.S., Zlotkin, S.H., 2000. Efficacy of meat and iron-fortified commercial cereal to prevent iron depletion in cow milk-fed infants 6 to 12 months of age: a randomized controlled trial. Canadian Journal of Public Health 91(4), 263–267.

Yevdokimova, N.Y., Yefimov, A.S., 2001. Effects of wheat germ agglutinin and concanavalin A on the accumulation of glycosaminoglycans in pericellular matrix of human dermal fibroblasts. A comparison with insulin. Acta Biochimica Polonica 48(2), 563–572.

Young, V.R., Pellett, P.L., 1994. Plant proteins in relation to human protein and amino acid nutrition. American Journal of Clinical Nutrition 59(5 Suppl), 1203S–1212S.

Zhu, S., Heymsfield, S.B., Toyoshima, H., Wang, Z., Pietrobelli, A., Heshka, S., 2005. Race-ethnicity-specific waist circumference cutoffs for identifying cardiovascular disease risk factors. American Journal of Clinical Nutrition 81(2), 409–415.

5. Hunting and Hunting Weapons of the Lower and Middle Paleolithic of Europe

Paola Villa
University of Colorado Museum
UCB 265
Boulder, CO 80309–0265 USA
villap@colorado.edu

and

Institut de Préhistoire et Géologie du Quaternaire
PACEA-UMR 5199 du C.N.R.S.
Université Bordeaux 1
33405 Talence, France

Michel Lenoir
Institut de Préhistoire et Géologie du Quaternaire
PACEA-UMR 5199 du C.N.R.S.
Université Bordeaux 1
33405 Talence, France
m.lenoir@ipgq.u-bordeaux1.fr

Keywords Hunting • Western Europe • Mousterian points • throwing and thrusting spears

Abstract This paper is a review of the state of our knowledge and ignorance on Early, Middle, and Late Pleistocene subsistence behavior in Western Europe. There are undoubtedly differences in subsistence behavior between early hominids and Upper Paleolithic humans in Europe. Yet recent research has shown that some of the most extreme statements about passive scavenging practiced on a regular basis at some Middle and Late Pleistocene sites are not supported by the evidence and must be rejected. We provide an overview of the hunting *versus* scavenging debate in African and European archaeology. We discuss the following issues: (1) evidence for hunting from the earliest sites in Europe, prior to 400–300,000 years ago; (2) evidence for hunting large mammals (elephants, rhinoceroses, large-size bovids) and for hunting/gathering very small vertebrates and invertebrates (leporids, birds, fish, shellfish) before the Upper Paleolithic; (3) evidence for the use of stone-tipped spears by Neanderthals in Western Europe.

Our analysis shows: (a) that for the period prior to OIS 12 (i.e. about 400 ka) very few generalizations can be made about the subsistence behavior of early humans in Europe because the informative sites are few and far between. Nevertheless, a good case can be made for hunting from two of the earliest sites in Europe, Gran Dolina TD 6 and Boxgrove; (b) even stronger evidence of hunting comes from sites such as Schöningen and later Middle Paleolithic sites where the topographic setting and the faunal accumulations indicate repeated episodes of hunting the same species of large-size mammals; (c) that Mousterian points were used to tip thrusting or throwing spears already by OIS 6 (i.e. between 186 and 127 ka), before the end of the Middle Pleistocene. Implications for a correlation between hunting weapons and Middle Paleolithic faunal remains are discussed.

J.-J. Hublin and M.P. Richards (eds.), *The Evolution of Hominin Diets: Integrating Approaches to the Study of Palaeolithic Subsistence,* 59–85.

Introduction

In African archaeology, the debate about hunting or scavenging by early hominids has come full circle. It started about 30 years ago with some cautious assessment of hunting of small prey and possibly scavenging of larger carcasses (such as elephants and hippos; Isaac, 1978; Isaac and Crader, 1981) and continued with detailed analysis of the Olduvai (Tanzania) and Koobi Fora (Lake Turkana, Kenya) faunal assemblages by Henry Bunn (1981, 1982) who underlined the complexities of site formation processes and the multiple origins of various assemblages. He suggested that scavenging could have played a role in the subsistence behavior of early hominids, but that additional research was required to demonstrate the hypothesis. Bunn (1981) also provided direct evidence of hominid butchering practices with the occurrence of abundant cutmarks and percussion marks on bones from archaeological sites in the Koobi Fora formation and at FLK Zinj (Olduvai Bed I). The FLK Zinj site, dated to about 1.75 Ma, has provided a large faunal assemblage (more than 3,500 identifiable bone specimens with well-preserved bone surfaces) and has been the subject of more debates than any other site of comparable age.

At the same time, Binford (1981) strongly asserted that early hominids at Olduvai were marginal scavengers of bone marrow at large carnivore kill sites and had little access to meat. Several other researchers contributed to the debate with analyses and contrasting interpretations of butchering marks, the role of carnivores in site formation processes, and the archaeological signatures of scavenging and hunting (e.g. Potts and Shipman, 1981; Blumenschine,1986, 1988; Blumenschine and Selvaggio, 1988). The debate intensified with an important paper by Bunn and Kroll (1986), which showed that abundant cutmarks and percussion marks occurred on meat-bearing limb bones (humerus, radio-ulna, femur, and tibia) of small and larger animal carcasses at FLK Zinj, proving that hominids had access to meaty elements. They concluded that carcasses had to be obtained either by hunting (hunting of smaller, gazelle-sized prey is consistent with trends in other primates) or by aggressive, confrontational driving away of primary predators or primary scavengers (cf. Bunn, 1996). According to Bunn and Kroll (1988), Binford's statements were wrong because they were based on preliminary and incomplete bone data that overrepresented lower limb elements, underestimated the abundance of meatier elements by excluding limb shafts, and did not take into account cutmark data.

The debate was not resolved because the evidence for early access to meat-bearing parts based on skeletal elements representation did not necessarily indicate who actually killed the animals. The debate was also complicated by discussions on the spatial distribution of bones and stones at FLK Zinj and other sites and on what behavioral model would best account for accumulation of artifacts and broken bones: the home base or central place foraging of Isaac (1984), the stone cache of Potts (1988), the refuge model to avoid carnivore competition of Blumenschine et al. (1994), or the near-kill model (O'Connell, 1997; Domínguez-Rodrigo, 2001).

The abundance of gnaw marks on the FLK Zinj bones (on about 400 bones, more than those with cutmarks, according to Bunn and Kroll, 1986) was an anomaly that would not fit actualistic data based on late access by carnivores. A multiple stage model was suggested by Blumenschine and colleagues, with large felids having primary access to carcasses, removing flesh partially or completely, followed by hominids who had secondary access and concentrated on marrow, and finally by other bone-crunching carnivores (hyenas) consuming the remaining epiphyses (Blumenschine, 1988, 1995; Domínguez-Rodrigo and Pickering, 2003 and references therein).

Experimental work by Domínguez-Rodrigo (1999) then showed that carcasses ravaged by lions provide very little edible tissue to hominid scavengers. Midshaft portions of upper limb bones (humerus and femur) display a complete lack of flesh scraps; flesh scraps are very poorly represented on midshafts of intermediate limb bones (radio-ulna and tibia). This is in contradiction with the evidence of abundant cutmarks on limb midshafts at FLK Zinj, as hominids would have no reason to use a cutting edge on a completely defleshed long bone (Pickering and Domínguez-Rodrigo, 2006). Recently a re-analysis of tooth marks on the FLK Zinj assemblage has provided new lower estimates of carnivore modifications (Domínguez-Rodrigo and Barba, 2006). Moreover, most researchers agree that hyenas ravaged the FLK Zinj fauna **after** hominid involvement. All these observations suggest that hominids had early access to fully-fleshed carcasses, and were not relegated to the role of passive scavengers of marginal scraps of flesh and marrow from carcasses primarily consumed by carnivores, as proposed by Binford as well as in the three-stage model of Blumenschine and colleagues. Analysis of tool marks on the fauna from FxJj 50 (Koobi Fora, Kenya), from the BK assemblage in Bed II at Olduvai, from the ST site complex at Peninj (Tanzania), and from Swartkrans Member 3 (South Africa) also indicate that humans were significant actors in the formation of bone accumulations (Domínguez-Rodrigo and Pickering, 2003; Pickering and Domínguez-Rodrigo, 2006). The re-analysis by Oliver of the FLK Zinj assemblage (1994) had previously confirmed that there are more upper and intermediate limb bones with cutmarks than lower limb bones, like metapodials, with cutmarks (see also Dominguez-Rodrigo et al., 2007).

Thus the overemphasis on passive, marginal scavenging is rejected, Bunn's hunting/aggressive scavenging hypothesis is resurrected, and the role of humans in faunal accumulations at these early archeological sites is confirmed. Although this is unlikely to be the last word on the interpretation of the human role in faunal accumulations at East and South African sites, it is safe to conclude that the case for active hunting or confrontational scavenging, if we exclude sites with single carcasses of very large mammals, is now much stronger, and a flexible adaptation combining hunting and scavenging is a viable scenario for early humans.

The Hunting Versus Scavenging Debate in European Archaeology

In European archaeology the idea that scavenging was an important pattern of subsistence behavior of early humans was also suggested by Binford (1985, 1987, 1988), based on sites such as Torralba, Hoxne, Swanscombe, Vaufrey, and Combe Grenal. According to him, the main activity represented at Torralba was scavenging of marginal parts left by carnivores of various non-elephant species. The predominance of head parts and feet (phalanges and lower foot bones such as metacarpals and metatarsals) was also cited as evidence of systematic scavenging at Middle Paleolithic French sites such as Vaufrey layer VIII (with an estimated age of OIS 7,[1] approximately 200 ka) and the Combe

[1] Throughout the text we give ages as in Oxygen Isotope Stages (OIS) whenever possible. However climato-stratigraphic terms, such as Early Glacial, Würm, or Last Interglacial, are extremely common in European prehistoric literature and we use them if they are embedded in the literature cited; correlation of those climato-stratigraphic terms to OIS stages is not always automatic. The dates of isotope stages are provided in Klein, 1999: 59.

Grenal Würm 1 (i.e. OIS 5a–d or Early Glacial, approximately 115–75 ka) layers. Binford thought that there was, through time, an increase in scavenging for meat; hunting became more common with the Early Glacial, but regular medium-to-large mammal hunting appeared (in Europe) only with fully modern man (Villa, 1990, 1991). This hypothesis of regular scavenging provided support to the idea that Neanderthals were behaviorally inferior to modern humans. Although both aggressive and passive scavenging are documented in modern hunter/gatherers (Bunn et al., 1988; O'Connell et al., 1988) scavenging by modern humans is not a systematic foraging strategy but an activity carried out in the context of hunting.

Later Stiner (1994) extended Binford's model, arguing that Neanderthals practiced a flexible and opportunistic passive scavenging mode, alternating with hunting. The case for scavenging was based on evidence from two Italian Middle Paleolithic sites, Grotta Guattari and Grotta dei Moscerini, both older than 55 ka; each site contains several layers, some with assemblages accumulated by hyenas and others with assemblages accumulated primarily by hominids. According to Stiner, the ungulate faunas in layers accumulated by hominids are dominated by head elements. These are elements of low utility because they have little flesh and provide little nutrition (they have relatively high fat levels; Stiner, 1994: 266), but they are those available to non-confrontational, passive scavengers.

This hypothesis was widely accepted, but in a series of important papers Curtis Marean and colleagues (Marean and Kim, 1998; Marean and Assefa, 1999) have shown that the interpretations of scavenging proposed by Binford and Stiner are based on faulty data and should be rejected. Assemblages dominated by head parts are the result of systematic biases in recovery or analysis, due to discard of long bone shafts after the excavation or their neglect by analysts because bone shafts are less easily identified. Their omission causes an underestimation of meat-rich long bones and creates an artificial head-dominated pattern. As documented by Zilhao (1998), head-dominated or head-and-foot dominated patterns are not only characteristics of Middle Paleolithic sites but can also be observed in Upper Paleolithic and Neolithic sites (where scavenging is not implicated), and are better interpreted as artifacts of taphonomy and excavation biases.

Mortality profiles with a bias toward old individuals have been described at Guattari and Moscerini. They have been considered by Stiner as complementing the skeletal element frequencies in favor of her scavenging hypothesis, because old individuals may have died a natural death, thus becoming available to passive, nonconfrontational scavengers. Since mortality profiles require a certain number of data points, we need to consider the size of the samples, their stratigraphic context, and accumulation process.

Several authors have pointed out that Guattari and Moscerini are very small assemblages, in addition to being heavily selected by post-excavation discard of long bone shafts (Klein, 1995; Gaudzinski, 1996a; Mussi, 2001). The stratigraphic sequence at Grotta dei Moscerini is 8 m thick and the excavators described 44 sedimentologically distinct layers but, to bolster sample size, Stiner lumped the layers in six units: M1 to M6. The units with faunas accumulated by hominids (M2, M3, M4 and M6) are extremely thick, between 90 and 290 cm, yet the samples are extremely small: M2 has yielded 39 NISP (Number of Identified Specimens), including bones and teeth; in M3 there are 361 NISP; in M4 159; and in M6 187. The density of NISP per liter of sediments, with an average of 32.4 lit. per one NISP (Table 5.1), is much lower than at other Mousterian sites. By comparison, in the Denticulate Mousterian of Saint-Césaire the density is 2.2 lit. of sediment per one NISP (Morin, 2004: 135); in layer 8 at La Quina the density of numbered bones (not counting the screen refuse) is 0.5 lit. of sediments per one bone (2,098/1000 lit.; Chase, 1999).

At Moscerini the excavators used dry-screening with a 5 mm mesh (Kuhn, 1990: 124), and apparently biases in selecting and keeping excavated items did not affect the stone artifacts. Yet the lithic assemblages at Moscerini (Table 5.1) are also quite small compared to other sites. The average density of artifacts in Middle Paleolithic cave sites of Northern Italy (e.g. Fumane, Tagliente) is between 5–10 lit. of sediments per one artifact; at La Chaise Abri Bourgeois-Delaunay (a cave site in the Charente region, France) where layers 10, 9, 8[1], and 8

TABLE 5.1. Grotta dei Moscerini and Grotta Guattari. Assemblages accumulated by hominids. Thickness of units measured from published section. NISP counts are for carnivore and ungulates (From Stiner, 1994: 50–56, 78, 247).

Site	Unit used by Stiner	Excavated area (m^2)	Thickness of the stratigraphic unit (cm)	Volume of excavated sediment (lit.)	Total NISP (bone and teeth)	Density (liters of sediment per one bone)	MNI of red and fallow deer	Stone artifacts	Density (liters of sediment per one artifact)
Moscerini	M2 (levels 11–20)	3	90	2,700	39	69.2	1	382	7.1
Moscerini	M3 (levels 21–36)	3	223	6,690	361	18.5	15	324	20.6
Moscerini	M4 (levels 37–43)	3	290	8,700	159	54.7	8	187	46.5
Moscerini	M6	<3	Unknown	Unknown	187	Unknown	7	196	Unknown
Guattari	G4	<25	Unknown	Unknown	53	Unknown	1 MNI of red and fallow deer and 2 MNI of aurochs G4–G5 combined	876	Unknown
Guattari	G5	<25	Unknown	Unknown	49	Unknown		482	Unknown

NISP = Number of identified specimens; MNE = Minimum number of elements; MNI = Minimum number of individuals.

dated to OIS 5 (hence comparable in age to Moscerini) were not screened, they have a density varying from 7.8–2.1 lit. per artifact (Porraz, 2005: 72; Delagnes, 1992: 23, 36).

The excavated area was only $3\,m^2$, corresponding to 5% of the preserved portion of the site; the very small size of the faunal and lithic assemblages compared to the thickness of the deposits indicate that the site was occupied only sporadically, and that these small assemblages are heterogeneous aggregates of materials accumulated over long intervals of time (Villa, 2004; Villa et al., 2005a).

The excavated area at Guattari is larger (Piperno and Giacobini, 1990–1991), but the assemblages from layers G4 and G5 (the assemblages accumulated by hominids) are even smaller, with 53 and 49 NISP respectively. Layers G4 and G5 occur at the cave mouth but thin out and disappear deep inside; their thickness and excavated surface are not provided in publications.

It is important to consider the minimum number of individuals (MNI) involved in Stiner's profiles in light of the problem of pooling of samples and cave sedimentation rates. At Guattari, in layers G4 and G5 the MNI of red and fallow deer is one, of aurochs is two. These numbers are too low and will not be considered further.

At Moscerini the MNIs used for mortality patterns (Stiner, 1994: 247, 302–305) are those red and fallow deer, the most common ungulates at the site. Stratigraphic unit M2 with only one MNI is 90cm thick; unit M3 with 15 MNI is 223cm thick; unit M4 with eight MNI is 290cm thick; the thickness of M6 is unknown. Cave sedimentation rates are generally very low; 1cm of deposit may represent from 5 to 167 years, with an average of 14 years (Speth and Johnson, 1976; Villa and Courtin, 1983). Clearly the Moscerini cave deposits represent very long time spans and probably many sporadic episodes of occupation. These very small and pooled samples of episodic behavior, spread over millennia and a large number of generations, cannot inform us on transport and subsistence patterns of the Moscerini inhabitants. It does not seem logical to treat aggregates of materials as if they were coherent assemblages with some measure of temporal and behavioral integrity. It is significant that in a recent paper Kuhn and Stiner (2006: 956) seem to agree that the question of scavenging by Neanderthals is a dead issue.

In conclusion, there is at present no evidence to support a hypothesis of systematic, regular scavenging activities by Neanderthals and earlier European hominids. In fact, the accumulated evidence from many Middle and Late Pleistocene Eurasian sites show that hunting, not scavenging, was the main method of meat procurement, and that in this respect Neanderthals and earlier humans in Europe did not differ from later, Upper Paleolithic humans (Villa et al., 2005a and references therein; Adler et al., 2006).

Reviewing the evidence for subsistence behavior in the Lower and Middle Pleistocene of Europe has lead us to examine in detail two subjects for which there is still limited empirical evidence, and which have not yet attracted the attention of many archaeologists. In the first part we examine the evidence for hunting from the earliest sites in Europe. In the second part we address a different question: if hunting, not scavenging, was the main method of meat procurement by Neanderthals, what kind of weapon did they use to dispatch their prey? Is there evidence for the use of stone-tipped spears in Western Europe?

The Earliest Sites

Only a small number of sites are securely dated to the Lower Pleistocene, just prior to the geomagnetic boundary between the Matuyama Reversed Chron and the Bruhnes Normal Chron dated to about 780ka ago (Klein, 1999: 51). Even fewer also contain faunas whose origins have been determined by taphonomic analysis. Some of the sites for which a Lower Pleistocene age has been claimed are either of uncertain age or contain no fauna (e.g., Monte Poggiolo), or the association of the lithic artifacts with the fauna is uncertain, and the role of humans in the accumulation of the faunal assemblage is unclear or undocumented. This is the case of Pakefield, Monte Peglia, Vallonet, Barranco León, Fuentenueva 3, and Sima del Elefante (Oms et al., 2000; Villa, 2001; Echassoux, 2004; Parés et al., 2006; Santonja and Villa, 2006 and references therein). The only exception is Gran Dolina TD 6 (Fig. 5.1, Table 5.2).

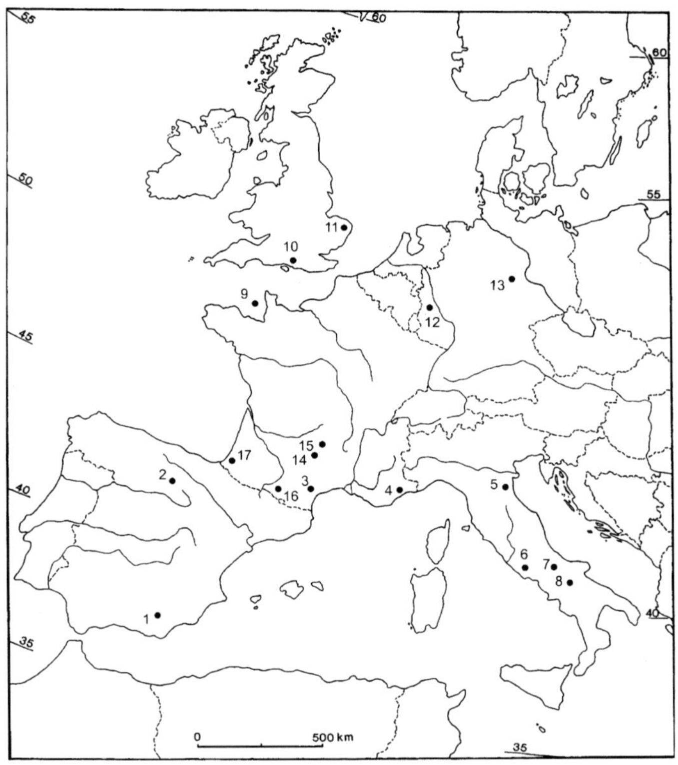

Fig. 5.1. Location of major European sites mentioned in the text. 1 = Barranco León and Fuentenueva 3; 2 = Atapuerca (Gran Dolina and Sima del Elefante); 3 = Arago; 4 = Vallonet; 5 = Monte Poggiolo; 6 = Ceprano; 7 = Isernia; 8 = Venosa Notarchirico; 9 = La Cotte de St. Brelade; 10: Boxgrove; 11 = Pakefield; 12 = Miesenheim; 13 = Schöningen; 14 = Coudoulous; 15 = La Borde; 16 = Mauran; 17 = Bouheben.

The 18 m thick karstic fill of Gran Dolina begins with a series of levels (TD 1 and 2) with sediments typical of a closed cave. Level TD 6 (2.5 m thick) belongs to a second phase that corresponds with an opening to the exterior through an entrance which no longer exists because it was destroyed by railway construction at the end of the nineteenth century (Fernández-Jalvo, 1998; Pérez-González et al., 2001). The Matuyama/Bruhnes boundary is in level TD 7, 1 m above the top of TD 6, which is thus of pre-Bruhnes age. ESR and uranium series dates support this conclusion and suggest that the age of TD 6 is between 860 and 780 ka (Falguères et al., 1999; Bermúdez de Castro et al., 2004). The top part of TD 6, called the Aurora stratum, is about 20–25 cm thick and is made of silty clay with clasts that come either from outside the cave or from the walls.

The total number of identified specimens of macrofauna from the Aurora stratum is 669, from an excavated area of 6 m^2, found together with 205 artifacts (Díez et al., 1999; Table 5.2). According to Díez et al. the NISP is 667, but we have corrected this value considering the NISP and MNI of carnivores identified by García and Arsuaga (1999) in excavation levels TD 38–43 corresponding to the Aurora stratum. Remains of six individuals of *Homo antecessor* are included in the macrofauna because the careful taphonomic analysis of Fernández-Jalvo et al. (1999) has shown that they had been cannibalized and used as food.

The minimum number of elements was provided in the publication, so it is possible to make a diagram with the skeletal completeness by regions and different size classes (Fig. 5.2). This diagram shows a good representation of limb bones, which suggests primary access to complete carcasses, supporting the indication of butchering provided by frequencies of cutmarks preferentially located on diaphyses (cutmarks located on diaphyses are defleshing marks and are considered an indication of primary access to a fully-fleshed carcass, see Introduction), and by frequencies of percussion marks (Table 5.3). Carnivore tooth marks are slightly less frequent than cutmarks; they have a small diameter (less than 5 mm) so it is suggested that they are from canids; in three cases they overlapped cutmarks indicating that canids scavenged bones abandoned by humans. Very few skeletal elements are intact (some phalanges and short bones); bones have fairly high frequencies of curved and V-shaped fractures, of oblique fracture angles and of bone splinters with shaft diameter less than half the original diameter (Díez et al., 1999: figs. 4–5). These patterns of bone breakage are clearly indicative of bones broken when fresh to extract marrow (Villa and Mahieu, 1991).

Some observations suggest a relatively short period of time for assemblage formation: (a) the bone accumulation and the stone artifacts have the same vertically restricted distribution, and there is no evidence of sorting by size or water transport; (b) 19 refitting links of human and animal bones have been found; some go across the depth of the level, supporting the idea of a short interval of time for deposition (Díez et al., 1999; Fernández-Jalvo et al., 1999: fig. 11).

Juveniles and medium to small sized animals predominate in this assemblage. The six human individuals (Fig. 5.3) are two infants (3–4 years old), two adolescents (11 and 14 years old), and two young adults (16–18 years old); the large to small size ungulates (*Mammuthus* and rhino are excluded) include eight infants and juveniles, four adults, and only two old individuals (one equid and one *Megaloceros*; Díez et al., 1999: table 3). In well-preserved assemblages accumulated as a result of normal mortality over a period of years, juvenile

TABLE 5.2. Stone artifacts and faunal remains in early sites in Spain. The total NISP in TD 6 includes bones anatomically identified to body size but not to taxon. The carnivore NISP remains belong to *Ursus* (3), *Canis mosbachensis* (2), *Vulpes praeglacialis* (3), *Crocuta* (2), *Lynx* (1), *Mustela palerminea* (1) and Canidae indet (1) (From Díez et al.,1999; García and Arsuaga, 1999; Fernández-Jalvo et al., 1999; Palmquist et al., 2005; Santonja and Villa, 2006).

Site	Stone artifacts	Taxon	NISP	MNE	MNI
Barranco León	295	*Equus* and many other species	>1,000	–	–
Fuentenueva 3	244	*Mammuthus meridionalis*, *Equus*, many other species	>1,400	–	–
Gran Dolina TD 6, Aurora stratum	205	*Mammuthus*	2	1	1
		Stephanorhinus	7	5	2
		Bison	55	35	2
		Equus	18	12	3
		Megaloceros	8	8	2
		Cervus elaphus	15	12	2
		Indet. large size cervids	94	34	1
		Dama	20	20	2
		Capreolus	5	5	2
		Sus	1	1	1
		Homo	92	69	6
		Possible *Homo*	90	30	–
		Carnivores	13	13	6
			Total 669		
Sima del Elefante	25	–	–	–	–

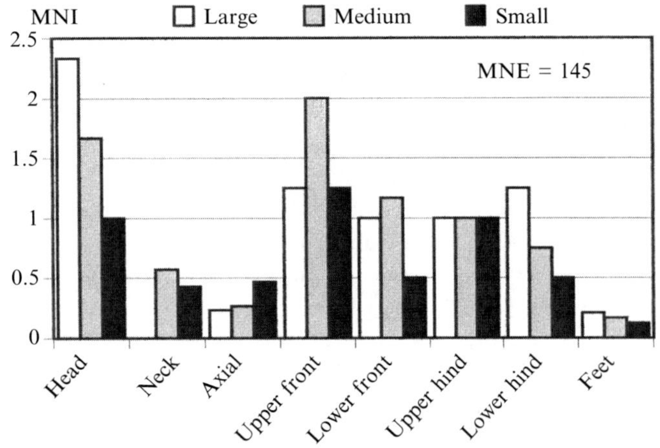

FIG. 5.2. Skeletal element abundance of faunal remains in TD 6 according to the anatomical regions defined by Stiner (1994). The total number of elements in any given region in a living animal (carnivores and Homo are excluded) is given in parenthesis. They are as follows: head (3) = 2 half-mandibles and 1 cranium; neck (7) = atlas, axis, cervical vertebrae 3–7; axial (49) = thoracic, lumbar, sacral vertebrae, 2 innominates and ribs; upper front (4) = scapula and humerus; lower front (6) = radius, ulna, metacarpal; upper hind (2) = femur; lower hind (8) = tibia, calcaneum, astragalus, metatarsal; feet (24) = 1st, 2nd and 3rd phalanx. Teeth, carpals, metapodials, patellas and long bone shafts are not included in Stiner's regions. Her first region (antler and horncores) is omitted because there were very few specimens in the Aurora stratum; various shed antlers occur in TD 6 but below the Aurora stratum (Made, 1999). Size classes were established by Díez et al. as follows: large includes *Bison, Equus,* and *Megaloceros*; medium includes red deer and indeterminate cervids of similar size; small includes *Dama, Capreolus,* and *Sus*. Values for each region are obtained by dividing the MNE of each region by the total number of elements in each region. The remains of *Mammuthus* and *Stephanorhinus* were included in the large class by Díez et al. (1999) due to their very scarce representation; it was not possible to exclude them from this diagram because MNE were calculated by classes. No information was provided for sacrum so the total number of elements used in this diagram for the axial region is 48, instead of 49.

TABLE 5.3. Bone modifications on animal and human bones at TD 6. Totals include indeterminate bone fragments. Cutmarks on limb bones are preferentially located on diaphyses (41/46), a few on near-epiphyses (4/46) and none on epiphises. Percussion marks are calculated on long bones. No marks were observed on proboscideans, rhinos and suids (mostly represented by teeth) (From Díez et al.,1999).

Type of marks	%	Total
Cutmarks	14.9	150/1,010
Percussion marks	12.7	42/330
Tooth marks	12.6	127/1,010

age groups are abundant relative to adults, reflecting the high mortality of young animals; beyond a certain age, very old individuals are also well-represented (Voorhies, 1969). Scavenging carcasses that died a natural death might seem a possibility. Yet assemblages accumulated by scavengers are unlikely to show a predominance of juveniles. The difficulty of locating small carcasses and the faster rates of biochemical deterioration due to low volume relative to surface area prevent scavengers from profiting of juvenile deaths, making carcasses from old animals more available to scavengers (Stiner, 1994: 300 and references therein).

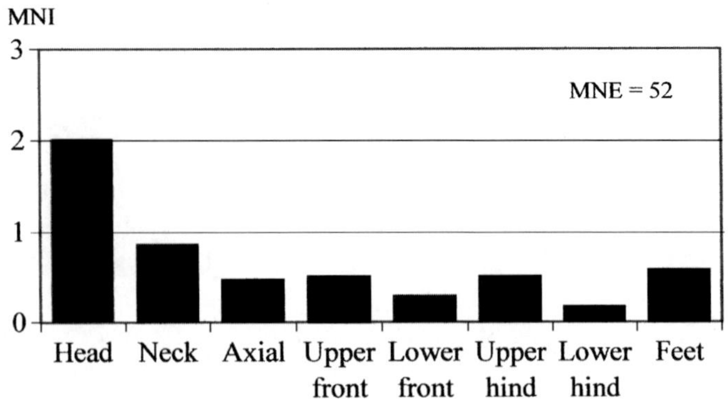

FIG. 5.3. Anatomical representation of human remains in TD 6. Regions as in Fig. 5.2 but the number of elements in each region is different from values of herbivores. The upper front region (scapula + humerus) includes clavicles. The total MNI is six and it was independently calculated taking into account the size, side and age of bones and teeth (From Fernández-Jalvo, 1999).

Scavenging from abandoned carnivore kills can be excluded because carnivores rapidly consume young carcasses (Vrba, 1980). In fact, archaeological profiles in which young animals are common are often seen as reflecting active human hunting (Klein and Cruz-Uribe, 1991). The occurrence of cutmarks, percussion marks and bone breakage on both human and animal remains, and the prevalence of defleshing marks on limb diaphyses (Table 5.3) clearly indicate human butchery of fully fleshed bones. The frequencies of skeletal elements (Fig. 5.2) refute a hypothesis of passive scavenging which would require a head-dominated or head-and-foot dominated pattern, and an age profile with a bias toward old individuals, as was suggested by Stiner (1994) in the case of Moscerini. The predominance of juveniles and of small to medium size animals might indicate hunters taking advantage of more vulnerable prey (Klein, 1978), though this kind of generalization cannot be based on one case only.

Sites from the Early Part of the Middle Pleistocene

In Western Europe there are only four sites older than OIS 12 (older than 400 ka) for which we have useful taphonomic data: Isernia, Venosa Notarchirico (level alpha), Boxgrove, and Miesenheim I. Arago cannot be included because there is, as yet, no complete faunal and taphonomic analysis of any single level, although we do have general papers (Moigne and Barsky, 1999; Lumley et al., 2004), paleontological papers, and doctoral theses on specific taxa (e.g. Monchot, 1996).

Isernia

Controversies about the age of Isernia (Villa, 1996) once dated by K-Ar to 730 ± 40 ka and by reversed polarity at or below the Matuyama-Bruhnes boundary (at the time the boundary was set at 730 ka but has since been redated at 788 ka; Klein, 1999: 51) have been settled by new $^{40}Ar/^{39}Ar$ dates on several sanidine crystals from tuff layers within the cross-bedded sands that cover level 3a, the main bone and artifact concentration at the site. The mean age is 606 ± 2 ka (Coltorti et al., 2005); this estimate matches the macrofaunal and the rodent faunas. *Arvicola terrestris cantiana* occurs at Isernia. This water vole with rootless molars, assumed to be a descendant of *Microtus savini* with rooted molars, appears in the second half or toward the end of the Cromerian Complex (for this climato-stratigraphic unit of the Middle Pleistocene see Klein, 1999: 28) i.e. between 0.6 and 0.5 Ma ago. The *Arvicola* from Isernia has been characterized as a primitive population, since a small proportion of the molars show evidence of root formation; a date of 0.6 Ma would be consistent with this observation. The presence of *Elephas* (*Palaeoloxodon*) *antiquus* and *Bison schoetensaki* also support a younger age for Isernia, again at approximately 0.6 Ma (see discussion in Villa, 2001 and references therein).

Layer 3a, excavated over a surface area of about 130 m^2, has yielded 334 artifacts of flint (average size is 2–4 cm) and limestone (some choppers and broken cobbles), and a spectacular concentration of macrofaunal remains (Peretto, 1994, 1996; Table 5.4) of mostly large-sized animals in a sandy matrix. The level has been interpreted by the excavators as an accumulation of essentially anthropic origin only partly displaced by a debris flow due to the volcanic eruption that provided the sanidine crystals.

The horizon contains many large blocks of travertine (Fig. 5.4), which may have acted as a barrier causing the grouping of materials already localized in ponds or abandoned channels. Bones are in variable states of preservation and many are affected by postdepositional compression fractures. Microscopic analysis of a sample of 40 bones indicates the occurrence of weathering cracks, abrasion striations, and loss of cortical surfaces due to postdepositional erosion. SEM analysis of the surfaces (carried out by G. Giacobini, in Anconetani et al., 1996) showed that abrasion striations mimicking cutmarks are probably due to friction by the numerous phenocrysts of volcanic origin present in the sedimentary matrix. Possible cutmarks were identified on three bones, all attributed to *Bison schoetensaki*, but they could not be verified with confidence due to their bad state of preservation. Some photos document rounding of fracture edges (Anconetani et al., 1996) although there is no systematic reporting on degrees of abrasion. Some bone long axes show preferred orientation (Giusberti et al., 1991), and the abrasion striations seem to us consistent with materials having been displaced by the viscous mixture of water, sand, silt, and rock debris which buried the paleosurface.

TABLE 5.4. Isernia. List of larger mammals found in layer 3a. The total number of macrofaunal remains is 4,240, of which 980 are unidentified shaft fragments. A few fragmented teeth of probable *Hippopotamus, Sus scrofa, Hemitragus* and one tooth of *Panthera leo fossilis* are not included. Cervids include antler fragments, teeth and a few postcranial bones of *Maegaceroides solilhacus, Cervus elaphus, Dama dama* cf. clactoniana and *Capreolus* sp. Note that the *Megaloceros* is mostly represented by shed antlers (After Peretto, 1996).

Taxon	NISP	MNI
Bison schoetensaki	744	61
Stephanorhinus hundsheimensis	358	31
Elephas (Paleoloxodon) antiquus	378	9
Ursus cf. deningeri	99	13
Cervids	46	–

Carnivore marks are said to be scarcely represented. Significantly, a number of bison long bones are reported to have percussion marks (67 of 572), and some of the published photos are convincing (Peretto, 1996: Figs. 6.11A, 6.13, 6.14, 6.16).

The excavators used a number of arguments to support the idea that the accumulation was the result of several episodes of hunting and butchering (Sala, 1996):

1. Anatomically connected units and complete bones are an important character of natural accumulations; at Isernia all bones are disarticulated and broken.
2. The material is not sorted, thus there is no evidence that the material was transported and concentrated by fluvial action like a flood.
3. The selectivity in skeletal representation, with a predominance of cranial parts from large size animals such as bison and rhinoceros (Fig. 5.5) is due to human action. The excavators thought that large bones were selected to consolidate a swampy area, unlikely as this may seem (Sala, 1996: 47).
4. The evidence of preferred orientation is limited and bones are horizontal, not jumbled up as might be expected in a volcanic mudflow.
5. There is no evidence of carnivore damage, except for small carnivores (Anconetani in Peretto, 1996: 128), so deaths are not due to carnivore predation.

As we will see, many of the features of the Isernia accumulation are not necessarily due to human action, throwing serious doubts on the interpretation of this occurrence as representing the remains of an ancient hunter/gatherer campsite.

There are at least two known cases of Miocene bone beds in Nebraska that show features identical or very similar to the bone concentration of Isernia. The Agate Fossil Beds National Monument is the locus of major mammalian bone beds preserved within fluvial sediments and near waterholes, containing the bones of hundreds of individuals of a small rhino (*Menoceras arikarense*), of 50–70 large-sized chalicotheres (*Moropus elatus*, a perissodactyl with claws on the feet

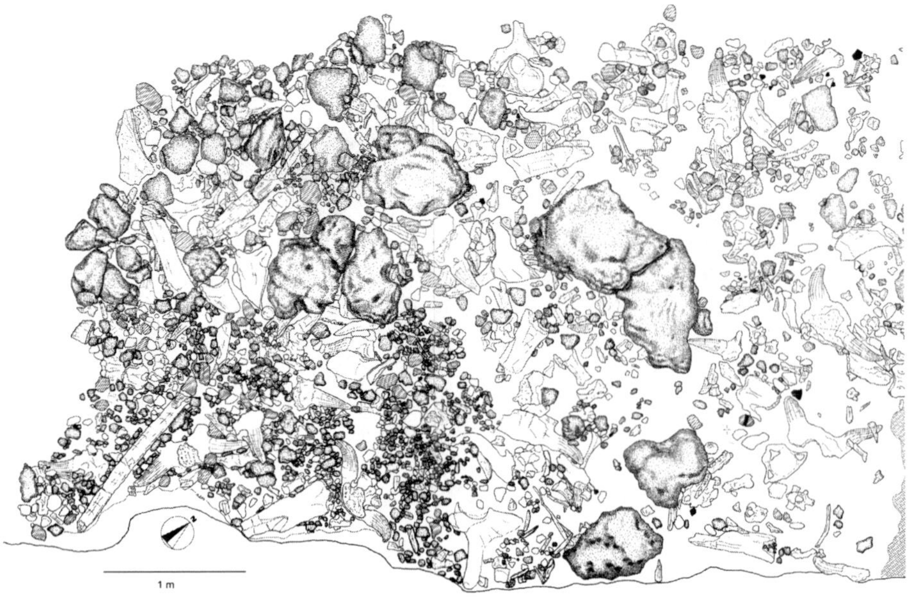

FIG. 5.4. Isernia. Plan of a portion of layer 3a. Note the large blocks of travertine forming a semicircle and many smaller natural cobbles, especially to the south of the distribution. Modified after Giusberti et al., 1983.

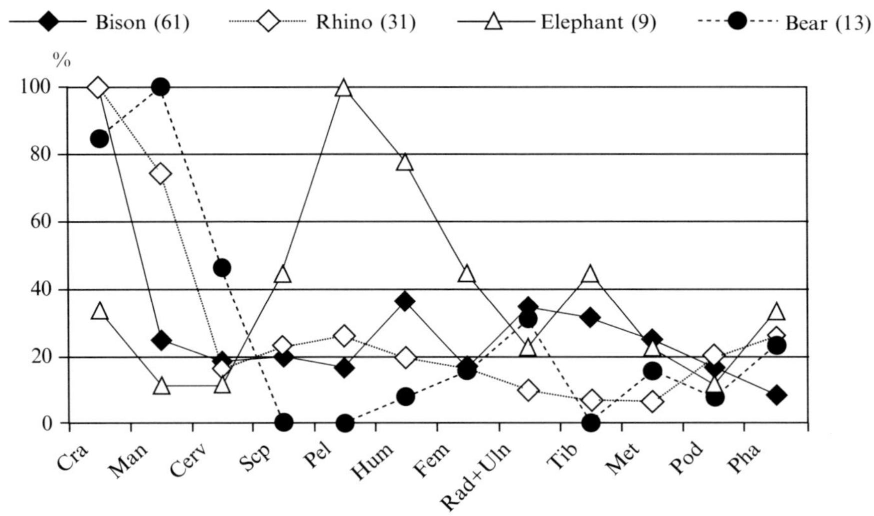

FIG. 5.5. Bodypart representation by percentage of MNI at Isernia, level 3a. Minimum number of individuals for each taxon in parenthesis (From Anconetani in Peretto, 1996).

rather than hooves, and weighing as much as 1,000 kg for adult males), a few bones of a very large primitive artiodactyl (the entelodont *Dinoyus hollandi* which was probably a scavenger), and extremely rarely, isolated and abraded bones of other mammalian species. These bone beds reflect natural deaths due to drought on river banks and in ponds; bones were displaced by river action. The density of bones is high comparable to Isernia, with as many as 100 bones per square meter, down to 40 bones at the periphery (Hunt, 1990).

The diagram in Fig. 5.6 compares the anatomical representation of the Isernia rhino (MNI = 31) with that of chalicotheres (MNI = 14) from one of the excavated localities in Agate Fossil Beds National Monument. The chalicothere bones were disarticulated and formed a single bed; the bones were not aligned (like at Isernia), showed moderate degrees of abrasion, no evidence of sorting, and there were a good deal of green bone fractures. We have chosen the Isernia rhino because its MNI and live size correspond better to that of the

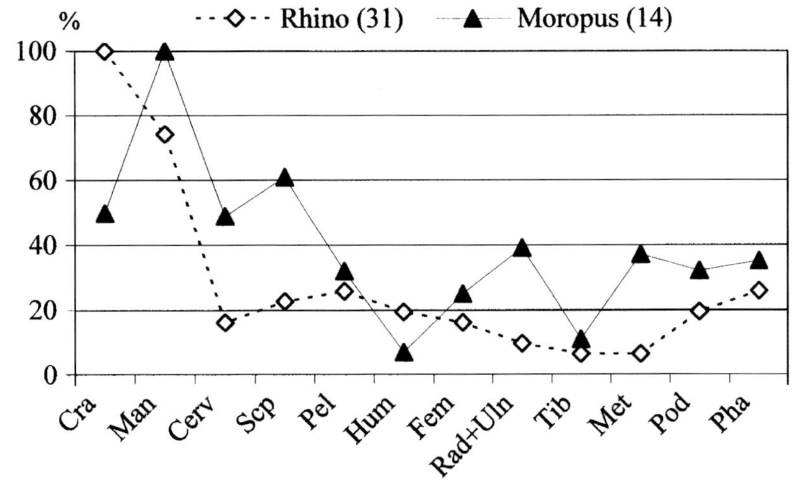

FIG. 5.6. Bodypart representation by percentage of MNI of the Isernia rhino in level 3a and of *Moropus elatus*, the chalicothere of the Agate Fossil Beds, Carnegie Quarry 2 (Miocene), Northwest Excavation. Minimum number of individuals for each taxon is given in parenthesis. The Northwest Excavation yielded dispersed and scattered bones at the periphery of a waterhole, a context comparable to that of Isernia (Data on *Moropus elatus* is from Hunt, 1990: Figure 22 top. Hunt's figure gives precise proportions of skeletal completeness for each of the bodyparts used in this diagram).

Agate chalicotheres; but a diagram with the bison would show the same pattern. It is clear that the predominance of cranial parts and many other features of the Isernia layer 3a are not necessarily due to human action since they also occur in the beds of Miocene age.

Likewise, the bone bed of Unit 3 in the Valentine quarry (Miocene, Nebraska), excavated over a surface of 1,300 m^2, contained an unsorted concentration of bones of various mammalian species (mainly horses) in a low velocity stream, with boulders up to 50 cm in diameter (like the travertine blocks of Isernia). A high percentage of bones was unweathered and had not undergone significant transport but disarticulation was complete; bones showed little evidence of preferred orientation and were lying horizontally, contained in a layer 5–80 cm thick. The variable degrees of weathering and abrasion suggested that bones were derived from a range of sources, that many originated in the immediate vicinity, and that depositional processes were not uniform during formation of the layer (McCool, 1988). Similar observations have been noted at the Lower Pleistocene paleontological site of Untermassfeld in Germany (Kahlke, 1999), which has no evidence of human action.

The only valid argument that associates the artifacts with the bones at Isernia is the presence of percussion marks on some bison bones. We conclude that human exploitation of the bones certainly occurred, but there is not enough evidence that the faunal accumulation was exclusively due to human action and we cannot say if the fauna was hunted or scavenged.

Venosa Notarchirico, Miesenheim I, and Boxgrove

For the other three sites for which good taphonomic and faunal analysis data are available (Table 5.5) only Boxgrove (the locus GPT 17 unit 4b, called "the horse butchery site") provides rather good evidence in favor of hunting. The other two sites were formed by a combination of natural processes and anthropic activities, and cannot be used to prove either hunting or scavenging. Miesenheim probably accumulated over a long period of time but it is difficult to disentangle the different processes that shaped the site (Turner, 1999). At Venosa, bone remains also seem to have multiple origins although the human presence is stronger (Piperno, 1999; Tagliacozzo et al., 1999).

The case for hunting or aggressive scavenging at Boxgrove is strong. It is based on a number of facts: (a) all skeletal elements of a single horse are represented. As noted before, passive or non-confrontational scavenging of carcasses that died natural, non-violent deaths, or were left from kills of other predators does not allow scavengers to exploit as much of the carcass as they encounter (O'Connell et al., 1988; Stiner, 1991: 465); (b) the frequencies of cutmarks are high and they occur on most of the larger bones, proving that the carcass was fleshed when butchered; (c) the frequency of refitting between lithic artifacts and some bones, the restricted vertical distribution of the horse bones, and the very fresh appearance of the artifacts indicate that a single episode of butchery is represented. The spread of lithic artifacts and bones and the fact that six flint nodules were brought to the site, and knapped to produce flakes and at least one handaxe (based on a void in the middle of a refitted nodule) suggest that a group of people took part in the butchery process (Roberts and Parfitt, 1999).

We should briefly mention a younger site, the Acheulian site of Ambrona in Spain, with a minimum age of 350 ka, possibly correlating with OIS 11 (Falguères et al., 2006). This is a deeply stratified site with faunal and lithic remains found in different sedimentary context. The most abundant species are

TABLE 5.5. Taphonomic and faunal data from sites in Western Europe dated between OIS 19 and 12. At Venosa Notarchirico level alpha is at the top of the sequence, so it is younger than 640 ± 70 ka, a TL date from a trachytic cinerite representing an ash fall of the Vulture volcano which occurs toward the base of the archeological sequence (Villa, 2001). The total excavation area at Miesenheim I is 436 m^2 but the main concentration of materials was about 300 m^2 (From Tagliacozzo et al., 1999; Turner, 1999; Roberts and Parfitt, 1999).

	Venosa Notarchirico, level alpha (20–30 cm thick)	Miesenheim I Upper part of layer G, and layer F (about 50 cm)	Boxgrove GTP 17, Unit 4b
Age	>OIS 12	ca. 600 ka	500 ka (OIS 13)
Excavated area (m^2)	62	About 300	150
Stone artifacts	950	100	>1,800 high frequencies of refitting
NISP	645	755	245
Main taxon	*Dama* (58% of NISP,10 MNI) Also *Bos/Bison*, elephant, other cervids	13 species of larger vertebrates, cervids are most common, some articulated remains.	*Equus ferus* (other remains are interpreted as background accumulation)
MNI	18	38	1, a single adult female
Cut marks	Not observable	No	46% (69 of 150)
Percussion marks	Some (two percussion marks and two bone flakes)	1	On humerus, radius, femur, mandible and indeterminate long bones
Gnaw marks	Two doubtful cases	Several	On five bones; on three pieces they overlap cutmarks
Carnivores in the assemblage	No	15.7%	No
Abrasion on bones	Present on all bones from slightly (45%) to abraded (20%) to very abraded (35%)	No	No
Abrasion on stone artifacts	6–40% are sligtly abraded to abraded, depending on raw material (flint, limestone)	No	No
Origin of the assemblage	Mixed human and natural processes	Mostly natural, some human presence	Human

elephant (*Palaeoloxodon antiquus*), cervids (*Cervus elaphus*, *Dama*), bovids (*Bos primigenius*), and equid (*Equus caballus torralbae*). The taphonomy of the Lower Complex from the central area of the site, with impressive finds of elephant bones and Acheulian bifaces and cleavers, has been extensively published (Villa et al., 2005a) and we briefly summarize it here. The analysis shows that the site is a complex mix of natural and human components and that some of the faunal remains of elephants and cervids represent natural occurrences without any clear evidence of hominid intervention. Evidence of human action on bones is provided by a few SEM verified cutmarks and anthropic bone fractures, documenting butchery of various animals, including elephants. A definitive interpretation of this controversial site (Freeman, 1975, 1978; Binford, 1987; Freeman, 1994) is limited by postdepositional disturbance processes, the loss of observable cortical surfaces on bones, and the fact that some materials probably derive from nearby locations now destroyed by erosion. However, the skeletal element representation and the lack or scarcity of carnivore modifications allow us to definitely reject Binford's idea of marginal scavenging of medium-size ungulates from carnivore kills. The site was regularly visited by hominids who transported some artifacts from non-local raw material sources and had an organized approach to meat acquisition. Whether meat was acquired through hunting or by taking advantage of natural deaths, we cannot say. It is only at late Middle Pleistocene sites like La Cotte de St. Brelade and the Late Pleistocene site of Lehringen that hunting of elephants (and rhinos) can be put forward as a valid hypothesis.

Late Middle and Late Pleistocene Sites

Hunting of large game is documented by a number of European sites dated to the second half of the Middle Pleistocene and the first part of the Late Pleistocene, beginning with well-known site of Schöningen in Germany (Thieme, 1997, 2000). Table 5.6 provides a list of sites later than Schöningen and dated between OIS 6 and 3, where the strongest evidence for hunting is provided by the topographic setting and the kind of accumulation. These sites show the use of cliff faces and rocky barriers associated with karstic systems for stampedes of one or two species of large mammals.

At La Cotte de Saint-Brelade (Jersey, English Channel Islands), layers 3 and 6, dated to OIS 6 by TL dates and stratigraphic data, are two separate accumulations of bones of woolly mammoths (total MNI = 18) and woolly rhinoceros (total MNI = 5). Only a small portion of the original site was excavated, approximately 12 m^2 in layer 3 and 18 m^2 in layer 6. Mammoths and rhinos comprise essentially the entire fauna of these two levels; a few bones of other species occur only at the base of each level. The site is at the base of a deep ravine (about 30 m at the time of deposition of layers 3 and 6), and there are several indications that these were rapid accumulations: some bones were found to rest vertically against other bones, a few bones were found in articulation, several mammoth scapulas were stacked in direct contact with each other without intervening sediment, and there was no evidence of subaerial weathering. The age distribution indicates a predominance of sub-adults and prime-age adults. It would have been impossible

TABLE 5.6. After Scott, 1986; Jaubert et al., 1990; Farizy et al.,1994. Gaudzinski, 1996b Jaubert et al., 2005; Brugal and Jaubert, 2006; also provides data on the bovid assemblages from Il'skaja (Russia) and Wallertheim (Germany) but we have not included them in this table as they have a more varied fauna with different accumulation histories.

Site	Age	Main fauna	MNI	Topographic setting	Evidence of hunting
La Cotte St. Brelade layers 3 and 6	OIS 6	*Mammuthus primigenius*, *Coelodonta antiquitatis*	7, 11 2, 3	Base of deep ravine	Rapid accumulation of only two very large species, almost no other species present
Coudoulous 1 layer 4	OIS 6 ca. 160 ka	*Bison priscus*	232	Karstic depression, open sky	Monospecific accumulation
La Borde single layer	OIS 7 or 5	*Bos primigenius*	40	Karstic depression, open sky	Monospecific accumulation
Mauran	OIS 3	*Bison priscus*	137	Base of an escarpment with rocky barrier	Monospecific accumulation

to kill a group of such dangerous animals without driving them off the cliff. Rhinos are frequently found on the fringes of herds of elephants, so they could have been driven together with the mammoths (Scott, 1986).

Like La Cotte, none of the other sites in Table 5.6 are really a cave or rock shelter in a strict sense. They are practically open-air sites in the sense that they were not protected by an overhanging roof; they were, however, physically constrained. La Borde and Coudoulous are accumulations in karstic depressions. At the time of accumulation Coudoulous had an open sky but was limited by a rocky wall on three sides. The archeological layer at La Borde was limited on two sides by a rock face. Mauran is at the base of an escarpment separated from the Garonne river by a rocky barrier. The three sites in SW France appear to have formed by a number of different episodes now occurring in the form of palimpsests (Jaubert et al., 1990; Farizy et al., 1994; Gaudzinski, 1996b; Joubert et al., 2005). Significantly, layers 3 and 6 at la Cotte preserve stratigraphic evidence strong enough to indicate two discrete episodes of mass killing.

Data from many other Middle Paleolithic sites show that Neanderthals hunted a wide range of prey, from dangerous animals such as brown bears, mammoths, and rhinos (Auguste, 1995; Bratlund, 1999) to large, medium, and small-size ungulates such as bison, aurochs, horses, red deer, reindeer, roe deer, and wild goats (Jaubert et al., 1990; Chase, 1999; Hoffecker, 1999; Hoffecker and Cleghorn, 2000; Conard and Prindiville, 2000; Roebroeks, 2001; Fernández and Legendre, 2003; Bar-Oz et al., 2004). Evidence of cutmarks on beaver bones suggest processing for pelts at Grotta Maggiore di San Bernardino in Northern Italy (Fiore et al., 2004). Cutmarks occur on ungual phalanges of the golden eagle (*Aquila chrysaetos*) in Mousterian levels at Pech de l'Azé I and Grotte de l'Hyène in France, and at Grotta di Fumane in Northern Italy (Fiore et al., 2004), suggesting removal of the claw for use as an ornament. Cutmarks on a swan phalanx have also been reported from a French Mousterian cave (Laroulandie, 2004). Aside from these few instances, however, hunting of birds for meat consumption remains uncertain; several sites in Spain, Italy, and France have yielded remains of aquatic and galliform birds, but we lack confirmation from taphonomic analyses (Villa and d'Errico, 2001; Fiore et al., 2004). Evidence for rabbit hunting is provided again by cutmarks, fresh bone breakage, and damage by disarticulation by flexion at just a few sites, e.g. Grotte de la Crouzade and Les Canalettes in Southern France (Cochard, 2004; Costamagno and Laroulandie, 2004).

Shellfish gathering is reported from some coastal caves in Italy and the Iberian peninsula, but the absolute quantities are small, not comparable with the shell middens of Mesolithic sites in Italy, coastal Spain, and Denmark (Stiner, 1993, 1994: 194; Kuhn, 1995: 177). Among the Mousterian sites in coastal Latium only Moscerini shows significant numbers of marine mollusks, but even there they are not especially abundant. The MNI of shell fish for the whole sequence at Moscerini (spanning the later part of OIS 5 and OIS 4 approximately 100–66 ka) is 613 (Stiner et al., 1999, table 1)[2] including mostly sand clams (*Callista chione* and *Glycimeris*), mussels (*Mytilus*) and a smaller number of other bivalves (*Cardium edulis*) and gastropods (*Patella*, *Monodonta turbinata*; Stiner, 1994). At Vanguard, the MNI of shellfish (mostly mussels) associated with a hearth dated between 49 and 45 ka is 73 (Barton, 2000). At Gruta da Figueira Brava (Raposo, 2000; Zilhao, 2001) significant numbers of *Patella* are reported from level 2, dated to about 30 ka; the site is close to a beach, like Moscerini and Vanguard caves. Other coastal Mousterian sites in the Iberian peninsula are reported to have mollusks (e.g. Gorham's Cave, Bajondillo Cave, sites on the Bay of Malaga; Barton, 2000; Cortés Sanchez, 2000), but detailed data are not yet available.

Mollusks can be overrated as a source of food. It is estimated that approximately 31,000 limpets (*Patella vulgata*) or 156,000 cockles (*Cardium edulis*) are required to supply the calorific equivalent of a single red deer carcass; 700 oysters or 400 limpets are needed to supply enough calories for one person

[2] The MNI of shellfish at Moscerini is based on percentage counts (93% of 660 total of small game and shellfish) provided in Table 5.1 of Stiner et al., 1999. We found difficult to calculate total shellfish MNI from numbers published in Stiner, 1994 (Tables 6.12–6.15) which provide NISP and hinge counts by individual layers. She recommends dividing hinge counts by two for bivalves to avoid the problem of fragmentation and this gives a MNI of 285 for bivalves such as *Callista chione*, *Cardium*, *Mytilus*, and *Glycimeris*. However Stiner's tables for gastropods (e.g. *Patella*, *Monodonta turbinata*) also contain hinge counts, which is confusing since gastropods do not have a hinge. The hinge is the structure at which the two valves are joined in a bivalve.

for 1 day if no other food is eaten (Bailey, 1978). At Mesolithic sites such as Meilgaard (Denmark), which was occupied for a few hundred years by a semi-sedentary community, 50 m^3 of deposits (representing only a small portion of the whole site) contained a minimum number of 100,000 oysters (Bailey, 1978). By comparison, the shellfish remains from Mousterian sites represent very small quantities. The limited evidence available for fishing in the Middle Paleolithic, including sparse remains of fish and marine mammals from Figueira Brava, also do not suggest an economically significant resource (Le Gall, 2000). Intensive use of marine food is documented only from the Late Upper Paleolithic onwards (Richards et al., 2005).

Use of shellfish as a raw material for scrapers has been reported for a number of Middle Paleolithic sites, in particular Moscerini, where Vitagliano (1984) described a fairly large number (about 100) of transverse and déjété scrapers on valves of *Callista chione* (sand clam). Other Middle Paleolithic coastal sites in Northern and Southern Italy also are reported to have tools on *Callista chione* (Vitagliano, 1984: fig. 2). This particular use of shells is not limited to Italian Neandertals. Three backed pieces made on limpets occur in the Howiesons Poort layer 11 of Klasies River Mouth (Singer and Wymer, 1982: Figs. 6.3.28–6.3.30 and P. V. personal observation, 2006). The Howiesons Poort at Klasies is dated to about 66 ± 5 ka or possibly between 55 and 60 ka (Soriano et al., 2007) so the occurrence might be just slightly younger than Moscerini, where ESR dates put the sequence roughly between 115 and 65 ka (Stiner, 1994: 35).

Remains of land tortoises (*Testudo graeca*) and aquatic turtles (*Emys orbicularis*) are reported by Stiner (1993, 1994) in unit M6 at Moscerini (NISP 39; MNI 5), with indications of human use such as impact damage and fresh break edges. At Gruta da Oliveira (Portugal), a late Mousterian site with levels dated to between 32 and 38–40 ka, tortoise remains are said to be common, and one scapula was cut-marked; at Gruta Nova da Columbeira (also in Portugal) levels with an estimated age of 54–61 ka have yielded 338 NISP of *Testudo hermanni* (another land tortoise) together with a large number (1,832) of Mousterian stone artifacts (Zilhao, 2001).

Undoubtedly, bovids, equids, and cervids are the most common prey species at Middle Paleolithic sites. Nevertheless the evidence of shellfish and small game hunting/gathering shows that Neanderthals were exploiting the same variety of resources as Upper Paleolithic humans (Zilhao, 2006). Sites such as La Cotte, Coudoulous, Mauran, and La Borde strongly suggest that hunting could be logistically organized as in the communal game drives and kill sites of late Upper Paleolithic and Paleoindian times.

Middle Paleolithic Hunting Weapons in Western Europe

If Neanderthals were capable of killing large game and prime-age animals, and hunting was their regular method of meat procurement, the next question to ask is: what kind of weapons did Neanderthals use to dispatch their prey? Did they use stone-tipped spears? This question is especially appropriate for the Western European record which has, until now, provided good evidence only for the use of wooden spears at the sites of Schöningen and Lehringen in Germany (Thieme and Veil, 1985; Veil and Plisson, 1990; Thieme, 1997, 2000). The six Schöningen wooden spears, dated to about 400–300 ka and associated with remains of at least 19 horses (*Equus mosbachensis*) have a diameter of 29–50 mm and a length of 1.8–2.5 m. Most of the spears were made from individual spruce trees, one was made from pine. The trees were felled, the bark, lateral twigs, and branches removed; the tips are worked from the hardest part of the wood at the base of the tree. The spears have been interpreted as javelins because their maximum thickness and weight is situated a third of the way from the tip, like in modern javelins. The Lehringen spear, also from Germany, made of yew wood and dated to OIS 5e, was found among the ribs of an elephant; its weight is concentrated on the proximal end, thus its use as a thrusting spear is a reasonable inference. The spear was 2.39 m long; its diameter was 3.1 cm at the base and 2.0 cm near the tip. The use of the Clacton "spear point," also made of yew, is less certain since it was recovered outside an archaeological context (Oakley et al., 1977).

Spear Points in South Africa and the Near East

In South Africa and in the Near East, several scholars have concentrated their attention on lithic points of the Middle Stone Age (MSA) and the Middle Paleolithic (MP) and their functional interpretation. Residue, microwear and impact scar analyses of a sample of 50 MSA (post-Howiesons Poort) unifacial points from several layers at the rock shelter of Sibudu in South Africa by Lombard (2005a) have shown that these artifacts were hafted and used as tips of hand-delivered spears. Technological and morphometric analyses of 138 points and distal tips from layer RSP of the same site (dated to ca. 50 ka by OSL) by Villa et al. (2005b) supported this interpretation.

In the analysis of the RSP points, we examined a number of variables used by different authors who have studied prehistoric weapons technology, i.e. the tip cross-sectional area (TCSA, obtained by the formula: 1/2 maximum width × maximum thickness, expressed in cm^2 or mm^2; Hughes, 1998), the penetrating angle, the maximum width (which is at least in part related to the width of the shaft), as well as the frequency of basal thinning (which is a way of accommodating the bases of the points to hafting procedures). Comparisons with archeological (North American and European Upper Paleolithic) and ethnographic data for the first three variables indicated to us that the Sibudu points were the tips of either thrusting or throwing spears. Shea (1988, 1998, 2003) had previously made the same suggestion for Levallois points from Levantine Mousterian sites, based on TCSA values similar to that of Sibudu RSP.

More recently Shea (2006) has conducted extensive statistical analyses of many Levallois and retouched unifacial and bifacial points from the Middle Paleolithic in the Levant. and from the Stillbay and post-Howiesons Poort phases of the South African

MSA, comparing the tip cross-sectional area of these MSA/MP artifacts with well-documented archeological and ethnographical examples of North American dart tips and arrowheads. The TCSA has been considered the best means to distinguish armatures of different weapon systems, whether arrows, spear-thrower darts, throwing, or thrusting spears (Hughes, 1998; Shea et al., 2002). Shea's statistical analyses strongly indicate that the Levant MP and the South African MSA points were used to tip hand-cast spears. His research is supported by the discovery of hafting traces (bitumen residues) on the proximal part of artifacts and of a Levallois point embedded in a wild ass cervical vertebra from the site of Umm el Tlel in Syria, dated to about 60 ka (Boëda et al., 1996, 1999).

Spear Points in Europe

The situation is different in Western Europe where most work has been concentrated on the study of Upper Paleolithic points and studies of Middle Paleolithic points have lagged behind. The strongest argument for the existence of spear points has been advanced by Callow (1986) for a few Mousterian points from the site of La Cotte de Saint-Brelade in the Channel Islands. Out of nine Mousterian points in layer 5, dated to OIS 6, four exhibit burin-like or flute impact scars; two others had been repaired after breaking and are less sure (Fig. 5.7). Still, the scarcity of detailed studies of pointed forms in Western Europe is striking; the study of possible spear points in the Middle Paleolithic record of Western Europe is a neglected topic, in clear contrast to research trends in the Near East and South Africa.

Three factors probably account for this lack of attention:

(a) The influence of Bordes' typology which gives great importance to scrapers and tends to lower the significance of pointed forms by merging them into the convergent scraper category. Thus pointed forms are thought to be rare in the European MP, which appears dominated by scrapers.
(b) Research by H. Dibble (1987a, b, 1988, 1995) on the effects of intensive reduction of tools and the idea that convergent scrapers are a reduced form of double scrapers.
(c) A few microwear analyses showing that convergent scrapers had been mainly used to work wood (Beyries, 1988a, b; Anderson-Gerfaud, 1990).

These issues are discussed in greater detail in Villa and Lenoir (2006). We note, however, that the impression of low frequency of points in Mousterian assemblages is at least in part due to different ways of counting artifacts. High frequencies of pointed forms do occur in some Mousterian assemblages, such as Biache, Vaufrey layer VIII, and Bérigoule in France, and Castelcivita in Italy (Villa et al., 2005b: Table 6; on Bérigoule see Richter et al., 2007). We expect that more cases will be recognized if we attract the attention of analysts to the subject.

Our analysis addresses this issue by providing information on points from Bouheben, a Middle Paleolithic/Final Acheulian site in SW France, and using comparisons with Middle Stone Age points from Sibudu and Rose Cottage; both of which contain long stratigraphic sequences with assemblages dated between 60 and 35 ka by OSL (Figs. 5.8–5.9). As indicated above, the Sibudu unifacial points have been previously identified as spear points (Lombard, 2005a; Villa et al., 2005b). The Howiesons Poort and post-Howiesons Poort lithic assemblages

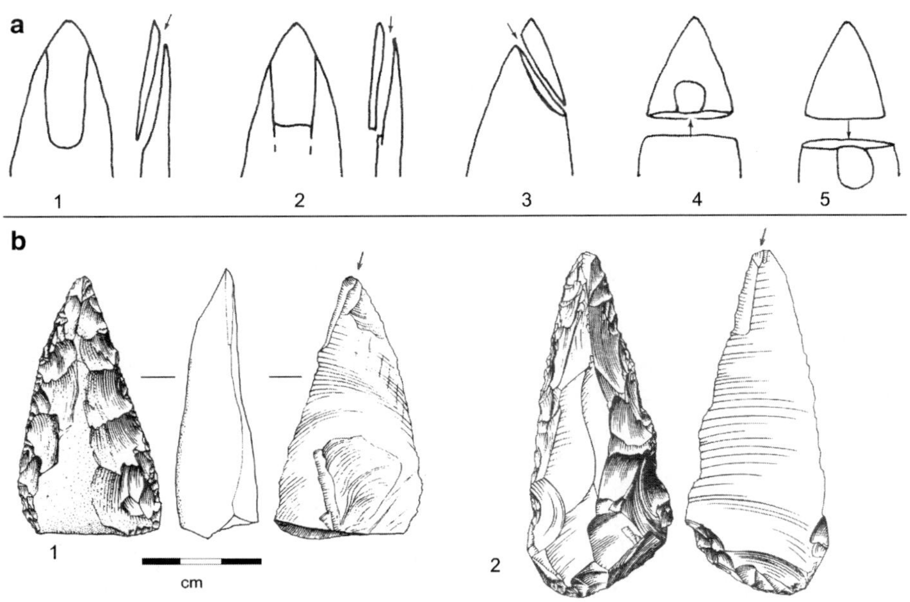

FIG. 5.7. (**a**) Schematic representation of impact scars on bifacial spear points from the Casper site (Wyoming, USA) a Paleoindian bison kill site, approximately 10,000 years old. 1 flute-like scar (i.e. bending fracture scar with feather termination in Fischer et al. terminology); 2 scar with step termination; 3 burin-like scar; 4–5 spin-off scars (Modified after Frison, 1974). (**b**) 1–2 Mousterian points from layer 5 at La Cotte de St. Brelade (Jersey, Channel Islands) with burin-like impact scars and thinned base (Modified after Callow, 1986).

FIG. 5.8. Map showing the location of the two Middle Stone Age sites mentioned in the text.

of Rose Cottage are the subject of another paper with more detailed information (Soriano et al., 2007).

The reason for selecting Bouheben, among so many other Middle Paleolithic assemblages, are simple: the assemblage was available for study, was well-excavated and has a high frequency of pointed forms (32% of formal tools in layer 2) comparable to those at Sibudu where layer RSP has 32.8% of unifacial points.

Bouheben (SW France)

Bouheben is an open air site in the department of Landes in the Aquitaine basin, located on a low plateau at about 117 m asl. Excavations were conducted by the late Claude Thibault in 1964 and in 1967–1969 on a surface of 43 m^2. According to Thibault (1970, 1976) the site was originally very large but was partly destroyed by a farm building, pathways, and digging of ponds. Bone was not preserved.

The upper part of the stratigraphic sequence contained two Mousterian levels (layer 1 and 1^1) and a Final Acheulian/ Mousterian level (layer 2). The top of layer 1 was marked by a recent soil and the upper part of layer 2 was also marked by a weathering horizon, interpreted as Riss-Würm (Last Interglacial, i.e. OIS 5e) soil. A pollen diagram for the base of layer 2 indicated a dry, cold climate. There are no absolute dates to confirm the original assignment to the end of Riss (i.e. OIS 6). The lithic assemblage provides some clues to the site age.

The excavated assemblage of layer 2 is large: the total number of artifacts from layer 2 was more than 4,500; the total number of flake tools is 312, and there is a small number of bifaces (n = 12). The occurrence of bifaces supports a typological attribution to the Upper or Final Acheulian. Assemblages with Acheulian bifaces persist in the region until the end of the Middle Pleistocene, as indicated by the site of Barbas I in the Dordogne region (Boëda et al., 2004) and by layer A at La Cotte de St. Brelade (Callow and Cornford, 1986). At Barbas layer $C^1 3$, dated to 147 ± 28 and 146 ± 29 ka by two TL dates, contained 167 bifaces, more than 40,000 flakes and fragments resulting from the making of bifaces, and about 2,000 flake tools. Layer A at La Cotte, dated to OIS 6, is also characterized by small flint bifaces (n = 70) larger quartzite and dolerite bifaces and cleavers (n = 20) and large numbers of small flake tools of Middle Paleolithic character (about 2,516 of flint and 137 of quartzite and dolerite). Barbas I and La Cotte show that assemblages without bifaces, which appear in Western Europe during OIS 8 and OIS 7, continued to coexist until the end of the Middle Pleistocene with assemblages with rare bifaces and a repertoire of flake types in many respects indistinguishable from Mousterian industries of Late Pleistocene age (Santonja and Villa, 2006).

Four of the 12 Bouheben bifaces from layer 2 are from large quartzite cobbles and resemble the typical Acheulian bifaces from the Garonne and Tarn terraces, also made on quartzite cobbles. A second set of six bifaces consists of much smaller bifaces with length from 5 to 7 cm. These flint bifaces are unlike the cordiform bifaces of the Mousterian of Acheulian Tradition which in SW France is dated to OIS 3, between 65 and 40 ka (Soressi, 2002).

In conclusion, the estimated age of Bouheben layer 2 as dating to the end of the Middle Pleistocene, although unconfirmed by quantitative dates, is strongly supported by its quartzite bifaces, typical of the regional Acheulian. The total number of pointed forms from layer 2 is 100.

Layer 1^1 contained a smaller assemblage of about 2,200 artifacts, very similar to layer 2 in terms of debitage and retouched pieces (mainly scrapers). In both levels the Levallois debitage is present but not predominant, and there are a few blades made by direct percussion by hard hammer. A comparison of this assemblage to that of layer 2 in terms of percentage frequencies of Bordes' types shows that the two assemblages are practically identical (cf. the two cumulative frequency curves in Thibault, 1970: fig. 98). We have included in our sample, mostly derived from layer 2, the 25 pointed forms from the overlying layer 1^1 because they are similar to points from layer 2 in terms of dimensions, frequencies of bulbar thinning, and impact scars. Thus the total sample is 125 pointed forms, including 11 broken tips or distal fragments.

Most artifacts from Bouheben are made of Senonian flint which outcrops at about 6 km from the site. Almost all artifacts have a white patina and some are slightly desilicified; thus the level of identification of microwear traces is very low and residue analysis is excluded. However morphometric studies and studies of impact scars are possible; they are some of the main methods for identifying use as spear tips.

We have analyzed all Mousterian points and elongated Mousterian points; other pointed forms such as convergent scrapers and *déjeté* (i.e. canted) scrapers, so classed following Bordes' typology (Bordes, 1961), have been included in the database, unless they had a round or blunt distal end or were too asymmetrical (see below observations on tip design). In Bordes' typology, distinctions between convergent scrapers and Mousterian points are based on the acuteness of the point, its thickness in profile, and bilateral symmetry. Thus, we followed traditional classificatory procedures for preliminary sorting, prior to detailed analysis of attributes expressing tool design (e.g. blank type, bulbar or lateral thinning, and kind of retouch) and morphometric features. We should make clear that our interpretations are not based on Bordes' typology but on morphometric and impact fracture analyses. Convergent and *déjeté* scrapers are excluded from all statistics, except in Table 5.8 for reasons explained in the caption. There are no unretouched Levallois points at Bouheben.

Sibudu (KwaZulu-Natal, South Africa)

This large rock shelter, approximately 40 km north of Durban and 15 km inland from the Indian Ocean, is under current excavation by Lyn Wadley. The stratigraphic sequence spans four MSA phases: the pre-Still Bay, the Still Bay, the Howiesons Poort and the Post-Howiesons Poort. Unifacial and bifacial points come from the post-Howiesons Poort layers, dated by OSL to ca. 60–36 ka. The excavated area (until 2005) varies from 2 m^2 for the lower part of the sequence to 18 m^2 for layer RSP and up to 6 m^2 for the top part of the sequence (Wadley, 2005; Wadley and Jacobs, 2006). Detailed analysis of the RSP lithic assemblage, one of the most extensively excavated post-Howiesons Poort layers at Sibudu, has been published by Villa et al. (2005b).

The analyzed sample consists of 272 specimens, which we have grouped in three subsamples: Final MSA layers from the East section (layers Ore to Co, dated by OSL between 50 and 36 ka), layers MOD-RSP (dated to ca. 50 ka), and layers below RSP from the North section (dated to ca. 60–50 ka; OSL dates and stratigraphy from Wadley, 2005; Wadley and Jacobs, 2006). All points from the first two subsamples have been analyzed. Points from layers below RSP represent artifacts that were found in preliminary sorting; thus the latter group may be incomplete (e.g. missing broken tips). For this reason the sample, which is quite large (70 pieces) is treated separately. There are, in addition, 41 pieces that would be classed as convergent scrapers in Bordes' typology. They are not included in the statistics unless specified. Our sample includes 36 bifacial points and 12 partly bifacial points; the rest are unifacial points. In the post-Howiesons Poort sequence, the two most common raw materials are hornfels and dolerite; the fine-grained hornfels is the preferred raw material for points (75.5%).

Rose Cottage (Free State, South Africa)

This large cave is located in the eastern Free State at 1,676 m elevation. It contains a long stratigraphic sequence with final MSA and Later Stone Age deposits (Wadley, 1997). The older MSA levels were excavated between 1989 and 1991 by Philip Harper under the guidance of Lyn Wadley (Harper, 1997). Our sample consists of all the points from the post-Howiesons Poort levels and comes from the 6 m^2 in the center of the cave plan. OSL dates place these levels between 57 and 33 ka; TL dates on burned lithics are broadly comparable though slightly younger (Valladas et al., 2005).

In the Post-Howiesons Poort levels there are 44 unifacial points, six tips of unifacial points, two bifacial and four partly bifacial points. The Howiesons Poort levels contained only one unifacial point from a layer in the middle of the sequence; it is included in our sample which thus totals 57 pieces. There are no convergent scrapers. The main raw material is opaline (equivalent terms are chalcedony and opal), a siliceous rock that formed as lenses 5–15 cm thick, or as roundish nodules, similar to geodes, within the Drakensberg basalt (Early Jurassic) to the east. Opaline nodules from the eroded outcrops were carried by tributaries to the Caledon river, which runs 8–10 km from the site. Opaline is a fine-grained raw material of variable colours; the knapping quality is generally very high and comparable to flint. Most blanks are water-rolled quadrangular blocks or slabs of small dimensions, generally less than 6 cm in size (Soriano et al., 2007).

Morphometric and Impact Scar Analyses

Analyses of stone points from Paleoindian kill sites have shown that the attributes necessary for the proper functioning of a device to kill a large animals are a sharp point to penetrate the hide and sharp side edges to open a hole for the remainder of the point and shaft (Frison, 1978: 337–338). Tip design is critical for the penetration of a low velocity weapon (Hughes, 1998). Thus, analysis of lithic points to test the hypothesis that the points were used as parts of hunting weapons, must show three things:

- There must be some evidence of hafting.
- The points must have a sharp tip, to penetrate the hide.
- Some should have impact scars, proving their use as killing weapons.

In this paper the term "projectile technology" is used exclusively in reference to high velocity weapons delivered by spearthrowers or by bows, as in Shea, 2006. This distinction is necessary since in the published literature the term "projectile point" has been used not only in reference to high-velocity weapons, but also for any point used for killing whether tipping thrusting spears, spears thrown by hand (i.e. javelins), spears thrown with a spearthrower, or arrowheads. This is a cause of some confusion. Following terminology common in North American literature (e.g. Frison, 1974; Thomas, 1978; Hughes, 1998) we use the term "dart" to indicate spears thrown with a spearthrower (spearthrowers are also called atlatl) and reserve the term spears and spear points to hand-delivered low velocity weapons. Of course, it is not always possible to identify stone points as part of one or another weapon system or to distinguish between stone points used to tip thrusting or throwing spears.

Hafting

Strong evidence of hafting has been found on some of the Sibudu unifacial points by microwear and residue analysis (Lombard, 2005a). Residue analyses cannot be carried out on the Rose Cottage and Bouheben materials. However, bulbar thinning is a good proxy. Thinning of the base by removal of the original striking platform and flaking of the ventral surface is generally considered a way to accommodate the bases of the points to hafting procedures, making sure that the haft bindings or adhesives do not project much above the stone, thus decreasing haft drag. Table 5.7 shows that all three sites have similar proportions of thinned bases.

Sharp Tip

The TCSA (tip cross-sectional area, obtained by the formula: 1/2 maximum width × maximum thickness) is one of the variables that influence penetration of a low velocity weapon, hence its killing power: the smaller the TCSA the better the penetration. Shea (2006) provides descriptive statistics for ethnographic and recent archeological (North American) hafted stone points (spear points, dart tips and arrowheads) based on data provided by Thomas (1978), Shott (1997), and Hughes (1998) (see also Shea, 2008; Churchill and Rhodes, 2008). Table 5.8 shows that the Sibudu and Bouheben TCSA mean values fall well within the range of throwing or thrusting spear. The Rose Cottage points instead have a smaller mean TCSA value (78) that may at first seem relatively close to the mean of dart tips (58). However a *t*-test shows that those values are significantly different ($t = 18.9$, $p < 0.001$). The reason for this clearly depends on the fact that the TCSA maximum value of the Rose Cottage points is much greater than those of dart tips (192 versus 94 mm^2), and that the Rose Cottage points are thicker than darts (mean is 7.1 ± 2.2 versus 5 ± 0.9 mm).

Table 5.9 shows that the Rose Cottage points are the smallest of the three sites. The small size of the original unworked raw material is very likely the reason for their small size. There was no deliberate reduction in size, just the opposite: the mean length of flakes (flakes < 20 mm are excluded) is 25.1 ± 5.8 mm; the mean length of retouched pieces, exclusive of points, is 31.2 ± 10.9; the mean length of points 36.6 ± 8.7. In other words, the Rose Cottage people consistently chose the largest available flakes for their points.

The statistics of the penetrating angle (the tip angle seen in plan view and measured in degrees) of all three sites are given in Table 5.10. This angle is related to the acuity of the point and its resistance to breakage (Peterkin, 1997). It was measured using the caliper method based on measurements of width at 1 cm from the tip; the angle is then calculated using a trigonometric formula (Dibble and Bernard, 1980; Villa et al., 2005b). All Upper Paleolithic points interpreted as projectile elements (dart tips or arrowheads) have smaller angles than our points: e.g. shouldered and tanged points of the Gravettian and Magdalenian have angle means of 49.2 ± 12.7 (n = 36) and 46.5 ± 4.57 (n = 40) respectively. The Solutrean and Magdalenian foliate points have angle means of 54.8 ± 12.5 (n = 92) and 46.5 ± 10.1 (n = 74). Of all foliate points, the Solutrean bifacial laurel leaves are the heaviest and it has been suggested that the more robust points may have been used as armatures for thrusting spears (Peterkin, 1997); yet their mean angle is significantly smaller ($p < 0.001$) than those of our three sites.

TABLE 5.7. Basal thinning on points from Bouheben, Sibudu and Rose Cottage. Counts exclude convergent scrapers, bifacial points (frequent at Sibudu but very rare or absent at the other two sites) and pieces with broken or damaged base. In the great majority of cases basal thinning is done by inverse retouch removing the bulb of percussion. Counts differ slightly from a similar table published in Villa and Lenoir, 2006 because here we also exclude bifacial pieces.

Site	Basal thinning (%)
Bouheben (*n* = 93)	25.8
Sibudu (*n* = 99)	25.3
Rose Cottage (*n* = 48)	18.8

TABLE 5.8. Tip cross-sectional area data in square millimeter for Bouheben, Sibudu, Rose Cottage. Data for ethnographic and recent archaeological hafted stone points (spear tips, dart tips and arrowheads) is from Shea (2006) and is based on ethnographic and North American archeological materials (Thomas, 1978; Shott, 1997; Hughes, 1998). For Bouheben three calculated means are given: Bouheben MP for Mousterian points only, Bouheben CS for convergent scrapers and forms intergrading between the two types, and Bouheben All Pointed Forms, which exclude convergent scrapers. The TCSA of convergent scrapers is given for Bouheben because the distinctiveness of convergent scrapers and Mousterian points has been questioned. Note that the TCSA of convergent scrapers at Bouheben exceeds those of ethnographic and historic spear tips and should be excluded from consideration.

Sites	Mean	SD	Min	Max	*n*
Bouheben (MP only)	165	67.2	50	322	70
Bouheben (CS)	232	94.4	70	420	31
Bouheben (All Pointed Forms except CS)	177	73	50	375	98
Sibudu, final MSA (East section, layers Ore to Co)	116.2	41.5	45	200	21
Sibudu, layers RSP–MOD (North section)	117.7	57.6	19.5	294	71
Sibudu, layers below RSP (North section)	139.4	60	54	320	42
Rose Cottage	78	33	19.5	192	47
Arrowheads	33	20	8	146	118
Dart tips	58	18	20	94	40
Spear tips	168	89	50	392	28

TABLE 5.9. Length (mm) of complete points (convergent scrapers are excluded).

	Mean	SD	Min	Max	*n*
Bouheben	57.4	14.5	28	96	95
Sibudu, final MSA (layers Ore to Co)	46.9	10.6	30	73	19
Sibudu, layers RSP–MOD	41.8	10.2	24	71	64
Sibudu, layers below RSP	45.3	11.2	28	74	42
Rose Cottage	36.6	8.7	23	60	43

TABLE 5.10. Penetrating angle, i.e. the tip angle seen in plan view and measured in degrees. Convergent scrapers are excluded.

Sites	Mean	SD	Min	Max	*n*
Bouheben	63.8	9.7	38.6	87.1	107
Sibudu, final MSA (layers Ore to Co)	62.3	13.2	43.6	106.9	36
Sibudu, layers RSP–MOD	68.3	12.7	38.6	95.5	126
Sibudu, layers below RSP	61.9	12.3	31.5	85.5	68
Rose Cottage	62.4	11.3	33	81	50

TABLE 5.11. Impact scars on the Bouheben, Sibudu and Rose Cottage points. Counts exclude convergent scrapers and pieces with broken or damaged distal end. At Bouheben three impact scars occur in layer 2 and three in layer 1^1. The percentage of imapct scars on the Rose Cottage points was given as 4.2% (2/47) in Villa and Lenoir, 2006 but it is now updated after a thorough revision of all points conducted in summer 2006; we have included one point with a scar 4 mm long in the Rose Cottage sample.

Site	Impact scars (%)
Bouheben (*n* = 113)	5.3
Sibudu, layer RSP (*n* = 101)	8.9
Rose Cottage (*n* = 48)	8.3

Impact Scars

Based on observations of impact scars on projectile points at Paleo-Indian bison kill sites (Fig. 5.7a; Frison, 1974) and experimental work by Fischer et al. (1984) and other researchers (e.g., Barton and Bergmann, 1982; Lombard, 2005a) step fractures, burin-like fractures, and spin-off fractures on the apex of a point, generally longer than 6 mm, are considered diagnostic of use as spear tips. Smaller scars at the tip can result from using the tip in a forceful motion. Scars with a negative bulb of percussion have also been excluded as they may simply represent retouch or resharpening of the tip and are uncommon in experiments using points as projectile elements (O'Farrell, 2005). Four tip scars at Bouheben were <6 mm and have not been considered in our counts. In the case of Rose Cottage, since points are smaller and length of impact scars is related to mass, we have considered scars 4–5 mm in length.

Table 5.11 shows that Sibudu layer RSP, Rose Cottage, and Bouheben have similar proportions of impact scars. Figure 5.10:1–3 and Fig. 5.11 show examples of diagnostic impact scars from Bouheben, Sibudu, and Rose Cottage. We used frequencies of the RSP assemblage at Sibudu because counts of artifacts from this major layer at the site are complete and thus fully comparable with the Rose Cottage and Bouheben assemblages that have been completely analyzed.

There are very few counts of frequencies of impact scars on spear points, based on complete assemblages from prehistoric sites, to compare with our sites. Published frequencies of impact scars on unifacial points from Sibudu are high (42%), but they are based on a sample selected from 11 post-Howiesons Poort assemblages from the site (Lombard, 2005a). Shea (1988) provides frequencies of impact scars of 7% on Levallois points and other artifacts from Kebara Units IX–XII, but these frequencies are based on counts of wear units on all artifacts, not comparable to our data which are based on counts of individual pointed forms. The only comparable statistics are those provided by Fischer et al. (1984) for Late Glacial, and Holocene assemblages in Northern Europe (Table 5.12). These assemblages consist of projectile points that were delivered by bows. It is not clear to us whether these statistics can be used to generate expectations concerning frequencies of impact scars on points tipping hand-delivered spears. They are, however, what we have. Table 5.12 shows assemblages with very variable proportions of impact scars, from high to quite low, as in our sites. Fischer and colleagues argued that the low proportions of diagnostic impact fractures could mean that the points were not used as weapons, or were made for that purpose but never used. The last explanation seemed more likely to him (see also Lombard, 2005b).

It is important to know whether the assemblage of points comes from a settlement or from a kill site where the main activity was the killing and butchering of animals. This is the case of Stellmoor, an open air site in northern Germany, with an Ahrensburgian layer dated to ca 10,000 BP, containing about 105 whole and fragmented arrows made of pinewood and more than 18,000 remains of reindeer, the results of at least four episodes of hunting drives. The minimum number of reindeer based on the scapula is 302 (Bratlund, 1996; Weinstock, 2000). The arrows consisted of a main shaft with a 20 cm long foreshaft. Some foreshafts had simple pointed wooden tips, others were armed with Ahrensburg points (small tanged and obliquely truncated tips), and impact scars are very common (42.2%; Table 5.12). Thirty identified bones, plus two antlers, have lesions with embedded pieces of flint (Bratlund, 1996).

At the Casper site, a Paleoindian bison kill site in Wyoming (USA) approximately 10,000 years old, a minimum of 74 bison were trapped against a parabolic sand dune and killed with spears tipped with a bifacial point (the so-called Hell Gap point) very probably thrown with a spearthrower. Of 60 bifacial points, Frison (1974) mentions that 26 (i.e., 43%) showed impact scars on the distal end, although few details are provided.

At our sites there is clear evidence of a variety of activities that included knapping and manufacturing many domestic (non-weapon) tools in addition to points. Some of the points we studied may have never been used and some, brought back

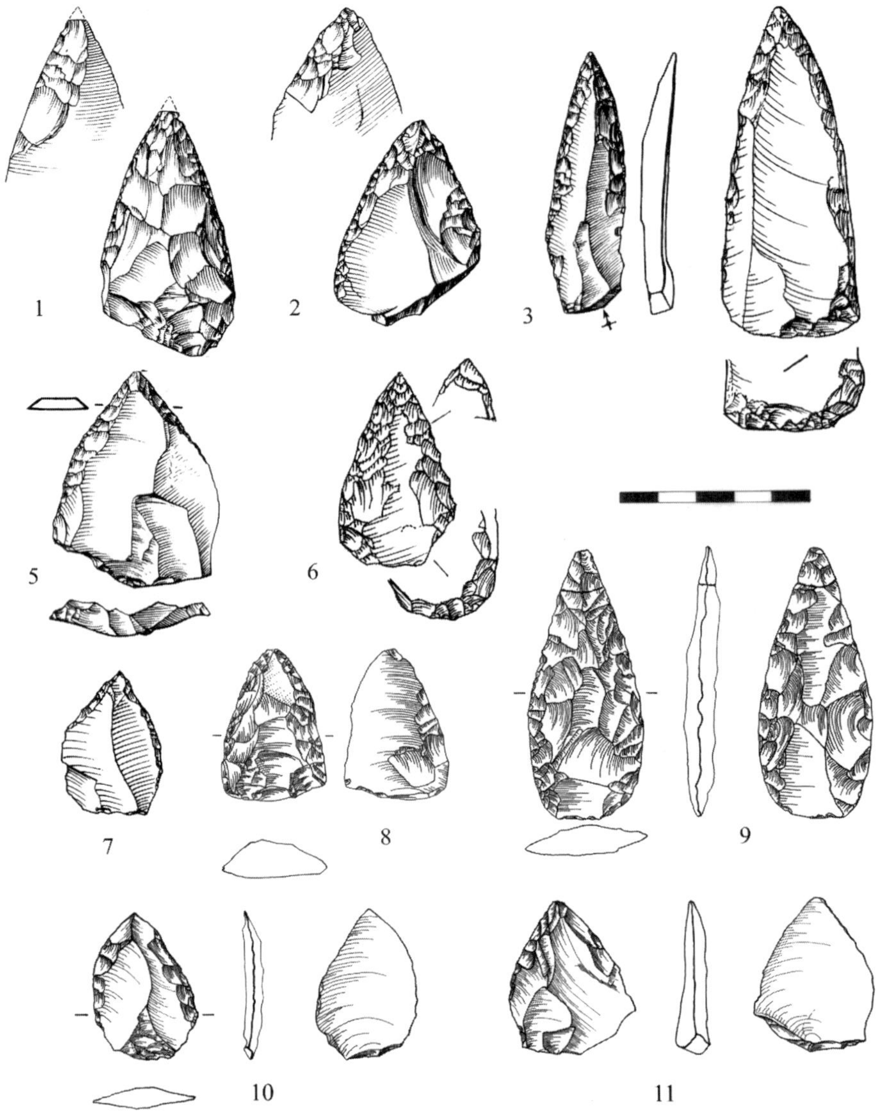

FIG. 5.9. 1–7 points from Bouheben, 3 is from layer 1^1, all others from layer 2; 8 point from Rose Cottage, layer Lou; 9 bifacial point from Sibudu, layer Ore; 10–11 unifacial points from Sibudu, layers Mod and RSP.

with the carcass or on their shaft, may have been recycled, thus removing impact scars (Fig. 5.11: 8).

In other words, we see no reason to expect a high percentage of impact scars on points found at a residential or a manufacturing site. A clear example is provided by one of the localities of the Agate Basin site, a Paleoindian occupation complex at the border of Wyoming and South Dakota. The Main Folsom component of Area II is a residential campsite dated to 10,780 ± 210 BP, which has yielded large amount of lithic debris related to weaponry manufacture together with the faunal remains of 11 bisons and five pronghorn antelopes. Based on the study of the typical byproducts of the Folsom point manufacture (channel flakes resulting from fluting), a minimum of 38 points were made at the site, yet only three were discarded (Sellet, 2004). This is not to say that the planning behavior and the technological organization of Middle Paleolithic people were in any way similar to those of the Paleoindian, only that we should keep in mind the possibility that used weapons may not have been discarded in equal proportions at any site.

Discussion

A visual assessment of the Bouheben assemblage would suggest that at least two main design shapes were in the minds of their makers: a broad, thin point with a wide front angle which would produce a greater wound area that would bleed

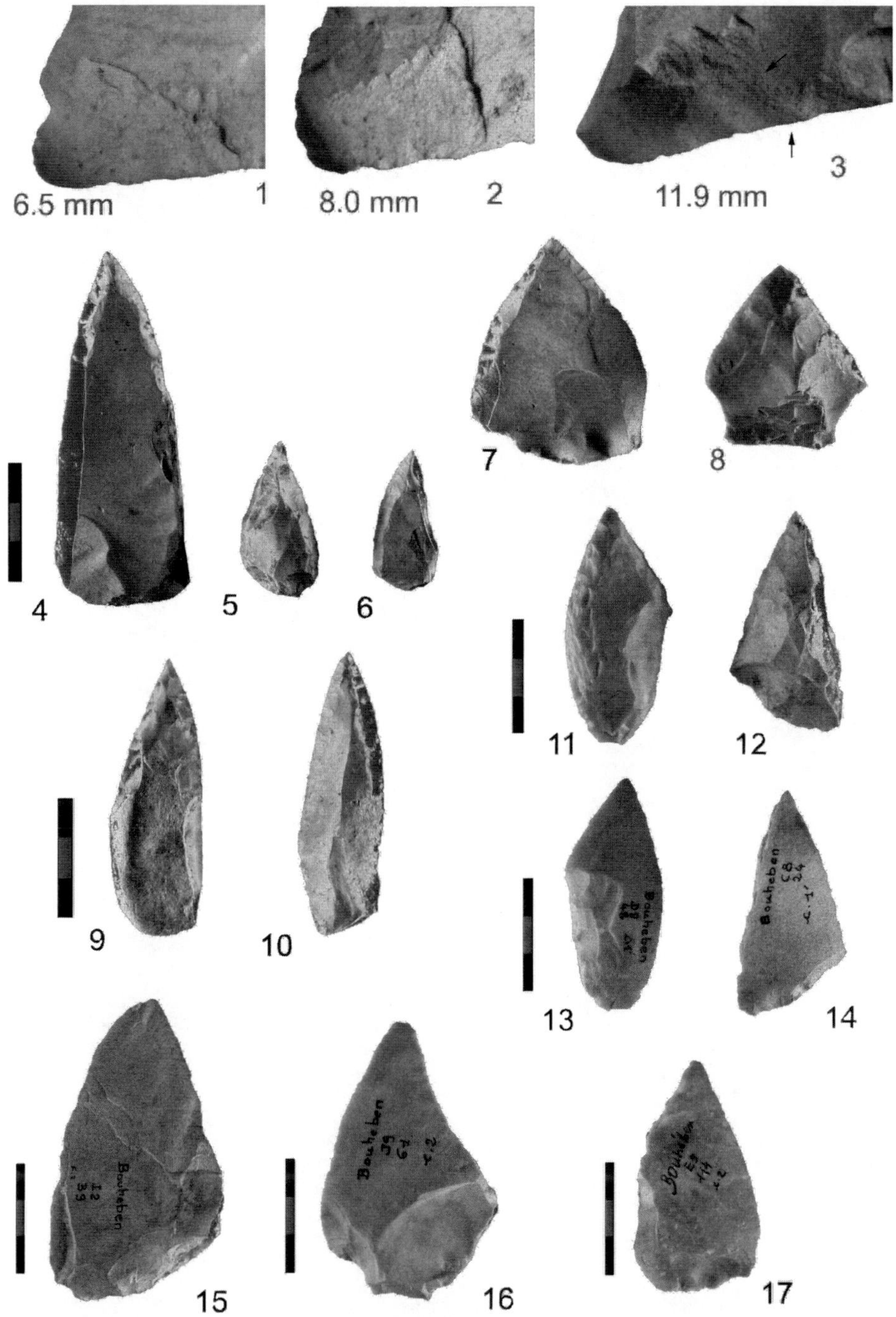

FIG. 5.10. Bouheben. 1–3 impact scars on three Mousterian point, length of scar is indicated on each micrograph: 1–2 step-terminating fractures from layer 2, 3 burin-like fracture, the black arrows indicate the termination of the scar, from layer 1^1; 4–12 Mousterian and elongated Mousterian points: 4–6, 9–11 from layer 2, 7–8, 12 from layer 1^1; 13–15 points with thinned base from layer 2. All points are on flint.

TABLE 5.12. Bronze Age, Mesolithic and Late Glacial assemblages with stone points in Northern Europe. Bromme and Ommelshoved belong to the Brommian culture dated to the Alleröd (about 13,000 BP); Stellmoor, upper level, belongs to the Ahrensburgian and is dated to the Younger Dryas (about 12,000 BP). Age estimates of these climatic periods vary between older and more recent ice core chronologies (Rasmussen et al., 2006). All of these points are interpreted as arrowheads (From Fischer et al., 1984).

Site, age and kinds of points	No. of points suitable for analysis	No. of impact scars	Percentage of impact scars
Muldbjerg about 2800 BC, transverse arrowheads	30	9	30.0
Præstelyng about 3200 BC, transverse arrowheads	56	8	14.3
Vejlebro, level 8 about 3500 BC, transverse arrowheads	24	5	20.8
Vejlebro, level 9 about 3500 BC, transverse arrowheads	42	2	4.8
Stellmoor, upper level, end of the last glacial, tanged points	45	19	42.2
Bromme, end of the last glacial, tanged points	47	3	6.4
Ommelshoved, end of the last glacial, tanged points	88	11	12.5

FIG. 5.11. Impact scars on Sibudu (1–4) and Rose Cottage (5–6) unifacial points. 1, 4 from layer RSP; 2–3 from layers above and below RSP, respectively; 5–6 from layer THO. Length of impact scar is indicated on each micrograph. 1, 2, 4 hornfels; 3 dolerite; 5–6 opaline; 7 three unifacial points from Rose Cottage layer THO, the scale is in centimeter; 8 tip retouch flake from Sibudu, layer YSP, showing a 3 mm impact scar at the tip. This flake was removed by direct percussion from a used point, proving that points were occasionally recycled.

easily (Fig. 5.9: 1–2, 5 and Fig. 5.10: 7–8), and a thick point with a slender head and a more obtuse leading edge angle (angle in profile) more resistant to breakage and with more stopping power (Fig. 5.9: 3–4 and Fig. 5.9–10). These shapes show the two contrasting requirements of stone spear points: the need to increase their mass to enhance impact and perhaps durability, and the need to create deep and lasting wounds that would hasten death (Cheshier and Kelly, 2006). A thin point would have the advantage of breaking more easily in the wound, thus protecting the haft from shock and breakage.

Experimental replication shows that the manufacture time of the Lehringen wooden spear was 5–20 times longer than that of the most sophisticated Middle Paleolithic stone implements (such as bifaces and Levallois cores). Depending on the implements used to cut off the twigs and branches from the stem, the time could vary from 4–5 h (if whittling is done by flakes) to about 1 h (if using a chopping implement like a biface). In other words, the making of a wooden spear shaft is much more time-consuming than the making of a Middle Paleolithic point (Veil and Plisson, 1990).

The Bouheben points have repetitive shapes but their variability and intergrading attribute states are such that it does not seem possible (and we have not attempted) to define different types by multivariate analysis. Their variability might be explained, at least in part, by Dibble's (1995) reduction model.

Although the morphometric features and impact scars diagnostic of spear tips are present in our three assemblages, their morphological variability (greater than that observable on the Hell Gap points of the Casper site or the tanged points of the Late Glacial of Northern Europe) suggests to us that some of these pieces may have had other functions, such as use as perforators, scrapers, or knife edges. This is the subject of another paper.

Thrusting or Throwing Spears?

Stone points of the Middle Stone Age/Middle Paleolithic are often interpreted as tips for thrusting spears. According to Churchill (1993), ethnographic sources indicate that the range of throwing spears is in the order of 8 m. This is considered a dangerously close distance for hunting large mammals (Shea, 2006). It is then suggested that Neanderthals used thrusting, not throwing, spears by placing a prey in a disadvantaged position and then killing it at close quarters.

In fact, the range of throwing spears is underestimated. Roman soldiers used javelins (*pila*) that were about 2 m long and weighted 2 kg to a maximum of 3–4 kg for the heavy variety. A *pilum* was made of a wooden shaft with a diameter of 2–3 cm and a length of 1.2–1.5 m; it was topped by a 60–90 cm long iron shank about 7 mm in diameter, leading to a small pyramidal or barbed point. The total length of a *pilum* was 2.0–2.3 m. The thin iron shank would easily pierce a shield or a cuirass and could bend on impact; the barbed point would make it difficult to withdraw from a shield so that the enemy was forced to drop it. Based on modern experiments, the *pilum* maximum range was in the order of 30 m, although the effective range (killing or wounding) was of 15 m. The legionaries threw their *pilum* (each soldier had two) after marching to within 30 m of the enemy, with a second volley at closer range (Goldsworthy, 2002). Some versions of the weapon were weighted by a lead ball to increase penetrative power.

There is no reason to suppose that Neanderthals were less strong than Roman soldiers; they had as much motivation to hit their target (a food animal) as Roman soldiers who had been drilled to obey their centurions and kill their opponents. The world record for throwing javelins in the Summer Olympics is 98.48 m (for a male athlete), but Olympic javelins weigh less than *pila* (800 g for male athletes; the javelin length is 2.6–2.7 m) and the throwing rules, which allow for a run-up of about 33 m, are hardly comparable to those of hunting defensive mammals in search of a caloric return.

In historical times (between the end of the eighth and the fourth century BC) the thrusting spears of Greek hoplites, advancing in phalanx formation to kill at close range, had the same diameter (2.5 cm) as the Roman throwing spear, were slightly longer (up to 2.7 m) but lighter (up to 2 kg); their shafts, made of dogwood or ash, were not weighted down (Hanson, 1989). In other words, thrusting spears are not necessarily heavier than throwing spears.

The experimental or ethnographic data for throwing spears is limited and does not allow distinguishing between thrusting or throwing spears when only the stone tip and impact scars are available to the analyst. We cannot define the morphometric features that would distinguish between stone points used to tip thrusting or throwing spears, both being used at close or relatively close distance, compared to darts thrown with a spearthrower. Based on data from Australia, darts thrown with a spearthrower in terrestrial environments for hunting or in warfare can reach a distance of 90–135 m; distances for accurate throws range from 45 to 55 m for immobile targets; animals ambushed near water sources are killed by spears usually thrown from 15 to 20 m. Throwing distances of hunting bows are much higher (Cattelain, 1997). Points used to tip darts or arrows have aerodynamic properties (are relatively light, thin, symmetrical, have a narrow penetrating angle and an elliptical cross-section) which do not seem to occur in the heavier points of the Middle Paleolithic and Middle Stone Age. In sum, we cannot say if the stone points of the Middle Paleolithic/MSA were used as tips for thrusting spears or throwing spears. We see, however, no reason for preferring one interpretation over the other; they could have been either.

Ethnographic and historic data suggest that thrusting and throwing spears were used mainly for hunting large and medium size mammals (Churchill, 1993; Hitchcock and Bleed, 1997); the Spartans used thrusting spears to hunt wild boars (Hanson, 1989). These data support a correlation between the preference for large and medium-size mammalian prey of Neanderthals and their weaponry. A similar correlation has been noted between the broadening of the subsistence base with the inclusion of smaller and more agile game in the Upper Paleolithic and the appearance of long-range weaponry in the form of spearthrowers and bows and arrows (Churchill, 1993). Nevertheless, the data available for the European Middle Paleolithic hunting weapons remain very limited and need to be strengthened by analyses of other assemblages, integrating faunal and lithic technology studies. Only then we will be able to fully test this hypothesis.

Conclusions

Our analysis shows that: (a) for the period prior to OIS 9 or 8 very few generalizations can be made about the subsistence behavior of early humans in Europe because the informative sites are few and far between. Nevertheless a good case can be made for hunting from two of the earliest sites in Europe, Gran Dolina TD 6 and Boxgrove; (b) even stronger evidence of hunting comes from sites such as Schöningen and later Middle Paleolithic sites where the topographic setting and the faunal accumulations indicate repeated episodes of hunting the same species of large-size mammals; (c) evidence for hunting/gathering of very small vertebrates and invertebrates (leporids, birds, fish and shellfish) before the Upper Paleolithic is limited; the available data indicate that Neanderthals relied primarily on herbivore meat and marrow as a dietary resource. This conclusion is supported by isotopic data from collagen extracted from Neanderthal bones (Bocherens et al., 2005). However limited, the evidence of shellfish and small game hunting/gathering shows that Neanderthals were capable of exploiting the same variety of resources as Upper Paleolithic humans.

Morphometric and impact scar analysis of the Bouheben assemblage show that at least some of the Mousterian points were used to arm thrusting or throwing spears. The evidence from La Cotte de St. Brelade suggests that stone-tipped spears were already in use by OIS 6 in Western Europe. Similar weapons were in common use in the MSA of South Africa and the Levant. According to Shea (2006) systematic production of projectile points, thrown with spearthrowers or bows, in the Levant, Europe and Africa dates to after 40–50 ka.

Acknowledgments. We thank Jean-Jacques Hublin and Michael Richards for the invitation to participate in the symposium on The Evolution of Hominid Diets at the Max Planck Institute. Research on the Sibudu and Rose Cottage lithics was funded by grants from the National Science Foundation (BCS 0314371 and BCS 0613319) and the Leakey Foundation to P.V. We are very grateful to Lyn Wadley of the University of Witwatersrand for letting us study materials from her excavations. Lucinda Backwell very graciously lent us her stereomicroscope (an Olympus SZX 9 with a Nikon Coolpix 990) to take photos of the Sibdu and Rose Cottage impact scars. The micrographs of the Bouheben impact scars were done using a Leica 216 APO and a Canon eo5 350D; we thank Eric Pubert of the Bordeaux Institute for his help in using the camera. The Bouheben artifact drawings were done by Pierre Laurent and by Michel Lenoir; those of Sibudu and Rose Cottage were done by Wendy Voorvelt of the University of Witwatersrand with help by Sylvain Soriano. The Bouheben materials are kept in the Institute of Prehistory and Quaternary Geology in Bordeaux; our research would not have been possible without the previous work of the late Claude Thibault, the very careful excavator of Bouheben. Grateful thanks also go to the friends of the Archaeology Lunch in Boulder, especially Jim Dixon, Payson Sheets, John Hoffecker, Steven Holen, Cathy Cameron and Steve Lekson for very valuable comments and discussions and to three anonymous reviewers for their suggestions. This paper was completed in January 2007 and does not include more recent references.

References

Adler, D.S., Bar-Oz, G., Belfer-Cohen, A., Bar-Yosef, O., 2006. Ahead of the game. Middle and Upper Palaeolithic hunting behaviors in the Southern Caucasus. Current Anthropology 47, 89–118.

Anconetani, P., Malerba, G., Giacobini, G., 1996. Studio micromorfologico di superficie dei reperti faunistici di Isernia La Pineta. Considerazioni tafonomiche. In: Peretto, C. (Ed.), I reperti paleontologici del giacimento paleolitico di Isernia la Pineta. Cosmo Iannone, Isernia, pp. 53–86.

Anderson-Gerfaud, P.A., 1990. Aspects of behaviour in the Middle Palaeolithic: functional analysis of stone tools from Southwest France. In: Mellars, P. (Ed.), The Emergence of Modern Humans. Edinburgh University Press, Edinburgh, pp. 389–418.

Auguste, P., 1995. Chasse et charognage au Paléolithique moyen: l'apport du gisement de Biache-Saint-Vaast (Pas-de-Calais). Bulletin de la Société Préhistorique Française 92, 155–167.

Bailey, G.N., 1978. Shell middens as indicators of postglacial economies: a territorial perspective. In: Mellars, P. (Ed.), The Early Postglacial Settlement of Northern Europe. University of Pittsburgh Press, Pittsburgh, pp. 37–64.

Bar-Oz, G., Adler, D.S., Vekua, A., Meshveliani, T., Tushabramishvili, N., Belfer-Cohen, A., Bar-Yosef, O., 2004. Faunal exploitation patterns along the southern slopes of the Caucasus during the Late Middle and Upper Palaeolithic. In: Mondini, M, Munõz, S., Wickler, S. (Eds.), Colonisation, Migration and Marginal Areas. Oxbow Books, Oxford, pp. 46–54.

Barton, R.N.E., 2000. Mousterian hearth and shellfish: Late Neanderthal activities on Gibraltar. In: Stringer, C.B., Barton, R.N.E., Finlayson, J.C. (Eds.), Neanderthals on the Edge. Oxbow Books, Oxford, pp. 211–220.

Barton, R.N.E., Bergman, C.A., 1982. Hunters at Hengistbury: some evidence from experimental archaeology. World Archaeology 14, 238–243.

Bermúdez de Castro, J.M., Martinón-Torres, M., Carbonell, E., Sarmiento, S., Rosas, A., Van der Made, J., Lozano, M., 2004. The Atapuerca sites and their contribution to the knowledge of human evolution in Europe. Evolutionary Anthropology 13, 25–41.

Beyries, S., 1988a. Etude tracéologique des racloirs du niveau IIA. In: Tuffreau, A., Sommé, J. (Eds.), Le Gisement Paléolithique Moyen de Biache-Saint-Vaast (Pas-de-Calais) Vol. 1. Mémoires de la Société Préhistorique Française 21, Paris, p. 215–230

Beyries, S., 1988b. Analyse tracéologique du matériel lithique de la couche VIII de la Grotte Vaufrey. In: Rigaud, J.P. (Ed.), La Grotte Vaufrey. Mémoires de la Société Préhistorique Française 19, Paris, p. 519–528.

Binford, L.R., 1981. Bones: ancient men and modern myths. Academic Press, New York.

Binford, L.R., 1985. Human ancestors: changing views of their behavior. Journal of Anthropological Archaeology 4, 292–327.

Binford, L.R., 1987. Were there elephant hunters at Torralba? In: Nitecki, M.H., Nitecki, D. (Eds.), The Evolution of Human Hunting. Plenum Press, New York, pp. 47–105.

Binford, L.R., 1988. Etude taphonomique des restes fauniques de la Grotte Vaufrey, couche VIII. In: Rigaud, J. Ph. (Ed.), La grotte Vaufrey. Mémoires de la Société Préhistorique Française 19, Paris, pp. 535–564.

Blumenschine, R.J., 1986. Early hominid scavenging opportunities. Implications of carcass availability in the Serengeti and Ngorongoro ecosystems. B.A.R. International Series 283, Oxford.

Blumenschine, R.J., 1988. An experimental model of the timing of hominid and carnivore influence on archaeological bone assemblages. Journal of Archaeological Science 15, 483–502.

Blumenschine, R.J., 1995. Percussion marks, tooth marks and the experimental determinations of the timing of hominid and carnivore access to long bones at FLK *Zinjanthropus*, Olduvai Gorge, Tanzania. Journal of Human Evolution 29, 21–51.

Blumenschine, R.J., Selvaggio, M.M., 1988. Percussion marks on bone surfaces as a new diagnostic of hominid behavior. Nature 333, 763–765.

Blumenschine, R.J., Cavallo, J.A., Capaldo, S.D., 1994. Competition for carcasses and early hominid behavioral ecology: a case study and a conceptual framework. Journal of Human Evolution 27, 197–213.

Bocherens, H., Drucker, D.G., Billiou, D., Patou-Mathis, M., Vandermeersch, B., 2005. Isotopic evidence for diet and subsistence pattern of the Saint-Césaire I Neanderthal: review and use of a multi-source mixing model. Journal of Archaeological Science 49, 71–87.

Boëda, E., Connan, J., Dessort, D., Muhesen, S., Mercier, N., Valladas, H., Tisnerat, N., 1996. Bitumen as hafting material on Middle Paleolithic artifacts. Nature 380, 336–338.

Boëda, E., Geneste, J.M., Griggo, C., 1999. A Levallois point embedded in the vertebra of a wild ass (*Equus africanus*): hafting, projectiles and Mousterian hunting weapons. Antiquity 73, 394–402.

Boëda, E., Soriano, S., Noël-Soriano, S., 2004. Fonction et fonctionnement d'un site à la fin du Pléistocène moyen. Le niveau Acheuléen C'3 base de Barbas I (Creysse, Dordogne). In: Bodu, P., Constantin, C. (Eds.), Approches fonctionnelles en Préhistoire. XXV Congrès Préhistorique de France, Nanterre 2000, Société Préhistorique Française, pp. 293–305.

Bordes, F., 1961. Typologie du Paléolithique ancien et moyen. Delmas, Bordeaux.

Bratlund, B., 1996. Hunting strategies in the Late Glacial of Northern Europe: a survey of the faunal evidence. Journal of World Prehistory 10, 1–48.

Bratlund, B., 1999. Anthropogenic factors in the thanotocoenose of the Last Interglacial travertines at Taubach (Germany). In: Gaudzinski, S., Turner, E. (Eds.), The Role of Early Humans in the Accumulation of European Lower and Middle Palaeolithic Bone Assemblages. Habelt, Bonn, pp. 255–262.

Brugal, J.P., Jaubert, J., 2006. A long Palaeolithic sequence in Southwest France: the karstic sites of Coudoulous (Lot, Quercy, France). Paper presented at the Paleoanthropology Society Meeting in San Juan, Puerto Rico, April 26, A 87.

Bunn, H.T., 1981. Archaeological evidence for meat-eating by Plio-Pleistocene hominids from Koobi Fora, Kenya. Nature 291, 574–577.

Bunn, H.T., 1982. Meat-eating and human evolution: studies on the diet and subsistence patterns of Plio-Pleistocene Hhminids in East Africa. Ph.D. Dissertation, University of California, Berkeley.

Bunn, H.T., 1996. Comments to Rose and Marshall. Current Anthropology 37, 321–323.

Bunn, H.T., Kroll, E.M., 1986. Systematic butchery by Plio-Pleistocene hominids at Olduvai Gorge, Tanzania. Current Anthropology 27, 431–452.

Bunn, H.T., Kroll, E.M., 1988. A reply to Binford. Current Anthropology 29, 123–149.

Bunn, T.H., Bartram, L.E., Kroll, E.M., 1988. Variability in bone assemblage formation from Hadza hunting, scavenging and carcass processing. Journal of Anthropological Research 7, 412–457.

Callow, P., 1986. The flint tools. In: Callow, P., Cornford, J. (Eds.), La Cotte de St. Brelade. Geo Books, Norwich, pp. 251–314.

Callow, P., Cornford, J. (Eds.), 1986. La Cotte de St. Brelade. Geo Books, Norwich, UK.

Cattelain, P., 1997. Hunting during the Upper Paleolithic: bow, spearthrower or both? In: Knecht, H. (Ed.), Projectile Technology. Plenum Press, New York, pp. 213–240.

Chase, P.G., 1999. Bison in the context of complex utilization of faunal resources. In: Brugal, J.P., David, F., Enloe, J.G., Jaubert, J. (Eds.), Le Bison: gibier et moyen de subsistance des hommes du Paléolithique aux Paléoindiens des Grandes Plaines. Editions APDCA, Antibes, pp. 159–184.

Cheshier, J., Kelly, R.L., 2006. Projectile point shape and durability: the effect of thickness: length. American Antiquity 71, 353–363.

Churchill, S.E., 1993. Weapon technology, prey size selection, and hunting methods in modern hunter-gatherers: implications for hunting in the Palaeolithic and Mesolithic. In: Peterkin, G.L., Bricker, H., Mellars, P.A. (Eds.), Hunting and animal exploitation in the Later Paleolithic and Mesolithic of Eurasia. Archaeological Papers of the American Anthropological Association, no. 4, pp.11–24.

Churchill, S.E., Rhodes, J.A., 2008. The evolution of the human capacity for "killing at a distance." In: Hublin, J.-J., Richards, M.P. (Eds.), The Evolution of Hominid Diets: Integrating Approaches to the Study of Palaeolithic Subsistence. Springer, Dordrecht.

Cochard, D., 2004. Les Leporidés dans la subsistence paléolithique du Sud de la France. Ph.D. Dissertation, Université Bordeaux 1.

Coltorti, M., Feraud, G., Marzoli, A., Peretto, C., Ton-That, T., Voinchet, P., Bahain, J.-J., Minelli, A., Thun Hohenstein, U., 2005. New $^{40}Ar/^{39}Ar$, stratigraphic and paleoclimatic data on the Isernia La Pineta Lower Paleolithic site, Molise, Italy. Quaternary International 131, 11–22.

Conard, N.J., Prindiville, T.J., 2000. Middle Palaeolithic hunting economies in the Rhineland. International Journal of Osteoarchaeology 10, 286–309.

Cortés Sanchez, M., 2000. Bajondillo Cave (Torremolinos, Malaga, Andalucia) and the Middle-Upper Palaeolithic transition in southern Spain. In: Stringer, C.B., Barton, R.N.E., Finlayson, J.C. (Eds.), Neanderthals on the Edge. Oxbow Books, Oxford, pp. 123–132.

Costamagno, S., Laroulandie, V., 2004. L'exploitation des petits vertebrés dans les Pyrénées françaises du Paléolithique au Mésolithique: un inventaire taphonomique et archéozoologique. In: Brugal, J.P., Desse, J. (Eds.), Petits animaux et sociétés humaines. Du complément alimentaire aux ressources utilitaires. Editions APDCA, Antibes, pp. 163–172.

Delagnes, A., 1992. L'organisation de la production lithique au Paléolithique moyen. Ph.D. Dissertation, University of Paris X.

Dibble, H., 1987a. Reduction Sequences in the Manufacture of Mousterian Implements of France. In: Soffer, O. (Ed.), The Pleistocene Old World: Regional Perspectives. Plenum Press, New York, pp. 33–45.

Dibble, H., 1987b. The interpretation of Middle Paleolithic scraper morphology. American Antiquity 52, 109–117.

Dibble, H., 1988. Typological Aspects of reduction and intensity of utilization of lithic resources in the French Mousterian. In: Dibble, H., Montet-White, A. (Eds.), Upper Pleistocene Prehistory of Western Eurasia. University Museum Monograph 54, University Museum Symposium Series, Philadelphia, vol. 1, pp. 181–197.

Dibble, H.L., 1995. Middle Paleolithic scraper reduction: background, clarification, and review of the evidence to date. Journal of Archaeological Method and Theory 2, 299–368.

Dibble, H.L., Bernard, M.C., 1980. A comparative study of basic edge angle measurement techniques. American Antiquity 45, 857–865.

Díez, J.C., Fernández-Jalvo, Y., Rosell, J., Cáceres, I., 1999. Zooarchaeology and taphonomy of Aurora stratum (Gran Dolina, Sierra de Atapuerca, Spain). Journal of Human Evolution 37, 623–653.

Domínguez-Rodrigo, M., 1999. Flesh availability and bone modifications in carcasses consumed by lions: palaeecological relevance in homind foraging patterns. Palaegeography, Palaeclimatology, Palaeecology 149, 373–388.

Domínguez-Rodrigo, M., 2001. A study of carnivore competition in riparian and open habitats of modern savannas and its implications for hominid behavioral modelling. Journal of Human Evolution 40, 77–98.

Domínguez-Rodrigo, M., Barba, R., 2006. New estimates of tooth mark and percussion mark frequencies at the FLK Zinj site: the carnivore-hominid-carnivore hypothesis falsified. Journal of Human Evolution 50, 170–194.

Domínguez-Rodrigo, M., Pickering, T. R, 2003. Early hominids, hunting and scavenging: a zooarcheological review. Evolutionary Anthropology 12, 275–282.

Domínguez-Rodrigo, M., Barba, R., Egeland, C.P. (Eds.), 2007. Deconstruction Olduvai: A Taphonomic Study of Bed I Sites. Springer, Dordrecht.

Echassoux, A., 2004. Etude taphonomique, paléoécologique et archéozoologique des faunes de grands mammifères de la seconde moitié du Pléistocène inférieur de la grotte du Vallonnet. L'Anthropologie 108, 11–53.

Falguères, C., Bahain, J.J., Yokohama, Y., Arsuaga, J.L., Bermúdez de Castro, J.M., Carbonell, E., Bischoff, J.L., Dolo, J.M., 1999. Earliest humans in Europe: the age of TD6, Gran Dolina, Atapuerca, Spain. Journal of Human Evolution 37, 343–352.

Falguères, C., Bahain, J.J., Pérez-González, A., Mercier, N., Santonja, M., Dolo, J.M., 2006. The Lower Acheulian site of Ambrona, Soria (Spain): ages derived from a combined ESR/U-series model. Journal of Archaeological Science 33, 149–157.

Farizy, C., David, F., Jaubert, J., 1994. Hommes et bisons du paléolithique moyen à Mauran (Haute Garonne). Editions CNRS, Paris.

Fernández-Jalvo, Y., 1998. Interpretación paleoambiental y etológica en Atapuerca a través de métodos tafonómicos. In: Aguirre, E. (Ed.), Atapuerca y la evolución humana. Fundación Ramón Areces, Madrid, pp. 111–152.

Fernández-Jalvo, Y., Díez, J.C., Rosell, J., Cáceres, I., 1999. Human cannibalism in the Early Pleistocene of Europe (Gran Dolina, Sierra de Atapuerca, Spain). Journal of Human Evolution 37, 591–622.

Fernández, Ph., Legendre, S., 2003. Mortality curves for horses from the Middle Palaeolithic site of Bau de l'Aubesier (Vaucluse, France): methodological, palaeo-ethnological and palae-ecological approaches. Journal of Archaeological Science 30, 1577–1598.

Fiore, I., Gala, M., Tagliacozzo, A., 2004. Ecology and subsistence strategies in the Eastern Italian Alps during the Middle Palaeolithic. International Journal of Osteoarchaeology 14, 237–286.

Fischer, A., Vemming Hansen, P., Rasmussen, P., 1984. Macro and micro wear traces on lithic projectile points. Experimental results and prehistoric examples. Journal of Danish Archaeology 3, 19–46.

Freeman, L.G., 1975. Acheulean sites and stratigraphy in Iberia and the Maghreb. In: Butzer, K.W., Isaac, G.Ll. (Eds.), After the Australopithecines. Mouton, The Hague. pp. 661–743

Freeman, L.G., 1978. The analysis of some occupation floors distributions from Earlier and Middle Paleolithic. In: Freeman, L.G. (Ed.), Views of the Past. Mouton, The Hague, pp. 57–93

Freeman, L.G., 1994. Torralba and Ambrona; a review of discoveries. In: Corruccini, R.S., Ciochon, R.L. (Eds.), Integrative Paths to the Past. Prentice Hall, Englewood Cliffs, pp. 597–638.

Frison, G.C., 1974. The Casper Site. Academic Press, New York.

Frison, G.C., 1978. Prehistoric Hunters of the High Plains. Academic Press, New York.

García, N., Arsuaga, J.L., 1999. Carnivores from the Early Pleistocene hominid-bearing Trinchera Dolina 6 (Sierra de Atapuerca, Spain). Journal of Human Evolution 37, 415–430.

Gaudzinski, S., 1996a. Review of M.C. Stiner, Honor among thieves. Antiquity 70: 227–229.

Gaudzinski, S., 1996b. On bovid assemblages and their consequences for the knowledge of subsistence patterns in the Middle Palaeolithic. Proceedings of the Prehistoric Society 62, 19–39.

Giusberti, G., Guerreschi, A., Peretto, C., 1983. Le strutture d'abitato dell'accampamento paleolitico di Isernia La Pineta. Prime considerazioni. In: Soprintendenza Archeologica del Molise (Ed.), Isernia La Pineta: un accampamento piu antico di 700,000 anni. Calderini Editore, Bologna, pp. 94–103.

Giusberti, G., Ferrari, M., Peretto, C., 1991. Tipologia, frequenza e distribuzione dei reperti paleontologici e paletnologici della paleosuperficie T. 3a del I° settore di scavo di Isernia La Pineta (Isernia, Molise). In: Peretto, C. (Ed.), Isernia La Pineta. Nuovi contributi scientifici. Istituto Regionale per gli Studi Storici del Molise, Firenze, pp. 5–42.

Goldsworthy, A., 2002. Roman warfare. Cassell, London.

Hanson, V.D., 1989. The western way of war. University of California Press, Berkeley.

Harper, P.T., 1997. The Middle Stone Age sequence at Rose Cottage Cave: a search for continuity and discontinuity. South African Journal of Science 93, 470–475.

Hitchcock, R., Bleed, P., 1997. Each according to need and fashion. In: Knecht, H. (Ed.), Projectile Technology. Plenum Press, New York, pp. 345–368.

Hoffecker, J.F., 1999. Neanderthals and modern humans in Eastern Europe. Evolutionary Anthropology 7, 129–141.

Hoffecker, J.F., Cleghorn, N., 2000. Mousterian hunting patterns in the Northwestern Caucasus and the ecology of the Neanderthals. International Journal of Osteoarchaeology 10, 368–378.

Hughes, S.S., 1998. Getting to the point: evolutionary change in prehistoric weaponry. Journal of Archaeological Method and Theory 5, 345–408.

Hunt, R.M., 1990. Taphonomy and sedimentology of Arikaree (lower Miocene) fluvial, eolian and lacustrine paleoenvironments, Nebraska and Wyoming: a paleobiota entombed in fine-grained volcaniclastic rocks. In: Lockley, M.G., Rice, A. (Eds.), Volcanism and Fossil Biotas. Geological Society of America, Boulder, Colorado, Special Paper 244.

Isaac, G.Ll., 1978. The food-sharing behavior of protohuman hominids. Scientific American 238, 90–108.

Isaac, G.Ll., 1984. The archaeology of human origins: studies of the Lower Pleistocene in East Africa 1971–1981. Advances in World Archaeology 3, 1–87.

Isaac, G.Ll., Crader, D.C., 1981. To what extent were early hominids carnivorous? An archaeological perspective. In: Harding, R., Teleki, G. (Eds.), Omnivorous Primates: Gathering and Hunting in Human Evolution. Columbia University Press, New York, pp. 37–103.

Jaubert, J., Lorblanchet, M., Laville, H., Slott-Moller, R., Turq, A., Brugal, J.Ph., 1990. Les chasseurs d'Aurochs de La Borde. Documents d'Archéologie Française, Editions Maison des Sciences de l' Homme, Paris.

Jaubert, J., Kervazo, B., Bahain, J.J., Brugal, J.P., Chalard, P., Falgueres, C., Jarry, M., Jeannet, M., Lemorini, C., Louchart, A., Maksud, F., Mourre, V., Quinif, Y, Thiebaut, C., 2005. Coudoulous 1 (Tour-de-Faure, Lot). Site du Pléistocène moyen en Quercy: bilan pluridiscipluinaire. In: Molines, N., Moncel, M.H., Morrier, J.L. (Eds.), Données récentes sur les modalités de peuplement et sur le cadre chronostratigraphique, géologique et paléogéographique des industries du Paléolithique ancien et moyen en Europe. B.A.R. International Series 1364, Archaeopress, Oxford, pp. 227–251.

Kahlke, R.D., 1999. Overview and first qualitative data on the taphonomy of the Lower Pleistocene fossil site of Untermassfeld (Thuringia, Germany). In: Gaudzinski, S., Turner, E. (Eds.), The Role of Early Humans in the Accumulation of European Lower and Middle Palaeolithic Bone Assemblages. Habelt, Bonn, pp. 7–20.

Klein, R.G., 1978. Stone Age predation on large African bovids. Journal of Archaeological Science 5, 195–217.

Klein, R.G., 1995. Neanderthal carnivory. Science 267, 1843–1844.

Klein, R.G., 1999. The Human Career. University of Chicago Press, Chicago and London.

Klein, R.G., Cruz-Uribe, K., 1991. The bovids from Elandsfontein, South Africa, and their implications for the age, palaeoenvironment, and origins of the site. The African Archaeological Review 9, 21–79.

Kuhn, S.L., 1990. Diversity within uniformity: tool manufacture and use in the Pontinian Mousterian of Latium (Italy). Ph.D. Dissertation, University of New Mexico.

Kuhn, S.L., 1995. Mousterian Lithic Technology. Princeton University Press, Princeton.

Kuhn, S.L., Stiner, M.C., 2006. What's a mother to do? The division of labor among Neandertals and modern humans in Eurasia. Current Anthropology 47, 953–980.

Laroulandie, V., 2004. Exploitation des ressources aviaires durant le Paléolithique en France: bilan critique et prespectives. In: Brugal, J.P., Desse, J. (Eds.), Petits animaux et sociétés humaines. Du complément alimentaire aux resspurces utilitaires. Editions APDCA, Antibes, pp. 163–172.

Le Gall, O., 2000. Les Moustériens étaient-ils pêcheurs ? Bulletin de la Société d' Anthropologie du Sud-Ouest 34, 3–11.

Lombard, M., 2005a. Evidence of hunting and hafting during the Middle Stone Age at Sibudu Cave, KwaZulu-Natal: a multianalytical approach. Journal of Human Evolution 48, 279–300.

Lombard, M., 2005b. A method for identifying stone age hunting tools. South African Archaeological Bulletin 60, 115–120.

Lumley, H. de et al. (22 other authors), 2004. Habitat et mode de vie des chasseurs paléolithicque de la Caune de l'Arago (600,000–400,000 ans). L'Anthropologie 108, 159–184.

Made, J. van der, 1999. Ungulates from Atapuerca. Journal of Human Evolution 37, 389–413.

Marean, C.W., Kim, S.Y., 1998. Mousterian large-mammal remains from Kobeh Cave. Current Anthropology 39 (Supplement), S79–S113.

Marean, C.W., Assefa, Z., 1999. Zooarcheological evidence for the faunal exploitation behavior of Neandertals and modern humans. Evolutionary Anthropology 8, 22–37.

McCool, K.E., 1988. Taphonomy of the Valentine railway quarry "B" bone bed (Late Barstovian), North-Central Nebraska. M.Sc. thesis, University of Nebraska.

Moigne, A.M., Barsky, D., 1999. Large mammal assemblages from Lower Palaeolithic sites in France: La Caune de l'Arago, Terra Amata, Orgnac 3 and Cagny l'Epinette. In: Gaudzinski, S., Turner, E. (Eds.), The Role of Early Humans in the Accumulation of European Lower and Middle Palaeolithic Bone Assemblages. Habelt, Bonn, pp. 219–236.

Monchot, H., 1996. Les petits Bovidés de la Caune de l'Arago (Tautavel, Pyrenées-Orientales). Ph.D. Dissertation, Université Aix-Marseille II.

Morin, E., 2004. Late Pleistocene population interaction in Western Europe and modern human origins: new insights based on the faunal remains from Saint-Césaure, Southwestern France. Ph.D. Dissertation, University of Michigan.

Mussi, M., 2001. Earliest Italy. An Overview of the Italian Paleolithic and Mesolithic. Kluwer Academic/Plenum, New York.

Oakley, K.P., Andrews, A., Keeley, L.H., Clark, J.D., 1977. A reappraisal of the Clacton spearpoint. Proceedings of the Prehistoric Society 43, 13–30.

O' Connell, J.F., 1997. On Plio-Pleistocene archaeological sites and central places. Current Anthropology 38, 86–88.

O' Connell, J.F., Hawkes, K., Blurton Jones, N., 1988. Hadza scavenging: implications for Plio/Pleistocene hominid subsistence. Currant Anthropology 29, 356–363.

O'Farrell, M., 2005. Étude préliminaire des éléments d'armature lithique de l'Aurignacien ancien de Brassempouy. In: Le Brun-Ricalens, F. (Ed.), Productions lamellaires attribuées à l'Aurignacien: Chaînes opératoires et perspectives technoculturelles, Archéologiques 1, Musée national d'Art et d'Histoire, Luxembourg, pp. 395–412.

Oliver, J.S., 1994. Estimates of hominid and carnivore involvement in the FLK *Zinjanthropus* fossil assemblage: some socioecological implications. Journal of Human Evolution 27, 267–294.

Oms, O., Parés, J.M., Martínez-Navarro, B., Agustí, J., Toro, I., Martínez-Fernández, G., Turq, A., 2000. Early human occupation of Western Europe: paleomagnetic dates for two Paleolithic sites in Spain. Proceedings of the National Academy of Sciences USA 97 (19), 10666–10670.

Palmquist, P., Martínez-Navarro, B., Toro, I., Patrocinio Espigares, M., Ros-Montoya, S., Torregrosa, V., Pérez-Claros, J., 2005. Réévaluation de la présence humain au Pléistocène inférieur dans le Sud de l'Espagne. L'Anthropologie 109, 411–450.

Parés, J.M., Pérez-González, A., Rosas, A., Benito, A., Bermúdez de Castro, J.M., Carbonell, E., Huguet, R., 2006. Matuyama-age lithic tools from the Sima del Elefante site, Atapuerca (northern Spain). Journal of Human Evolution 50, 115–218.

Pérez-González, A., Parés, J.M., Carbonell, R.E., Aleixandre, T., Ortega, A.I., Benito, A., Merino, M.A.M., 2001. Géologie de la Sierra de Atapuerca et stratigraphie des remplissages karstiques de Galería et Dolina (Burgos, Espagne). L'Anthropologie 105, 27–43.

Peretto, C. (Ed.), 1994. Le industrie litiche del giacimento paleolitico di Isernia la Pineta. Cosmo Iannone, Isernia.

Peretto, C. (Ed.), 1996. I reperti paleontologici del giacimento paleolitico di Isernia la Pineta. Cosmo Iannone, Isernia.

Peterkin, G.L., 1997. Upper Palaeolithic hunting technology and prey selection in Southwest France. Ph.D. Dissertation, Tulane University, UMI Microform 9732439.

Pickering, T.R., Domínguez-Rodrigo, M., 2006. The acquisition and use of large mammal carcasses by Oldowan hominids in Eastern and Southern Africa: a selected review and assessment. In: Toth, N., Schick, K. (Eds.), The Oldowan: case studies into the Earliest Stone Age. Stone Age Institute Press, Bloomington (Indiana), pp. 113–128.

Piperno, M., (Ed.), 1999. Notarchirico. Un sito del Pleistocene medio iniziale nel bacino di Venosa. Edizioni Osanna, Venosa.

Piperno, M., Giacobini, G., 1990&NDASH;1991. A taphonomic study of the paleosurface of Guattari Cave (Monte Circeo, Latina, Italy). Quaternaria Nova, 1, 143–161.

Porraz, G., 2005. En marge du milieu alpin. Dynamique de formation des ensembles lithiques et modes d'occupation des territoires au Paléolithique moyen. Ph.D. Dissertation, University of Provence.

Potts, R., 1988. Why the Oldowan? Plio-Pleistocene toolmaking and the transport of resources. Journal of Anthropological Research 47, 153–176.

Potts, R., Shipman, P., 1981. Cutmarks made by stone tools on bones from Olduvai Gorge. Nature 291, 577–580.

Raposo, L., 2000. The Middle-Upper Palaeolithic transition in Portugal. In: Stringer, C.B., Barton, R.N.E., Finlayson, J.C. (Eds.), Neanderthals on the Edge. Oxbow Books, Oxford, pp. 95–110.

Rasmussen, S.O. et al. (15 other authors), 2006. A new Greenland ice core chronology for the last glacial termination. Journal of Geophysical Research 111, X 1–15.

Richards, M.P., Jacobi, R., Cook, J., Pettitt, P.B., Stringer, C.B., 2005. Isotope evidence for the intensive use of marine foods by Late Upper Palaeolithic humans. Journal of Human Evolution 49, 390–394.

Richter, D., Mercier, N., Valladas, H., Jaubert, J., Texier, P.J., Brugal, J.B., Kervazo, B., Reyss, J.L., Joron, J.L., Wagner, G.A., 2007. Thermoluminescence dating of heated flint from the Mousterian site of Bérigoule, Murs, Vaucluse, France. Journal of Archaeological Science 34, 532–539.

Roebroeks, W., 2001. Hominid behaviour and the earliest occupation of Europe: an exploration. Journal of Human Evolution 41, 437–461.

Roberts, M.B., Parfitt, S.A., 1999. Boxgrove. A Middle Pleistocene hominid site at Eartham Quarry, Boxgrove, West Sussex. English Heritage, London.

Sala, B., 1996. Gli animali del giacimento di Isernia La Pineta. In: Peretto, C. (Ed.), I reperti paleontologici del giacimento paleolitico di Isernia la Pineta. Cosmo Iannone, Isernia, pp. 25–49.

Santonja, M., Villa, P., 2006. The Acheulian of Western Europe. In: Goren-Inbar, N., Sharon, G. (Eds.), Axe Age: Acheulian Toolmaking-from Quarry to Discard. Equinox, London, pp. 429–478.

Scott, K., 1986. The bone assemblages of layers 3 and 6. In: Callow, P., Cornford, J. (Eds.), La Cotte de St. Brelade. Geo Books, Norwich, pp. 159–192.

Sellet, F., 2004. Beyond the point: projectile manufacture and behavioral inference. Journal of Archaeological Science 31, 1553–1566.

Shea, J.J., 1988. Spear points from the Middle Paleolithic of the Levant. Journal of Field Archaeology 15, 441–450.

Shea, J.J., 1998. Neandertal and early modern human behavioral variability. Current Anthropology 39 (Supplement) S45–S78.

Shea, J.J, 2003. The Middle Paleolithic of the East Mediterranean Levant. Journal of World Prehistory 17, 313–394.

Shea, J.J., 2006. The origins of lithic projectile point technology: evidence from Africa, the Levant and Europe. Journal of Archaeological Science 33, 823–846.

Shea, J.J., 2008. The impact of projectile weaponry on late Pleisotcene hominin evolution. In: Hublin, J.-J., Richards, M.P. (Eds.), The Evolution of Hominid Diets: Integrating Approaches to the Study of Palaeolithic Subsistence. Springer, Dordrecht.

Shea, J.J., Brown, K.S., Davis, Z.J., 2002. Controlled experiments with Middle Paleolithic spear points: Levallois points. In: Mathieu, J.R. (Ed.), Experimental Archaeology: Replicating Past Objects, Behaviors and Processes. BAR Intern. Series 1035, Archeopress, Oxford, pp. 55–72.

Shott, M.J., 1997. Stones and shafts redux: the metric discrimination of chipped-stone dart and arrow points. American Antiquity 62, 86–101.

Singer, R., Wymer, J., 1982. The Middle Stone Age at Klasies River Mouth in South Africa. Chicago, Chicago University Press.

Soressi, M., 2002. Le Moustérien de tradition acheulénne du sud-ouest de la France. Ph.D. Dissertation, University of Bordeaux 1.

Soriano, S., Villa, P., Wadley, L., 2007. Blade technology and tool forms in the Middle Stone Age of South Africa: the Howiesons Poort and post-Howiesons Poort at Rose Cottage Cave. Journal of Archaeological Science 34, 681–703.

Speth, J.D., Johnson, G.A., 1976. Problems in the use of correlation for the investigation of tool kits and activity areas. In: Cleland, Ch.E. (Ed.), Cultural Change and Continuity: Essays in Honor of J.G. Griffin. Academic Press, New York, pp. 35–76.

Stiner, M.C., 1991. Food procurement and transport by human and non-human predator. Journal of Archaeological Science 18, 455–482.

Stiner, M.C., 1993. Small animal exploitation and its relation to hunting, scavenging and gathering in the Italian Mousterian. In: Peterkin, G.L., Bricker, H., Mellars, P.A. (Eds.), Hunting and Animal Exploitation in the Later Paleolithic and Mesolithic of Eurasia. Archaeological Papers of the American Anthropological Association, no. 4, pp. 107–125.

Stiner, M., 1994. Honor among thieves. A zooarchaeological study of Neandertal ecology. Princeton University Press, Princeton.

Stiner, M.C., Munro, N.D., Surovell, T.A., Tchernov, E., Bar-Yosef, O., 1999. Paleolithic population growth pulses evidenced by small animal exploitation. Science 283, 190–194.

Tagliacozzo, A., Cassoli, P.F., Curci, A., Fiore, I., 1999. Analisi tafonomica dei resti ossei del livello Alfa. In: Piperno, M. (Ed.), Notarchirico. Un sito del Pleistocene medio iniziale nel bacino di Venosa. Edizioni Osanna, Venosa, pp. 455–520.

Thibault, C., 1970. Recherches sur les terrains quaternaires du bassin de l'Adour. Thèse de Doctorat d'Etat, University of Bordeaux.

Thibault, C., 1976. Les civilisations du Paléolithique inférieur dans le Sud-Ouest. In: de Lumley, H. (Ed.), La Préhistoire Française, vol. 2. Editions CNRS, Paris, pp. 1048–1052.

Thieme, H., 1997. Lower Palaeolithic hunting spears from Germany. Nature 385, 807–810.

Thieme, H., 2000. Lower Palaeolithic hunting weapons from Schöningen, Germany – The Oldest Spears in the World. In: Dong, W. (Ed.), Proceedings of the 1999 Beijing International Symposium on Paleoanthropology, Acta Anthropologica Sinica 10 (Supplement) 136–143.

Thieme, H., Veil, S., 1985. Neue Untersuchungen zum eemzeitlichen Elefanten-Jagdplatz Lehringen, Ldkr. Verden. Die Kunde, N.F. 36, 11–58.

Thomas, D.H., 1978. Arrowheads and atlatl darts: how the stones got the shaft. American Antiquity 43, 461–472.

Turner, E., 1999. Lithic artefacts and animal bones in flood-plain deposits at Miesenheim I (Central Rhineland, Germany). In: Gaudzinski, S., Turner, E. (Eds.), The Role of Early Humans in the Accumulation of European Lower and Middle Palaeolithic Bone Assemblages. Habelt, Bonn, pp. 103–120.

Valladas, H., Wadley, L., Mercier, N., Froget, L., Tribolo, Ch., Reyss, J.L., Joron, J.L., 2005. Thermoluminesence dating on burnt lithics from Middle Stone Age layers at Rose Cottage Cave. South African Journal of Science 101, 169–174.

Veil, S., Plisson, H., 1990. The elephant kill-site of Lehringen near Verden on Aller, Lower Saxony (Germany). Unpublished manuscript in possession of S. Veil, Niedersächsisches Landesmuseum, Hanover, Germany.

Villa, P., 1990.Torralba and Aridos: elephant exploitation in Middle Pleistocene Spain. Journal of Human Evolution 19, 299–309.

Villa, P., 1991. Middle Pleistocene prehistory of Southwestern Europe. Journal of Anthropological Research 47 (2), 193–218.

Villa, P., 1996. Book review. The first Italians. Le industrie litiche del giacimento paleolitico di Isernia La Pineta. Lithic Technology 21, 71–79.

Villa, P., 2001. Early Italy and the colonization of Western Europe. Quaternary International 75, 113–130

Villa, P., 2004. Taphonomy and stratigraphy in European prehistory. Before Farming 1, 1–20.

Villa, P., Courtin, J., 1983. The interpretation of stratified sites: a view from underground. Journal of Archaeological Science 10, 267–281.

Villa, P., d'Errico, F., 2001. Bone and ivory points in the Lower and Middle Palaeolithic of Europe. Journal of Human Evolution 41, 69–112.

Villa, P., Lenoir, M., 2006. Hunting weapons of the Middle Stone Age and the Middle Palaeolithic: spear points from Sibudu, Rose Cottage and Bouheben. Southern African Humanities 18 (1), 89–122.

Villa, P., Mahieu, E., 1991. Breakage patterns of human long bones. Journal of Human Evolution 21, 27–48.

Villa, P., Soto, E., Pérez-González, A., Santonja, M., Mora, R., Parcerisas, Q., Sese, C., 2005a. New data from Ambrona (Spain): closing the hunting *versus* scavenging debate. Quaternary International 126–128, 223–250.

Villa, P., Delagnes, A., Wadley, L., 2005b. A late Middle Stone Age artifact assemblage from Sibudu (KwaZulu- Natal). Comparisons with the European Middle Paleolithic. Journal of Archaeological Science 32, 399–422.

Vitagliano, S., 1984. Nota sul Pontiniano della Grotta dei Moscerini, Gaeta (Latina). Atti della XXIV Riunione Scientifica dell'Istituto Italiano di Preistoria e Protostoria nel Lazio, Firenze, pp. 155–164.

Voorhies, M.R., 1969. Taphonomy and population dynamics of an early Pliocene vetrtebarte fauna, Knox County, Nebraska. Contributions to Geology Special Paper no. 1, University of Wyoming, Laramie.

Vrba, E.S., 1980. The significance of bovid remains as indicators of environment and predation patterns. In: Beherensmeyer, A.K., Hilla, A.P. (Eds.), Fossils in the Making. University of Chicago Press, Chicago, pp. 247–271.

Wadley, L., 1997. Rose Cottage Cave: Archaeological work 1987 to 1997. South African Journal of Science 93, 439–444.

Wadley, L., 2005. A typological study of the final Middle Stone Age stone tools from Sibudu Cave, KwaZulu-Natal. South African Archaeological Bulletin 60, 51–63.

Wadley, L., Jacobs, Z., 2006. Sibudu Cave: background to the excavations, stratigraphy and dating. Southern African Humanities 18 (1), 1–26.

Weinstock, J., 2000. Osteometry as a source of refined demographic information: sex-ratios of reindeer, hunting strategies and herd control in the Late Glacial site of Stellmoor, Northern Germany. Journal of Archaeological Science 27, 1187–1195.

Zilhao, J., 1998. Comments to Marean and Kim. Current Anthropology 39 (Supplement), S104–S105.

Zilhao, J., 2001. Middle Paleolithic settlement patterns in Portugal. In: Conard, N.J. (Ed.), Settlement Dynamics of the Middle Paleolithic and Middle Stone Age. Kerns, Tübingen, pp. 597–608.

Zilhao, J., 2006. Neandertals and moderns mixed, and it matters. Evolutionary Anthropology 15, 183–195.

6. Neanderthal and Modern Human Diet in Eastern Europe

John F. Hoffecker
Institute of Arctic and Alpine Research
University of Colorado at Boulder
Boulder, CO, USA
John.Hoffecker@colorado.edu

Keywords Neanderthals • modern humans • diet • Eastern Europe

Abstract The Late Pleistocene environments of Eastern Europe – especially the cool and dry environments of the East European Plain – offer a unique setting for comparison of Neanderthal and modern human diet. There are some taphonomic factors specific to this setting, however, including those related to the scarcity of natural shelters and distribution of woody plants on the central plain, which complicate the analysis of Late Pleistocene human diet. Neanderthals occupied southern upland areas, and sometimes the southwest and central plain, probably sustaining themselves primarily on the hunting of large mammals, which often included steppe species such as *Bison priscus* and *Saiga tatarica*. It is difficult to understand how recent hunter-gatherers in some of these environments (e.g., southwest plain during the Middle Pleniglacial) could have sustained themselves throughout the year on the basis of large mammal hunting without food storage. The key to Neanderthal survival in such habitats may have been the hunting of mammoth and rhinoceros – very large herbivores not available to recent hunter-gatherers in northern interior settings. Modern humans – who are present on the East European Plain as early as anywhere in Europe – broadened the diet to include small mammals, birds, and/or fish during the Middle Pleniglacial. Evidence of this shift, which seems to have been achieved through the design of novel food-getting technologies, is derived both from the faunal remains and stable isotope analysis of human bone. Modern humans occupied the loess-steppe habitat on the central plain during the Late Pleniglacial and often used their food debris as fuel.

Introduction

Eastern Europe provides a unique setting for the comparative study of diet between Neanderthals and modern humans. The environments of Eastern Europe – especially those of the immense East European Plain – are more continental than those of Western Europe. Winters are colder and biological productivity is generally lower. A similar pattern apparently prevailed during the Late Pleistocene and presented special challenges to both taxa. The contrasting responses of Neanderthals and modern humans to these challenges offer some insights that may not be accessible in other parts of western Eurasia.

Clockwise currents in the North Atlantic bring warm and moist air to Western Europe creating an *oceanic effect* on climate and biota (cf. Gamble, 1995). Western Europe enjoys the mildest winters in northern Eurasia. Mean January temperature in Paris is 3°C, but at the same latitude (48° North) in Eastern Europe (Volgograd) it falls to −7°C. Mean annual precipitation declines from 650 mm in Paris to 370 mm in Volgograd (http://weatherbase.com). Primary productivity also falls significantly in Eastern Europe, especially on the arid southern plain (cf. Archibold, 1995). Various lines of evidence, including paleopedology and paleobiology, suggest that the contrasts between Western and Eastern Europe were also present during the Pleistocene (Hoffecker, 2002a).

As a result of the oceanic effect, climate and productivity in northern Eurasia exhibit a west-east – as well as a south-north – gradient. The pattern seems to be reflected in the distribution of human settlement during the Pleistocene. The earliest occupation of areas above latitude 45° North appears to be confined to Western and Central Europe (Roebroeks and van Kolfschoten, 1995). The cold-adapted Neanderthals subsequently expanded eastward (not further north) into the East European Plain, and apparently as far as the Altai region of southwestern Siberia (Hublin, 1998; Goebel, 1999). However, their ability to occupy the East European Plain may have been restricted to warmer periods by winter temperatures and/or productivity levels (Hoffecker, 2005a: 55–59).

The landscapes of Eastern Europe introduce some local taphonomic factors that must be accounted for in the analysis of archaeological sites and faunal remains. The scarcity of natural shelters on the central East European Plain exercises a significant effect on the preservation and discovery of sites, which are confined to open-air localities and often recognized by the exposure of large mammal bones (typically mammoth). Although the loess-derived colluvium in which the

J.-J. Hublin and M.P. Richards (eds.), *The Evolution of Hominin Diets: Integrating Approaches to the Study of Palaeolithic Subsistence*, 87–98.

bones are usually buried provides a generally favorable geochemical medium, weathering rates are higher than in cave sediments and assemblages are often biased towards larger taxa and body parts. Also, the periodic scarcity of woody vegetation on the central plain has sometimes encouraged or necessitated the use of bone for fuel. Much of the large mammal food debris in these sites probably was consumed for fuel (e.g., Tarasov, 1979: 33).

East European Neanderthal Sites in Space and Time

The geographic distribution of Neanderthal sites in Eastern Europe seems to have varied during the Late Pleistocene. The pattern stands in sharp contrast to that of southwestern Europe, where Neanderthal settlement was more or less continuous throughout the later Middle and late Pleistocene (cf. Mellars, 1996). The pattern of spatial and temporal variation in the East European sites may provide important clues to Neanderthal climate tolerance and habitat preference (Hoffecker, 2005a).

Two problems present potential complications to the reconstruction of spatial and temporal variation in East European settlement. The first of these is that the absence of occupation in specific times and places is based on negative evidence and assumes that sites have not been obscured or destroyed. As noted above, however, sites on the East European Plain may be eroded or concealed by various factors that might have varied in intensity according to time or place (e.g., accumulation of mammoth bones at occupations). Secondly, with the exception of an isolated tooth at Rozhok I on the Sea of Azov coast (Praslov, 1968: 83–84) and scapula from riverine sediment on a tributary of the Don River (Shevyrev and Khrisanfova, 1984), Neanderthal skeletal remains are unknown on the East European Plain. Bones and teeth that may be firmly assigned to *Homo neanderthalensis* are confined to Crimea and Northern Caucasus (Hoffecker, 2002a: 88–91). Thus, Neanderthal presence on the East European Plain is based entirely on the classification of artifacts.

Assuming that most or all of the sites in Eastern Europe that contain assemblages of artifacts assigned to the Middle Paleolithic were occupied by Neanderthals, the earliest appearance of the latter seems to be during one of the late Middle Pleistocene interglacials (possibly MIS 9 age equivalent) near the confluence of the Severskii Donets and Don rivers (Praslov, 1968, 1995; Hoffecker, 2002a: 48–51). This is broadly consistent with estimates of a late Middle Pleistocene divergence (350–150 ka) of East European Neanderthals from the West European population on the basis of fossil DNA extracted from a partial skeleton in the Northern Caucasus (Ovchinnikov et al., 2000). Most traces of Neanderthal settlement in Eastern Europe post-date the Middle Pleistocene (Fig. 6.1).

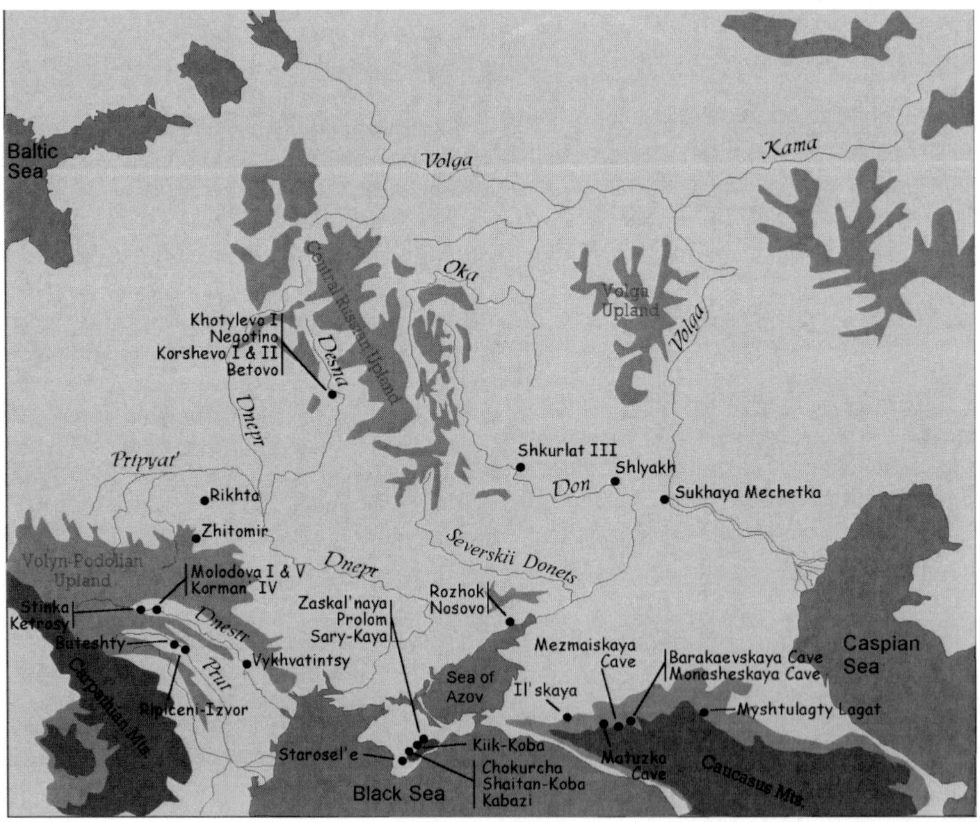

FIG. 6.1. Map of Eastern Europe illustrating the locations of Middle Paleolithic sites and Neanderthal fossil finds of the Late Pleistocene.

Many of the more reliably dated Middle Paleolithic sites on the central and southern East European Plain were occupied during the climatic optimum of the Last Interglacial (MIS 5e substage equivalent). The artifacts are associated with faunal remains that indicate climates warmer than those of the present day (e.g., Khotylevo I on the Desna River [Motuz, 1967; Zavernyaev, 1978]) or with a buried soil assigned to the Last Interglacial (e.g., Sukhaya Mechetka on the Lower Volga [Zamyatnin, 1961: 9–12]). The scapula from the Don River area also was recovered from deposits of the Last Interglacial climatic optimum (Shevyrev et al., 1979).

Some occupations on the central plain appear to date to cooler periods, such as the "Early Glacial" interval (MIS 5d-5a age equivalent). An example is Korshevo I & II on the Middle Desna River, where associated pollen-spore samples indicate boreal and mixed forest that is restricted today to more northern latitudes (Tarasov, 1989). And at least one occupation may date to the Middle Pleniglacial (MIS 3 age equivalent) on the Desna River at Betovo (Tarasov, 1989: 174). Overall, the Neanderthals seem to have been scarce or absent during very cold periods, when periglacial loess-steppe conditions prevailed on the central plain (especially during the Lower Pleniglacial or MIS 4 age equivalent) (Hoffecker, 2002a: 71–75).

A somewhat different picture is evident on the southwest region of the East European Plain. Sites apparently were occupied during MIS 5a-5d, such as Ripiceni-Izvor (Prut Valley) and Ketrosy (Dnestr Valley) (Carciumaru, 1980; Anisyutkin, 1981). The well known site at Molodova 5 contains occupation levels dating to both the end of the Early Glacial (MIS 5a?) and the cooler interstadial period correlated with MIS 3 (Haesaerts et al., 2003). Although traces of occupation are lacking in the Lower Pleniglacial loess at Molodova, the Neanderthals were clearly present in the Dnestr Valley during the generally cool Middle Pleniglacial. Winters in this region are milder today than those of the central plain, and trees seem to have survived in the southwest plain during even the coldest glacial periods (Hoffecker, 2002a).

In contrast to the pattern of fluctuating settlement on the plain, the southern upland margins of Eastern Europe were occupied during both warm and cold periods. The lengthy sequence at Matuzka Cave suggests a more or less continuous Neanderthal presence in the northwestern Caucasus from the end of the Middle Pleistocene onwards (Golovanova et al., 1990). A comparable sequence of occupations is reported from the open-air locality of Kabazi II in southwest Crimea (Chabai and Ferring, 1998), and the appearance of arctic fox (*Alopex lagopus*) in several Neanderthal occupations of this region indicates glacial conditions (Baryshnikov, 2006).

Reconstructing Neanderthal Diet in Eastern Europe

Information on Neanderthal diet from mammal remains in East European Plain sites is limited due to the problematic relationship between many of the remains and the occupants of the sites. The situation is better in the Northern Caucasus and Crimea, where the remains are less weathered and have been subject to some taphonomic analyses. The analyses indicate that here – as elsewhere in western Eurasia – the Neanderthals were consuming significant quantities of meat from large mammals. To date, there are no published results of the isotopic analysis of Neanderthal bone from Eastern Europe.

Mammoth (*Mammuthus primigenius*) dominates faunal assemblages from many sites in the southwest plain, including Ripiceni-Izvor, Ketrosy, Molodova 1, and Molodova 5 (Paunescu, 1965; David, 1981: 136; Chernysh, 1982: 88; Alekseeva, 1987: 154). Mammoth is also common at Khotylevo I and Sukhaya Mechetka on the central and southern plain (Vereshchagin and Kolbutov, 1957: 84; Zavernyaev, 1978: 29), and is present in most other site faunas. The percentage of mammoth bones and teeth in these sites has probably been inflated by weathering and differential preservation (combined with the high identifiability of even small fragments of mammoth).

Although the faunal remains at Khotylevo I may have accumulated by flowing water (cf. Zavernyaev, 1978: 31–35), most of these open-air sites are found on low terraces buried in loess and/or colluvium, where large bones are unlikely to have been concentrated by natural processes (Hoffecker, 2002a: 113). It is unclear, however, if they represent food debris – either hunted or scavenged. The bones and tusks might have been brought to sites like Molodova for other purposes (Alekseeva, 1987: 156). Few taphonomic data are available to address the problem.

Several other large mammals are relatively common in the East European Plain sites, including steppe bison (*Bison priscus*), horse (*Equus caballus*), and red deer (*Cervus elaphus*). Less common taxa include woolly rhinoceros (*Coelodonta antiquitatis*), giant deer (*Megaceros giganteus*), and reindeer (*Rangifer tarandus*) (cf. Hoffecker, 2002a: 113–115). It is widely assumed that these remains represent food debris; although – as in the case of mammoth – there is little supporting evidence (some tool cut-marks are reported on the highly fragmented bones from Ketrosy [David, 1981: 135]).

Sites in the southern plain typically contain high percentages of bison; this apparently reflects the local predominance of steppe habitat in this region. Steppe bison predominate at Sukhaya Mechetka, Kodak (Lower Dnepr Basin), and Rozhok, as well as at Il'skaya in the foothills of the northwestern Caucasus (Vereshchagin and Kolbutov, 1957: 84; Boriskovskii and Praslov, 1964: 22; Praslov, 1968: 71; Hoffecker et al., 1991: 121). At Il'skaya I, prime-age individuals dominate the age profile for adults, and the bones exhibit intense breakage in green condition. Prime-age adults also predominate among the small sample of teeth from Sukhaya Mechetka (Hoffecker, 2002a: 117), while juveniles are reportedly common at Rozhok I (Praslov, 1968: 84).

Most of our conclusions about Neanderthal diet on the East European Plain are deduced from the environmental setting and a reconstruction of Neanderthal physiology. The caloric demands of the Neanderthals are thought to have been high – comparable to those of modern peoples living in the Arctic. In the relatively cool and dry habitats of the East European Plain, they are likely to have been heavily dependent on a diet high in protein and fat derived from animals. The isotopic analysis of Neanderthal bones from Western and Central Europe indicates that – even in milder climate settings where plant foods were probably more abundant – they derived the overwhelming proportion of protein from animal foods (Bocherens et al., 1999; Richards et al., 2001). At the same time, the isotope data suggest that – unlike most modern forgers in northern environments – the Neanderthals derived little protein from freshwater aquatic sources (i.e., fish and/or waterfowl) (Richards et al., 2001).

In the southern upland margins of Eastern Europe, the analysis of well preserved faunal assemblages from natural shelters provides more direct evidence for Neanderthal diet. On the higher slopes of the Northern Caucasus, Neanderthals hunted sheep (*Ovis orientalis*) and goat (*Capra caucasica*) at Mezmaiskaya Cave (1,300 m asl). The bones are heavily fractured with percussion marks and tool cuts, and prime-age adults predominate (Baryshnikov et al., 1996; Cleghorn, 2005). In Crimea, wild ass (*Equus hydruntinus*), horse (*Equus caballus*), saiga (*Saiga tatarica*), and giant deer (*Megaceros giganteus*) are the most common large mammal taxa in Neanderthal occupations (Vereshchagin and Baryshnikov, 1980: 32; Kolosov et al., 1993: 73–74). It is interesting to note that here, as in other parts of Europe (cf. Straus, 1982), the cave and rockshelter occupations yield much evidence for carnivore activity (Enloe et al., 2000; Cleghorn, 2005). To a greater extent than modern humans of the later Upper Paleolithic, Neanderthals shared their sites with competitors.

The Problem of Neanderthal Diet in Eastern Europe

The data from Eastern Europe conform to the pattern seen in other parts of western Eurasia. The Neanderthals appear to have met their high caloric requirements by consuming large quantities of animal protein and fat, derived primarily from the hunting of large terrestrial mammals (cf. Chase, 1986; Gaudzinski, 1996; Mellars 1996). Compared to low-latitude environments, scavenging opportunities were limited. Although evidence for consumption of smaller vertebrates and plant foods is scarce (Madella et al., 2002), Neanderthals sometimes supplemented their diet with other foods (e.g., shellfish, carnivores [Stiner, 1994]). They were central-place foragers who brought food back to their occupation areas; the short distances over which they transported raw materials (rarely exceeding 100 km) suggest that their foraging range may have been small relative to recent hunter-gatherers in northern interior settings.

A few Neanderthal sites are dominated by the remains of one species (Gaudzinski, 1996). The most striking examples from Eastern Europe are Starosel'e in Crimea (wild ass) and Il'skaya (steppe bison) in the Northern Caucasus (Hoffecker et al., 1991; Burke, 2000). Although such sites have sometimes been interpreted as evidence of mass-kill hunting tactics (e.g., Vereshchagin, 1967: 373), none of them appears to represent a bone bed created by a herd drive or surround. A natural catastrophe – such as a flash flood in a narrow ravine or canyon – may explain these occurrences (Burke, 2000: 330), and the use of mass-kill techniques by the Neanderthals remains to be demonstrated.

The East European sites provide a perspective on Neanderthal diet in a cooler and drier environment. Steppe taxa are more common in these sites – most notably bison in the southern plain and saiga in Crimea. To these may be added horse, wild ass, mammoth, and woolly rhinoceros. The cold winters and reduced productivity of habitats on the East European Plain probably were a challenge to the Neanderthals, and they may have abandoned the central plain during very cold periods. While they presumably could have tolerated summer conditions, their apparently limited mobility might have precluded even seasonal use of the central plain during these periods (i.e., requiring movements of more than 250 km). The ability of the Neanderthals to cope with very cold winters and comparatively low productivity is perhaps best illustrated by their occupation of the southwest plain during all but the coldest periods (e.g., Lower Pleniglacial).

In fact, Neanderthal occupation of areas like the Dnestr Valley during the Middle Pleniglacial poses a problem for the study of their diet. The southwest plain currently experiences a January mean temperature of about −5°C and annual precipitation of 600 mm. During interstadial phases of the Middle Pleniglacial, winter temperatures and moisture were probably lower than today. Paleoenvironmental data indicate open woodland dominated by pine that supported a variety of large mammals, including both steppic and woodland forms. Recent hunter-gatherers in northern interior settings practiced a broad-based economic strategy that coped with seasonal fluctuations in resource availability by harvesting waterfowl in early summer, exploiting fish runs in the autumn, constructing trap-lines for small mammals in winter, storing various foods in the fall, and other foraging tactics. Most hunter-gatherers in latitudes above 50° North derive at least one third of their diet from fishing (Kelly, 1995: 66–73).

It is difficult to imagine how a group of recent foragers could have sustained themselves in a setting like the Dnestr Valley solely on the basis of large mammal hunting. The Neanderthals do not seem to have systematically exploited small mammals, birds, or fish. Their sites also lack evidence for any food storage facilities. The problem is exacerbated by their apparent lack of technology for effective cold protection (i.e., heated shelters, tailored fur clothing) and its implications for energy loss. Furthermore, the faunal remains from caves and rockshelters suggest intense competition with carnivores for large mammals (cf. Anisyutkin, 2001).

The solution to this problem may lie in the fact that the Neanderthals had access to two very large mammals that were not available to recent hunter-gatherers in northern latitudes – mammoth and rhinoceros. Bocherens et al. (2005) compared isotopic signatures for Neanderthal and contemporaneous hyena specimens from Western Europe, and concluded that the former probably consumed more mammoth and rhinoceros and less reindeer than the latter (presumably reflecting niche separation between the two consumers of large mammals). Mammoth and rhinoceros remains are present in Neanderthal occupations of Western Europe, but typically in low numbers (Patou-Mathis, 2000). Nevertheless, there is some evidence for the hunting of mammoth and rhinoceros in this part of Europe (e.g., Cavarretta et al., 2001), most notably at La Cotte de St. Brelade on the island of Jersey (Scott, 1989). The low quantities of mammoth and rhinoceros in the West European sites – many of which probably represent habitation sites in natural shelters – may reflect comparatively limited retrieval of the massive bones of these taxa from kill-butchery locations (Bocherens et al., 2005: 82).

By contrast, mammoth remains are often abundant in East European sites, as already noted (Fig. 6.2); rhinoceros remains are not abundant, but are usually present in modest numbers. Published information on mammoth remains from selected Middle Paleolithic sites of Eastern Europe is presented in Table 6.1. It may be noted that high numbers of mammoth bones and tusks are not confined to open-air sites (which might represent locations where carcasses were butchered), but also are found in some natural shelters, such as Chokurcha I in Crimea (Vereshchagin and Baryshnikov, 1980: 32). Fragmented mammoth remains sometimes are associated with large stones that might have been used to break the bones (e.g., Anisyutkin, 1981: Fig. 6; Kolosov, 1983: 45–50). Nevertheless, compelling evidence for the consumption of either taxon by Neanderthals in Eastern Europe is thus far lacking – this is a task for future research. The problem is similar to that of investigating whale bones in late prehistoric sites of the Arctic coast (cf., McCartney, 1979).

Mammoth and rhinoceros may have been an important component of the Neanderthal niche and the key to explaining how they survived cold winters in regions like the Dnestr Valley during the Middle Pleniglacial. While evidence of storage facilities in Neanderthal sites is absent, mammoth and rhinoceros carcasses could have been allowed to freeze in the autumn, providing several months' supply of protein and fat for a Neanderthal group. At the same time, an emphasis on these animals would have reduced resource competition with hyaenas (Bocherens et al., 2005). An emphasis on mammoth and rhinoceros may be especially evident in an East European context because of their higher visibility in the archaeological record of the region (i.e., preponderance of open-air sites) and/or an intensified reliance on these taxa in colder and drier habitats.

Modern Human Colonization of Eastern Europe

Modern humans appeared in northern Eurasia (above latitude 45° North) prior to 40 cal ka (40,000 calibrated years ago), and they may have occupied the central East European Plain as early as 45 cal ka. On the Don River at Kostenki, the lowest occupation levels underlie a volcanic tephra now identified

FIG. 6.2. Mammoth remains in a Middle Paleolithic open-air site: Excavation of Layer 6 at Il'skaya 2 in 1987 (photograph by G.F. Baryshnikov).

TABLE 6.1. Remains of Mammoth (*Mammuthus primigenius*) in selected Middle Paleolithic sites/levels of Eastern Europe (From Bonch-Osmolovskii, 1940: 69; Vereshchagin and Kolbutov, 1957; Zavernyaev, 1978: 29; Vereshchagin and Baryshnikov, 1980: 32; Chernysh, 1982; Kolosov, 1983: 45–50; Alekseeva, 1987; Kolosov et al., 1993; Baryshnikov and Hoffecker, 1994; Anisyutkin, 2001: 121–129).

Site	Level	NISP	MNI	Skeletal part representation	Demographic profile	Breakage/damage
Buzduzhany 1 rockshelter	All layers	64	3	No data	No data	Heavily fragmented
Chokurcha I cave	–	3,041	20	Crania, tusks, pelvis, limb bones	All ages represented	Limb bones fractured by stones (?)
Il'skaya 2	Layer 4b	135	4	Mandibles, tusks, vertebrae (cervical, caudal), ribs (often complete), scapulae, limb bones, extremities	Sub-adults and adults present	Traces of carnivore gnawing (hyaena?)
Il'skaya 2	Layer 5	101	3	Most skeletal parts present, except for skulls and teeth; metapodials common	No data	Limb bones fragmented
Il'skaya 2	Layer 6	81	3	Mandible, tusks, vertebrae, rib fragments, limb-bone epiphyses	Adult female present	Most bones intact
Ketrosy	–	192	4	Tusks (numerous), mandibles, vertebrae, ribs, pelves, limb bones, extremities	Primarily sub-adults and adults	Bones are heavily fragmented
Khotylevo I	–	Common	?	No data	No data	Most bones fragmented
Kiik-Koba rockshelter	Upper	42	2	Teeth, other parts present	No data	No data
Molodova 1	Layer 4	Abundant	38	Crania, tusks, limb bones	Young predominate	Limb bones fragmented
Molodova 5	Layer 11	Abundant	5	Crania, tusks, ribs, limb bones, extremities	No data	Limb bones and pelves intact; epiphyses missing
Molodova 5	Layer 12	Abundant	?	Tusks, scapulae, vertebrae, pelvis, rib fragments (numerous)	No data	Heavily fragmented
Prolom 2 rockshelter	Layer 1	51	19	No data	7 adults, 1 young, 8 juvenile, 3 embryonic	No data
Prolom 2 rockshelter	Layer 2	28	11	Tusks, vertebra, others	3 adults, 4 young, 3 juveniles	No data
Starye Duruitory rockshelter	Layer 3	34	4	No data	No data	No data
Sukhaya Mechetka	–	51	?	1 tooth, 3 vertebrae, 2 ribs, 5 limb bone epiphyses, 40 shaft fragments	No data	Limb bones fragmented
Zaskal'naya 5 rockshelter	Layer 2	122	7	Tusks, mandibles, limb bones	Young present	Generally fragmented (associated with large hammer-stones?)

NISP = Number of Identified Specimens
MNI = Minimum Number of Individuals

at the Campanian Ignimbrite Y5 (ca. 40 cal ka) (Anikovich et al., 2007). These levels may be correlated with several minor warm and cold oscillations of the Middle Pleniglacial between 45 and 40 cal ka. Other traces of modern human settlement in Eastern Europe in this time range are scarce and problematic (e.g., Molodova 5, Layer 10a [Chernysh, 1987: 25; Haesaerts et al., 2003: 183]), and human skeletal remains are confined to two isolated teeth at Kostenki, but modern humans (*Homo sapiens*) are assumed to be present on the basis of the artifacts (i.e., stone bladelets, bone and ivory implements, ornaments, possible figurative art, and materials imported from more than 500 km [cf. Sinitsyn, 2002]) (Fig. 6.3).

The scarcity of sites in this time range is due in part to their low visibility on the East European Plain. The occupations at Kostenki and Molodova are deeply buried and lack any surface clues. Their existence is known only because later Upper Paleolithic occupations (i.e., less than 28 cal ka) comprising large quantities of debris overlie the older remains. The debris of the younger occupations – including mammoth bone – provided for their discovery. After years of excavation, archaeologists eventually reached the deeply buried levels containing the early occupations.

In some respects, it seems odd that modern humans initially occupied the central East European Plain, given their origins

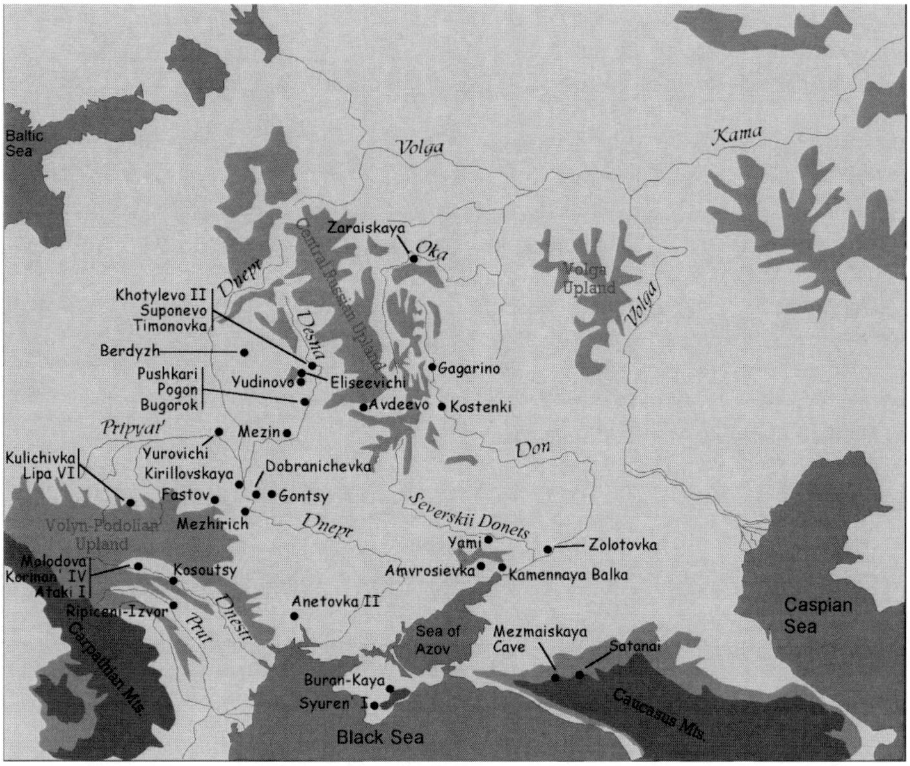

FIG. 6.3. Map of Eastern Europe illustrating locations of Upper Paleolithic sites and modern human fossil finds.

in the lower latitudes. More complete skeletal remains dating to 40–30 cal ka exhibit body proportions typical of modern peoples of the tropical zone (Trinkaus, 1981; Holliday, 1999). The modern human population of the Middle Pleniglacial must have been highly susceptible to cold injury in northern Eurasia, and the East European Plain probably saw the coldest winters in mid-latitude Europe (the current January mean temperature at Kostenki is −8°C). One factor that might have encouraged initial settlement here is the scarcity or absence of Neanderthals on the central plain at this time (Hoffecker, 2005a: 76–82). There would have been significant niche overlap and competition for resources between the two taxa.

If modern humans were capable of exploiting an "empty niche" on the central East European Plain during the Middle Pleniglacial, they must have possessed a means of coping with the low temperatures and/or low productivity of the region lacked by the Neanderthals. This presumably lay in the ability of modern humans to create novel and sometimes complex technologies – a pattern that emerges in the African Middle Stone Age and becomes strikingly apparent during the subsequent course of the Upper Paleolithic in Eurasia (Hoffecker, 2005b).

If *recursion* – the ability to generate potentially infinite combinations from a finite set of elements – is a "core property" of language (cf. Hauser et al., 2002: 1571), the same property is evident in the archaeological record of modern humans over time. The pattern is evident not only with respect to art, but also is manifested in technology. The East European archaeological record contains early traces of novel technologies (or their effects) for cold protection and expansion of dietary breadth among modern humans (e.g., eyed needles [Hoffecker, 2002a: 160–161]).

Modern Human Diet in Eastern Europe: Middle Pleniglacial

Much of the evidence for modern human diet in Eastern Europe during the Middle Pleniglacial is derived from Kostenki. The sites here have yielded many thousands of vertebrate bones and teeth – many of them relatively well preserved for an open-air setting – and they indicate hunting of horse, reindeer, and other large mammals, as well as procurement of smaller mammals, and possibly some birds and fish (Vereshchagin and Kuz'mina, 1977). Some isotope data also are available on human bone from a late Middle Pleniglacial context, and suggest heavy consumption of freshwater aquatic foods (Richards et al., 2001).

The Kostenki sites are located along large tributary ravines incised into the limestone bedrock of the high west bank of the Don River. Occupations of Middle Pleniglacial age are buried in slope and spring deposits that overlie alluvium of the second terrace. Springs are active today in the ravines, and probably were a major attraction for animals and people during the Middle Pleniglacial. In places where spring activity was limited (e.g., Kostenki 1), weak soil profiles developed during this period (Holliday et al., 2007).

Most of the large mammal bones in the Middle Pleniglacial occupations are horse. At Kostenki 14, more than 2,000 horse bones from a minimum of 19 individuals were recovered from Layer II (ca. 32–28 cal ka). All age groups were represented, and many bones were found in anatomical sequences (vertebrae and extremities) and/or with cut marks (Rogachev, 1957: 78–80). Similar concentrations of horse remains have been excavated at other Kostenki localities (cf. Hoffecker, 2002a: 181–183), including a recently discovered bone bed composed of both horse and reindeer bones in Layer III at Kostenki 12 (>40 cal ka). These bones are weathered, heavily fragmented, and seem to have been partially dispersed and sorted by slopewash (Hoffecker et al., 2005).

The concentrations of horse bone at Kostenki probably are the remains of hunted prey, and the occurrence of anatomical groups suggests that at least some of them represent kill-butchery locations. The distribution of ages reported from Kostenki 14 may reflect the collective hunting of groups of horses. It remains unclear if the Neanderthals were capable of hunting groups of large mammals (Hoffecker 2002a: 117); this might represent a change in foraging tactics – although not one necessarily tied to technological innovation on the part of modern humans.

Although information on fauna from other Middle Pleniglacial sites on the East European Plain is limited, horse also dominates the assemblage from Layer 10 at Molodova 5 (ca. 30 cal ka) in the Dnestr Valley (Alekseeva, 1987: 154), and from Layer 3 at Brynzeny I in the Prut Valley (David and Ketraru, 1970: 23–25). Like Kostenki, Molodova is associated with spring activity (N.K. Anisyutkin, 2006, personal communication), and might have been used in a similar way. Other large mammals turn up in the Middle Pleniglacial sites, and reindeer is especially common in some assemblages. Mammoth and bison are present, but not common (Hoffecker, 2002a: 181).

More interesting than the large mammals in modern human occupations of the Middle Pleniglacial is the evidence for procurement of small mammals, birds, and fish. Virtually all of this evidence has been collected from Kostenki. The large Don hare (*Lepus tanaiticus*) is the most abundant mammal in several assemblages, including Layer II at Kostenki 8 and Layer III at Kostenki 14 (Vereshchagin and Kuz'mina, 1977: 104–107). Arctic fox and wolf are also common in several occupation levels, and wolf actually dominates the assemblage from Layer II (below the tephra) at Kostenki 17 (Vereshchagin and Kuz'mina, 1977: 108). In 2003, I observed cut marks on specimens of all three taxa from an occupation level associated with the Y5 tephra at Kostenki 14 (see Fig. 6.4).

The remains of birds and fish are also reported from some Middle Pleniglacial occupation levels at Kostenki, but they are comparatively rare. A modest quantity of bird remains was recovered from Kostenki 8 (Layer II), Kostenki 14 (Layer IV), Kostenki 15, and Kostenki 16 (Layer II), and isolated bird bones have been reported from several other occupations

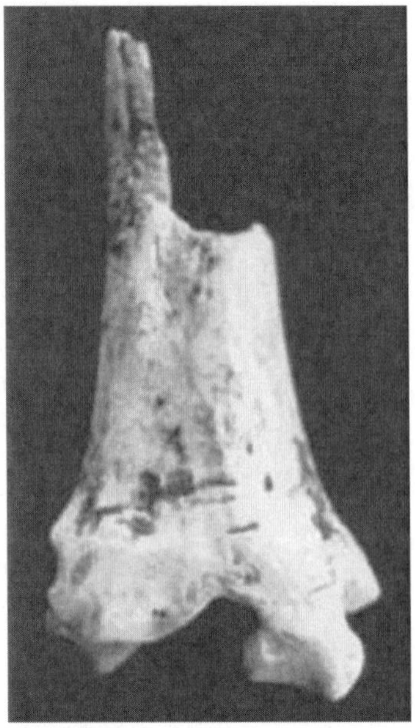

FIG. 6.4. Small mammal remains from an early Upper Paleolithic context: distal tibia of Don hare (*Lepus tanaiticus*) exhibiting stone tool cut marks recovered from the volcanic tephra layer at Kostenki 14 on the Don River (dating to 39 cal ka) (photograph by J.F. Hoffecker).

(Vereshchagin and Kuz'mina, 1977: 104–108). As many as 36 fish bones were identified in Layer II at Kostenki 8 (ca. 30 cal ka), and isolated fish remains also were found in other occupations (Vereshchagin and Kuz'mina, 1977: 104). In general, birds and fish are likely to be under-represented at Kostenki due to the effects of weathering in open-air settings and recovery bias (i.e., lack of sediment-sieving in earlier excavations).

The Kostenki data probably indicate systematic harvesting of hares, which may have been important as both a source of food and of pelts. Arctic fox and wolf also seem to have been harvested in significant numbers, and may have been used primarily for furs. The remains of birds and fish are more problematic – at least some of them might have been introduced to the sites by natural processes. Isotopic analysis of a human bone from Layer III at Kostenki 1 (ca. 30 cal ka) nevertheless indicated a heavy dependence on freshwater aquatic foods (Richards et al., 2001), which might have included waterfowl and/or fish.

In sum, there is good evidence for modern human expansion of dietary breadth to a variety of smaller vertebrates not systematically exploited by the Neanderthals (cf. Stiner et al., 1999). This expansion seems to have begun with the initial settlement of modern humans in Eastern Europe, but becomes more evident after 40 cal ka. As noted above, it may have been essential to modern human colonization of environments like the central plain during the Middle Pleniglacial – where the Neanderthals seem to have been scarce or absent at this time. Ethnographic data indicate that hares, waterfowl, and fish often play a major role in the diet and economy of northern interior peoples (e.g., Nelson, 1973; Kelly, 1995). Their procurement in significant numbers, however, requires a variety of specialized and sometimes complex technology in the form of nets, snares, traps, throwing darts, and other gadgetry. The ability of modern humans to design such technology probably was a critical advantage over the Neanderthals (Stiner et al., 1999).

Modern Human Diet in Eastern Europe: Upper Pleniglacial

As in other parts of northern Eurasia, modern humans exhibit a significant expansion in the size and complexity of their habitation areas during the Upper Pleniglacial (MIS 2 age equivalent) on the East European Plain (Klein, 1969). This almost certainly reflects an increase in residential group size, but also probably reflects an increase in population density and carrying capacity (cf. Jochim, 2002). The archaeological record of the Upper Pleniglacial illustrates – more effectively than that of the Middle Pleniglacial – the process of cumulative technological change among modern humans, and its impact on economy and population (Hoffecker, 2005b).

The Upper Pleniglacial record of the East European Plain also provides a stark contrast with that of the Neanderthals during the Lower Pleniglacial. Conditions were similar during the two glacial intervals (i.e., MIS 2 and 4 age equivalents), and both are represented by loess units. Instead of the complex Upper Pleniglacial settlements of modern humans, the Neanderthals appear to be scarce or absent altogether.

A quantum jump in site size and complexity is evident in Central and Eastern Europe at the end of the Middle Pleniglacial (cf. Svoboda et al., 1996). The change seems to take place somewhat later in Western Europe (Straus, 1995). Sites on the central east European Plain are especially large, and include massive feature complexes assigned to the *Eastern Gravettian* technocomplex dating to ca. 25 cal ka (e.g., Grigor'ev, 1993; Amirkhanov, 1998; Hoffecker, 2002b), and sites comprising multiple structures of mammoth bone (*Epigravettian*) that postdate the cold maximum of the last glacial (e.g., Pidoplichko, 1976; Abramova, 1993). Sites from both periods contain large storage pits, and some evidence for both warm and cold season occupation. New technologies include fired ceramics and basketry (Vandiver et al., 1989; Soffer, 2004) and – by 18 cal ka – evidence for the domestication of canids (Sablin and Khlopachev, 2002).

The Upper Pleniglacial sites of the central plain contain abundant evidence for the use of bone fuel – presumably reflecting a largely treeless landscape – and this phenomenon introduced an unusual taphonomic agent into the analysis of faunal assemblages. Much of the large mammal foods debris accumulated at these sites probably was consumed as fuel (e.g., Grigor'ev, 1967: 348). The most common large mammal bones in most sites belong to mammoths – apparently often collected from natural accumulations for use as material for constructing shelters (Klein, 1973: 53–54; Pidoplichko, 1976; Soffer, 1985). In the southwest plain, where wood fuel was available, the large mammal remains are probably more representative of the diet; they are dominated by reindeer and horse, although some mammoth remains are present (Chernysh, 1959; Borziyak, 1993; Hoffecker, 2002a: 238–239).

During the Upper Pleniglacial, modern humans apparently continued to pursue a northern interior economy based heavily on smaller mammals, birds, and probably fish. Small mammals such as hare and fox are often abundant in faunal assemblages. Roughly 15,000 bones of arctic fox were recovered from Eliseevichi on the Sudost' River (Vereshchagin and Kuz'mina, 1977). Bird remains are more common than in the Middle Pleniglacial occupations, and typically represent willow ptarmigan (*Lagopus lagopus*), black grouse (*Lyrurus tetrix*), teal (*Anas* sp.), and goose (*Anser* sp.) (Klein, 1973: 57). The remains of fish are not common, but isotope data on a human bone from Kostenki 18 (ca. 25 cal ka) suggest heavy consumption of freshwater foods, which may include fish (Richards et al., 2001). The southwest plain sites have yielded numerous grinding stones, thought to reflect significant plant use (Borziac et al., 1997).

Conclusions

1. The Neanderthals were the first hominins to occupy the East European Plain (characterized by the coldest winters and lowest productivity in mid-latitude Europe), apparently as a consequence of their tolerance for low winter temperatures and a diet high in protein and fat.
2. As documented in other parts of Eurasia, the Neanderthals in Eastern Europe probably derived most of their protein and fat from the hunting of large mammals. Scavenging opportunities probably were rare, relative to low latitudes. Although they may have supplemented their diet with small mammals, birds, and other items, large mammals seem to have been the primary protein source.
3. It is difficult to explain how recent hunter-gatherers could have occupied environments comparable to those of the southwest East European Plain – where Neanderthals were present during the cool Middle Pleniglacial – with an economy based primarily on the hunting of large mammals. The Neanderthals may have depended heavily on the hunting of mammoth and rhinoceros – two massive herbivores not available to modern hunter-gatherers in northern interior settings. Significant hunting of mammoth and rhinoceros has yet to be documented in analyses of faunal remains from Neanderthal sites, but new isotope data suggest that these taxa may indeed have played an important role in the diet (Bocherens et al., 2005).
4. Modern humans occupied the central plain of Eastern Europe as early as 45–40 cal ka (Middle Pleniglacial), and exhibit an expansion of dietary breadth relative to the Neanderthals. Modern humans systematically harvested small mammals – probably both for food and pelts – and may have also consumed significant numbers of birds and

fish. Modern humans probably achieved this expansion of diet primarily through the development of new technologies (e.g., nets, snares, throwing darts).

5. During the Upper Pleniglacial, modern human settlements – especially on the central plain – increased significantly in size and complexity, and probably reflect greater population density and higher carrying capacity. As during the preceding period, their diet was based on a broad array of large and small mammals, birds, and probably some fish. At least some of the apparent increase in population density may be attributable to further technological innovation (e.g., spear-throwers, hunting dogs).

Acknowledgments. I am grateful to J.-J. Hublin and M. Richards for inviting me to present this paper at a conference on hominid diet held in May 2006 at the Max-Planck Institut fur evolutionare Anthropologie in Leipzig. My thanks also to H. Bocherens for sharing his thoughts with me on Neanderthal diet. Helpful comments on an earlier draft were received from C. Cameron, A. Joyce, S. Holen, S. Lekson, P. Sheets, P. Villa, and three anonymous reviewers.

References

Abramova, Z.A., 1993. 2 examples of terminal Paleolithic adaptations. In: Soffer, O., Praslov, N.D. (Eds.), From Kostenki to Clovis, Plenum Press, New York, pp. 85–100.

Alekseeva, L.I., 1987. Teriofauna mnogosloinoi stoyanki Molodova V. In: Ivanova, I.K., Tseitlin, S.M. (Eds.), Mnogosloinaya paleoliticheskaya stoyanka Molodova V: lyudi kamennogo veka i okruzhayushchaya sreda. Nauka, Moscow, pp. 153–162.

Amirkhanov, Kh.A. (Ed.), 1998. Vostochnoi Gravett. Nauchnyi Mir, Moscow.

Anikovich, M.V., Sinitsyn, A.A., Hoffecker, J.F., Holliday, V.T., Popov, V.V., Lisitsyn, S.N., Forman, S.L., Levkovskaya, G.M., Pospelova, G.E., Kuz'mina, I.E., Burova, N.D., Goldberg, P., Macphail., R.E., Giaccio, B., Praslov, N.D., 2007. Early Upper Paleolithic in eastern Europe and implications for the dispersal of modern humans. Science 315, 223–226.

Anisyutkin, N.K., 1981. Arkheologicheskoe izuchenie must'erskoi stoyanki Ketrosy. In: Praslov, N.D. (Ed.), Ketrosy: Must'erskaya Stoyanka na Srednem Dnestre, Nauka, Moscow, pp. 7–53.

Anisyutkin, N.K., 2001. Must'erskaya epokha na Yugo-Zapade Russkoi ravniny. Evropeiskii Dom, St. Petersburg.

Archibold, O.W., 1995. Ecology of World Vegetation. Chapman & Hall, London.

Baryshnikov, G., 2006. Late Pleistocene arctic fox (*Alopex lagopus*) from Crimea, Ukraine. Quatern. Int. 142–143, 208–217.

Baryshnikov, G., Hoffecker, J.F., 1994. Mousterian hunters of the NW Caucasus: preliminary results of recent investigations. J. Field Archaeol. 21, 1–14.

Baryshnikov, G., Hoffecker, J.F., Burgess, R.L., 1996. Zooarchaeology and palaeontology of Mezmaiskaya Cave, northwestern Caucasus. J. Archaeol. Sci. 23, 313–335.

Bocherens, H., Billiou, D., Mariotti, A., Patou-Mathis, M., Otte, M., Bonjean, D., Toussaint, M., 1999. Palaeoenvironmental and palaeodietary implications of isotopic biogeochemistry of Last Interglacial Neanderthal and mammal bones in Scladina Cave (Belgium). J. Archaeol. Sci. 26, 599–607.

Bocherens, H., Drucker, D.G., Billiou, D., Patou-Mathis, M., Vandermeersch, B., 2005. Isotopic evidence for diet and subsistence pattern of the Saint-Cesaire I Neanderthal: review and use of a multi-source mixing model. J. Hum. Evol. 49, 71–87.

Bonch-Osmolovskii, G.A., 1940. Paleolit Kryma, vol.1, Grot Kiik-Koba. Akademiya nauk SSSR, Moscow.

Boriskovskii, P.I., Praslov, N.D., 1964. Paleolit bassein Dnepra i Priazov'ya. Svod Arkheologicheskikh Istochnikov A 1–5.

Borziac, I.A., Allsworth-Jones, P., French, C., Medyanik, S.I., Rink, W.J., Lee, H.K., 1997. The Upper Palaeolithic site of Ciuntu on the Middle Pruth, Moldova: a multidisciplinary study and reinterpretation. Proc. Prehist. Soc. 63, 285–301.

Borziyak, I.A., 1993. Subsistence practices of Late Paleolithic groups along the Dnestr River and its tributaries. In: Soffer, O., Praslov, N.D. (Eds.), From Kostenki to Clovis, Plenum Press, New York, pp. 67–84.

Burke, A., 2000. The view from Starosele: faunal exploitation at a Middle Palaeolithic site in western Crimea. Int. J. Osteoarchaeol. 10, 325–335.

Carciumaru, M., 1980. Mediul Geografic in Pleistocenul Superior si Culture Paleolitice din Romania. Editura Academiei Republicii Socialiste Romania Bucharest.

Cavarretta, G., Giola, P., Mussi, M., Palambo, M.R. (Eds.), 2001. The world of Elephants. Proceedings of the first international Congress, Roma.

Chabai, V.P., Ferring, C.R., 1998. Kabazi II: introduction. In: Marks, A.E., Chabai, V.P. (Eds.), The Middle Paleolithic of Western Crimea, vol. 1, ERAUL 84, pp. 167–200.

Chase, P.G., 1986. The Hunters of Combe Grenal: Approaches to Middle Paleolithic Subsistence in Europe. British Archaeological Reports International Series S-286.

Chernysh, A.P., 1959. Pozdnii paleolit Srednego Pridnestrov'ya. Trudy Komissii po Izucheniyu Chetvertichnogo Perioda 15, 5–214.

Chernysh, A.P., 1982. Mnogosloinaya paleoliticheskaya stoyanka Molodova I. In: Goretskii, G.I., Ivanova, I.K. (Eds.), Molodova I: Unikal'noe Must'erskoe Poselenie na Srednem Dnestre, Nauka, Moscow, pp. 6–102.

Chernysh, A.P., 1987. Etalonnaya mnogosloinaya stoyanka Molodova V. arkheologiya. In: Ivanova, I.K., Tseitlin, S.M. (Eds.), Mnogosloinaya Paleoliticheskaya Stoyanka Molodova V: Lyudi Kamennogo Veka i Okruzhayushchaya Sreda, Nauka, Moscow, pp. 7–93.

Cleghorn, N., 2005. A zooarchaeological perspective on the Upper to Middle Paleolithic transition at Mezmaiskaya Cave, NW Caucasus, Russia. Ph.D. dissertation, State University of New York at Stony Brook.

David, A.I., 1981. Ostatki mlekopitayushchikh iz raskopok paleoliticheskoi stoyanki Ketrosy. In: Praslov, N.D. (Ed.), Ketrosy: Must'erskaya Stoyanka na Srednem Dnestre, Nauka, Moscow, pp. 135–142.

David, A.I., Ketraru, N.A., 1970. Fauna mlekopitayushchikh paleolita Moldavii, in Fauna Kainozoya Moldavii, AN MSSR, Kishinev, pp. 3–53.

Enloe, J.G., David, F., Baryshnikov, G., 2000. Hyenas and hunters: zooarchaeological investigations at Prolom II Cave, Crimea. Int. J. Osteoarchaeol. 10, 310–324.

Gamble, C., 1995. The earliest occupation of Europe: the environmental background. In: Roebroeks, W., van Kolfschoten, T. (Eds.), The Earliest Occupation of Europe, University of Leiden, Leiden, pp. 279–295.

Gaudzinski, S., 1996. On bovid assemblages and their consequences for the knowledge of subsistence patterns in the Middle Palaeolithic. Proc. Prehist. Soc. 62, 19–39.

Goebel, T., 1999. The Pleistocene colonization of Siberia and peopling of the Americas: an ecological approach. Evol. Anthropol. 8, 208–227.

Golovanova, L.V., Levkovskaya, G.M., Barychnikov, G.F., 1990. Le nouveau site mousterien en Grotte de Matouzka, Caucase Septentrional (resultats des fouilles de 1985–1987). L'Anthropologie 94, 739–762.

Grigor'ev, G.P., 1967. A new reconstruction of the above-ground dwelling of Kostenki. Curr. Anthropol. 8, 344–349.

Grigor'ev, G.P., 1993. The Kostenki-Avdeevo archaeological culture and the Willendorf-Pavlov-Kostenki-Avdeevo cultural unity. In: Soffer, O., Praslov, N.D. (Eds.), From Kostenki to Clovis, Plenum Press, New York, pp. 51–65.

Haesaerts, P., Borziak, I., Chirica, V., Damblon, F., Koulakovska, L., van der Plicht, J., 2003. The East Carpathian loess record: a reference for the middle and late Pleniglacial stratigraphy in central Europe. Quaternaire 14, 163–188.

Hauser, M.D., Chomsky, N., Fitch, W.T., 2002. The faculty of language: what is it, who has it, and how did it evolve? Science 298, 1569–1579

Hoffecker, J.F., 2002a. Desolate Landscapes: Ice-Age Settlement of Eastern Europe. Rutgers University Press, New Brunswick.

Hoffecker, J.F., 2002b. The Eastern Gravettian 'Kostenki Culture' as an arctic adaptation. Anthropol. Pap. Univ. Alaska, NS 2(1), 115–136.

Hoffecker, J.F., 2005a. A Prehistory of the North: Human Settlement of the Higher Latitudes. Rutgers University Press, New Brunswick.

Hoffecker, J.F., 2005b. Innovation and technological knowledge in the Upper Paleolithic of northern Eurasia. Evol. Anthropol. 14, 186–198.

Hoffecker, J.F., Baryshnikov, G.F., Potapova, O.R., 1991. Vertebrate remains from the Mousterian site of Il'skaya I (northern Caucasus, USSR): new analysis and interpretation. J. Archaeol. Sci. 18, 113–147.

Hoffecker, J.F, Kuz'mina, I.E., Anikovich, M.V., Popov, V.V., 2005. Taphonomy of an early Upper Paleolithic bone bed at Kostenki 12. In: Anikovich, M.V. (Ed.), Early Upper Paleolithic Research at Kostenki, Russian Academy of Sciences, St. Petersburg, pp. 161–176.

Holliday, T.W., 1999. Brachial and crural indices of European Late Upper Paleolithic and Mesolithic humans. J. Hum. Evol. 36, 549–566.

Holliday, V.T., Hoffecker, J.F., Goldberg, P., Macphail, R.I., Forman, S.L., Anikovich, M., Sinitsyn, A., 2007. Geoarchaeology of the Kostenki-Borshchevo sites, Don River, Russia. Geoarchaeol. Int. J. 22, 183–230.

Hublin, J.-J., 1998. Climatic changes, paleogeography, and the evolution of the Neandertals. In: Akazawa, T., Aoki, K., Bar-Yosef, O. (Eds.), Neandertals and Modern Humans in Western Asia, Plenum Press, New York, pp. 295–310.

Jochim, M., 2002. The Upper Palaeolithic. In: Milisauskas, S. (Ed.), European Prehistory: A Survey, Kluwer Academic/Plenum Publishers, New York, pp. 55–113.

Kelly, R.L., 1995. The Foraging Spectrum: Diversity in Hunter-Gatherer Lifeways. Smithsonian Institution Press, Washington, DC.

Klein, R.G., 1969. Man and Culture in the Late Pleistocene: A Case Study. Chandler, San Francisco.

Klein, R.G., 1973. Ice-Age Hunters of the Ukraine. University of Chicago Press, Chicago.

Kolosov, Yu.G., 1983. Must'erskie stoyanki raiona Belogorska. Naukova Dumka, Kiev.

Kolosov, Yu.G., Stepanchuk, V.N., Chabai, V.P., 1993. Rannii Paleolit Kryma. Naukova Dumka, Kiev.

Madella, M., Jones, M.K., Goldberg, P., Goren, Y., Hovers, E., 2002. The exploitation of plant resources by Neanderthals in Amud Cave (Israel): the evidence from phytolith studies. J. Archaeol. Sci. 29, 703–719.

McCartney, A.P. (Ed.), 1979. Archaeological Whale Bone: A Northern Resource. Univ. Arkansas Anthropol. Pap. No. 1.

Mellars, P., 1996. The Neanderthal Legacy: An Archaeological Perspective from Western Europe. Princeton University Press, Princeton.

Motuz, V.M., 1967. Chetvertichnye mollyuski iz Khotylevskogo nizhnepaleoliticheskogo mestonakhozhdeniya Bryanskoi Oblasti. Byulleten' Komissii po Izucheniyu Chetvertichnogo Perioda 33, 150–154.

Nelson, R.K., 1973. Hunters of the Northern Forest. University of Chicago Press, Chicago.

Ovchinnikov, I.V., Gotherstrom, A., Romanova, G.P., Kharitonov, V.M., Lidon, K., Goodwin, W., 2000. Molecular analysis of Neanderthal DNA from the Northern Caucasus. Nature 404, 490–493.

Patou-Mathis, M., 2000. Neandertal subsistence behaviours in Europe. Int. J. Osteoarchaeol. 10, 379–395.

Paunescu, A., 1965. Sur la succession des habitats paleolithiques et postpaleolithiques de Ripiceni-Izvor. Dacia 9, 5–31.

Pidoplichko, I.G., 1976. Mezhirichskie Zhilishcha iz Kostei Mamonta. Naukova Dumka, Kiev.

Praslov, N.D., 1968. Rannii Paleolit Severo-Vostochnogo Priazov'ya i Nizhnego Dona. Nauka, Leningrad.

Praslov, N.D., 1995. The earliest occupation of the Russian Plain: A short note. In: W. Roebroeks and T. van Kolfschoten (Eds.), The Earliest Occupation of Europe, University of Leiden, Leiden, pp. 61–66.

Richards, M.P., Pettitt, P.B., Stiner, M.C., Trinkaus, E., 2001. Stable isotope evidence for increasing dietary breadth in the European mid-Upper Paleolithic. Proc. Natl. Acad. Sci. 98, 6528–6532.

Roebroeks, W., van Kolfschoten, T. (Eds.), 1995. The Earliest Occupation of Europe. University of Leiden, Leiden.

Rogachev, A.N., 1957. Mnogosloinye stoyanki Kostenkovsko-Borshevskogo raiona na Donu i problema razvitiya kul'tury v epokhy verkhnego paleolita na Russkoi Ravnine. Materialy i Issledovaniya po Arkheologii SSSR 59, 9–134.

Sablin, M.V., Khlopachev, G.A., 2002. The earliest Ice Age dogs: evidence from Eliseevichi I. Curr. Anthropol. 43(5), 795–799.

Scott, K., 1989. Mammoth bones modified by humans: evidence from La Cotte de St. Brelade, Jersey, Channel Islands. In: Bonnichsen, R., Sorg, M.H. (Eds.), Bone Modification, Center for the Study of the First Americans, Orono, pp. 335–346.

Shevyrev, L.T., Khrisanfova, E.N., 1984. Nakhodka ostankov iskopaemogo cheloveka v verkhnem pleistotsene tsentra Russkoi Ravniny (mestonakhozhdenie Shkurlat III). Voprosy Antropologii 73, 69–71.

Shevyrev, L.T., Raskatov, G.I., Alekseeva, L.I., 1979. Shkurlatskoe mestonakhozhdenie fauny mlekopitayushchikh mikulinskogo vremeni. Byulleten' Komissii po Izucheniyu Chetvertichnogo Perioda 49, 39–48.

Sinitsyn, A.A., 2002. Nizhnie kul'turnye sloi Kostenok 14 (Markina gora) (raskopki 1998-2001 gg.). In: A.A. Sinitsyn, V. Ya. Sergin, and J.F. Hoffecker (Eds.), Kostenki v kontekste paleolita Evrazii, Russian Academy of Sciences, St. Petersburg, pp. 219–236.

Soffer, O., 1985. The Upper Paleolithic of the Central Russian Plain. Academic Press, San Diego.

Soffer, O., 2004. Recovering perishable technologies through use wear on tools: preliminary evidence for Upper Paleolithic weaving and net making. Curr. Anthropol. 45, 407–413.

Stiner, M.C., 1994. Honor Among Thieves: A Zooarchaeological Study of Neandertal Ecology. Princeton University Press, Princeton.

Stiner, M.C., Munro, N.D., Surovell, T.A., Tchernov, E., Bar-Yosef, O., 1999. Paleolithic population growth pulses evidenced by small animal exploitation. Science 283, 190–194.

Straus, L.G., 1982. Carnivores and cave sites in Cantabrian Spain. J. Anthropol. Res. 38, 75–96.

Straus, L.G., 1995. The Upper Paleolithic of Europe: an overview. Evol. Anthropol. 4(1), 4–16.

Svoboda, J., Lozek, V., Vlcek, E., 1996. Hunters Between East and West: The Paleolithic of Moravia. Plenum Press, New York.

Tarasov, L.M., 1979. Gagarinskaya stoyanka i ee mesto v paleolite Evropy. Nauka, Leningrad.

Tarasov, L.M., 1989. Periodizatsiya paleolita basseina Verkhnei Desny, in Chetvertichnyi Period: Paleontologiya i Arkheologiya, Shtiintsa, Kishinev, pp. 166–175.

Trinkaus, E., 1981. Neanderthal limb proportions and cold adaptation. In: Stringer, C. (Ed.), Aspects of Human Evolution, Taylor & Francis, London, pp. 187–224.

Vandiver, P., Soffer, O., Klima, B., Svoboda, J., 1989. The origins of ceramic technology at Dolni Vestonice, Czechoslovakia. Science 246, 1002–1008.

Vereshchagin, N.K., 1967. Primitive hunters and Pleistocene extinction in the Soviet Union. In: Martin, P., Wright, H.E. (Eds.), Pleistocene Extinctions, Yale University Press, New Haven, pp. 365–398.

Vereshchagin, N.K., Baryshnikov, G.F., 1980. Mlekopitayushchie predgornogo severnogo Kryma v epokhu paleolita. Trudy Zoologicheskogo Instituta AN SSSR 93, 26–49.

Vereshchagin, N.K., Kolbutov, A.D., 1957. Ostatki zhivotnykh na must'erskoi stoyanke pod Stalingradom i stratigraficheskoe polozhenie paleoliticheskogo sloya. Trudy Zoologicheskogo Instituta AN SSSR 22, 75–89.

Vereshchagin, N.K., Kuz'mina, I.E., 1977. Ostatki mlekopitayushchikh iz paleoliticheskikh stoyanok na Donu i Verkhnei Desne. Trudy Zoologicheskogo Instituta AN SSSR 72, 77–110.

Zamyatnin, S.N., 1961. Stalingradskaya paleoliticheskaya stoyanka. Kratkie Soobshcheniya Instituta Arkheologii 82, 5–36.

Zavernyaev, F.M., 1978. Khotylevskoe Paleoliticheskoe Mestonakhozhdenie. Nauka, Leningrad.

7. Hominin Subsistence Patterns During the Middle and Late Paleolithic in Northwestern Europe

Sabine Gaudzinski-Windheuser
Paleolithic Research Unit
Römisch-Germanisches Zentralmuseum
Schloss Monrepos
56567 Neuwied Germany
gaudzinski@rgzm.de

and

Laura Niven
Paleolithic Research Unit
Römisch-Germanisches Zentralmuseum
Schloss Monrepos
56567 Neuwied Germany
and Max Planck Institute for Evolutionary Anthropology
Department of Human Evolution
Deutscher Platz 6, 04103 Leipzig, Germany
laura.niven@eva.mpg.de

Keywords Middle/Upper Paleolithic • large mammal hunting • reindeer • diet breadth

Abstract The aim of this paper is to more clearly classify Middle Paleolithic subsistence tactics by considering this evidence against an Upper Paleolithic background, where we discern a clearer picture of human subsistence tactics. Therefore, a diachronous comparative analysis of reindeer assemblages from northwestern European archaeological sites was undertaken. Differences in exploitation strategies become clearly visible for the late Upper Paleolithic, which can be interpreted to partly reflect the demands of elaborate settlement dynamics, evidence of which we especially lack for the Middle Paleolithic. Because of these differences in social networking strategies between Middle and Upper Palaeolithic groups, it seems highly likely that subsistence behavior involving the careful selection of large mammal resources was particularly crucial to maintaining high foraging return rates during the Middle Palaeolithic.

J.-J. Hublin and M.P. Richards (eds.), *The Evolution of Hominin Diets: Integrating Approaches to the Study of Palaeolithic Subsistence*, 99–111.

Introduction

It is widely accepted that as early as the Lower Paleolithic of northwestern Europe, hominins regularly hunted large mammals. The German site of Schöningen, dating to 400–300 ka, yielded ample proof of this in the form of wooden spears, fashioned from spruce and pine. Interpreted as javelins, these artifacts were found *in situ* amidst the remains of at least 20 horses (Thieme, 1999; Rieder, 2000).

Although the excellent preservation at Schöningen is unique in the archaeological record for this period, the site reflects a general trend seen in these early faunal assemblages: the almost exclusive exploitation of large game and the generally scarce presence – or even absence – of small herbivores, birds, and fish in faunal assemblages from Lower and Middle Paleolithic sites. In the early days of Pleistocene zooarchaeological research, Neanderthals were perceived to have focused on small- and medium-sized herbivores, since their acquisition strategies were presumably opportunistic; cooperative group hunting was not considered to be part of Neanderthals' behavioral repertoire (Binford, 1984, 1989). However, following more than 20 years of research methods developed for closely analyzing faunal assemblages from archaeological sites, a different picture emerges. Evidence from Middle Paleolithic sites instead points to Neanderthal foraging strategies directed towards hunting of large game that involve a variety of tactics, but always focus on "prime" animals and parts thereof.

In the following paper, this perspective is presented in a general and synthetic overview of faunal assemblages from Middle Paleolithic sites that are characteristic of this period in northwestern Europe. Different large mammal hunting tactics used by hominins are discussed, and evidence for the exploitation of small game is also considered.

Evidence from the Middle Paleolithic is then reviewed in comparison to the Upper Paleolithic record, where we discern a clearer picture of human subsistence. There is general consensus that a definite broadening of human diet breadth becomes visible during the Upper Paleolithic; during this time, the exploitation of small game and fish played an important role in human subsistence strategies (e.g., Richards et al., 2001; Stiner, 2005). The character of the subsistence changes from the Middle to the Upper Paleolithic is still unclear. For example, future research should focus on whether Upper Paleolithic humans' hunting strategies of large mammals changed along with the increased exploitation of small game, fish, and birds.

Lastly, we consider changes in subsistence behaviors from the Middle to Upper Paleolithic by evaluating hominin hunting strategies and utilization of reindeer (*Rangifer tarandus*). Because this taxon is ubiquitous in the archaeological record in question and generally represented by large sample sizes, a diachronic comparison of results from zooarchaeological studies of late Middle Paleolithic as well as early and late Upper Paleolithic reindeer assemblages allows us to assess a variety of issues regarding large mammal utilization, including whether the focus on energy-rich large mammal resources that characterizes Middle Paleolithic subsistence is maintained in the same way into and throughout the Upper Paleolithic. We hope to contribute to the ongoing pursuit of a better understanding of the Neanderthal dietary niche.

Summarizing the Middle Paleolithic Faunal Record

Middle Paleolithic subsistence in northwestern Europe has been well covered in a number of recent regional syntheses (e.g., Boyle, 2000; Conard and Prindiville, 2000; Patou-Mathis, 2000; Grayson and Delpech, 2006). Many of the faunal assemblages considered in these contexts are characterized by broad species diversity, revealing no clear patterning that could be translated into particular behavioral patterns. Moreover, the lack of a seasonal signal in the timing of hunting events is apparent, perhaps emphasizing the high flexibility in Middle Paleolithic subsistence. However, one pattern that does emerge among Middle Paleolithic faunal assemblages from numerous cave and open-air sites is the predominance of a single prey species, often represented by a minimum number of individuals (hereafter MNI) of up to 100 animals or more. These faunal assemblages are associated with lithic artifacts, and chronologically cover the earlier part of the Middle Paleolithic, their numbers increasing significantly from MIS 5 onwards (Gaudzinski, 2005).

The dominant taxon varies from large herbivores that live in herds such as bovids, equids, and cervids (Chase, 1986; Gaudzinski, 1996; Valensi, 2000) to solitary animals such as rhinoceros (Bratlund, 1999). Hominin exploitation of the remains is shown by unambiguous cutmarks and hammerstone-induced impact notches on bones. What is particularly interesting is that although there are clearly similarities between a variety of thanatocoenoses, the exploitation strategies used by hominins are not necessarily the same. A selection of examples will better illustrate this point.

Numerous large bovid-dominated faunas are known from the earlier part of the Weichselian throughout Europe, for example Mauran (Farizy et al., 1994) and La Borde (Jaubert et al., 1990) in France, and Wallertheim in Germany (Gaudzinski, 1995). The associated lithic assemblages mainly consist of unmodified flakes, denticulates, and simple scrapers produced from locally available raw materials. All of these localities have been interpreted as kill sites. The age structure of bovids at these sites indicates a dominance of individuals at the height of their reproductive life, i.e., prime-aged animals. Where data are available, the sex composition reflects a typical herd during the rutting season. Hominin exploitation of bovid carcasses follows a uniform strategy, involving the breaking of bones for marrow extraction (Gaudzinski, 1996).

Data from bovid ethology, as well as the prime adult-dominated age composition of the thanatocoenoses, suggest that these assemblages might reflect selective, controlled, and systematic exploitation of these animals, presumably over a long period of time. In turn, this suggests repeated, communal hunting events involving the participation of several hominin individuals (following Driver, 1990). Overall, there are no indications for seasonally restricted hunting, nor an apparent preference for a particular topographic position of the sites (Gaudzinski, 1996).

In contrast to these repeated, selective, and controlled communal hunting episodes documented at bovid-dominated sites, the reindeer-dominated site of Salzgitter-Lebenstedt (Germany) saw short-term mass death scenarios characterized by *unselective* killing of entire reindeer herds. At Salzgitter-Lebenstedt, considered to be ca. 50 ka, the largest group of animals in the assemblage – adult reindeer males – died around September, as discerned from the development stage of male antlers, as well as by means of tooth eruption stages in young individuals. The reindeer remains display abundant evidence of butchery, including filleting of meat and systematic processing of bones for marrow. During marrow extraction, a clear selection against subadults, as well as against bones with low marrow content occurred. For example, of the 38 reindeer metacarpals examined, 34 bones survived in a complete condition. In contrast, only 11 metatarsals from 163 metatarsal fragments altogether represent complete bones. In summary, the evidence from Salzgitter-Lebenstedt reflects the seasonal, yet unselective killing of reindeer, followed by the selective exploitation of nutritionally-rich skeletal parts (Gaudzinski and Roebroeks, 2000, 2003).

Based on the examples presented above, a large game hunting strategy aimed at better quality animals was clearly characteristic of Neanderthals. This is also underlined by isotopic studies in which Neanderthals have been described as top-level carnivores (e.g., Richards et al., 2000; Bocherens et al., 2001). In order to obtain these high quality nutritional resources, different hunting tactics involving considerable flexibility were employed, which could have been influenced by animal ethology.

For the Mediterranean Basin, it appears that exploitation of small game constituted a key element of Middle and Upper Paleolithic hominin subsistence (Stiner et al., 2000). The general consensus is that with the exception of easily captured game, such as tortoises and shellfish, Middle Paleolithic foragers only seldom obtained small game. This pattern changes in the Upper Paleolithic, when a wide variety of small game, as well as birds, increasingly became a larger component of human diets. A range of factors potentially influenced these economic decisions, including (1) the declining availability of higher ranked prey as well as human population growth (Stiner, 2005); (2) the nutritional benefits of small mammalian, avian, and marine resources for human reproductive fitness (Hockett and Haws, 2003; Haws and Hockett, 2004); and (3) in the case of rabbit (*Oryctolagus cuniculus*), its periodic abundance in some areas and the ease of its capture (Hockett and Haws, 2002; Jones, 2006).

For the Middle Paleolithic of northwestern Europe, the inclusion of small mammals and birds in the hominin diet remains largely invisible, a situation that has several possible explanations. First, in most cave and open-air sites, we see an underrepresentation of small animals and birds in archaeofaunas; this could be a factor of taphonomic or excavation biases, or alternatively, it is a true reflection of the faunal spectrum. In either scenario, the low numbers of these animals is unexpected, considering the number of excavated sites. In other cases, small mammalian and avian remains are present in the assemblage, but attributing these remains to the human dietary spectrum, as opposed to the natural background fauna, is problematic.

Until recently (e.g., Brugal and Desse, 2004), many such small mammal and bird assemblages have not received attention due to these interpretive challenges. Only in cases where small animal, bird and fish remains were particularly numerous could one argue for an anthropic origin, e.g., the bird accumulation at Orgnac (France, OIS 9) (Mourer-Chauviré, 1979; Moigne and Barsky, 1999); or Grotte du Lazaret, layer UA 25 (France, MIS 6), where the remains of at least 25 individual *Oryctolagus cuniculus* (rabbit) were found among the remains of large mammals. Although the rabbit bones lacked clear anthropogenic traces, the assemblage is characterized by a dominance of burned long bone diaphyses whose spatial distribution corresponds to zones interpreted as centers of anthropic activities. This evidence, coupled with a lack of carnivore involvement suggests that the rabbits were part of the human prey assemblage at Grotte du Lazaret. Whether the similarly-large bird assemblage from the same find horizon (n = 960, MNI = 42) is also anthropogenic remains unclear (de Lumley et al., 2004).

Accumulations of bird remains, as well as large amounts of broken egg shells, are also reported from the late Middle Paleolithic site of Buhlen (Germany), attributed to MIS 5c. In level III, 23 different avian species could be determined from a total of 102 bones; 60% of this assemblage represents Galliformes, or large ground birds such as grouse (Eastham, 1998). Unambiguous anthropogenic traces are lacking, as are those from carnivores or predatory birds. Skeletal element representation shows a predominance of meat-rich proximal limbs and portions of the axial skeleton, having been dismembered without leaving butchering traces (Eastham, 1998).

Fish remains have been recovered from numerous Middle Paleolithic sites, e.g., Grotte Vaufrey and Pair non Pair in France (Rigaud, 1989; Cleyet-Merle, 1990); Taubach and Sesselfelsgrotte in Germany (Soergel, 1922; Richter, 1997) and Tata in Hungary (Kretzoi, 1964). Only at Grotte Vaufrey (France), a site probably dating to the penultimate Interglacial, have 71 fish remains, mostly vertebrae from six different species, been interpreted as opportunistically exploited by hominins (Le Gall, 1988).

Tortoises constituted an important part of hominin diet during the Levantine Middle Paleolithic (Stiner et al., 2000; Speth and Tchernov, 2002). However, in northwestern Europe indications for foraging on lower vertebrates are lacking, as terrapins found in limnic and travertine deposits of the Last Interglacial are not directly associated with lithic remains, and seem to have died naturally. Only in rare cases is a direct association of *Emys orbicularis* (turtle) with lithic remains evident, such as in the Last Interglacial sites Grotte Trou Félix and Caverne des Grands Malades in Belgium (Wenzel, 1998).

By the late Middle Paleolithic, hominin involvement with small game, birds, and fish becomes more apparent, as seen in the increasing frequency of anthropogenically modified bones from these species within the archaeological record. For example, definitive evidence for exploitation of birds was found at Salzgitter-Lebenstedt, where cutmarks on a right metacarpus of *Cygnus* sp. (swan) as well as a humerus of *Anas* sp. (indeterminate duck) could be identified. The Mousterian horizons from Grotta du Fumane, Italy (Fiore et al., 2004) contain a wide variety of avian taxa. Although clear evidence of butchery is mostly lacking on the bird remains, many of the species could have potentially served as rich sources of meat and fat, particularly waterfowl and grouse. Raptors are also numerous, and a talon from *Aquila chrysaetos* (golden eagle) represents the sole example of cutmarks by hominins at the site (Fiore et al., 2004: Fig. 7.2a). Interestingly, the talon of a medium-sized raptor from the lowest Mousterian level at Pech de l'Azé IV, France (Niven, in press, 2009) exhibits cutmarks in the exact location of the Fumane specimen – on the articular surface – possibly indicating removal of the keratin sheath from this phalanx bone (Fiore et al., 2004: 280). Fiore et al. (2004) propose that the cutmarked talon from Fumane was used as an ornament, an intriguing suggestion that awaits additional support from more samples. The analyses of faunal remains from the extensive Mousterian deposits at Pech de l'Azé IV are still underway, though preliminary results indicate that birds were exploited in small numbers by Neanderthals at this site.

As we can see from the discussion above, the number of sites reflecting unambiguous evidence of Neanderthals' utilization of small game and birds has increased in recent years, and ongoing research will undoubtedly contribute more to our understanding of these activities. However, it remains unclear what role these animals actually played in the diet of Neanderthals; i.e., whether they fulfilled nutritional needs as well as practical ones (e.g., feathers, bone for ornaments and tools). Moreover, can we assume that some birds such as waterfowl and ground birds represent hominin prey, while raptors do not?

The challenge in understanding the role of small game and birds in prehistoric diets is due in part to generally small sample sizes, difficulty in interpreting modifications and butchery strategies of humans versus other predators, and the use of small faunal remains to fashion personal ornaments and tools. For example, it has recently been argued, for sites in the French Pyrénées, that abundant ornaments and tools made from bones and teeth of small prey do not necessarily reflect a broader dietary spectrum for humans (Costamagno and Laroulandie, 2004). It is certainly plausible that small game and birds served both nutritional and practical purposes, considering that the remains of large ungulates as well as woolly mammoths were exploited for a wide range of uses other than nutritional ones, such as skins, tendons, bone, ivory, and antler.

Examples of extensive small game and bird utilization during the early and middle Upper Paleolithic are similarly scarce, and hence reflect a comparable pattern to what we see for the late Middle Paleolithic. The evidence for a true transformation in the human dietary spectrum becomes more robust in the Magdalenian period, approximately 15,000 calendar years BP. At this time, fish (Le Gall, 2003), birds (Laroulandie, 2003), and small game (e.g., Hockett and Bicho, 2000; Fontana, 2003; Cochard and Brugal, 2004) become important components of the human diet and in turn, visible in the archaeological record across much of continental Europe. Synthetic studies show that subsistence strategies without seasonal focus on both small and large mammals, as well as fish and birds, were embodied in a broader settlement system (Costamagno, 2003; Gaudzinski and Street, 2003).

In summary, there remains much to understand about the role of small game and birds in the diets of Neanderthals and the first anatomically modern humans in northwestern Europe. Although the outline above has shown that the dietary niches of Middle and Upper Paleolithic hominins were not strictly limited to large mammals, it is these prey that represent the best samples in the archaeological record with which to evaluate subsistence in the past.

Material and Methods

In this paper, we evaluate reindeer assemblages from both cave and open-air settings, representing a variety of time periods (Table 7.1). We focus on *Rangifer tarandus* since it is a common member of Upper and Middle Paleolithic assemblages, although examples are rarer in the latter period; only three late Middle Paleolithic assemblages could be included here.

Criteria for choosing the assemblages rested primarily on whether MNI (minimum number of individuals) and MNE (minimum number of skeletal elements) values were available.

TABLE 7.1. Summary of sites, relevant stratigraphic layers and details about the reindeer assemblages discussed in text.

	S-L	PoF2	PoF3	Cés	Vog	Ren	Flag	Sol	Verb	RB	P27
Cultural attribution	MP	MP	MP	Aurig	Aurig	Aurig	Grav	Magd	Magd	Magd	Magd
Setting	Open	Cave	Cave	Abri	Cave	Cave	Cave	Open	Open	Cave	Open
Reindeer NISP	2,130	2,532	1,461	2,823	1,633	1,509	1,801	271	4,522	446	896
Reindeer MNI	86	48	20	42	28	12	43	9	40	10	19
Season of death	Fall	Fall-wi	Fall-wi	Fall-wi-spr	Fall	Fall-wi	Winter	Wi-spr	Fall	nd	Fall
Dominant age	Adult	Juv + adult	Juv + adult	Juv + adult	Adult	Juv + adult	Subad-adu	Subad-adu	Juv-subad	nd	Juv + adult
Dominant sex	Male	Male	Male	Female	Female	nd	Female	nd	Male	nd	Even
Marrow exploitation	Extensive	Extensive	Extensive	Extensive	Extensive	Extensive	Extensive	Minimal	Extensive	Extensive	Extensive
Preservation bias	Minimal	Middle	Middle	Middle	Middle	nd	Minimal	Middle	Minimal	Middle	Middle

Site abbreviations and stratigraphy: S-L (Salzgitter-Lebenstedt, Germany); PoF2 (Le Portel Couche F2, France); PoF3 (Le Portel Couche F3, France); Cés (St. Césaire EJF, France); Vog (Vogelherd, AH IV/V, Germany); Ren (Grotte du Renne VII [Arcy-sur-Cure], France); Flag (Abri du Flageolet Couche V, France); Sol (Solutré P16, France); Verb (Verberie II-1, France); RB (Rond-du-Barry Couche F2, France); P27 (Pincevent Section 27 – IV-20, France). Cultural attributions: MP (Middle Palaeolithic); Aurig (Aurignacian); Grav (Gravettian); Magd (Magdalenian).

References for data: S-L (Gaudzinski and Roebroeks, 2000); PoF2–PoF3 (Gardeisen, 1997); Cés (Morin, 2004); Vog (Niven, 2006); Ren (David and Poulain, 2002); Flag (Enloe, 1993); Sol (Turner, 2002); Verb (Enloe, 2004); RB (Costamagno, 1999); P27 (David and Enloe, 1993; Enloe et al., 1994; Enloe, 2003). No data other than NISP and MAU available for P36.

nd = data not available.

From MNE data we are able to derive MAU (minimal number of animal units) values, which are a useful analytical unit. Although we agree with arguments that NISP (number of identified specimens) can be as analytically valuable as MNE for discerning patterns of human subsistence (e.g., Grayson, 1984), we chose to include only those assemblages with published raw MNE and/or MAU data for reasons that are explained below. In other words, the assemblages we have chosen for our focused study are by no means the only ones available. In fact, we regret the exclusion of samples from the French Middle Paleolithic site of Les Pradelles (Costamagno et al., 2006), for which there are no raw data; the German Magdalenian sites of Stellmoor (Spiess, 1979; Bratlund, 1994), Lommersum (Hahn, 1989), and Schussenquelle (Schuler, 1994), or the rich cave assemblages spanning the Middle to Late Upper Paleolithic of southwestern France (Grayson and Delpech, 2003) from which we were unable to derive MNE values.

Additional data considered to be important to our study include season of death, mortality profile, and herd ratio, although in some cases these data were not available (e.g., Pincevent 36). Lastly, although adequate sample size was deemed important, some smaller samples were included in our study based on the availability of the other datasets.

Among our assemblages, reindeer either dominates the faunal spectrum or is second in frequency following horse (Table 7.2). Although mammoth is nearly equal in number to reindeer and horse in the assemblage from Vogelherd, there

TABLE 7.2. Summary of fauna from sites discussed in text. Microfauna and unidentifiable material excluded in all cases; antler excluded from St. Césaire, Rond du Barry and Pincevent. NISP of taxa other than reindeer not available for Pincevent. Site name abbreviations defined in Table 7.1.

Taxon/family NISP	S-L	PoF2	PoF3	Cés	Vog	Ren	Flag	Sol	Verb	RB	P27	P36
Lepus sp. (hare)	–	–	–	4	27	39	–	–	–	222	–	–
Marmota marmota (marmot)	–	–	–	–	–	3	–	–	–	–	–	–
Canis lupus (wolf)	1	111	31	9	38	9	–	30	–	68	–	–
Alopex lagopus (arctic fox)	–	–	–	3	–	–	–	–	–	200	–	–
Vulpes vulpes (red fox)	–	428	119	–	–	–	–	–	–	–	–	–
Vulpes/Alopex (fox)	–	–	–	23	20	223	–	7	–	–	–	–
Mustelidae	–	–	1	1	–	–	–	–	–	3	–	–
Ursus spelaeus (cave bear)	–	5	2	–	120	415	–	–	–	–	–	–
Ursus arctos (brown bear)	–	–	–	–	2	–	–	–	–	–	–	–
Ursus sp. (indeterminate bear)	–	–	–	–	–	–	–	–	–	2	–	–
Crocuta crocuta spelaea (cave hyaena)	–	37	–	–	17	30	–	–	–	–	–	–
Crocuta crocuta (spotted hyena)	–	–	–	1	–	–	–	–	–	–	–	–
Meles meles (badger)	–	2	4	1	–	–	–	–	–	27	–	–
Gulo gulo (wolverine)	–	–	–	–	1	–	–	4	–	–	–	–
Panthera leo spelaea (cave lion)	–	2	–	2	4	1	–	–	–	–	–	–
Felis silvestris (wild cat)	–	–	–	–	3	–	–	–	–	–	–	–
Felis sp. (indeterminate small felid)	–	–	–	–	–	–	–	–	–	5	–	–
Felis lynx (lynx)	–	–	–	1	–	–	–	–	–	–	–	–
Indeterminate carnivore	–	–	–	–	–	–	–	–	–	6	–	–
Mammuthus primigenius (mammoth)	410	–	–	16	3,540	40	–	–	–	3	–	–
Equus ferus (horse)	227	1,522	622	385	1,423	2,133	20	3,577	–	1,556	–	–
Equus hydruntinus (ass)	–	–	–	–	–	–	–	–	–	6	–	–
Coelodonta antiquitatis (woolly rhino)	8	–	–	8	124	3	–	–	–	–	–	–
Sus scrofa (boar)	–	–	–	–	8	–	1	–	–	35	–	–
Cervus elaphus (red deer)	–	502	80	10	19	3	22	–	–	110	–	–
Megaloceros giganteus (giant deer)	–	3	–	7	–	–	–	–	–	–	–	–
Capreolus capreolus (roe deer)	–	12	4	–	–	–	–	–	–	23	–	–
Rangifer tarandus (reindeer)	2,130	2,532	1,461	2,823	1,633	1,509	1,801	271	4,522	446	896	2,850
Bison priscus (bison)	79	792	160	164	–	–	–	–	–	–	–	–
Bos/Bison (aurochs/bison)	–	–	–	–	61	8	9	142	–	136	–	–
Rupicapra rupicapra (chamois)	–	25	5	–	2	10	6	–	–	152	–	–
Capra ibex (ibex)	–	60	37	–	–	–	4	–	–	709	–	–
Homo sp. (hominin)	5	21	3	–	–	–	–	–	–	6	–	–
Birds	6	–	–	16	13	24	1	–	–	–	–	–
Fishes	5	–	–	–	–	–	–	–	–	–	–	–
Total identified to taxon/family	2,781	6,054	2,529	3,474	7,055	4,450	1,864	4,031	4,522	3,715	896	2,850
Indeterminate	1	6,033	2,526	nd	6,227	4,143	36	93	1,889	6,472	nd	nd
Total NISP/N	2,782	12,087	5,055	3,474	13,282	8,593	1,900	4,124	6,411	10,187	896	2,850

References for data: S-L (Gaudzinski and Roebroeks, 2000: Table 2; data on file); PoF2–PoF3 (Gardeisen, 1997: Table 80); Cés (Morin, 2004: Table 14, 17); Vog (Niven, 2006: Table 5.1); Ren (David and Poulain, 2002: 51–95; Mourer-Chauviré, 2002: 105); Flag (Enloe, 1993: 104); Sol (Turner, 2002: 12, Table 1); Verb (Enloe, 2004: 152); RB (Costamagno, 1999: Table 7–26, 7–22); P27 (Enloe et al., 1994: 106); P36 (Enloe, 1991: Table 4.5). Site abbreviations and stratigraphy: see Table 7.1.

is a great deal of mammoth ivory and for a variety of reasons, this taxon is not considered to be a subsistence taxon like reindeer or horse (Niven, 2006).

Within the different assemblages considered, the MNI value for reindeer ranges between 86 (Salzgitter-Lebenstedt) and nine (Solutré), with NISP counts between 4,522 (Verberie) and 271 (Solutré). Reindeer skeletal remains exhibit evidence of exploitation for meat and marrow resources.

All of the sites show a predominance of juvenile and adult individuals. Several methods were used by the analysts in aging reindeer, including tooth eruption and crown height measurements (Salzgitter-Lebenstedt, Le Portel, Vogelherd, St. Césaire, Le Flageolet, Grotte du Renne, Solutré, Verberie, Pincevent), and epiphyseal fusion of bones (Le Portel, Vogelherd, Le Flageolet). To facilitate comparison of age profiles among the assemblages, we group reindeer into juvenile, prime adult, and old age categories (following Stiner, 1990). The *juvenile* category covers samples from the age at birth to the loss of its deciduous teeth; *prime adult* encompasses the reproductive years of life; and *old* are those individuals past reproductive age and whose tooth crown is >50% worn away. It is worth noting that each age group does not represent an equal proportion of longevity, but instead correlates to life-history stages of the taxon in question (Stiner, 1990). In the case of reindeer, we grouped individuals between 0–2 years into juvenile; 3–9 years into prime adult; and 9+ years into old.

Hunting events of reindeer took place in the cold seasons of the year (Table 7.1). Season of death estimates on reindeer are derived from one or more of the following datasets: tooth eruption and wear (Salzgitter-Lebenstedt, Le Portel, Vogelherd, Solutré, Verberie, Pincevent), and shed antler (St. Césaire, Solutré,Verberie, Pincevent). When samples are present, epiphyseal fusion of bones and development stage of fetal bone can also be used for estimating season of death. Such methods work because most ungulates, including reindeer, have seasonally restricted mating and birthing seasons – for reindeer, mating occurs in autumn and birth in late May to early June (Spiess, 1979).

Sex ratio of the assemblages could be discerned in most assemblages (Table 7.1). For reindeer, sex determination can be done by means of osteometrics and/or the size of antler. The former method yielded sex determination in Vogelherd, Le Flageolet, Verberie, and Pincevent, while the latter one was applied to the assemblages from Salzgitter-Lebenstedt and St. Césaire.

One method we used for evaluating the exploitation of reindeer by people involves the overall standardized skeletal element representation (%MAU) of the archaeofaunas. Raw MAU values are derived by dividing the comprehensive MNE (minimum number of elements) of each skeletal element by the number of that element in a living animal, while %MAU involves dividing the MAU value for each element by the highest MAU value(s) in the assemblage overall and multiplying by 100 in order to scale all values between 0 and 100 (Binford, 1984). Percent MAU facilitates comparison of skeletal element frequencies by taxon to a standard, the complete animal skeleton (Binford, 1984), with the goal of illustrating patterns in human decision-making in terms of transport and processing of prey. %MAU values are then compared to the meat (Metcalfe and Jones, 1988: Table 7.3) and marrow (Binford, 1978: Table 1.7) indices established for *Rangifer*. Statistical relationships between %MAUs of each assemblage provide further clarification of the datasets. Although other data such as cutmark and fracture frequencies must also be considered in an evaluation of human processing decisions, the method described above can clearly illustrate whether people butchered reindeer beyond the level of meat removal, and invested time and labor in extracting nutrient- and fat-rich marrow resources.

An essential step in our analysis included assessing the degree of post-depositional bone destruction among the assemblages by such agents as carnivores or chemical deterioration, which better facilitates our ability to discern economic choices of hominins from skeletal element frequencies in an archaeofauna. A standard method for such evaluation entails comparison of bone density values of skeletal elements with their abundance in an archaeofauna. In this study, we employ the bone mineral density (BMD) values of reindeer elements measured by computed tomography (CT) provided by Lam et al. (1999: Table 7.1). Generally, low density elements such as vertebrae and ribs are more easily affected by density-mediated attrition or carnivore ravaging than high density bones such as those from the appendicular skeleton. For the correlation with %MAU values, we used the highest density values provided for long bones.

Results

Comparison of skeletal element abundances from each reindeer assemblage reveals different patterns. As visible in Fig. 7.1, two distinct groups can be distinguished. Group 1 (Fig. 7.1a) comprises the assemblages from Salzgitter-Lebenstedt, Vogelherd, St. Césaire and Le Flageolet. Skeletal element representation within this group is characterized by a moderate presence of skulls, an average representation of the forelimbs and a high representation of lower hindlimbs, with only a minor representation of phalanges.

Samples from Pincevent 27 and 36 as well as Verberie are summarized in Group 2 (Fig. 7.1b). Compared to the skeletal element representation for Group 1, skulls are more often represented, and a definite dominance of forelimb elements is obvious in contrast to the moderate abundance of lower hindlimb elements. In addition, phalanges are better represented than in Group 1. Comparison of five additional assemblages (Le Portel F2, F3, Grotte du Renne, Solutré, Rond-du-Barry) representing most of the time periods in question reveals a more ambiguous pattern, but one showing clusters of similar element abundances (Fig. 7.1c).

The role of density-mediated attrition on survivorship of long bones varies among our assemblages (Table 7.3).

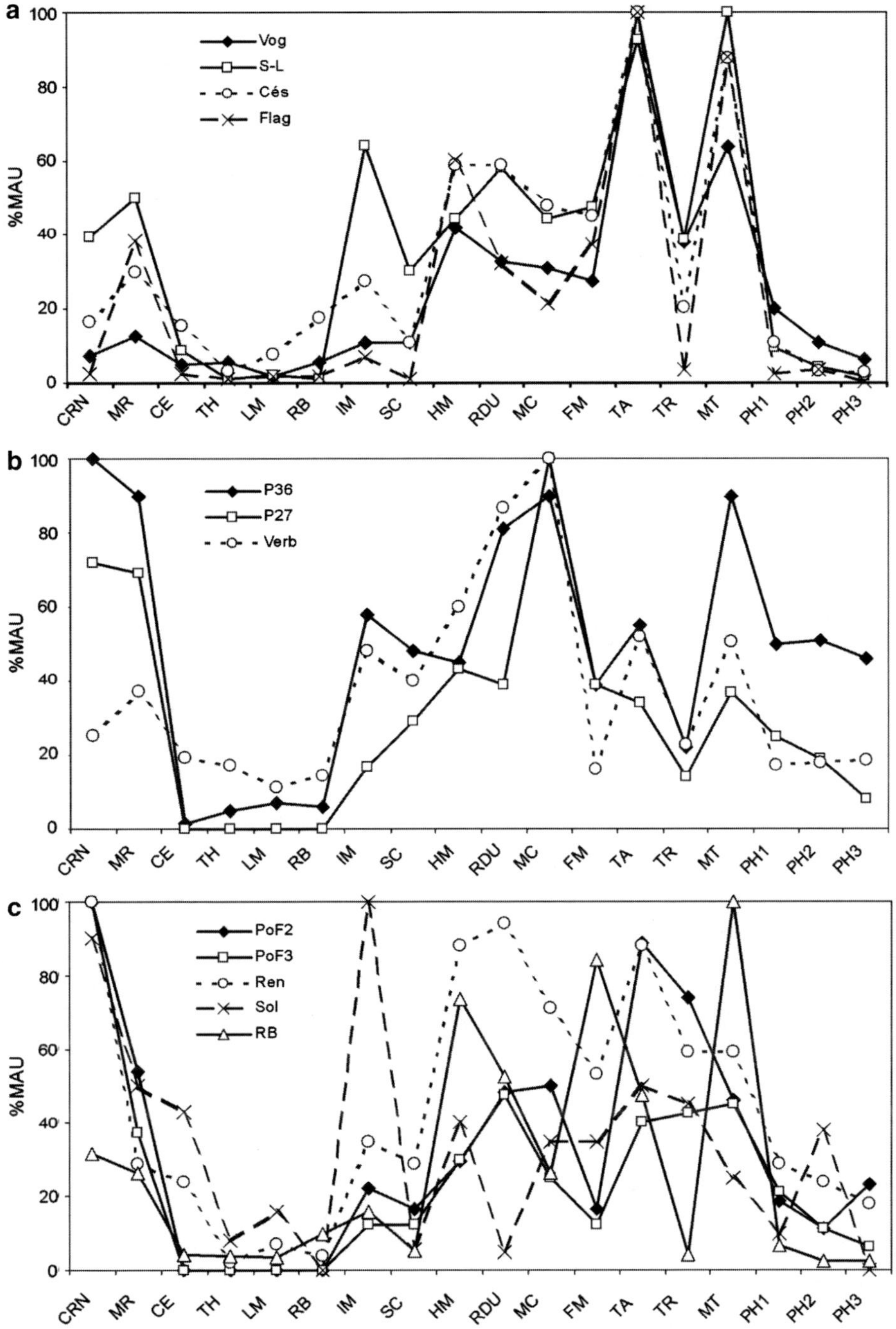

FIG. 7.1. Graphs illustrating skeletal element representation of reindeer, expressed as standardized MAU (%MAU) values. Skeletal part abbreviations: CRN cranium, MR mandible, CE cervical vertebrae, TH thoracic vertebrae, LM lumbar vertebrae, RB ribs, IM pelvis, SC scapula, HM humerus, RDU, radius/ulna, MC metacarpus, FM femur, TA tibia, TR tarsals, MT metatarsus, PH1-PH3 phalanges. Site name abbreviations defined in Table 7.1.

We see correlation coefficients between r = 0.69 and r = −0.89. The coefficients of determination (r^2) scatter mainly between $r^2 = 0.01$ and $r^2 = 0.79$. The correlation coefficient and the coefficient of determination (r^2) for the majority of the assemblages suggest that the survival of skeletal elements was governed by the influence of bone attrition prior to burial (e.g., carnivore destruction) or post-burial (e.g., chemical processes, sediment compaction) only to a minor degree. Only in the case of Le Portel F3 can we discern considerable impact of density-mediated attrition on the assemblage. Based on these results, we conclude that long bone element abundances of reindeer at the majority of sites considered here were the result of economic decisions of hominins.

TABLE 7.3. Statistical correlations between standardized MAU values versus bone mineral density (BMD) values of long bones from Lam et al. (1999: Table 7.1). Highest BMD values are used. Site name abbreviations defined in Table 7.1.

	%MAU versus BMD	
Site	r	r^2
S-L	−0.12	0.01
PoF2	−0.21	0.04
PoF3	−0.69	0.48
Cés	−0.05	0.00
Vog	0.14	0.02
Ren	−0.35	0.13
Flag	0.18	0.03
Sol	0.61	0.37
Verb	−0.83	0.69
RB	0.27	0.07
P27	−0.32	0.11
P36	−0.89	0.79

In order to evaluate this conclusion in further detail, we compared skeletal element representation among the assemblages with two utility indices established for reindeer: the standardized food utility index, which involves all skeletal parts (Metcalfe and Jones, 1988: Table 7.3); and the marrow index, based on marrow cavity volume of all appendicular elements as well as the pelvis, mandible, and first two phalanges (Binford, 1978: Table 1.7). These indices measure average values of meat, and marrow, and the overall economic utility of skeletal elements for each given prey taxon (Binford, 1978) and are one method for exploring human decisions regarding prey utilization.

What we see is that within the late Upper Paleolithic assemblages from Solutré and Rond-du-Barry, the meatiest portions of the skeleton are best represented (Fig. 7.2, Tables 7.4–7.5). Comparison of the %MAU values with the marrow index highlights two different groups of assemblages: (1) those from the Middle Paleolithic/Early- and Middle Upper Paleolithic (i.e., Salzgitter-Lebenstedt, Vogelherd, St. Césaire, Le Flageolet) indicate a clear focus on bones high in marrow content; and (2) two of the Late Upper Paleolithic reindeer samples (Pincevent 27, 36, Verberie) show much lower number of bones high in marrow content. Although these patterns are visible in the basic comparison of %MAU values illustrated in Fig. 7.1, they are confirmed by the use of the marrow index.

It is striking that with one exception, all sites showing a focus on marrow acquisition are characterized by a dominance of reindeer within the overall faunal assemblage composition (Table 7.2). The exception is Vogelherd, which has an equal representation of horse, though this taxon was exploited extensively for marrow similar to the reindeer.

Another notable pattern among the assemblages involves the mortality profiles of reindeer (Fig. 7.3). As seen in Table 7.1, the younger age classes are certainly represented in these sites, yet it is prime-aged adults that were most frequently hunted across all time periods in question. Prime adult ungulates such as reindeer would have provided the best yield in terms of nutrition (meat, marrow, organs) and other resources such as skins, tendons, and antler. Humans are unique in their focus on this age class, in contrast to carnivore predators who target the vulnerable young and old individuals (Stiner, 1990; Steele, 2003), and this aspect of the human hunting strategy appears to have already been in place during the early Middle Paleolithic and Middle Stone Age of Africa (Steele, 2003). With the exception of Le Portel, where the young age classes might be underrepresented due to density-mediated attrition, the adult-dominated age profiles in our reindeer assemblages

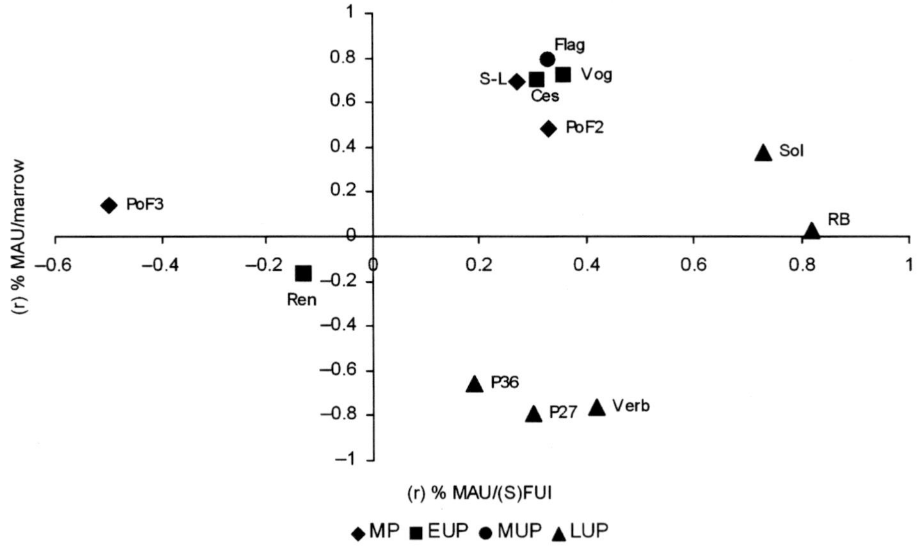

FIG. 7.2. Comparison of reindeer %MAU values to: (*Y* axis) marrow cavity volume (ml) (reference data from Binford, 1978: Table 1.7); (*X* axis) standardized food utility index (SFUI) Site name abbreviations defined in Table 7.1 (From Metcalfe and Jones, 1988: Table 7.3).

TABLE 7.4. Standardized MAU values for reindeer assemblages discussed in text. Site name abbreviations defined in Table 7.1.

	S-L	PoF2	PoF3	Cés	Vog	Ren	Flag	Sol	Verb	RB	P27	P36
Cranium	39.5	**100.0**	**100.0**	16.4	7.3	**100.0**	2.4	90.0	25.3	31.6	72.0	**100.0**
Mandible	50.0	53.7	37.5	30.1	12.7	29.0	38.5	5.0	37.3	26.3	69.0	90.0
Cervical	8.7	0.0	0.0	15.6	5.1	24.0	2.4	43.0	19.2	4.2	0.0	1.5
Thoracic	1.2	0.0	0.0	3.1	5.5	2.0	0.9	8.0	17.1	4.0	0.0	5.0
Lumbar	2.2	0.0	0.0	7.9	1.8	7.0	1.6	16.0	11.2	3.5	0.0	7.0
Rib	1.2	0.0	0.0	17.5	5.5	4.0	1.6	0.0	14.5	9.7	0.0	6.0
Pelvis	64.2	22.2	12.5	27.4	10.9	35.0	7.2	**100.0**	48.0	15.8	17.0	58.0
Scapula	30.2	16.6	12.5	11	10.9	29.0	1.2	55.0	40.0	5.3	29.0	48.0
Humerus	44.4	29.6	30.0	58.9	41.8	88.0	60.2	40.0	60.0	73.7	43.0	45.0
Radius-ulna	58.0	48.1	47.5	58.9	32.7	94.0	32.5	25.0	86.7	52.6	39.0	81.0
Metacarpal	44.4	50.0	25.0	47.9	30.9	71.0	21.6	35.0	**100.0**	26.3	**100.0**	90.0
Femur	47.5	16.6	12.5	45.2	27.3	53.0	37.3	35.0	16.0	84.2	39.0	39.0
Tibia	92.6	88.8	40.0	**100.0**	**100.0**	88.0	**100.0**	50.0	52.0	47.4	34.0	55.0
Tarsal	38.9	74.0	42.5	20.5	47.3	59.0	3.6	45.0	22.7	4.2	14.0	22.0
Metatarsal	**100.0**	46.2	45.0	87.7	63.6	59.0	87.9	25.0	50.7	**100.0**	37.0	90.0
Ph1	9.4	18.5	21.2	11.0	20.0	29.0	2.4	10.0	17.3	6.6	25.0	50.0
Ph2	4.2	11.1	11.2	3.0	10.9	24.0	3.6	38.0	18	2.6	19.0	51.0
Ph3	1.7	23.1	6.25	2.7	6.5	18.0	0.3	0.0	18.7	2.6	8.0	46.0

References for data: S-L (Gaudzinski and Roebroeks, 2000: Table 2); PoF2–PoF3 (Gardeisen, 1997: Table 39); Cés (Morin, 2004: Table 3.8); Vog (Niven, 2006: Table 5.1); Ren (David and Poulain, 2002: Table V); Flag (Enloe, 1993: Table 2); Sol (Turner, 2002: Table 13); Verb (Enloe, 2004: Table 1); RB (Costamagno, 1999: 10–100); P27 (Enloe et al., 1994: Fig. 7.1); P36 (Enloe, 1991: Table 4.5).

TABLE 7.5. Statistical correlations between %MAU (see Table 7.3) versus (S)FUI (Metcalfe and Jones, 1988: Table 7.3); and %MAU versus marrow cavity volume (Binford, 1978: Table 1.7). Site name abbreviations defined in Table 7.1.

Site	%MAU/(S)FUI	%MAU/marrow
S-L	0.27	0.69
PoF2	0.33	0.48
PoF3	−0.5	0.14
Cés	0.31	0.7
Vog	0.36	0.72
Ren	−0.13	−0.17
Flag	0.33	0.79
Sol	0.73	0.38
Verb	0.42	−0.76
RB	0.82	0.03
P27	0.19	−0.66
P36	0.3	−0.79

reflect hominin hunting strategies targeting animals that would provide the best returns.

Discussion

Reindeer assemblages from all the sites considered here reflect hunting activities during the cold period of the year, and the overwhelming majority of the assemblages indicate extensive exploitation of meat and marrow, as seen by the presence of cutmarks, impact notches, and bone breakage. Originating from caves and open-air sites, the individual taphonomic pre-burial histories of the assemblages are obviously different; i.e., in regard to assemblages originating from caves (e.g., Le Flageolet), we presume the animal remains were transported to the site by people, while open-air sites such as Salzgitter-Lebenstedt might be closer to, or located on the kill location. Whereas the accumulation at Le Flageolet is interpreted in terms of repeated individual encounter hunting spaced throughout winter (Enloe, 1993), Salzgitter-Lebenstedt represents mass kill encounters during the fall (Gaudzinski and Roebroeks, 2000). In addition, some of the reindeer assemblages stem from highly specialized reindeer (e.g., Pincevent and Verberie) and horse (e.g., Solutré) hunting camps that represent either short-term accumulations of numerous individuals procured in a single kill event; or long-term accumulations of repeated, single animal kill events.

We presume that these different hunting scenarios required a variety of hunting tactics that should become visible by means of analyzing mortality profiles. However, as seen in Fig. 7.3, the influence of hunting tactics on age representation appears to have been minimal or alternatively, simply not expressed in this dataset. The reindeer assemblages cluster in a homogeneous group in the lower right corner of the triangular graph, representing what is generally labeled "catastrophic" and "prime-dominated" mortalities. We conclude that either considerable differences in hunting tactics did not occur; or that the methods employed in analyzing reindeer age structure result in an ambiguous picture.

Although reindeer population structure within the different assemblages did not reveal any clear patterns across time, our analysis of skeletal element representation provides insight to different behaviors involving carcass exploitation. These behaviors appear to be diagnostic of faunal assemblages in which reindeer is the primary large mammal taxon exploited;

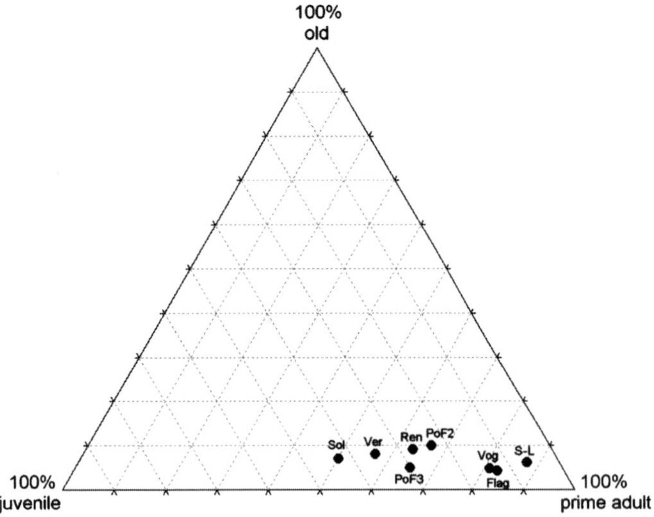

FIG. 7.3. Mortality profiles of reindeer, classified as juvenile, prime adult and old age groups (following Stiner, 1990). Age data not available for RB, P36, P37. Site name abbreviations defined in Table 7.1.

or where reindeer is of equal importance to horse as a prey species, as in the case of Vogelherd.

Reindeer assemblages from Middle Paleolithic, early-, and middle Upper Paleolithic sites show a moderate to high presence of skulls, an average representation of elements of the forelimbs in contrast to a high representation of the lower hindlimbs, and a minor presence of phalanges. Late Upper Paleolithic reindeer assemblages reveal a rather contrasting pattern in that skulls are better represented, while forelimb elements and phalanges are more numerous than lower hindlimb elements. Based on our reading of the various studies involved here, density-mediated attrition did not have considerably more impact on Middle and Upper Paleolithic samples in comparison to late Upper Paleolithic assemblages. In addition, carnivore predators were not a significant influence on the assemblages in question, supporting our conclusion that hominins were the primary collectors of bone in all of these sites. Thus, what we see in the samples from the Middle Paleolithic and early to middle Upper Paleolithic is a focus on bones high in marrow content.

For the late Upper Paleolithic samples, two aspects should be emphasized. We see a lower representation of bones high in marrow content within assemblages where reindeer is the main prey species. In addition, reindeer assemblages that are part of faunal accumulations dominated by horse reveal a definite focus on carcass parts particularly rich in meat value, as seen in Rond-du-Barry and Solutré. It is quite interesting to note in this context that marrow procurement clearly did not play a role in subsistence behaviors at Solutré, since only two impact notches associated with marrow bone breakage were documented from Solutré area P16 (Turner, 2002).

These results show that exploitation of reindeer was consistently directed towards acquisition of marrow resources, and it can be assumed that the additional exploitation of meat was also of great importance. Only during the late Upper Paleolithic does the sequencing of carcass exploitation tactics become visible, showing that humans could afford to only focus on part of the animal resources available, while other major resources such as marrow remained either untouched or were transported and used elsewhere. It is difficult to evaluate whether this shift towards more focused resource exploitation is also accompanied by intensification in meat and marrow exploitation.

The intensity of carcass exploitation has been measured among other aspects in establishing the nutritional sources which have been regularly targeted. In this scenario, intensity of carcass exploitation should increase with the amount of energy required for exploitation (Munro and Bar-Oz, 2005). Following this argument, high intensity exploitation should become evident in the extraction of the smallest marrow stores of marrow-poor elements such as phalanges, whereas breakage of only the richest marrow bones might account for low intensity exploitation. Against this background, fragmentation of phalanges in gazelles has been taken to measure intensity of carcass exploitation (Munro and Bar-Oz, 2005).

In evaluating the fragmentation of phalanges among our reindeer assemblages, we see only a weak pattern. Data on fragmentation of phalanges were available for all sites except Pincevent, Verberie and St. Césaire. Only at Salzgitter-Lebenstedt and Solutré did phalanges remain complete; for the former, the focused exploitation of high nutritional sources in reindeer was emphasized (Gaudzinski and Roebroeks, 2000). Within all other assemblages, the first and second phalanges have been broken for marrow extraction. Phalanges occur more often in late Upper Paleolithic assemblages, but quantitative analysis of fragmentation patterns based on data available is not possible (e.g., Costamagno, 1999).

Overall, exploitation of within-bone nutrients among the reindeer assemblages discussed here appears to have been limited to marrow extraction by means of breakage. In other words, we see no evidence of processing for bone grease and fats. Exploitation of grease is difficult to demonstrate except in cases where we have evidence of bone in pits used for cooking (Bosinski, 1979), and it is not thought to have been a regular part of subsistence behaviors until the Gravettian (Stiner, 2003).

Concerning meat and marrow exploitation, we demonstrated that within the earlier assemblages a definite focus on marrow exploitation occurred. These arguments could initially be interpreted in terms of an increase in resource intensification. However the question is much more complex, as we have to take into account the increasingly complex settlement systems during the late Upper Paleolithic (Street et al., 2006). Evidence from sites focusing on meat exploitation only (e.g., Solutré) shows low intensity carcass exploitation, a pattern presumably governed by site use. Thus it appears that the reduced pattern of carcass exploitation we demonstrated for the late Upper Paleolithic, which undoubtedly depends on the number of prey species involved, was influenced by complex settlement system organization. It could be assumed that due to settlement dynamics humans could afford economization in subsistence tactics. The broadening of the dietary niche with the inclusion of cost effective small game hunting, as well as the shift in large mammal hunting strategies, reflect in part the demands of settlement dynamics.

Conclusions

This brings us back to our original question: whether Middle Paleolithic subsistence represents only one aspect of the much broader subsistence pattern that we see during the late Upper Paleolithic. If we accept the above- formulated hypothesis that settlement dynamics shaped the character of late Upper Paleolithic large mammal exploitation patterns, it seems quite clear that Middle Paleolithic subsistence was different in character. Middle Paleolithic hominins have been described as "hominids without homes" (Kolen, 1999). Based on archaeological evidence, we lack good examples of Middle Paleolithic home bases and dwellings. Moreover, indications for a cultural geography where space was differentiated according to the societies' cosmology in terms of material culture are rare (Kolen, 1999). Studies such as these have emphasized the situational character of Neanderthals' spatial organization.

It has been illustrated above that Neanderthals were very selective in their mammalian diet, in contrast to later modern humans. In addition, Neanderthals tended to focus on resources high in energy content, as they occupied a high trophic level in the food chain (e.g., Richards et al., 2000; Bocherens et al., 2001). It seems highly likely that careful selection of resources would have been crucial to maintain a high foraging return rate necessary to cover their high metabolic rate and energy requirements (Sorensen and Leonard, 2001; Aiello and Wheeler, 2003). Thus it seems that late Upper Paleolithic modern humans could afford to diversify their mammalian diet (Gaudzinski and Street, 2003) due to their organization in extended settlement and foraging systems, which would have allowed compensation in times of short supply.

Acknowledgments. We are grateful to the conference organizers Prof. Jean-Jacques Hublin and Prof. Michael Richards for inviting us to participate in their meeting. Thanks to Matthew G. Hill for sending us relevant literature and data. LN thanks Harold Dibble, Shannon McPherron and Paul Goldberg for their collaboration on the fauna from Pech de l'Azé IV. Lastly, we are grateful for the helpful comments and suggestions of three anonymous reviewers.

References

Aiello, L., Wheeler, P., 2003. Neanderthal Thermoregulation and the Glacial Climate. In: van Andel, T.H., Davies, W. (Eds.), Neanderthals and Modern Humans in the European Landscape During the Last Glaciation: Archaeological Results of the Stage 3 Project. McDonald Institute for Archaeological Research, Cambridge, pp. 147–166.

Binford, L.R., 1978. Nunamiut Ethnoarchaeology. Academic Press, New York.

Binford, L.R., 1984. Faunal Remains from Klasies River Mouth. Academic Press, New York.

Binford, L.R., 1989. Isolating the transition to cultural adaptations: an organizational approach. In: Trinkaus, E. (Ed.), The Emergence of Modern Humans. Biocultural Adaptations in the Late Pleistocene. Cambridge University Press, Cambridge, pp. 18–41.

Bocherens, H., Billiou, D., Mariotti, A., Toussaint, M., Patou-Mathis, M., Bonjean, D., Otte, M., 2001. New isotopic evidence for dietary habits of Neanderthals from Belgium. Journal of Human Evolution 40, 497–505.

Bosinski, G., 1979. Die Ausgrabungen in Gönnersdorf 1968–1976 und die Siedlungsbefunde der Grabung 1968. Steiner, Wiesbaden.

Boyle, K.V., 2000. Reconstructing Middle Palaeolithic subsistence strategies in the South of France. International Journal of Osteoarchaeology 10, 336–356.

Bratlund, B., 1994. A survey of the subsistence and settlement pattern of the Hamburgian culture in Schleswig-Holstein. Jahrbuch des Römisch-Germanischen Zentralmuseums Mainz 41, 59–93.

Bratlund, B., 1999. Taubach revisited. Jahrbuch des Römisch-Germanischen Zentralmuseums Mainz 46, 67–174.

Brugal, J-P., Desse, J., 2004. Petits Animaux et Sociétés Humaines. Éditions APDCA, Antibes.

Chase, P.G., 1986. The Hunters of Combe Grenal: Approaches to Middle Palaeolithic Subsistence in Europe. BAR International Series 286, Oxford.

Cleyet-Merle, J.-J., 1990. La prehistoire de la pêche. Editions Errance, Paris.

Cochard, D., Brugal, J.P., 2004. Importance des functions de sites dans les accumulations paléolithiques de léporidés. In: Brugal, J-P., Desse, J. (Eds.), Petits Animaux et Sociétés Humaines. Éditions APDCA, Antibes, pp. 283–296.

Conard, N.J., Prindiville, T.J., 2000. Middle Palaeolithic hunting economies in the Rhineland. International Journal of Osteoarchaeology 10, 286–309.

Costamagno, S., 1999. Stratégies de Chasse et Fonction des Sites au Magdalénien dans le Sud de la France. Ph.D. Dissertation, L'Université Bordeaux I.

Costamagno, S., 2003. L'éxploitation des ongulés au Magdalénien dans le sud de la France. In: Costamagno, S., Laroulandie, V. (Eds.), Mode de Vie au Magdalénien: Apports de L'Archéozoologie. BAR International Series 1144, Oxford, pp. 73–88.

Costamagno, S., Laroulandie, V., 2004. L'exploitation des petits vertébrés dans les Pyrénées françaises du Paléolithique au Mésolithique: un inventaire taphonomique et archéozoologique. In: Brugal, J-P., Desse, J. (Eds.), Petits Animaux et Sociétés Humaines. Éditions APDCA, Antibes, pp. 403–416.

Costamagno, S., Meignen, L., Beauval, C., Vandermeersch, B., Maureille, B., 2006. Les Pradelles (Marillac-le-Franc, France): a Mousterian reindeer hunting camp? Journal of Anthropological Archaeology 25, 466–484.

David, F., Enloe, J.G., 1993. L'exploitation des animaux sauvages de la fin du Paléolithique moyen au Magdalénien. In: Desse, J., Audouin, F. (Eds.), Exploitation des Animaux Sauvages à travers le Temps. Editions APDCA, Juan-les-Pins, pp. 29–47.

David, F., Poulain, T., 2002. Les Mammifères. In: Schmider, B. (Ed.), L'Aurignacien de la Grotte du Renne: Les Fouilles d'André Leroi-Gourhan à Arcy-sur-Cure (Yonne). CNRS Editions, Paris, pp. 51–95.

Driver, J., 1990. Meat in due season: the timing of communal hunts. In: Davis, L.B. (Ed.), Hunters of the Recent Past. Unwin Hyman, London, pp. 68–86.

Eastham, A., 1998. Buhlen Upper Cave: the avifauna. Jahrbuch des Römisch-Germanischen Zentralmuseums Mainz 45, 251–266.

Enloe, J.G., 1991. Subsistence organization in the Upper Paleolithic: carcass refitting and food sharing at Pincevent. Ph.D. Dissertation, University of New Mexico.

Enloe, J.G., 1993. Subsistence organization in the Early Upper Paleolithic: reindeer hunters of the Abri du Flageolet, Couche V. In: Knecht, H., Pike-Tay, A., White, R. (Eds.), Before Lascaux: The Complex Record of the Early Upper Paleolithic. CRC Press, Boca Raton, pp. 101–115.

Enloe, J.G., 2003. Acquisition and processing of reindeer in the Paris basin. In: Costamagno, S., Laroulandie, V. (Eds.), Mode de Vie au Magdalénien: Apports de l'Archéozoologie, BAR International Series 1144, Oxford, pp. 23–31.

Enloe, J.G., 2004. Equifinality, assemblage integrity and behavioral inferences at Verberie. Journal of Taphonomy 2, 147–165.

Enloe, J.G., David, F., Hare, T.S., 1994. Patterns of faunal processing at Section 27 of Pincevent: the use of spatial analysis and ethnoarchaeological data in the interpretation of archaeological site structure. Journal of Anthropological Archaeology 13, 105–124.

Farizy, C., David, F., Jaubert, J., 1994. Hommes et Bisons du Paléolithique Moyen á Mauran (Haute Garonne). CNRS Editions, Paris.

Fiore, I., Gala, M., Tagliacozzo, A., 2004. Ecology and subsistence strategies in the Eastern Italian Alps during the Middle Palaeolithic. International Journal of Osteoarchaeology 14, 273–286.

Fontana, L., 2003. Characterization and exploitation of the Arctic Hare (*Lepus timidus*) during the Magdalenian: surprising data from Gazel Cave (Aude, France)? In: Costamagno, S., Laroulandie, V. (Eds.), Mode de Vie au Magdalénien: Apports de L'Archéozoologie. BAR International Series 1144, Oxford, pp. 101–118.

Gardeisen, A., 1997. La Grotte Ouest du Portel, Ariège, France: Restes Fauniques et Stratégies de Chasse dans le Pléistocene Supérieur Pyrénéen, Oxford.

Gaudzinski, S., 1995. Wallertheim revisited: a re-analysis of the fauna from the Middle Palaeolithic site of Wallertheim (Rheinessen/Germany). Journal of Archaeological Science 22, 51–66.

Gaudzinski, S., 1996. On Bovid assemblages and their consequences for the knowledge of subsistence patterns in the Middle Palaeolithic. Proceedings of the Prehistoric Society 62, 19–39.

Gaudzinski, S., 2005. Monospecific or species-dominated faunal assemblages during the Middle Paleolithic in Europe. In: Hovers, E., Kuhn, S.L. (Eds.), Transitions Before the Transition: Evolution and Stability in the Middle Paleolithic and Middle Stone Age. Springer, New York, pp. 137–147.

Gaudzinski, S., Roebroeks, W., 2000. Adults only: reindeer hunting at the Middle Palaeolithic site Salzgitter-Lebenstedt, northern Germany. Journal of Human Evolution 38, 497–521.

Gaudzinski, S., Roebroeks, W., 2003. Profile analysis at Salzgitter-Lebenstedt. A reply to Munson and Marean. Journal of Human Evolution 44, 275–281.

Gaudzinski, S., Street, M., 2003. Reconsidering hunting specialization in the German Magdalenian faunal record. In: Costamagno, S., Laroulandie, V. (Eds.), Mode de Vie au Magdalénien: Apports de L'Archéozoologie. BAR International Series 1144, Oxford, pp. 11–22.

Grayson, D.K., 1984. Quantitative Zooarchaeology: Topics in the Analysis of Archaeological Faunas. Academic Press, Orlando.

Grayson, D.K., Delpech, F., 2003. Ungulates and the Middle-to-Upper Paleolithic transition at Grotte XVI (Dordogne, France). Journal of Archaeological Science 30, 1633–1648.

Grayson, D.K., Delpech, F., 2006. Was there increasing diet specialization across the Middle-to-Upper Paleolithic transition in France? In: Conard, N.J. (Ed.), When Neanderthals and Modern Humans Met. Kerns, Tübingen, pp. 377–417.

Hahn, J., 1989. Genese und Funktion einer jungpaläolithischen Freilandstation: Lommersum im Rheinland. Rheinland, Köln.

Haws, J.A., Hockett, B.S., 2004. Theoretical perspectives on the dietary role of small mammals in human evolution. In: Petits Animaux et Sociétés Humaines, Éditions APDCA, Antibes, pp. 533–544.

Hockett, B.S., Bicho, N.F., 2000. The rabbits of Picareiro Cave: small mammal hunting during the late Upper Palaeolithic in the Portugese Estremadura. Journal of Archaeological Science 27, 715–723.

Hockett, B., Haws, J.A., 2002. Taphonomic and methodological perspectives of leporid hunting during the Upper Paleolithic of the western Mediterranean Basin. Journal of Archaeological Method and Theory 9, 269–302.

Hockett, B., Haws, J.A., 2003. Nutritional ecology and diachronic trends in Paleolithic diet and health. Evolutionary Anthropology 12, 211–216.

Jaubert, J., Lorblanchet, M., Laville, H., Slott-Moller, R., Turq, A., Brugal, J.-P., 1990. Les Chasseurs d'Aurochs de La Borde: un Site du Paléolithique Moyen. Maison des Sciences de l'Homme, DAF No. 27, Paris.

Jones, E.L., 2006. Prey choice, mass collecting, and the wild European rabbit (*Oryctolagus cuniculus*). Journal of Anthropological Archaeology 25, 275–289.

Kolen, J., 1999. Hominids without homes: on the nature of Middle Palaeolithic settlement in Europe. In: Roebroeks, W., Gamble, C. (Eds.), The Middle Palaeolithic Occupation of Europe. University of Leiden Press, Leiden, pp. 139–175.

Kretzoi, M., 1964. Die Wirbeltierfauna des Travertinkomplexes von Tata. In: Vertes, L. (Ed.), Tata. Eine mittelpaläolithische Travertin-Siedlung in Ungarn. Archaeologica Hungarica, Series Nova 43, Budapest, pp. 105–126.

Lam, Y.M., Chen, X., Pearson, O.M., 1999. Intertaxonomic variability in patterns of bone density and the differential representation of bovid, cervid, and equid elements in the archaeological record. American Antiquity 64, 343–362.

Laroulandie, V., 2003. Exploitation des Oiseaux au Magdalénien en France: Etat des Lieux. In: Costamagno, S., Laroulandie, V. (Eds.), Mode de Vie au Magdalénien: Apports de L'Archéozoologie. BAR International Series 1144, Oxford, pp. 129–138.

Le Gall, O., 1988. Analyse palethnologique de l'ichtiofaune. In: Rigaud, J.P. (Ed.), La Grotte Vaufrey. Paléoenvironnement – Chronologie – Activités humaines. Mémoires de la Société Préhistorique Française XIX, pp. 565–568.

Le Gall, O., 2003. Des Magdaléniens et … des Poissons. In: Costamagno, S., Laroulandie, V. (Eds.), Mode de Vie au Magdalénien: Apports de L'Archéozoologie. BAR International Series 1144, Oxford, pp. 119–128.

Lumley, de H., Echassoux, A., Bailon, S., Cauche, D., de Marchi, M.P., Desclaux, E., Guennouni, K., Khatib, S., Lacombat, F.,

Roger, T., Valensi, P. (Eds.), 2004. Le sol d'occupation acheuléen de l'unite archéostratigraphique UA 25 de La Grotte du Lazaret, Nice, Alpes-Maritimes. Edisud, Aix-en-Provence.

Metcalfe, D., Jones, K.T., 1988. A reconsideration of animal body-part utility indices. American Antiquity 53, 486–504.

Moigne, A.-M., Barsky, D.R., 1999. Large mammal assemblages from Lower Palaeolithic sites in France: La Caune de lÀrago, Terra Amata, Orgnac 3 and Cagny l'Epinette. In: Gaudzinski, S., Turner, E. (Eds.), The Role of Early Humans in the Accumulation of European Lower and Middle Palaeolithic Bone Assemblages. RGZM, Bonn, pp. 219–235.

Morin, E., 2004. Late Pleistocene population interaction in western Europe and modern origins: new insights based on the faunal remains from Saint-Césaire, Southwestern France. Ph.D. Dissertation, University of Michigan.

Mourer-Chauviré, C., 1979. La chasse aux oiseaux pendant la préhistoire. La Recherche 106, 1202–1210.

Munro, N.D., Bar-Oz, G., 2005. Gazelle bone fat processing in the Levantine Epipalaeolithic. Journal of Archaeological Science 32, 223–239.

Niven, L., 2006. The Palaeolithic Occupation of Vogelherd Cave: Implications for the Subsistence Behavior of Late Neanderthals and Early Modern Humans. Kerns, Tübingen.

Patou-Mathis, M., 2000. Neanderthal subsistence behaviors in Europe. International Journal of Osteoarchaeology 10, 379–395.

Richards, M., Pettitt, P.B., Stiner, M.C., Trinkaus, E., 2001. Stable isotope evidence for increasing diet breadth in the European mid-Upper Paleolithic. Proceedings of the National Academy of Sciences 98, 6258–6532.

Richards, M., Pettit, P., Trinkaus, E., Smith, F., Paunovic, M., Karavanic, I., 2000. Neanderthal diet at Vindija and Neanderthal predation: the evidence from stable isotopes. Proceedings of the National Academy of Sciences 97, 7663–7666.

Richter, J., 1997. Der G-Schichtenkomplex der Sesselfelsgrotte. Ein Beitrag zum Verständnis des Micoquien. Saarbrücker Verlag, Saarbrücken.

Rieder, H., 2000. Die altpaläolithischen Wurfspeere von Schöningen, ihre Erprobung und ihre Bedeutung für die Lebensumwelt des Homo erectus. Praehistoria Thuringica 5, 68–75.

Rigaud, J.-P. (Ed.), 1989. La Grotte Vaufrey. Paléoenvironments, chronologie et activites humaines. Mémoires de la Société Préhistorique Française 19, Paris.

Schuler, A., 1994. Die Schussenquelle. Eine Freilandstation des Magdalénien in Oberschwaben. Theiss Verlag, Stuttgart.

Soergel, W., 1922. Die Jagd der Vorzeit. Gustav Fischer, Jena.

Sorensen, M.V., Leonard, W.R., 2001. Neandertal energetics and foraging efficiency. Journal of Human Evolution 40, 483–495.

Speth, J.D., Tchernov, E., 2002. Middle Paleolithic tortoise use at Kebara Cave (Israel). Journal of Archaeological Science 29, 471–483.

Spiess, A.E., 1979. Reindeer and Caribou Hunters: An Archaeological Study. Academic Press, San Francisco.

Steele, T.E., 2003. Using mortality profiles to infer behavior in the fossil record. Journal of Mammalogy 84, 418–430.

Stiner, M.C., 1990. The use of mortality patterns in archaeological studies of hominid predatory adaptations. Journal of Anthropological Archaeology 9, 305–351.

Stiner, M.C., 2003. Zooarchaeological evidence for resource intensification in Algarve, southern Portugal. Promontoria 1, 27–58.

Stiner, M.C., 2005. Middle Paleolithic subsistence ecology in the Mediterranean Region. In: Hovers, E., Kuhn, S.L. (Eds.), Transitions Before the Transition: Evolution and Stability in the Middle Paleolithic and Middle Stone Age. Springer, New York, pp. 213–231.

Stiner, M.C., Munro, N.D., Surovell, T.A., 2000. The tortoise and the hare: small-game use, the Broad-Spectrum Revolution, and Paleolithic demography. Current Anthropology 41, 39–73.

Street, M., Gelhausen, F., Grimm, S., Moseler, F., Niven, L., Sensburg, M., Turner, E., Wenzel, S., Jöris, O., 2006. L' occupation de bassin de Neuwied (Rhénanie centrale, Allemagne) par les Magdaléniens et les groupes à Federmesser (aziliens). Bulletin de la Société préhistorique française 103, 753–780.

Thieme, H., 1999. Altpaläolithische Holzgeräte aus Schöningen, Lkr. Helmstedt. Bedeutsame Funde zur Kulturentwicklung des frühen Menschen. Germania 77, 451–487.

Turner, E., 2002. Solutré: An Archaeozoological Analysis of the Magdalenian Horizon. Habelt, Bonn.

Valensi, P., 2000. The archaeozoology of Lazaret Cave (Nice, France). International Journal of Osteoarchaeology 10, 357–367.

Wenzel, S., 1998. Die Funde aus dem Travertin von Stuttgart-Untertürkheim und die Archäologie der letzten Warmzeit in Mitteleuropa. Dr. Rudolf Habelt, Bonn.

8. Late Pleistocene Subsistence Strategies and Resource Intensification in Africa

Teresa E. Steele
Department of Human Evolution
Max Planck Institute for Evolutionary Anthropology
Deutscher Platz 6
Leipzig 04103 Germany

and

Department of Anthropology
University of California – Davis
One Shields Avenue
Davis, CA 95616 USA
testeele@ucdavis.edu

Richard G. Klein
Program in Human Biology
Stanford University
Stanford, CA 94305 USA
rklein@stanford.edu

Keywords Middle Stone Age • Later Stone Age • modern humans • Late Pleistocene • subsistence • hunting • ecology • intensification • diet breadth • prey choice

Abstract Studies in southern Africa, western Europe, and the Mediterranean Basin have documented changes in subsistence strategies and technologies during the Late Pleistocene, and have often related them to differences between the Middle and Upper Paleolithic of Europe and the Middle and Later Stone Age of Africa. However, few explicit comparisons have been made between these geographic regions to create a global view of the evolution of human diet. In this paper, we begin by documenting the state of our knowledge in coastal South Africa, where Middle Stone Age paleoecology is best documented. The lines of evidence that we consider here follow those identified by previous researchers: large game exploitation as indicated by prey mortality profiles and processing intensity; changes in the relative abundance and increasing diversity of species included in the diet, especially the addition of small, fast game such as hares, birds, and fish relative to small, slow game such as shellfish and tortoises; and changes in shellfish size reflecting intensity of collection. We then expand the discussion to include all of Africa, highlighting similarities and differences in the record and the areas that require further research. Two of the major limitations to this research are the lack of zooarchaeological assemblages that span the "transition" from the Middle to Later Stone Age (60,000 to 20,000 years ago) and the lack of zooarchaeological data from most regions of Africa. Finally, we consider how the documented subsistence changes relate to demographic changes, the modern human expansion out of Africa, and models of modern human origins.

J.-J. Hublin and M.P. Richards (eds.), *The Evolution of Hominin Diets: Integrating Approaches to the Study of Palaeolithic Subsistence,* 113–126.

Introduction

Many researchers investigating changes in human diet during the Late to Terminal Pleistocene have concentrated on Europe and the eastern Mediterranean, with the focus usually on the "transition" from the Middle Paleolithic, associated with Neanderthals, to the Upper Paleolithic, associated with modern humans. Initially, most investigations focused on larger game exploitation, but few differences have been found (Grayson and Delpech, 2003; Steele, 2003; Stiner, 2005a). The key difference may have been an increase in the rate or degree of hunting success, but currently it is not possible to directly measure this with archaeological records. The most significant documented shift is the appearance of small, quick, difficult to catch game, such as fish, birds, hares and small carnivores, in Upper Paleolithic assemblages. In contrast, these prey species are rare or absent in earlier time periods (Straus, 1977; Chase, 1986; Stiner, 2001; Costamagno and Laroulandie, 2004; Laroulandie, 2004; Pérez Ripoll, 2004; Stiner, 2005a). This difference is further supported by the isotopic signatures preserved in Neanderthal and modern human bones, which indicate that some modern humans had a more diverse diet than Neanderthals (Richards et al., 2001). Additional changes during the Terminal Pleistocene, just before the advent of agriculture and domestication in the Near East, include the addition of more small game, a decrease in the ages of the large ungulates that were hunted, and an increase in how intensively these ungulates were exploited for resources such as marrow and grease (Munro, 2004; Davis, 2005).

However, paleontological and genetic data point to Africa as the place where modern humans originated, either during the Middle Stone Age (MSA, whose makers were Neanderthal contemporaries/near-modern humans living about 250,000 to 50,000 years ago) or at its end with the "transition" the to Later Stone Age (LSA associated with fully modern humans from 50,000 years ago to ethnohistoric times). It was only with the beginning of the LSA that modern humans spread from Africa, eventually replacing the Neanderthals in Eurasia. This means that the Middle to Upper Paleolithic, or the Neanderthal to modern human "transition" in Eurasia was not an *in situ* "transition." To understand the origins of modern human subsistence, we must examine the African record.

The modern human expansion out of Africa would have involved increases in human population sizes, and probably also in local population densities. This increasing population size comes at or after the "transition" from the MSA to LSA, and therefore followed technological changes that allowed modern humans to extract more resources from the environment. Following optimal foraging theory (Stephens and Krebs, 1986), this increased demand could be reflected in the archaeological record in a variety of ways (Munro, 2004). Modern humans could have increased their exploitation of existing high-ranked prey resources, assumed here to be ungulates, by obtaining more of these individuals. If they could, hunters would have first targeted prime-aged animals, because these animals maximize the caloric return for energy spent hunting them. Then, if there was strong hunting pressure on these prime individuals, which could happen if the technological changes of the LSA increased the hunting success rate of modern humans, the typical ages of the animals in the prey herds would have declined; so that by necessity, more young individuals would be taken. The modern humans could also have processed each animal more intensively to extract more calories and nutrients in the forms of marrow and grease. Additionally, decreasing abundance of high-ranked prey due to over-exploitation could cause people to increase their exploitation of existing low-ranked resources, here assumed to be small animals. As a result, small game would become more abundant in the diet, and the median size of slow-growing and long-lived mollusks and tortoises would decline as exploitation increased. Finally, the expanding populations may have added new low-ranked species to the diet. The addition of these previously low-ranked species to the LSA diet could also be the result of technological changes that made it possible or more efficient to capture these resources, and therefore, they would have been included in the diet.

Extensive research in southern Africa has identified a number of differences in human subsistence between the MSA and the LSA, including increased exploitation of dangerous game, such as buffalo and wild pigs; increased exploitation of young seals; increased exploitation of marine mollusks, tortoises, and airborne birds; as well as the addition of fish to the diet (Klein, 1975; Klein and Cruz-Uribe, 1983, 1996; Klein, 1998; Klein et al., 1999, 2004; Steele and Klein, 2005/2006).

However, despite the abundance of publications discussing the Eurasian and African changes in human diet during the Late Pleistocene, only a few formal comparisons of the two records have been made (e.g., Marean and Assefa, 1999; Marean, 2005). Our goal here is to begin making these comparisons. Specifically, we analyze the southern African record using similar methods to those typically employed by researchers working in Eurasia (e.g., Stiner and Munro, 2002; Munro, 2004; Stiner, 2005b). We highlight similarities and differences between the records and suggest areas that require further research. In doing so, we hope to encourage a more global view of the evolution of human diet.

Samples

The faunal remains from nine sites found along the coast of South Africa are considered here (Fig. 8.1, Table 8.1). We chose these sites b ecause they have provided the best-documented assemblages available, and most of them were included in earlier investigations into MSA and LSA ecology. Although there is some inconsistency in reporting, the assemblages have been excavated and analyzed by related teams of researchers, making the data comparable between assemblages. The recovery and recording of small animals, such as hare (*Lepus* sp.) and dune molerat (*Bathyergus suillus*) dental remains indicates that excavation and analytical procedures were sufficient for consideration of small game exploitation.

The assemblages span two environmental zones. This is important because it increases the ecological variability under consideration, making it more likely that any identified patterns reflect human behavior and not climatic or environmental variation. The ocean waters off the western coast of South Africa are cool and nutrient-rich, while the land experiences long, warm, dry summers and short, cool, wet winters, limiting the available land resources. In contrast, the waters off the southern coast are more variable, and the land receives rainfall during the summer months and in larger amounts than on the western coast. As a result, the southern coast has a richer terrestrial environment, and these environmental differences are also reflected in the animal community. Overall, the contrast between western coast and southern coast faunal assemblages is present throughout the Late Pleistocene and Holocene, indicating that this environmental variation has great time depth, but the boundary between the two environmental zones has fluctuated with the glacial cycles.

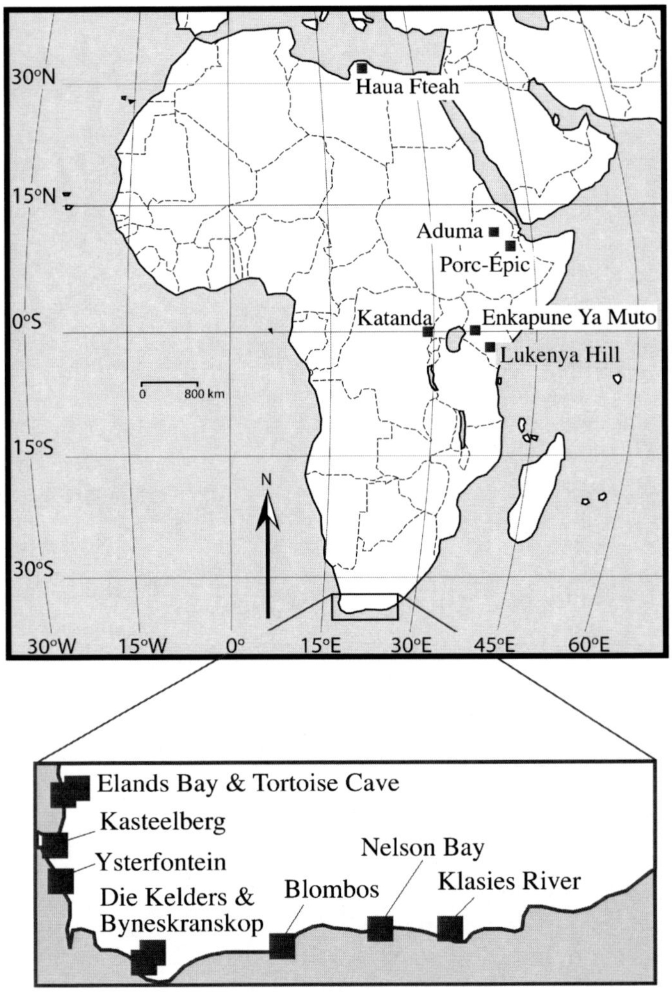

FIG. 8.1. Map of the sites discussed in the text. Table 8.1 lists the references for each sample that provided the data presented here.

Increased Exploitation of Existing High Ranked Prey Resources

Previous studies of large game exploitation in South Africa have investigated the relative species abundances of some the largest prey, as well as the associated mortality profiles. Eland (*Taurotragus oryx*) were more common in MSA assemblages (KRM, DK1, and BBC), while Cape buffalo (*Syncerus caffer*) and bushpig (*Potamochoerus larvatus*) were more abundant in LSA assemblages (NBC and BNK1) (Klein, 1994; Klein and Cruz-Uribe, 2000). This pattern holds across two geographic areas and multiple environmental regimes (Fig. 8.2). Cape buffalo and bushpig are among the most difficult to capture of those animals identified in the assemblages because of their strong predator-defense capabilities. Their greater abundance in the LSA assemblages could reflect the introduction of projectile technology, which allowed hunters to strike from a safer distance. When the mortality profiles for the bovids are compared, the profiles for Cape buffalo from KRM MSA and NBC LSA are similar to what might be expected from attritional mortality factors like predation and endemic disease, which disproportionately affect the weakest members of the population – the youngest and oldest. The mortality profiles constructed for eland from KRM MSA and DK1 MSA more closely resemble what would be expected if individuals of different ages were being taken in proportion to their live abundance (Klein, 1978, 1994; Klein and Cruz-Uribe, 1996). These patterns were found to be consistent across two techniques of mortality profile analysis: histograms of age-at-death estimates and boxplots of tooth crown heights (see Steele, 2005 for a detailed comparison of the methods). Unfortunately, small sample sizes do not allow for an examination of eland mortality profiles from LSA assemblages.

Following analyses conducted in the Mediterranean Basin, mortality profiles were further investigated via triangular graphs, which plot the percentages of juveniles, prime, and old individuals in a faunal sample (Stiner, 1990). The method was modified to add 95% confidence limits to the triangular graphs (Steele and Weaver, 2002; Steele, 2005). When plotted this way, the eland samples from three MSA assemblages are statistically indistinguishable from each other, and show an abundance of prime individuals; however attritional hunting cannot be ruled out with statistical confidence for DK1 MSA (Fig. 8.3a; Tables 8.2 and 8.3). When Cape buffalo and giant or long-horned buffalo (*Pelorovis antiquus*) are added to the graph, they are also statistically indistinguishable from each other and indicate attritional or juvenile dominated mortality. In addition, with the exception of the DK1 MSA eland sample, the eland profiles are significantly different from the

TABLE 8.1. References that provided the data used in this paper.

Site	Abbreviation	References
Blombos Cave	BBC	Henshilwood et al., 2001
Byneskranskop 1	BNK1	Klein, 1981; Schweitzer and Wilson, 1982
Die Kelders 1	DK1	Schweitzer 1979; Klein and Cruz-Uribe, 2000; Avery, 1990; G. Avery, 2003, personal communication
Elands Bay Cave	EBC	Klein and Cruz-Uribe, 1987; Avery, 1987; Avery, 1990
Kasteelberg	KBA, B	Klein and Cruz-Uribe, 1989
Klasies River Main	KRM1, 1A, 5	Klein, 1976b; Singer and Wymer, 1982; G. Avery, 2003, personal communication
Nelson Bay Cave	NBC	Klein, 1972a, b; Avery, 1990
Tortoise Cave	TC	Klein and Cruz-Uribe, 1987
Ysterfontein 1	YFT1	Klein et al., 2004; Halkett et al., 2003; G. Avery, 2006, personal communication, continuing research

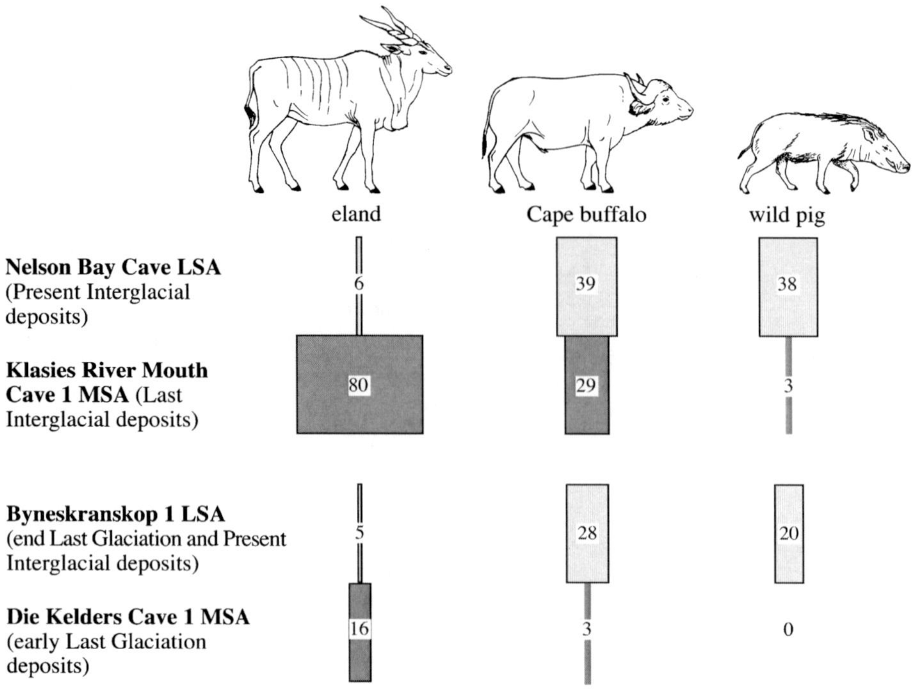

FIG. 8.2. Top: The minimum numbers of eland, Cape buffalo, and bushpig in samples from the Last Interglacial deposits of KRM Cave 1 and in the Present Interglacial deposits of nearby NBC. Bottom: The minimum number of individuals from the same species in the early Last Glaciation deposits of DK1 and in the end Last Glacial/Present Interglacial deposits of nearby BNK1. The relatively greater abundance of buffalo and bushpig at NBC and BNK1 probably reflects the more sophisticated technology of local LSA hunters over their MSA predecessors.

buffalo profiles. These results are consistent with earlier studies based on histograms and boxplots.

We examined a few additional species to further investigate the question of MSA and LSA hunting patterns, and the possibility that large game experienced over-exploitation with the "transition" from the MSA to the LSA. The mortality profiles of the extinct blue antelope (*Hippotragus leucophaeus*) from KRM MSA and NBC LSA show a similar pattern to the buffalo from the same sites (Fig. 8.3b). The samples are statistically indistinguishable from each other and most similar to what is expected for attritional hunting, although the hunting of more prime individuals cannot be excluded, particularly for NBC LSA.

All of the bovids we have discussed so far are large bodied (large-medium or large body-size classes), so they would provide the most calories and potentially be the most highly ranked in a forager's diet. Their slower reproductive rates also mean that they should be sensitive indicators to over-exploitation. However, when comparisons are available, we find no differences in mortality profiles between the MSA and LSA samples. Rather, the differences we do find are between species. During the MSA, eland may have been individually hunted in equal proportion to their abundance on the landscape or entire herds may have been taken at once (as would occur during a drive), creating a catastrophic mortality profile. In contrast, the buffalo and blue antelope samples consistently fall into the attritional zone, indicating that they were probably hunted singly, and only the most vulnerable individuals were taken during both the MSA and LSA. Advances in LSA technology, such as projectile weapons, should have allowed hunters to successfully hunt buffalo more frequently, but because they were taking only the most vulnerable individuals, the age structure of the living population would not have been affected, and hunting was sustainable. The abundance of juvenile remains in the buffalo and blue antelope samples suggests that these animals were not the result of scavenging, because the initial predator would be expected to consume juvenile carcasses more or less completely. These results contrast with those from the Mediterranean Basin where very few samples fall within the attritional zone and none within the juvenile dominated zone. Additionally, old individuals dominate a few Mediterranean Basin samples but no MSA or LSA samples (Stiner, 1990).

To further examine predator-prey dynamics, we investigated the mortality profiles of steenbok (*Raphicerus campestris*) and grysbok (*R. melanotis*), small bovids that are much more abundant in the faunal assemblages than the species discussed above (Fig. 8.3c). The two MSA samples we consider contain only grysbok, while the LSA samples contain a mix of grysbok and steenbok (which are distinguishable skeletally mainly by their horncores and the shape of the inferior margin of the mandible (Klein, 1976a)). The steenbok/grysbok mortality profiles show a variety of distributions. The two MSA assemblages are significantly different from each other, with KRM having more prime adults and DK1 having more juveniles. The DK1 LSA, BNK LSA, and EBC LSA samples

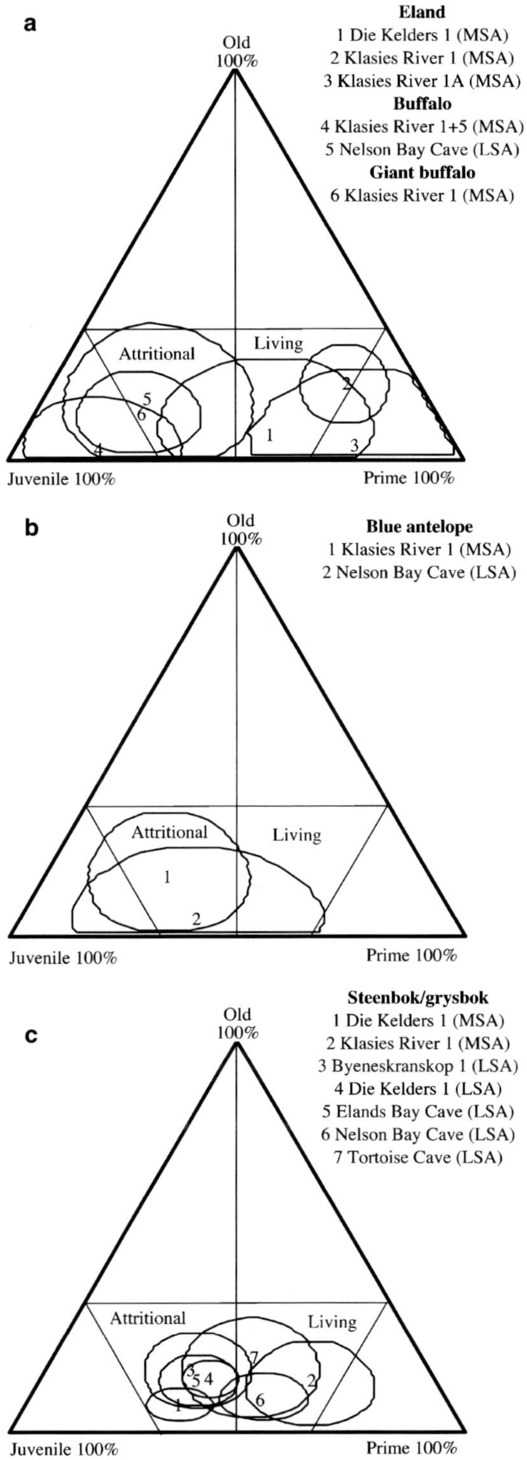

FIG. 8.3. Modified triangular graphs showing the percentage of young, prime, and old individuals in each sample, along with the 95% confidence interval around each point (Steele and Weaver, 2002). (**a**) MSA eland hunters were able to hunt many adult eland, while both MSA and LSA hunters could only take the youngest and oldest buffalo. (**b**) Both MSA and LSA people hunted only the youngest and oldest blue antelope. (**c**) There is a wide variety in the ages of steenbok and grysbok that were taken by MSA and LSA hunters. Together, these graphs show that prey age selection is more closely related to species than to technology.

are all similar to the DK1 MSA sample, and fall within the zone for attritional mortality on the graph. These results are somewhat surprising given the expectation that neither MSA nor LSA people should have had problems capturing these small-bodied bovids. There are a few possible explanations. First, steenbok and grysbok are fast-breeding (continuously through the year with minor birth peaks) and short-lived species with relatively low-crowned teeth (Klein, 1976a; Skinner and Chimimba, 2005). Juveniles are abundant and old individuals are naturally rare in these populations, so it may not be possible to distinguish attritional and catastrophic shaped mortality profiles unless samples are very large. Furthermore, reliably separating the few old individuals from prime adults can be difficult in species with low-crowned teeth. As a result, the variation identified here may have a number of causes, and larger sample sizes and more data on the age structures of living *Raphicerus* populations and on age estimation techniques are necessary for fuller examination. Stiner (2005b: 207–208) offers an alternate explanation for some samples that fall within the attritional zone of the triangular graph. These samples may reflect hunting pressure, because the age structure of the herd has been altered so that a larger proportion of juveniles is present than expected in a typical living structure, and there are fewer old individuals. In this case, the proportion of old individuals in the sample can help distinguish attritional mortality from hunting pressure, because more old individuals should be present in the attritional profile. A final explanation is that some of the specimens, especially those in select DK1 MSA samples, were not the result of human hunting at all, but were in fact brought into the cave by large raptors (Klein and Cruz-Uribe, 2000).

Modern humans may have extracted more calories from large game by processing carcasses more intensely. Unfortunately, few studies on MSA and LSA faunal assemblages have directly addressed this issue, perhaps because marrow processing has been securely documented in earlier Plio-Pleistocene faunal assemblages (Blumenschine, 1995; de Heinzelin et al., 1999). For the MSA of South Africa, only a few samples of the DK1 MSA fauna have been coded for traits characteristic of marrow extraction, including fracture outlines and angles and percussion marks, and the samples show evidence for marrow extraction (Marean et al., 2000). Only one similar analysis has been conducted on an LSA sample, and it also concludes that the bones had been exploited for marrow (Dewar et al., 2006). To the best of our knowledge, no LSA assemblage has been examined for evidence of boiling for grease extraction (as in Outram, 2001; Munro and Bar-Oz, 2005). Further research will hopefully shed more light on this issue.

In sum, the primary contrast between MSA and LSA large game exploitation is in relative species abundance. The variation in mortality profiles appears to be more species-specific and not directly related to differences in technology. Advances in technology may have allowed LSA hunters to more frequently take buffalo and pigs, but not so frequently that the age structures of the herds were impacted.

TABLE 8.2. Data used to calculate mortality profiles on the modified triangular graph (Steele and Weaver, 2002). Young = number of dp4s with measurable crown heights; Prime = number of measurable m3s with crown heights less than 50% worn; and Old = number of measurable m3s with crown heights more than 50% worn (following Klein et al., 2007). The degree of wear on each m3 was assessed by dividing each crown height by maximum unworn m3 crown height (mm) for that species: eland = 47.1[a], buffalo = 53.0[b], giant buffalo = 70.2[b], blue antelope = 40.0[b], and steenbok/grysbok = 14.7[a].

Species	Site	Industry	Young	Prime	Old	Total
Eland	DK1	MSA	8	11	2	21
Eland	KRM1	MSA	15	59	17	91
Eland	KRM1A	MSA	3	11	1	15
Buffalo	KRM1 + 5	MSA	18	4	1	23
Buffalo	NBC	LSA	12	5	3	20
Giant buffalo	KRM1	MSA	29	10	5	44
Blue antelope	KRM1	MSA	15	7	4	26
Blue antelope	NBC	LSA	9	6	1	16
Steenbok/grysbok	BNK1	LSA	40	26	12	78
Steenbok/grysbok	DK1	LSA	140	106	37	283
Steenbok/grysbok	DK1	MSA	119	68	13	200
Steenbok/grysbok	EBC	LSA	75	50	18	143
Steenbok/grysbok	KRM1	MSA	12	26	5	43
Steenbok/grysbok	NBC	LSA	48	62	10	120
Steenbok/grysbok	TC	LSA	20	23	8	51

[a]Unworn m3 crown height taken from original data.
[b]Unworn m3 crown height taken from Klein (1978).

TABLE 8.3. The minimum number of individuals (MNI) used to calculate the relative abundances presented in Figs. 8.3 and 8.4. Presence and absence of other taxa are also provided.

	DK1	KRM	BBC	NBC	EBC	KBA	KBB	BNK1	DK1	KRM	BBC	YFT1
	LSA	LSA	LSA	LSA	LSA	LSA	LSA	LSA	MSA	MSA	MSA	MSA
Small game sub-total	2,811	2	108	148	6,361	99	2,217	2,943	8,848	100	943	109
Hares	5	0	4	3	61	1	4	10	392	2	9	4
Cape dune molerat	1,766	0	60	0	334	1	5	95	3,646	0	34	33
Hyrax	28	2	12	145	86	0	0	9	170	98	37	0
Tortoise[a]	1,012	0	32	0	5,880	97	2,208	2,829	4,640	0	863	72
Carnivore sub-total	140	35	9	204	329	32	182	41	98	184	28	21
Cape fur seal	108	28	4	160	216	24	133	10	38	134	7	11
Other carnivores	32	7	5	44	113	8	49	31	60	50	21	10
Ungulate sub-total	330	59	238	369	481	43	102	237	444	634	66	33
Bird sub-total[b]	99	56		509	445	Rare	Rare	Y	66	174		28
Penguin	16	6	?	106	65				26	124	?	12
Airborne marine birds	73	50	?	443	380				40	50	?	16
Ostrich egg shell	Y	?	?	Y	?	?	?	Y	?	Y	Y	Y
Mollusks	Y	Y	Y	Y	Y	Y	Y	Y	Y[c]	Y	Y	Y
Fish	Y	Y	Y	Y	Y	Rare	Rare	Y	N	Rare	Y	N

[a]Number of identifiable humeri.
[b]Freshwater and terrestrial birds are not considered here because the data are not available for all sites.
[c]Poorly preserved.

Increased Exploitation of Existing but Low Ranked Resources

Following the work conducted around the Mediterranean, we next investigate the relative abundance of prey taxa, because foragers can extract more calories from the environment by increasing their exploitation of lower-ranked resources that are already part of their diet. This is done either when the higher ranked resources became harder to find, or when technological changes make the lower ranked resources more profitable. In the case of coastal South Africa, significant new technologies are found in the LSA that were not present in the MSA, and so technological change can be implicated in many of the dietary changes. We follow Munro (2004), and begin by comparing the proportions of ungulates, carnivores, and small game in a number of MSA and LSA assemblages (Fig. 8.4). As with Mediterranean Basin samples (Stiner and Munro, 2002), this analysis shows little patterning, although KRM LSA and MSA and NBC show a low abundance of small game. We note that Cape fur seals (*Arctocephalus pusillus*) are commonly classified as Carnivora (Skinner and Chimimba, 2005; Wilson and Reeder, 2005) and therefore are included as carnivores here, but their ecology is very different from other members of this order; they were frequently exploited for human subsistence, unlike other members of this order. This underscores the importance of examining prey ecologically, not always taxonomically.

To further examine relative abundances in more detail, we calculated the reciprocal of Simpson's Index (as described in Stiner, 2005b: 167) to examine the evenness in the representation of genera in the samples (approximately 60 genera are represented between the samples; 40 more than in the Mediterranean Basin series). Evenness is consistently low in all samples (<4.2), except that again the KRM LSA, KRM MSA, and NBC differ (8.7, 12.6 and 9.0 respectively). Furthermore, there is no significant trend through time ($r^2 = 0.15$, $p = 0.22$) or between the MSA and LSA medians (Wilcoxon 1-way Test, ChiSquare Approximation $p = 0.61$). The KRM and NBC samples are distinctive because they lack tortoises, which are extremely abundant in all the other samples. These sites are further east on the South African coast than the other samples considered here, and are located in an environment that must not have ever been favorable for tortoises.

Much of the recent research into changing diets in the Middle and Upper Paleolithic has emphasized the importance of small game, and especially the need to consider small game in terms of its ease of capture and reproductive potential instead of taxonomic classification (Stiner et al., 2000; Stiner and Munro, 2002; Munro, 2004). These studies have contrasted the exploitation of slow versus quick game, namely tortoises and mollusks versus hares and game birds, although other taxa are included. Middle Paleolithic samples contain primarily slow, easy-to-capture prey, and quick game do not become more abundant until the Upper Paleolithic, increasing into the Epipaleolithic (Stiner et al., 1999, 2000; Stiner and Munro, 2002; Stiner, 2005a, b).

As with the large game in South Africa, comparable comparisons are not straightforward. Slow, easy to capture resources include tortoises (*Chersina angulata*), mollusks, and ostrich (*Struthio camelus*) eggs. However, these remains are inconsistently quantified, making comparisons difficult. Tortoise abundance may be compared using either the minimum number of humeri or femora, or the number of identified specimens (which can be artificially inflated because of the high fragmentation potential of the carapace and plastron). Mollusks are weighed and where possible, non-duplicated parts (mussel hinges and limpet apices) are counted for minimum numbers of individuals. Ostrich eggshell can be quantified by weight or by counting the number of fragments, but some pieces of eggshell may enter the site as water canteens, blanks for making beads, or already-made beads. Nevertheless, if these data are consistently reported across samples, comparisons can be made.

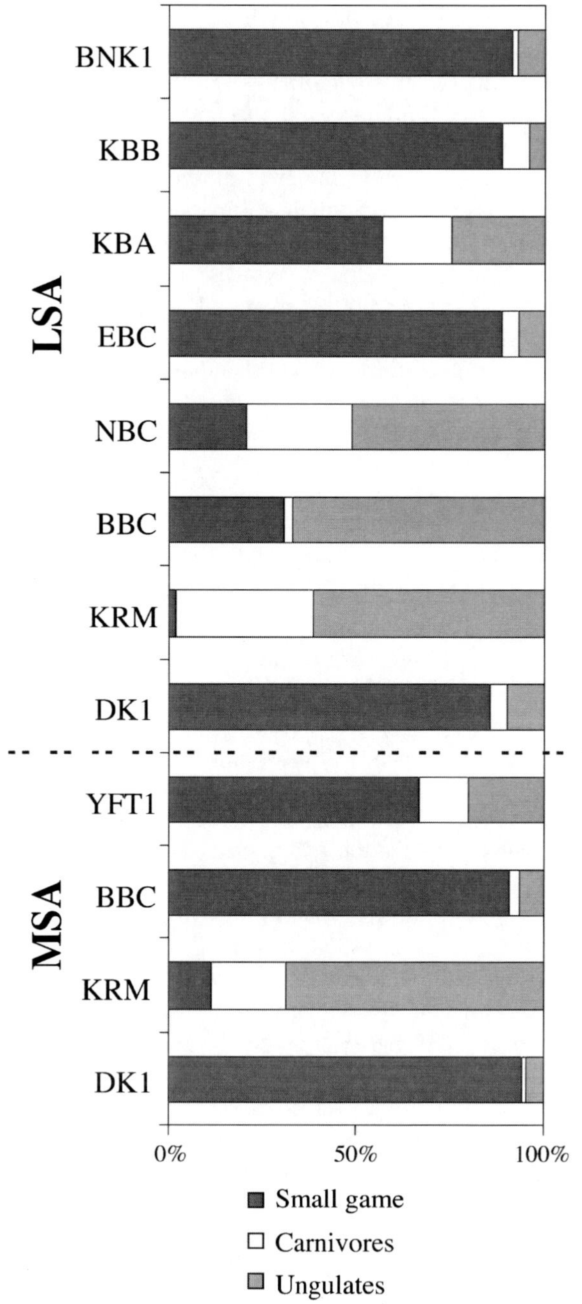

FIG. 8.4. Relative abundance of ungulates (bovids, equids, rhinos, suids), carnivores (small and large, seals), and small game (hare, molerat, hyrax, tortoise) in MSA and LSA assemblages. Birds are not included here because data are not available for all assemblages. All abundance indices are based on MNI counts. Abbreviations for sites follow Table 8.1.

Small, quick game also create some complications. Hare are rare in both MSA and LSA faunal assemblages (<1.5% in the samples considered here), although their consistent but sparse presence indicates that they existed on the landscape, at least in small numbers, when environments were suitable. Hyraxes (*Procavia capensis*) and dune molerats are also present in the majority of assemblages considered here, often in high numbers. Like hares, they probably should be considered quick, hard-to-capture game, given their ability to rapidly return to their shelters or burrows. The more important question, however, is ensuring that the hares, hyraxes, and molerats were brought into the sites as human food and not by non-human predators, especially raptors. Klein and Cruz-Uribe (2000) concluded that the extremely high abundance of molerats in the MSA at DK1 (over 90% of the fauna) was a result of eagle owl hunting, not human exploitation. However, in the same study, they showed that humans probably collected the molerats in the LSA layers, because unlike the MSA specimens, some of these were burnt or had cut-marks. Henshilwood (1997) documented another example of humans consuming molerats in the LSA layers at BBC, identified by a distinct burning pattern on their premaxillas, mandibles and incisors; he supported this inference with ethnographic examples. Because of these factors, detailed taphonomic studies are needed to determine the dominant mode of accumulation for each sample. Unfortunately, these detailed analyses have not yet been documented for the majority of samples considered here, but the general pattern appears to be that molerats were more frequently exploited for food in the LSA than they were in the MSA, supporting the proposal that quick game were more frequently taken in the LSA than in the MSA.

Cruz-Uribe and Klein (1998) determined that skeletal part representation can be useful for distinguishing hares and hyraxes accumulated by humans from those accumulated by eagles, because eagle roost assemblages often contain an abundance of hind-limb elements from these small animals. When they examined hare and hyrax skeletal part representation in many of the archaeological sites considered here, they concluded that humans, not raptors, were responsible for introducing the majority of these small animals into the sites. When they further examined the ages of the hyraxes in the samples, they documented that MSA samples (KRM and DK1) contained few young hyraxes compared to the LSA samples (NBC and DK1). They argued that this difference could reflect differences in hunting technology, which allowed LSA foragers to more readily capture quick game.

The final type of quick game is birds. These can be divided into three groups along the South African coast: marine, freshwater, and terrestrial, although marine birds (primarily cormorants [*Phalacrocorax* sp.] and gannets [*Morus capensis*]) provide the majority of specimens (Avery, 1987). However, the South African coast has one species that deserves special consideration – African or jackass penguins (*Spheniscus demersus*). These flightless birds are probably easier to take than their airborne cousins, but unless they are scavenged, they are not gatherable like tortoises. Because they do not fall neatly into a particular category, we consider them separately here. The ethnohistoric record provides a few descriptions of techniques for capturing coastal birds: young penguins on the beach were chased and captured, juvenile cormorants were clubbed, and gulls were shot with bows and arrows (Avery, 1984 in Avery, 1987). In inland regions snares, bows and arrows, and sticks were used (Steyn, 1971 in Avery, 1987).

We start by comparing slow, easy-to-capture game to quick game, with penguins in their own category. Because of the taphonomic and quantification issues described above, our comparison includes tortoises, penguins, and airborne marine birds, plus hares and hyraxes only for assemblages where marine bird abundances are available (Fig. 8.5a). When present, tortoises dominate all assemblages, masking any difference between the MSA and LSA, so relative abundances of tortoises versus fast game do not seem to be informative. All of these samples also contain varying abundances of mollusks (although they were too poorly preserved in DK1 MSA to quantify), and if they were included, slow game would further dominate the assemblages. Therefore, we must examine the other components in more detail. A comparison of just penguins to airborne marine birds reveals that the LSA samples all contain significantly more airborne marine birds than penguins (Fig. 8.5b). These differences could reflect different seasons of site occupation (Avery and Underhill, 1986; Avery, 1987), but because they are consistent between so many assemblages, we think they are more likely to reflect differing subsistence behaviors. Furthermore, only LSA sites contain artifacts similar to ethnohistoric fowling gear. Although Avery (1987) argues that these could actually be fishing implements, this further supports the case for an increased exploitation of airborne birds in the LSA. Comparisons are not available for inland sites, and inland sites contain fewer birds overall (Avery, 1987).

Despite the consistently high abundance of tortoises and mollusks in most assemblages, these taxa may still reveal evidence for increased human exploitation. These animals have slow and continuous growth, and a relatively extended time until sexual maturity; therefore, under intense exploitation the median body size of the available animals will decline (Klein and Cruz-Uribe, 1983; Stiner et al., 2000; Mannino and Thomas, 2002; Steele and Klein, 2008). In the Late Pleistocene Mediterranean Basin, tortoise (*Testudo graeca*) size may decline from the Middle to Upper Paleolithic, and limpet (*Patella caerula*) size declines within the Upper Paleolithic (Stiner et al., 1999, 2000; Stiner, 2005b). In South Africa, smaller LSA mollusks (relative to their MSA counterparts) have been clearly documented for both the western (limpets [*Cymbula* sp. and *Scutellastra* sp.]) and southern (Cape Turban shells [*Turbo sarmaticus*]) coasts, indicating more intense mollusk exploitation in the LSA (Fig. 8.6) (Voigt, 1982; Parkington, 2003; Steele and Klein, 2005/2006, 2008). Because of the large number of samples considered from a variety of environments, climate is probably not the causal factor, making it more likely that higher human population

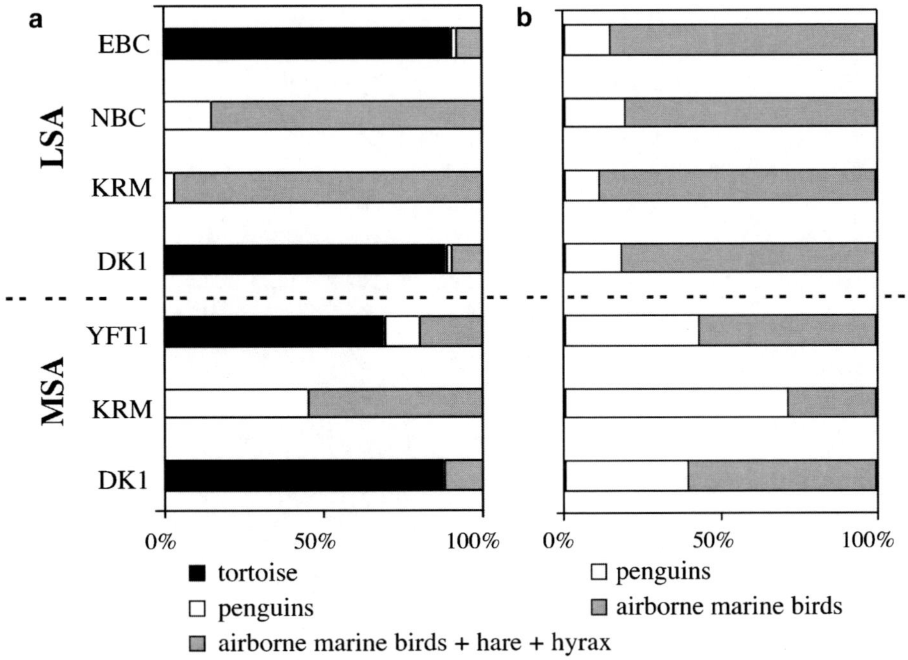

FIG. 8.5. (**a**) Relative abundance of slow (tortoise, penguins) versus quick (airborne marine birds, hare, hyrax) prey and (**b**) penguins versus airborne marine birds. LSA samples contain a significantly larger abundance of airborne marine birds compared to penguins (Likelihood Ratio or *G*-test of Independence [Sokal and Rohlf, 1995: 731–732] on pair-wise comparisons: YFT1 MSA/EBC LSA $p = 0.0087$, KRM MSA/NBC LSA $p < 0.0001$, KRM MSA/KRM LSA $p < 0.0001$, DK1 MSA/DK1 LSA $p = 0.0030$). Abbreviations for sites follow Table 8.1.

densities, and hence the collecting of more mollusks, caused the diminution. Declines in median tortoise sizes, as determined by the maximum breadth of distal humeri, from the MSA to LSA samples are not as clear as in the mollusks, but in general LSA tortoises are smaller than MSA ones (Klein and Cruz-Uribe, 1983; Steele and Klein, 2005/2006).

Exploitation of New Prey

Intensifying human groups may add new species to their diets to obtain more calories from the environment. Along the coast of South Africa, we see the inclusion of many more marine components in the LSA diet (Klein et al., 2004; Steele and Klein, 2008). The most significant addition to the LSA diet that is rarely seen in the MSA is large numbers of coastal fish. As with fowling implements, only LSA assemblages have provided artifacts that resemble recent fishing equipment, such as fish gorges. On the west coast, rock lobsters (*Jasus islandii*) were also added to the diet in the LSA (Buchanan, 1988; Jerardino et al., 2001; Orton et al., 2005). While a variety of mollusks were gathered during the MSA, exploitation concentrated on a few species, especially limpets and black mussels (*Choromytilus meridionalis*). In contrast, in the LSA, other taxa, including sand mussels (*Donax serra*) and whelks (*Burnupena* sp.), were regularly gathered, occasionally in large quantities (Buchanan et al., 1978; Klein et al., 2004, current research). This higher diversity of mollusks in the LSA samples reflects the LSA people's wider diet breadth, which was probably necessary to support their larger population sizes. It is unknown if LSA technological changes made it more efficient to gather a higher diversity of mollusks, but these advances almost certainly were critical for supporting larger human population sizes.

Discussion

Our results indicate that there were significant changes in subsistence between the MSA and LSA, and that LSA foragers consumed a wider variety of resources. We propose that this increased diet breadth was able to support larger and more dense human populations, and that these changes must have occurred close in time to when modern humans were expanding out of Africa.

However, there is one major limitation to our analysis: the MSA samples included here all accumulated more than 60,000 years ago, while the LSA samples are all younger than 20,000 years old. Unfortunately, coastal South Africa experienced extreme aridity in the midst of the Last Glaciation (OIS 3), and as a result, local human populations were very sparse during this time. Therefore, it may not be possible to fully test our proposals using the coastal record from South Africa. We must turn to other parts of Africa where climatic conditions were more favorable to human occupation during this time.

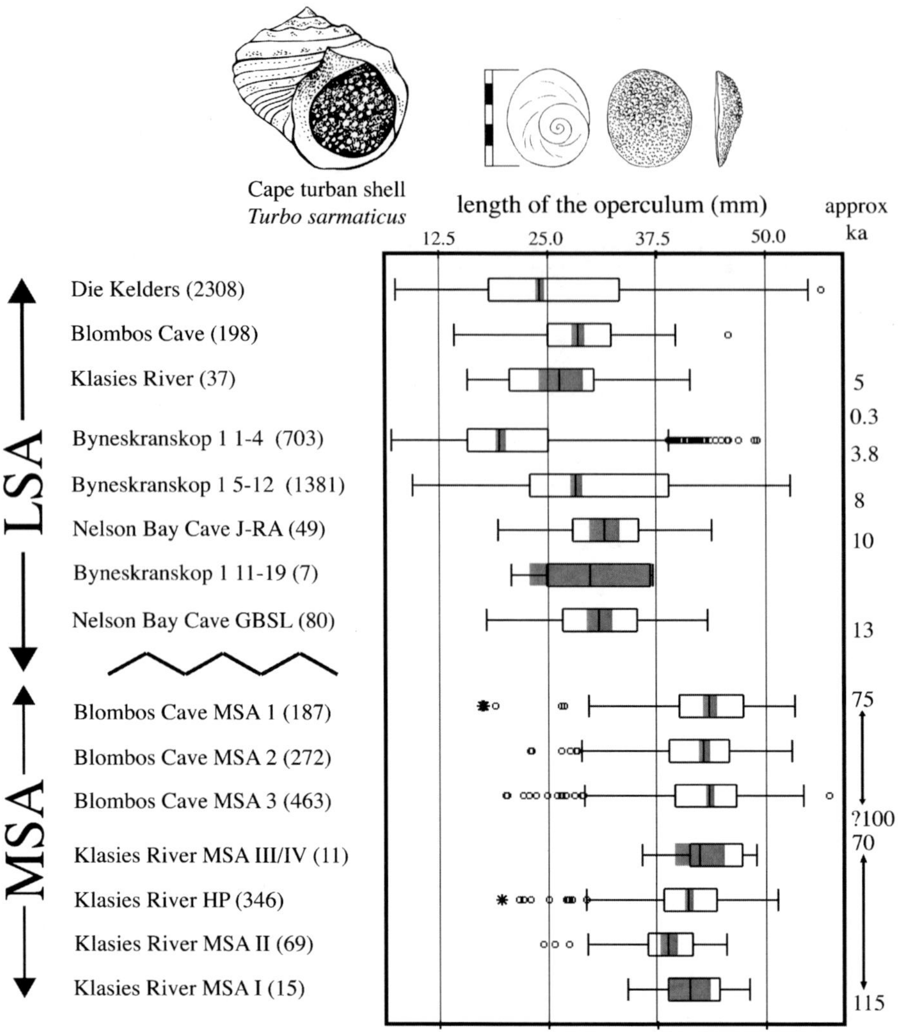

FIG. 8.6. Box plots summarizing the maximum diameter of Cape turban shell opercula from successive units of many of the sites considered here. The number in parentheses after each sample name is the number of specimens in that sample. The sample median is indicated by the vertical line near the center of each plot; the 95% confidence limits for the median is indicated by the shaded rectangle around each median; and the number of measured shells is shown in parentheses. Medians whose 95% confidence limits do not overlap are significantly different. The numbers along the right margin designate the approximate ages of the samples in thousands of years before present. The LSA opercula tend to be significantly smaller than the MSA ones, suggesting that LSA people were more intensively collecting turban shells, likely because LSA people were living at higher population densities than MSA people.

After southern Africa, northern Africa has been subject to the most archaeological investigations into Late Pleistocene deposits. Unfortunately, despite this attention, little is known about the details of the subsistence strategies. This region also probably suffered from the same aridity as southern Africa, and therefore human populations were probably equally sparse during the interval of interest. Bone preservation is generally poor in the northeast corner of the continent, so only a few faunal analyses have been possible, and they have necessarily been limited. The western part of the region contains more cave and rockshelter deposits, so bone preservation and relative chronological control is better. However, much of the work done in this area has been paleontological and biochronological in nature. Although some preliminary work has been done (Thomas, 1981; Amani, 1991; Aouraghe, 2004), more detailed zooarchaeological studies are needed. One exception is Klein and Scott's (1986) re-analysis of the fauna from Haua Fteah and a few nearby sites from Cyrenaican, Libya, which span from the Middle Paleolithic to historic times. However, their analysis was hampered by a lack of stratigraphical control over the assemblages, because the excavation was conducted in spits that crossed natural stratigraphic levels. Nonetheless, they discuss mollusk remains from this site and others from coastal northwestern Africa, tortoises that were present in variable numbers through the sequence, and abundances of large mammals that varied with climate. Unfortunately, sample sizes were not large enough to examine variation in mollusk or tortoise body size through time. Birds were preserved throughout the sequence, along with ostrich eggshells. Although bird bones are sparse in the Middle Paleolithic levels, some dove (*Columba* sp.) and partridge (*Alectoris barbara*) bones were burnt (6% of the total), indicating that humans occasionally exploited them (MacDonald, 1997).

Because of the extreme aridity at the northern and southern ends of the African continent during the Last Glaciation, the most likely place to find faunal assemblages that span the MSA to LSA "transition" is in East Africa. However, until recently, few zooarchaeological studies have been conducted, in part because few suitable assemblages from the relevant sites are known. One exception is Porc-Epic, Ethiopia, where Assefa (2006) recently completed a detailed analysis of the MSA faunal remains. The fauna supports the interpretation of accumulation during the Last Glaciation because more grazers are present in the MSA assemblage than in the historic fauna, indicating a drier, more open environment. The sample size of dental remains is too small for mortality profile analysis, but the high frequency of green fractures and of incomplete shaft circumferences, especially on the larger bovid remains, shows extensive marrow extraction. However, the assemblage also shows a high degree of carnivore tooth-marks relative to cut-marks, but Assefa (2006) argues that the overall pattern still indicates first access by hominids, not by carnivores. For small, quick game, both hare and hyrax are present in the assemblage; slow game are either not present or not considered. Many of the hare and hyrax bones had butchery marks and traces of burning, but the skeletal parts found in the assemblage were similar to those that Cruz-Uribe and Klein (1998) documented in eagle assemblages, causing Assefa (2006) to conclude that both humans and raptors played a role in accumulating these small animals. Unfortunately, no studies on nearby LSA remains are available for further comparison. Elsewhere in Ethiopia, preliminary work at sites in Aduma, Middle Awash Valley, suggest that MSA people exploited large, easily predated catfish (*Clarias* sp.) (Yellen et al., 2005). Catfish (*Clarias* sp. and *Synodontis* sp.) may also have been exploited at Katanda, Upper Semliki Valley, Democratic Republic of the Congo (Brooks et al., 1995).

The contrasts between MSA and LSA faunal assemblages are greater than the variation found among MSA assemblages. This indicates that significant dietary changes occurred with the "transition" from the MSA to LSA, and had a major role in the expansion of modern humans out of Africa. Specialists agree that humans were fully modern in their behavior with the "transition" from the MSA to the LSA, but they disagree on whether the behavioral shift occurred gradually within the MSA (Gradual Accumulation Model) or abruptly at its end (Abrupt and Late Model) (Henshilwood and Marean, 2003; Steele and Klein, 2005/2006). Our results argue that intensification did not occur until the LSA, and could not occur until the changes in LSA technology took place. Therefore, modern human behavior more likely arose at the beginning of the LSA and not during the MSA. However, this proposal remains to be tested with more assemblages spanning the geographic and temporal gaps in the African Late Pleistocene archaeological record.

Conclusions

There are many subsistence differences between MSA and LSA foragers of coastal South Africa. LSA people more intensively extracted resources from the environment. These results are broadly consistent with those found in the Mediterranean Basin contrasting Middle Paleolithic and Upper Paleolithic foragers, although the specifics are different. Prey mortality profiles remain consistent between the MSA and LSA, and they vary more with species than with technology. However, relative species abundances are different between MSA and LSA assemblages. LSA people exploited more bushpigs and buffalo, while MSA people took more eland. LSA people consumed more molerats, young hyraxes, and airborne birds. They added fish and rock lobster to their diet, as well as regularly consumed a wider variety of mollusks. Finally, the median mollusk size is smaller in LSA samples than in their MSA counterparts, indicating that more LSA people were more heavily exploiting mollusks. All of these lines of evidence point to a larger diet breadth for LSA foragers, which supported larger human population densities. The challenge for understanding modern human origins is finding faunal assemblages from throughout the African continent that span the 60,000 to 20,000 year gap in coastal South African assemblages, so that we can more closely document the subsistence changes that co-occur with the expansion of modern humans out of Africa.

Acknowledgments. We would like to thank Jean-Jacques Hublin and Mike Richards for inviting us to participate in this symposium, funded by the Max Planck Society, and Allison Cleveland for patiently handling the manuscript. Tim Weaver and conference participants provided helpful comments.

References

Amani, F., 1991. La faune de la grotte à Hominidé du Jebel Irhoud (Maroc). University of Rabat, Rabat, p. 229.

Aouraghe, H., 2004. Les populations de mammifères atériens d'El Harhoura 1 (Témara, Maroc). Bulletin d'Archéologie Marocaine 20, 83–104.

Assefa, Z., 2006. Faunal remains from Porc-Epic: paleoecological and zooarchaeological investigations from a Middle Stone Age site in southeastern Ethiopia. Journal of Human Evolution 51(1), 50–75.

Avery, G., 1987. Coastal birds and prehistory in the western Cape. In: Parkington, J.E., Hall, M. (Eds.), Papers in the Prehistory of the Western Cape, South Africa. British Archaeological Reports, Oxford, pp. 164–191.

Avery, G., 1990. Avian fauna, Palaeoenvironments and Palaeoecology in the Late Quaternary of the Western and Southern Cape, South Africa. Department of Archaeology, University of Cape Town, Cape Town, p. 206.

Avery, G., Underhill, L.G., 1986. Seasonal exploitation of seabirds by Late Holocene coastal foragers: analysis of modern and archaeological data from the western Cape, South Africa. Journal of Archaeological Science 13, 339–360.

Blumenschine, R.J., 1995. Percussion marks, tooth marks, and experimental determinations of the timing of hominid and carnivore access to long bones at FLK *Zinjanthropus*, Olduvai Gorge, Tanzania. Journal of Human Evolution 29(1), 21–51.

Brooks, A.S., Helgren, D.M., Cramer, J.S., Franklin, A., Hornyak, W., Keating, J.M., Klein, R.G., Rink, W.J., Schwarcz, H.P., Leith Smith, J.N., Stewart, K., Todd, N.E., Verniers, J., Yellen, J.E., 1995. Dating and context of three Middle Stone Age sites with bone points in the Upper Semliki Valley, Zaire. Science 268, 548–553.

Buchanan, W.F., 1988. Shellfish in Prehistoric Diet: Elands Bay, S. W. Cape Coast, South Africa. British Archaeological Reports International Series, Oxford.

Buchanan, W.F., Hall, S.L., Henderson, J., Olivier, A., Pettigrew, J.M., Parkington, J.E., Robertshaw, P.T., 1978. Coastal shell middens in the Paternoster area, South-Western Cape. South African Archaeological Bulletin 33, 89–93.

Chase, P.G., 1986. The hunters of Combe-Grenal: approaches to Middle Paleolithic subsistence in Europe. BAR International Series, Oxford.

Costamagno, S., Laroulandie, V., 2004. L'exploitation des petits vertébrés dans les Pyrénées françaises du Paléolithique au Mésolithique: un inventaire taphonomique et archéozoologique. In: Brugal, J.-P., Desse, J. (Eds.), Petits Animaux Sociétés Humaines: du Complément Alimentaire aux Ressources Utilitaires. Éditions APDCA, Antibes, pp. 403–416.

Cruz-Uribe, K., Klein, R.G., 1998. Hyrax and hare bones from modern South African eagle roots and the detection of eagle involvement in fossil bone assemblages. Journal of Archaeological Science 25(2), 135–147.

Davis, S.J.M., 2005. Why domesticate food animals? Some zoo-archaeological evidence from the Levant. Journal of Archaeological Science 32, 1408–1416.

de Heinzelin, J., Clark, J.D., White, T., Hart, W., Renne, P., WoldeGabriel, G., Beyene, Y., Vrba, E., 1999. Environment and behavior of 2.5-million-year-old Bouri hominids. Science 284 (23 April 1999), 625–629.

Dewar, G., Halkett, D., Hart, T., Orton, J., Sealy, J., 2006. Implications of a mass kill site of springbok (Antidorcas marsupialis) in South Africa: hunting practices, gender relations, and sharing in the Later Stone Age. Journal of Archaeological Science 33, 1266–1275.

Grayson, D.K., Delpech, F., 2003. Ungulates and the Middle-to-Upper Paleolithic transition at Grotte XVI (Dordogne, France). Journal of Archaeological Science 30(12), 1633–1648.

Halkett, D., Hart, T., Yates, R., Volman, T.P., Parkington, J.E., Orton, J., Klein, R.G., Cruz-Uribe, K., Avery, G., 2003. First excavation of intact Middle Stone Age layers at Ysterfontein, Western Cape Province, South Africa: implications for Middle Stone Age ecology. Journal of Archaeological Science 30, 955–971.

Henshilwood, C.S., 1997. Identifying the collector: evidence for human processing of the Cape Dune Mole-Rat, *Bathyergus suillus*, from Blombos Cave, southern Cape, South Africa. Journal of Archaeological Science 24(7), 659–662.

Henshilwood, C.S., Marean, C.W., 2003. The origin of modern behavior: critique of the models and their test implications. Current Anthropology 44(5), 627–651.

Henshilwood, C.S., Sealy, J.C., Yates, R., Cruz-Uribe, K., Goldberg, P., Grine, F.E., Klein, R.G., Poggenpoel, C., Van Niekerk, K., Watts, I., 2001. Blombos Cave, Southern Cape, South Africa: preliminary report on the 1992–1999 excavations of the Middle Stone Age levels. Journal of Archaeological Science 28(4), 421–448.

Jerardino, A., Navarro, R., Nilssen, P., 2001. Cape rock lobster (*Jasus islandii*) exploitation in the past: estimating carapace length from mandible sizes. South African Journal of Science 97, 59–62.

Klein, R.G., 1972a. A preliminary report on the June through September 1970 excavations at Nelson Bay Cave (Cape Province, South Africa). Palaeoecology of Africa 6, 177–208.

Klein, R.G., 1972b. The late Quaternary mammalian fauna of Nelson Bay Cave (Cape Province, South Africa): its implications for megafaunal extinctions and for cultural and environmental change. Quaternary Research 2, 135–142.

Klein, R.G., 1975. Middle Stone Age man-animal relationships in southern Africa: evidence from Die Kelders and Klasies River Mouth. Science 190, 265–267.

Klein, R.G., 1976a. The fossil history of *Raphicerus* H. Smith, 1827 (Bovidae, Mammalia) in the Cape Biotic Zone. Annals of the South African Museum 71, 169–191.

Klein, R.G., 1976b. The mammalian fauna of the Klasies River Mouth sites, southern Cape Province, South Africa. South African Archaeological Bulletin 31(123&124), 75–96.

Klein, R.G., 1978. Stone Age predation on large African bovids. Journal of Archaeological Science 5, 195–217.

Klein, R.G., 1981. Later Stone Age subsistence at Byeneskranskop Cave, South Africa. In: Harding, R.S.O., Teleki, G. (Eds.), Omnivorous Primates: Gathering and Hunting in Human Evolution. Columbia University Press, New York, pp. 166–190.

Klein, R.G., 1994. Southern Africa before the Iron Age. In: Corruccini, R.S., Ciochon, R.L. (Eds.), Integrative paths to the past: paleoanthropological advances in honor of F Clark Howell. Prentice Hall, Englewood Cliffs, pp. 471–519.

Klein, R.G., 1998. Why anatomically modern people did not disperse from Africa 100,000 years ago. In: Akazawa, T., Aoki, K., Bar-Yosef, O. (Eds.), Neandertals and Modern Humans in Western Asia. Plenum Press, New York, pp. 509–521.

Klein, R.G., Cruz-Uribe, K., 1983. Stone Age population numbers and average tortoise size at Byneskranskop Cave 1 and Die Kelders Cave 1, Southern Cape Province, South Africa. South African Archaeo logical Bulletin 38, 26–30.

Klein, R.G., Cruz-Uribe, K., 1987. Large mammal and tortoise bones from Elands Bay Cave and nearby sites, Western Cape Province, South Africa. In: Parkington, J.E., Hall, M. (Eds.), Papers in the Prehistory of the Western Cape, South Africa. British Archaeological Reports International Series, Oxford, pp. 132–163.

Klein, R.G., Cruz-Uribe, K., 1989. Faunal evidence for prehistoric herder-forager activities at Kasteelberg, Vredenburg Peninsula, western Cape Province, South Africa. South African Archaeological Bulletin 44, 82–97.

Klein, R.G., Cruz-Uribe, K., 1996. Exploitation of large bovids and seals at Middle and Later Stone Age sites in South Africa. Journal of Human Evolution 31, 315–334.

Klein, R.G., Cruz-Uribe, K., 2000. Middle and Later Stone Age large mammal and tortoise remains from Die Kelders Cave 1, Western Cape Province, South Africa. Journal of Human Evolution 38, 169–195.

Klein, R.G., Scott, K., 1986. Re-analysis of faunal assemblages from the Haua Fteah and other Late Quaternary archaeological sites in Cyrenaican Libya. Journal of Archaeological Science 13, 515–542.

Klein, R.G., Cruz-Uribe, K., Skinner, J.D., 1999. Fur seal bones reveal variability in prehistoric human seasonal movements on the southwest African coast. ArchaeoZoologia 10, 181–188.

Klein, R.G., Avery, G., Cruz-Uribe, K., Halkett, D., Parkington, J.E., Steele, T.E., Volman, T.P., Yates, R., 2004. The Ysterfontein 1 Middle Stone Age site, South Africa, and early human exploitation of coastal resources. Proceedings of the National Academy of Sciences 101(16), 5708–5715.

Klein, R.G., Avery, G., Cruz-Uribe, K., Steele, T.E., 2007. The mammalian fauna associated with an archaic hominin skullcap and later Acheulean artifacts at Elandsfontein, Western Cape Province, South Africa. Journal of Human Evolution 52, 164–186.

Laroulandie, V., 2004. Exploitations des ressources aviaires durant le Paléolithique en France: bilan critique et perspectives. In: Brugal, J.-P., Desse, J. (Eds.), Petits Animaux Sociétés Humaines: du Complément Alimentaire aux Ressources Utilitaires. Éditions APDCA, Antibes, pp. 163–172.

MacDonald, K.C., 1997. The avifauna of the Haua Fteah (Libya). ArchaeoZoologia 9, 83–102.

Mannino, M.A., Thomas, K.D., 2002. Depletion of a resource? The impact of prehistoric human foraging on intertidal mollusc communities and its significance for human settlement, mobility and dispersal. World Archaeology 33(3), 452–474.

Marean, C.W., 2005. From the tropics to the colder climates: contrasting faunal exploitation adaptations of modern humans and Neanderthals. In: d'Errico, F., Backwell, L. (Eds.), From Tools to Symbols: From Early Hominids to Modern Humans. Wits University Press, Johannesburg, pp. 333–371.

Marean, C.W., Assefa, Z., 1999. Zooarcheological evidence for the faunal exploitation behavior of Neanderthals and early modern humans. Evolutionary Anthropology 8(1), 22–37.

Marean, C.W., Abe, Y., Frey, C.J., Randall, R.C., 2000. Zooarchaeological and taphonomic analysis of the Die Kelders Cave 1 layers 10 and 11 Middle Stone Age larger mammal fauna. Journal of Human Evolution 38(1), 197–233.

Munro, N.D., 2004. Zooarchaeological measures of hunting pressure and occupation intensity in the Natufian: implications for agricultural origins. Current Anthropology 45, S5–S33.

Munro, N.D., Bar-Oz, G., 2005. Gazelle bone fat processing in the Levantine Epipalaeolithic. Journal of Archaeological Science 32, 223–239.

Orton, J., Hart, T., Halkett, D., 2005. Shell middens in Namaqualand: two Later Stone Age sites at Rooiwalbaai, Northern Cape Province, South Africa. South African Archaeological Bulletin 60, 24–32.

Outram, A.K., 2001. A new approach to identifying bone marrow and grease exploitation: why the "indeterminate" fragments should not be ignored. Journal of Archaeological Science 28, 401–410.

Parkington, J.E., 2003. Middens and moderns: shellfishing and the Middle Stone Age of the Western Cape, South Africa. South African Journal of Science 99, 243–247.

Pérez Ripoll, M., 2004. La consommation humaine des lapins pendant le Paléolithique dans la région de Valencia (Espagne) et l'étude des niveaux gravétiens de La Cova de les Cendres (Alicante). In: Brugal, J.-P., Desse, J. (Eds.), Petits Animaux Sociétés Humaines: du Complément Alimentaire aux Ressources Utilitaires. Éditions APDCA, Antibes, pp. 191–206.

Richards, M.P., Pettitt, P.B., Stiner, M.C., Trinkaus, E., 2001. Stable isotope evidence for increasing dietary breadth in the European mid-Upper Paleolithic. Proceedings of the National Academy of Sciences 98(11), 6528–6532.

Schweitzer, F.R., 1979. Excavations at Die Kelders, Cape Province, South Africa: the holocene deposits. Annals of the South African Museum 78(10), 101–233.

Schweitzer, F.R., Wilson, M.L., 1982. Byneskranskop 1, a late Quaternary living site in the southern Cape Province, South Africa. Annals of the South African Museum 88, 1–203.

Singer, R., Wymer, J.J., 1982. The Middle Stone Age at Klasies River Mouth in South Africa. University of Chicago Press, Chicago.

Skinner, J.D., Chimimba, C.T., 2005. The Mammals of the Southern African Subregion, 3rd edition. Cambridge University Press, Cambridge.

Sokal, R.R., Rohlf, F.J., 1995. Biometry: the principles and practice of statistics in biological research, 3rd edition. W.H. Freeman, New York.

Steele, T.E., 2003. Using mortality profiles to infer behavior in the fossil record. Journal of Mammalogy 84(2), 418–430.

Steele, T.E., 2005. Comparing methods for analyzing mortality profiles in zooarchaeological and paleontological samples. International Journal of Osteoarchaeology 15, 404–420.

Steele, T.E., Klein, R.G., 2005/2006. Mollusk and tortoise size as proxies for stone age population density in South Africa: implications for the evolution of human cultural capacity. Munibe (Antropologia – Arkeologia) 57/1, 221–237.

Steele, T.E., Klein, R.G., 2008. Intertidal shellfish use during the Middle and Later Stone Age of South Africa. Archaeofauna 17, 63–76.

Steele, T.E., Weaver, T.D., 2002. The modified triangular graph: a refined method for comparing mortality profiles in archaeological samples. Journal of Archaeological Science 29(3), 317–322.

Stephens, D.W., Krebs, J.R., 1986. Foraging theory. Princeton University Press, Princeton.

Stiner, M.C., 1990. The use of mortality patterns in archaeological studies of hominid predatory adaptations. Journal of Anthropological Archaeology 9, 305–351.

Stiner, M.C., 2001. Thirty years on the "Broad Spectrum Revolution" and paleolithic demography. Proceedings of the National Academy of Sciences 98(13), 6993–6996.

Stiner, M.C., 2005a. Middle Paleolithic subsistence ecology in the Mediterranean region. Transitions Before the Transition: Evolution and Stability in the Middle Paleolithic and Middle Stone Age. Springer, New York, pp. 213–231.

Stiner, M.C., 2005b. The Faunas of Hayonim Cave, Isreal: a 200,000-Year Record of Paleolithic Diet, Demography, and Society. Peabody Museum of Archaeology and Ethnology. Harvard University, Cambridge.

Stiner, M.C., Munro, N.D., 2002. Approaches to prehistoric diet breadth, demography, and prey ranking systems in time and space. Journal of Archaeological Method and Theory 9(2). 181–214.

Stiner, M.C., Munro, N.D., Surovell, T.A., Tchernov, E., Bar-Yosef, O., 1999. Paleolithic population growth pulses evidenced by small animal exploitation. Science 283, 190–194.

Stiner, M.C., Munro, N.D., Surovell, T.A., 2000. The tortoise and the hare: small-game use, the Broad-Spectrum Revolution, and Paleolithic demography. Current Anthropology 41(1), 39–73.

Straus, L.G., 1977. Of deerslayers and mountain men: Paleolithic faunal exploitation in Cantabrian Spain. In: Binford, L.R. (Ed.), For Theory Building in Archaeology. Academic Press, New York, pp. 41–76.

Thomas, H., 1981. La faune de la Grotte à Néandertaliens du Jebel Irhoud (Maroc). Quaternia 23, 191–217.

Voigt, E.A., 1982. The molluscan fauna. In: Singer, R., Wymer, J. (Eds.), The Middle Stone Age at Klasies River Mouth in South Africa. The University of Chicago Press, Chicago, pp. 155–186.

Wilson, D.E., Reeder, D.M. (Eds), 2005. Mammal Species of the World. Johns Hopkins University Press, Baltimore.

Yellen, J., Brooks, A.S., Helgren, D.M., Tappen, M., Ambrose, S.H., Bonnefille, R., Feathers, J., Goodfriend, G., Ludwig, K., Renne, P., Stewart, K.M., 2005. The archaeology of Aduma Middle Stone Age Sites in the Awash Valley, Ethiopia. PaleoAnthropology 10, 25–100.

9. Seasonal Patterns of Prey Acquisition and Inter-group Competition During the Middle and Upper Palaeolithic of the Southern Caucasus

Daniel S. Adler
University of Connecticut
Department of Anthropology
U-2176
Storrs, CT 06269 USA
daniel.adler@uconn.edu

and

Guy Bar-Oz
University of Haifa, Zinman Institute of Archaeology
Mount Carmel
31905 Haifa, Israel
guybar@research.haifa.ac.il

Keywords Zooarchaeology • taphonomy • hunting • diet breadth • competitive exclusion

Abstract Zooarchaeological and taphonomic analyses provide an essential backdrop to discussions of Late Middle Palaeolithic and Early Upper Palaeolithic patterns of mobility, land-use, and hunting, and the degree and manner(s) of Neanderthal and modern human competition within the southern Caucasus. Recent research at Ortvale Klde has documented the hunting of prime-age adult *Capra caucasica* and the organization of hunting activities according to this species' migratory behaviors, which made them locally abundant on a seasonal basis. Our analyses suggest that Neanderthals and modern humans occupied the same ecological niche and were equally capable of learning and exploiting key biogeographic information pertaining to the feeding, mating, migratory, and flight behaviors of this species. In these respects there appear to have been few alterations in medium/large game hunting practices between the Late Middle Palaeolithic and Early Upper Palaeolithic, with ungulate species abundance in the entire stratigraphic sequence of Ortvale Klde reflecting seasonal fluctuations in food supply rather than specialization, differences in diet breadth, hunting ability, or technology. Attention is paid to faunal data from neighboring sites to test whether patterns identified at Ortvale Klde are in any way representative of larger regional subsistence behaviors. We find that such patterns are only replicable at sites that have experienced similar zooarchaeological and taphonomic study. We conclude that Neanderthal and modern human populations occupied and exploited the same ecological niches, at least seasonally, and that the regional archaeological record documents a clear spatial and temporal disruption in Neanderthal settlement resulting from failed competition with expanding modern human groups. In terms of niche and resource preference, we suggest that Neanderthals and modern humans were sympatric to the point of exclusion.

J.-J. Hublin and M.P. Richards (eds.), *The Evolution of Hominin Diets: Integrating Approaches to the Study of Palaeolithic Subsistence*, 127–140.

Neanderthal-Modern Human Subsistence and Competition

The data presented in this paper are considered in light of ongoing questions concerning Neanderthal and modern human foraging behaviors and inter-group competition. At its core, the following discussion centers on the relative degree of niche overlap, prey choice, and population size between indigenous Neanderthals and expanding modern humans. Consideration of the primary literature relating to human behavioral ecology, the specific ethnographic data employed, and its application in archaeological contexts is beyond the scope of this paper, but detailed summaries are provided elsewhere (e.g., Winterhalder and Smith, 2000; Shennan, 2002; Bird and O'Connell, 2006). Particularly relevant to the main issues addressed here are several recent studies that consider the issue of forager competition from an ecological and evolutionary perspective. These focus on resource exploitation and diet breadth, and the attendant influences these factors have on foraging efficiency and individual and group fitness (e.g., Finlayson, 2004; Kuhn and Stiner, 2006; O'Connell, 2006).

Finlayson (2004) proposes that competition between Middle and Upper Palaeolithic populations, if it occurred at all, was ephemeral and played little if any role in the extinction of the Neanderthals. This assessment is based on theoretical associations between population size and resource exploitation that suggest competition occurs only among populations at environmental carrying capacity who exploit similar resources at similar times (see Pianka, 1988). As Neanderthal populations are presumed to have been small (effective population size of 10,000–20,000), and thus below carrying capacity, the chance for competition was limited. In other words, there was ample territory to accommodate both populations within a given geographic zone, especially those containing heterogeneous environments.

Neanderthal-modern human coexistence is documented elsewhere (e.g., the Levant; Shea, 1998, 2003), which Finlayson attributes to the fact that both populations existed below carrying capacity and exploited different landscapes and geographic areas. Here he seems to suggest of a degree of allopatry, even though data highlighting niche overlap between the two populations are common during the Middle–Upper Palaeolithic "transition" (e.g., Grayson and Delpech, 2002, 2003, 2006). In those instances where Neanderthal territories were encroached upon by expanding modern humans, Finlayson argues, based on research by Wang et al. (2000), that native populations were competitively superior owing to their long-term, in-depth knowledge of local resource availability and distribution. While expansion into new territories entails risk (Kelly, 2003; Meltzer, 2003), Finlayson's conclusion implies that newcomers (modern humans) were unable to obtain and exploit such critical biogeographic information in a timely fashion. Ultimately, Finlayson attributes the cause of demographic change and Neanderthal extinction to climate change and the attendant natural contractions experienced by Neanderthal populations, especially those outside the Mediterranean zone. The potential impact of such climate change on contemporaneous modern human populations and the persistence of Neanderthal populations in refugia outside the Mediterranean zone (e.g., the Caucasus) require further consideration. That important medium–large size prey species remain viable and significant resources throughout the Upper Palaeolithic of the southern Caucasus argues against the general idea that Neanderthals may have been just another large-bodied mammal driven to extinction by Middle Pleniglacial climate fluctuations (see Stewart et al., 2003; Stewart, 2004).

Expanding forager populations who exploit a wider array of resources, and so operate a more expensive subsistence economy, typically occupy a wider array of habitats and may do so at higher densities (O'Connell, 2006). In the context of competition, such a foraging strategy can lead to the elimination of foraging opportunities among indigenous populations. Work by Stiner et al. (2000) provides evidence of increased diet breadth among modern humans as compared to their Neanderthal contemporaries, with the former exploiting not only the same medium–large game upon which Neanderthals relied so heavily, but also smaller game that required increased effort to harvest relative to nutritional gain. Such resources are rarely encountered in Neanderthal assemblages; instead, Neanderthals appear to have focused much of their energies on the procurement of high-ranked terrestrial prey (see Stiner, 2002; Kuhn and Stiner, 2006; Speth and Clark, 2006). Under such conditions, and in combination with related factors such as specialized technologies, lower energy budgets, and larger populations, O'Connell (2006) argues that modern humans had a considerable competitive advantage over Neanderthals. Why Neanderthals did not simply adjust their economies to meet increasing subsistence pressures caused by competition can be linked to fitness-related opportunity costs that varied between males and females. Concerning subsistence, O'Connell (2006) argues that Neanderthal males had to balance efforts in the realms of alliance building and mating opportunities, perhaps via male–male competition in the arena of big-game hunting, with the need to maintain high return rates and provision their families. Shifting to a broader diet in the face of modern human competition would have required increased foraging effort, resulting in a reduction in time spent in the pursuit of other fitness-related activities, namely competition for mates.

But this proposition, which is not currently possible to test archaeologically, hinges on whether Neanderthals divided labor between the sexes in a manner similar to that commonly documented among historic and current foragers (see Kelly, 1995; Winterhalder and Smith, 2000; Bird and O'Connell, 2006). The danger of projecting recent ethnographic observations into the distant past has long been recognized (e.g., Wobst, 1978) and the question of sexual divisions of labor in the Middle and Upper Palaeolithic has recently come under greater scrutiny. For example, Kuhn and Stiner (2006) claim that, because of their narrow diet breadth, sex-based divisions of labor were untenable among Neanderthals as there were too few subsistence roles to be filled.

Consequently, because Neanderthals pursued economies more narrowly focused than their modern human counterparts (i.e., medium–large game), Kuhn and Stiner argue that females participated closely in time and space with activities more commonly associated with men (i.e., hunting). Only during the Upper Palaeolithic did a sex-based division of labor emerge in response to an increased range of subsistence and social roles, as evidenced, for example, by an increase in diet breadth, the development of specialized technologies (e.g., cordage and basketry, non-utilitarian organic items), and the production of elaborate clothing, structures, and items of personal adornment. While Kuhn and Stiner do not speculate as to the sex or age of those who took on the new roles indicated by the technological and subsistence data, they do stress the point that, unlike in the Middle Palaeolithic, diverse roles were there to be filled.

Again, it is necessary to ask why the Neanderthals did not simply broaden their resource base in the face of increasing subsistence pressures. Kuhn and Stiner (2006) argue that the risks associated with a shift to more diversified subsistence, similar to that documented for the Upper Palaeolithic, may have been too great for the Neanderthals, who are argued to have maintained populations below carrying capacity through the exploitation of high-yield but risky resources. The inability to adapt to the new socioeconomic reality created by the appearance and perhaps growth of modern human populations put Neanderthals at a competitive disadvantage, tipping the reproductive and demographic balance in favor of modern humans.

In the following pages we consider these recent studies from the perspective of Ortvale Klde, a late Middle Palaeolithic (LMP) and early Upper Palaeolithic (EUP) rockshelter located in the southern Caucasus, where recent excavations and zooarchaeological and taphonomic studies document seasonal hunting behaviors prior to and following the Middle–Upper Palaeolithic "transition." Using data on species representation, age structure, life history, sex ratios, and processing behaviors we investigate (a) whether Neanderthals and modern humans were sympatric or allopatric; (b) whether diet breadth increased from the LMP to EUP; and (c) what, if any, competitive advantage either population may have enjoyed. We believe that Neanderthals and modern humans were sophisticated, flexible foragers capable of adapting to a wide variety of stable and changing environmental conditions in response to immediate and anticipated needs, a perspective based on the well recognized ability of foragers to shift easily between different modes of subsistence and mobility on a daily, seasonal, or yearly basis (see Butzer, 1982; Kelly, 1995; Kuhn, 1995). Therefore, the data presented here should be considered dependent upon local, situational factors, and thus not necessarily reflective of all populations, in all places, at all times. We do not attempt to build a broad-based geographic model of Neanderthal and modern human behavioral variability and competition, but confine our observations and interpretations to the data currently available from the southern Caucasus.

The Faunal Assemblage of Ortvale Klde

Ortvale Klde currently represents the only stratified LMP and EUP site within the southern Caucasus to have experienced careful excavation and detailed zooarchaeological and taphonomic analyses (Bar-Oz and Adler, 2005; Adler et al., 2006a; but see Cleghorn, 2006 for an example form the northern Caucasus). The site is situated in western Georgia (Imereti region) in the Cherula river valley, approximately 530 m above sea level (m.a.s.l.) and 35 m above the gorge (Fig. 9.1). It is a karstic rockshelter comprised of two chambers opening to the east (~300 m^2). D. Tushabramishvili first investigated the rockshelter between 1973 and 1992, and excavated roughly 40 m^2 at the front of the southern chamber (Tushabramishvili et al., 1999; Adler and Tushabramishvili, 2004). Faunal remains from this excavation were selectively collected and analyzed by the eminent palaeontologist, Prof. A. Vekua (Georgian Academy of Sciences), for the purpose of taxonomic identification based on teeth and selected long bone epiphyses.

Excavations resumed from 1997 to 2001 under the direction of Nicholas Tushabramishvili (Georgian State Museum) and Daniel S. Adler; a sample of 6 m^2 was excavated at the rear of the southern chamber where stratified LMP and EUP deposits were preserved. A total of 3,209 EUP and 12,541 LMP faunal specimens were recovered. All of the excavated sediments were carefully dry-screened through 2 mm mesh and sorted for small bones; all bone fragments were collected and processed according to their spatial and stratigraphic location. The faunal remains discussed here are associated with three EUP horizons (Layers 2–4; ~19.5–34.3 ka Uncal BP) and three LMP horizons (Layers 5–7; ~38/36.5–42 ka Uncal BP) and thus span the regional "transition" from the Middle to Upper Palaeolithic (Adler et al., 2006a; Adler et al., 2008).

In the following section we summarize several of the major features of the LMP and EUP faunal assemblages from Ortvale Klde (e.g., species representation; survivorship and mortality; bone modification) while also highlighting similarities with results obtained from previously excavated material analyzed by Vekua. These data allow us to consider the diachronic change in the relative importance of species-specific hunting practices at the site in relation to faunal processing behaviors, seasonality, and mobility.

Species Representation

Our analyses support the main conclusions of Vekua (Tushabramishvili et al., 1999), with 95% of the combined 1997–2001 ungulate assemblage, based on the number of identified specimens (NISP), dominated by Caucasian tur (*Capra caucasica*; Table 9.1), suggesting that diet breadth at the site was relatively low during both the LMP and EUP. Steppe bison (*Bison priscus*) constitutes a small proportion of the assemblage (4%), and other large ungulates, such as red deer (*Cervus elaphus*) and roe deer (*Capreolus capreolus*) are nearly absent (<1% in each layer). Noteworthy are the even smaller proportions (≤0.5% of total assemblage) of carnivores (e.g., *Ursus* sp. and fox [*Vulpes vulpes*]). It is not currently possible to find such species-specific hunting behaviors mirrored at other Palaeolithic sites in the region or in the northern Caucasus. Within the EUP faunal assemblage of Dzudzuana Cave, located in an adjacent river valley approximately 4.6 km to the east (see Fig. 9.1; Meshveliani et al., 1999, 2004), the proportions of steppe bison and aurochs (*Bos primigenius*) increase considerably (over 40% in all levels) at the expense of Caucasian tur (Bar-Oz et al., 2007). Similarly, Barakaevskaïa Cave, located roughly 350 km northwest of Ortvale Klde in the northern Caucasus, contains a faunal

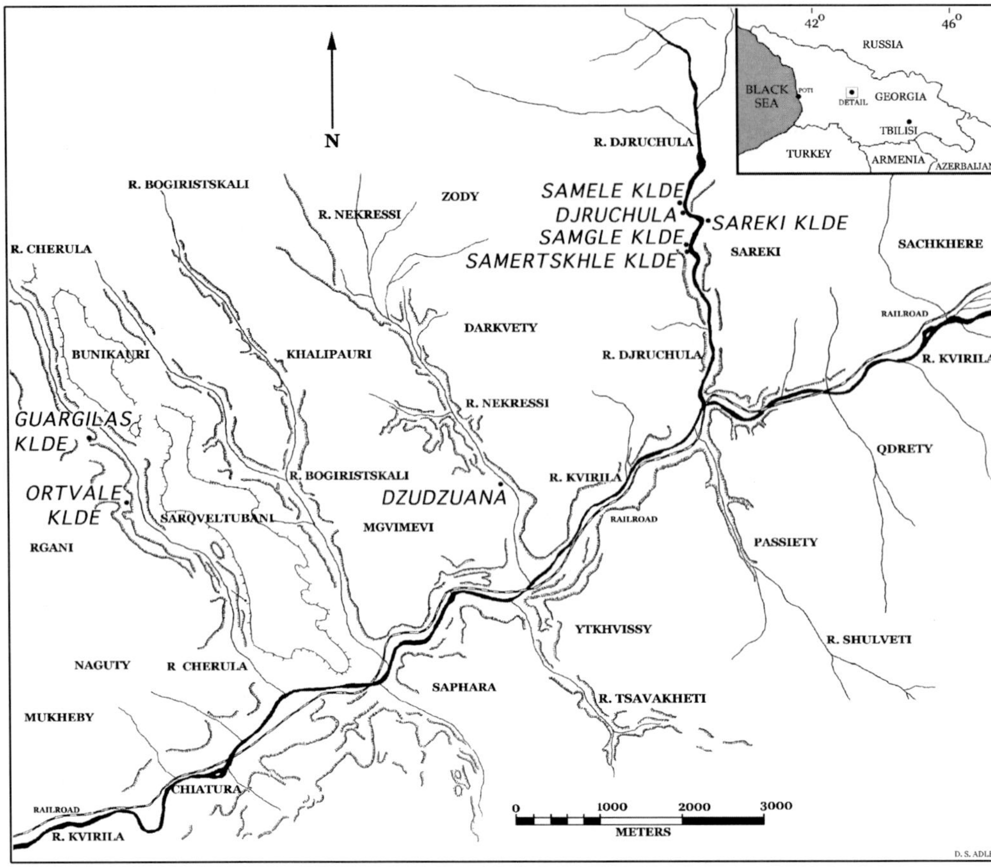

FIG. 9.1. Kvirila River Valley, Imereti region, Republic of Georgia. Archaeological sites are uppercase and italicized. Bronze Cave is located approximately 35 km due southwest (Modified after Adler, 2002).

assemblage with one of the highest frequencies of Caucasian tur (28.2%; Liubin, 1998). The recently analyzed fauna from Mezmaiskaya Cave (Cleghorn, 2006) indicates that while the frequency by NISP of *Capra/Ovis* is not as large as that for Bos/Bison, the very large number of size 2 bovid/cervids and ungulates likely represent sheep and goat. Thus, with only a few exceptions, Caucasian tur is generally poorly represented at most LMP and EUP sites in the Caucasus (e.g., Hoffecker et al., 1991; Baryshnikov and Hoffecker, 1994; Baryshnikov et al., 1996; Hoffecker and Cleghorn, 2000; Hoffecker, 2002).

Survivorship and Mortality

The age structure of the Caucasian tur recovered from Ortvale Klde was analyzed on the basis of tooth wear (see Bar-Oz and Adler, 2005; Adler et al., 2006a). Mortality profiles, expressed as 10% increments of potential lifespan (see Klein, 1976; Klein et al., 1981; Munson and Marean, 2003), of Caucasian tur for the combined LMP assemblage, determined according to dental wear height of the dP4 and M3, show that 32% of the specimens (n = 18) were hunted as juveniles (<20% lifespan), while 66% of the specimens (n = 38) were hunted as prime-age adults (between 20–70% lifespan); old adults were taken in very low frequencies (n = 1). These data highlight a bias toward prime-age individuals (Table 9.1) and resemble LMP and EUP ungulate profiles obtained at numerous sites in Eurasia on samples of similar size (e.g., Stiner, 1994; Speth and Tchernov, 1998; Steele, 2004; Speth and Clark, 2006).

The hunting of prime-age individuals may indicate the use of sophisticated weapons technology, group coordination and anticipatory behavior, and/or an intimate knowledge of prey behavior. In fact, it is likely that such knowledge, combined with the ability to identify and utilize key natural features that afforded cover and took advantage of the varied terrain, were important factors enabling LMP hunters at Ortvale Klde to cull economically important prime-age adults. As with the species-specific hunting practices outlined above, the procurement of prime-age adults is documented elsewhere in Eurasia during the LMP and EUP (e.g., Levine, 1983; Jaubert et al., 1990; Stiner, 1990, Hoffecker et al., 1991; Stainer, 1994; Baryshnikov and Hoffecker, 1994; Speth and Tchernov, 1998; Baryshnikov and Hoffecker, 1999; Pike-Tay et al., 1999; Hoffecker and Cleghorn, 2000; Stainer, 2002; Miracle, 2005), as well as earlier periods (Gaudzinski and Roebroeks, 2000, 2003; Steele, 2004; Stiner, 2005; Speth and Clark, 2006). This situation is in line with Stiner's (2002) proposal that carnivory, or more specifically prime-adult-biased hunting, among hominins dates to at least 250 ka and was a common adaptive feature among archaic populations. If true, then it is not surprising to find well-developed hunting skills among both Neanderthals and modern humans, especially if one considers the projected daily caloric requirements of Neanderthals (Aiello and Wheeler, 2003) or recent stable isotope studies (Bocherens et al., 1999; Richards et al., 2000, 2001; Bocherens et al., 2005; Bocherens and Drucker, 2006). In terms of hunting and niche preference it appears that Neanderthals and modern humans in the southern Caucasus were sympatric and thus in direct competition for the same limited resources.

Human Modifications

Faunal Transport, Butchery, Bone Fragmentation, and Burning

Examination of element abundance in relation to associated food utility index (FUI; Metcalfe and Jones, 1988) reveals that Caucasian tur skeletal profiles are not altered by selective transport decisions (Table 9.1). Detailed analysis of LMP skeletal part abundance shows some bias against axial elements (vertebrae and ribs) and higher representations of toes (Bar-Oz and Adler, 2005: Fig. 7). The lack of evidence for selective transport, coupled with the low occurrence of axial elements suggests that prey was subject to extensive butchery, either in the field prior to import to the site or within the site itself, and that hunters at Ortvale Klde returned large portions of carcasses to the site.

Butchery marks on Caucasian tur document all stages of carcass processing (n = 53, combined EUP and LMP assemblage): dismemberment, filleting, and skinning (Table 9.1). The overall proportion of butchery marks on Caucasian tur is low for all archaeological layers (<3%). However, the presence of cut marks from all stages of the butchering process suggests that the full range of butchering activities is represented within each layer. The most obvious similarity in butchery patterns from each layer is the location of cut marks on and around the major limb joints. We also identify higher frequencies of cut marks on head and limb bones, and fewer on axial elements and foot bones. These data indicate that Caucasian tur carcasses underwent thorough dismemberment and preparation following arrival on site, and that carcass processing included both low and high utility elements.

The intensity of bone fragmentation and the analysis of bone breakage patterns (fracture angle, fracture outline, and fracture edge) of long bone shaft fragments (Villa and Mahieu, 1991) point to the prevalence of oblique, V-shaped, and jagged bone fractures, indicating that the majority of bones from each layer were broken while fresh (Table 9.1). These data demonstrate that the faunal assemblage results from intensive bone processing, possibly related to the extraction of bone marrow. Such bone processing behaviors produce high frequencies of fresh/green fractures and likely account for the high number of small, unidentified shaft fragments within each layer. The presence of percussion fractures (Blumenschine and Selvaggio, 1988) close to the fracture edge of long bone shaft fragments (Table 9.1) and average fragment length further support this interpretation. In addition, the high rate of fresh breakage among the first phalanges (>60% in each layer) (Bar-Oz and Adler, 2005) and the low relative frequency of complete bones (22.0% of MNE; following Munro and Bar-Oz, 2005) suggest that these elements were also systematically split open for marrow extraction.

Burned specimens are observed in all of the archaeological horizons and appear randomly distributed. Layers 2 and 3 (EUP) are characterized by low rates of burning (2–3%) while within Layer 4 (EUP) and Layers 5–7 (LMP) the rate of burning increases (Table 9.1). The frequencies of burned bones among the identified Caucasian tur and bison remains are too small to analyze by anatomical unit. Burning was most common on small, unidentified fragments (n = 1,030, mean = 25 mm, SD = 11 mm) from the combined LMP sample. As Speth argues (in Bar-Yosef et al., 1992) the relationship between burning and specimen size support the view that burning is related to food preparation; if bones were accidentally burned by later activities, larger bones with bigger surface areas should display high rates of burning (Stiner et al., 1995).

Non-human Modifications

Carnivore activity (e.g., chewing, gnawing, and scratch marks) is infrequent (<10%, excluding teeth) among the identifiable elements of Caucasian tur and declines in frequency from Layers 4–7 (Table 9.1). Likewise, no signs of digestion are found on any bones in the assemblage. The non-identifiable elements (fragments >40 mm), mainly mid-shafts fragments, bear even lower rates of carnivore tooth marks (<2% in all levels), demonstrating that carnivores had only secondary access to bones (see discussion in Dominguez-Rodrigo, 2002) and that the assemblage was not accumulated or significantly altered by carnivores. Examples of carnivore marks on the inner surface of Caucasian tur bones (n = 4) further indicate that carnivores had access to discarded bones only following site abandonment by hominins.

The low frequency of carnivore damage at Ortvale Klde likely results from long-term human habitation, perhaps on the order of weeks–months, during which carnivores could not gain access to discarded remains (i.e., Bunn, 1993) or, the intensive and thorough processing of faunal resources occurred, after which little remained for carnivores to ravage (Lupo, 1995). Our results are in accordance with natural experiments and actualistic studies with captive spotted hyena (*Hyaena hyaena*) that show few carnivore tooth marks on assemblages processed by humans (Marean and Spencer, 1991; Marean et al., 1992). Decreasing rates of carnivore damage on identifiable bone fragments within the LMP layers may reflect intensive use of the site by Neanderthals.

The Seasonal Exploitation of Caucasian Tur at Ortvale Klde

Life History Characteristics

Given the economic importance of Caucasian tur to the inhabitants of Ortvale Klde, we summarize the life history characteristics of this species and discuss how such behaviors may have structured mobility and land-use patterns during LMP and EUP. Based on the data provided by Brown and Burton (1974), Heptner et al. (1988, 1989), Parker (1990), and Nowak (1991) several general observations may be made regarding species size and weight, reproduction and life cycle, diet, habitat, seasonality, and group structure (Table 9.2). These data are derived from recent observations of extant Caucasian tur populations located within small, isolated nature reserves in the Georgian Republic and on the northern side of the Caucasus, mainly in Daghastan. Although these populations have undergone some interbreeding with other closely related species, mainly East Caucasian tur (*Capra cylindricornis*), they remain important sources of information regarding Caucasian tur (Vereshchagin, 1967) and we assume that the behaviors of Pleistocene Caucasian tur were not significantly different from those of recent herds.

Body Size, Sexual Dimorphism, and Breeding

Caucasian tur, also known as the west Caucasian tur, is closely related to the east Caucasian tur in morphology and behavior. On average, adult males exhibit a body length of 150–165 cm, a shoulder height of 95–109 cm, and weigh between 65 and 100 kg. Females are on average smaller: 120–140 cm long, 78–90 cm tall, and weigh 50–60 kg. Both males and females have horns that in cross-section approximate rounded triangles. Male horns are scimitar-shaped and exhibit heavy ridges; these average 75 cm in length. The horns of females are considerably shorter and narrower, and exhibit fewer and smaller ridges. Therefore, sexual dimorphism is expressed in horn morphology and overall body size.

Breeding occurs from late November to early January, with the birth of one to two offspring in May and June, representing a 150–160 day gestation period. Weaning occurs gradually, and sexual maturity among females is reached following the second year after which they may breed annually. Among males sexual maturity is not reached until the fourth or fifth year and it is at this time that adult males compete vigorously for females. Life expectancy of both males and females is 12–13 years.

Home Range, Seasonality, Diet, and Group Size

The Caucasian tur has one of the smallest natural ranges of any ungulate today, spanning an approximately 4,500 km^2 region in the western Caucasus that includes elevations of 800–4,200 m.a.s.l. They follow a seasonal migration cycle, moving upslope in May to take advantage of summer pastures and avoid insects (e.g., ticks, horseflies, deerflies, and other blood-sucking pests), and downslope in October for mating and feeding (Heptner et al., 1989). They thrive in alpine meadows, barren areas, or forests, and their diet is composed of a wide variety of plants and grasses in summer and the leaves of trees and shrubs in winter. During the summer, Caucasian tur can cover a daily vertical distance of up to 1,500–2,000 m, thereby taking advantage of numerous resources at different elevations. During spring and summer they feed periodically throughout the late afternoon, night, and morning, and spend the hottest portions of the day resting in shaded places. During winter months, herds often remain in open pastures, alternately grazing and resting. Adult males (>4–5 years old) generally stay at higher altitudes than females, who are usually accompanied by young. Maternal herds of approximately 12 animals are joined by otherwise solitary males (in particular those in their reproductive prime [6–8 years old]) in the breeding season (late November–early January), during which competition for females is often violent (Heptner et al., 1989). Unfortunately, it is not clear from the literature to what extent the behaviors of sexually immature males vary from those of adult males.

During winter, the home range is much reduced (Heptner et al., 1989) and within the range Caucasian tur have the habit of utilizing the same trails, sometimes several kilometers long, for many generations. Heptner et al. (1989) note that rocks in these trails have been literally polished by the repeated passage of herds. The average population density of Caucasian tur is currently estimated at 50–160 per 1,000 ha, but due to overhunting and habitat lose related to farming and herding, these numbers likely underestimate Pleistocene population densities.

Sex Ratio

Since males are generally solitary and therefore more difficult to hunt throughout much of the year, especially the summer months, their representation within an archaeological assemblage is a sensitive indicator of season of procurement. Alternatively, an over-representation of females within the assemblage might indicate hunting prior to and/or following the breeding season when males disperse and are therefore more difficult to locate and hunt.

The ability to differentiate between males and females in zooarchaeological assemblages can provide information on sex-based strategies of animal exploitation (Klein and Cruz-Uribe, 1984). As discussed above the Caucasian tur is sexually dimorphic, with adult males being larger and heavier than adult females. The difference in weight is reflected in the breadth and width of portions of some elements. Complete astragali and, to a lesser extent, distal humeri are among the most abundant measurable skeletal elements in the assemblage. Only adult specimens (i.e., fused epiphyses and non-porous astragali) are included in the analysis. Table 9.3 presents data showing that the Caucasian tur recovered from the LMP and EUP of Ortvale Klde are larger than the mean values obtained for recent specimens of both sexes of *Capra caucasica* collected in the Caucasus at the beginning of the twentieth century (three females and four males from St. Petersburg Zoological Institute and the Humboldt Zoologische Museum, Berlin). A similar pattern of decreased body size has been observed among numerous Late Pleistocene Levantine mammals (Kurten, 1965; Davis, 1981; Bar-Oz et al., 2004a) and is in accordance with Bergmann's rule (1847). However, since we lack specific information on the geographic origin of recent Caucasian tur specimens, the observed size difference may also be affected by spatial rather than temporal differences. The broad range of bone measurements shows that both large and small specimens are present (Table 9.3), thus it appears that both males and females are represented at Ortvale Klde.

Modern adult male and female Caucasian tur live separately, in independent groups, with mixed herds forming only during the period of estrus (end of November to early January) and for 1 or 2 months thereafter (Heptner et al., 1989). Caucasian tur herds encountered in the Caucasian preserves of alpine meadows exhibit a sex ratio among adult animals close to 1:1, however females predominant in the forest, located at lower elevations, constituting between 60–84% of the entire population (Heptner et al., 1989). Based on the limited available bone sample we suggest that Caucasian tur does and bucks were hunted in the same frequencies as they occur in recent herds at different elevations and thus within different environmental settings. Pleistocene hunters did not cull herds according to sex-based preferences. As one might expect of pre-pastoral groups, the occupants of Ortvale Klde ate what was immediately available, in this case Caucasian tur, probably in sex frequencies equal to their natural distribution within particular environments during particular seasons (see also Marks and Chabai, 2001; Bar-Yosef, 2004; Marks and Chabai, 2006).

Discussion

The zooarchaeological and taphonomic results from Ortvale Klde indicate that LMP and EUP hunters focused their foraging efforts on Caucasian tur, a high-ranked species that remained a predictable seasonal resource throughout the Middle and Late Pleniglacial. We find no evidence for the preferential culling of herds by sex and propose that the observed sex ratios reflect seasonally determined frequencies of males and females. Caucasian tur age classes demonstrate that Neanderthals exploited all age groups, with a bias toward prime-age adults, suggesting ambush or intercept hunting rather than some form of encounter hunting. As Caucasian tur can be particularly elusive and difficult to hunt given their predilection for rocky heights and steep crags, terrain to which they often retreat after being disturbed (Lay, 1967; Roberts, 1977; Marean and Kim, 1998), the hunting of Caucasian tur was no simple matter. However, the migratory and mating behaviors discussed here likely lowered search cost and enhanced encounter rates, establishing this taxon as a high-rank food item on a seasonal basis.

The seasonal exploitation of Caucasian tur is strongly suggested by several independent lines of evidence. As mentioned above, Caucasian tur migrate upslope in May and downslope in October. While Ortvale Klde lies below the historic lower range (800 m.a.s.l.) of Caucasian tur distribution, the topography surrounding the site is quite steep and the plateau immediately above Ortvale Klde quickly reaches an elevation of 800 m. During Oxygen Isotope Stage 3 (OIS 3) the sub-alpine zone was periodically as much as 1,000 m lower than the current level due to climatic oscillations, although compared to surrounding regions, such oscillations were muted by the ameliorating affects of the Black Sea (Adler et al., 2006a, b). It is assumed that Caucasian tur populations responded to this environmental shift by reorganizing their migratory behaviors to take advantage of resources available at lower elevations. This would have placed Ortvale Klde well within the lower, late fall–early spring range of their seasonal movements and established the site as a strategic position within the landscape, serving as a central place adjacent to rich resources from which foragers radiate and to which they return (Fig. 9.2).

The bias at Ortvale Klde towards prime-age adults suggests that prey densities were high enough to allow LMP hunters to target the largest, most nutritious members of the herd. Still, bone fusion and dental wear data indicate that some juveniles (<2 years) are present in the assemblage. This pattern again suggests late fall–early spring occupation of the site when otherwise solitary males and small maternal herds with young aggregated for mating at lower elevations. Hunting efficiency would be maximized at this time due to the increased density and vulnerability of prey in and around the site. At least seasonally then, both LMP and EUP diet breadth appears to have been equivalently low.

The processing of Caucasian tur long bones indicates, not unexpectedly, that hominins extracted and consumed large amounts of fat-rich marrow. The presence of Caucasian tur heads at Ortvale Klde may further support habitation during late fall–early spring when animals were leaner and long bones contained less marrow. Brains may have been exploited at this time in order to maximize fat intake during such periods when Caucasian tur were experiencing some degree of dietary stress. If so, the most effective way of maximizing the nutritional value of brains would be through roasting (Stiner, 1994), a practice for which little direct evidence (e.g., burned cranial fragments) exists at Ortvale Klde. While the process of fat mobilization among ungulates has been outlined (Speth, 1983; Bar-Oz and Munro, 2007), the timing, actual extent, and effects of dietary stress on the fat levels of Caucasian tur remain unknown.

Speth and Spielmann (1983), and Speth (1987, 1989, 1990) demonstrate that, with respect to available ethnographic data, hunter-gatherer groups who rely on high levels of lean-meat consumption, usually in the spring, require very high kill

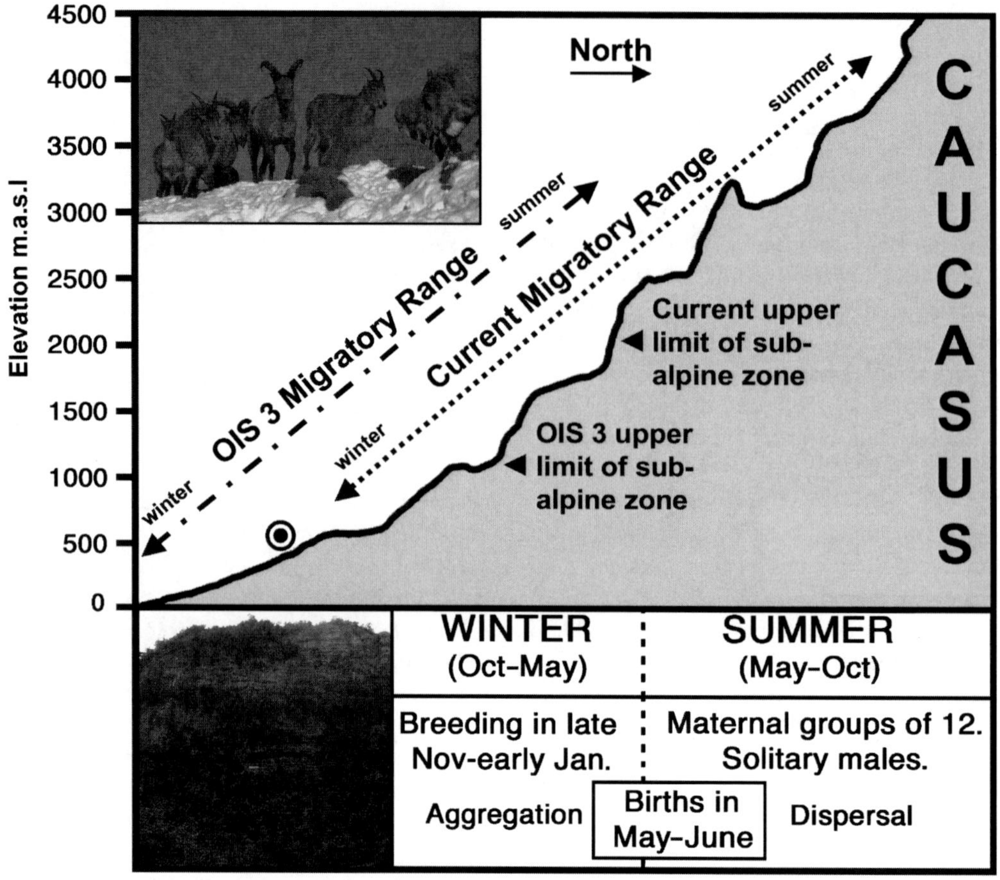

FIG. 9.2. Schematic cross-section of the Southern Caucasus (not to scale), illustrating current migratory range, and reproductive and social behaviors of Caucasian tur (*Capra caucasica*). OIS 3 migratory range is proposed. The nested circle indicates the relative elevation of Ortvale Klde. Upper left inset is a photograph of a mixed herd of west Caucasian tur (© Brent Huffman, Ultimate/Ungulate); lower left inset is a view of Ortvale Klde from the east (Photograph by D.S. Adler; Figure modified after Adler, 2002).

rates that are difficult to maintain over extended periods of time. Speth suggests that such examples likely reflect serious shortfalls in other non-protein foods, and that this subsistence approach reflects a stop-gap measure rather than a long-term solution to seasonal caloric shortages (1987: 16). Therefore, the targeting of prime-age adults, the thorough processing of marrow-rich elements, and the consumption of heads may actually indicate a strategy whereby LMP hominins at Ortvale Klde sought to maintain body fat as a hedge against predicted dietary shortfalls later in the season. According to Speth:

> When under stress (i.e., losing weight and subsisting on marginal calorific intakes), hunter-gatherers would avoid high intakes of lean meat and other high-protein food sources that are low in fat or carbohydrate. Thus male ungulates of prime age or females of similar age that are neither pregnant nor lactating become principal targets of spring hunts. (Speth, 1987: 20)

The use of brains as an agent in hide tanning should also be considered. The late fall–early spring exploitation of Caucasian tur for meat and fat may have also been accompanied by the removal and preparation of skins in anticipation of colder weather and/or in association with the general need to service or replace clothing. In this respect the seasonal availability of Caucasian tur in the Cherula valley may have marked one of the most anticipated events in the yearly subsistence cycle of hominins, during which meat, fat, and hides could be readily procured, consumed, and prepared prior to the onset of winter conditions and the anticipated declines in food availability and quality.

Species-specific hunting from late fall–early spring is a hominin behavior attested to in many parts of Eurasia (Stiner, 1994, 2002), but one that must have been predicated on an intimate knowledge of the specified prey's life history characteristics. For example, since Caucasian tur is "fixed" as a resource seasonally, it stands to reason that hominins, given appropriate weapons technology and/or organizational skills, would maximize their food gathering efforts by learning and exploiting this prey species' seasonal behaviors. If carnivores can map onto and exploit seasonally mobile terrestrial prey, why couldn't Neanderthals or modern humans (see Henshilwood and Marean, 2003)? Chasing solitary animals or small, dispersed, fast moving herds through the forests and along the steep slopes of the southern Caucasus probably did not qualify as an efficient expenditure of time or energy. Instead, large numbers of animals or specific size/age/sex could be exploited by small groups of hunters during periods of Caucasian tur population aggregation. Such behaviors may have also resulted in the aggregation of hominins, which served multiple social and economic purposes, the exact form and function of which can only be speculated for the LMP and EUP, but which likely included the exchange of resources, knowledge, and mates.

During late spring and summer, populations of Caucasian tur disperse throughout higher elevations, thus prey densities within the immediate vicinity of Ortvale Klde were lowered, and the increased costs associated with the search for and transport of kills far from their place of procurement likely excluded the site from intensive use. To date no LMP or EUP sites have been identified at higher elevations, but it is possible that Neanderthals and modern humans followed migrating herds of into the mountains, in which case we would expect to find faunal assemblages dominated by females and young, as males are solitary at this time. It is equally plausible that hominins shifted to the exploitation of larger areas by smaller, perhaps family units during the late spring and summer. Many resources would still have remained readily available during this part of the year while others, such as large ungulates, would be dispersed and more difficult to hunt in the overgrown forests and summer vegetation. Under such conditions the reorganization of foraging behaviors around the intensive exploitation of seasonally abundant floral resources, and an increase in residential mobility may have been the best way to minimize dietary risk (Kelly, 1995). If either population continued the committed pursuit of Caucasian tur during the late spring and summer, they were likely faced with higher search costs and relatively lower returns.

Regional Perspective on Seasonality in the Southern Caucasus

Based on detailed zooarchaeological and taphonomic analyses we have established a working understanding of seasonal exploitation patterns and mobility at Ortvale Klde. But how representative are these data of LMP and EUP hunting patterns at the larger regional scale? Is Ortvale Klde a special locality or is it part of a wider system of seasonally based mobility? In the following section we summarize the main zooarchaeological characteristics of three neighboring Paleolithic sites in an attempt to answer these questions. Unfortunately, in most instances the lack of secure chronometric estimates and methodological variability make such comparison difficult (see Adler and Tushabramishvili, 2004).

Faunal assemblages from most neighboring Paleolithic sites have been studied from a palaeontological perspective only and are of limited comparative value. For example, Djruchula Cave, an early Middle Paleolithic (late Middle Pleistocene–early Upper Pleistocene) site located approximately 7.5 km northeast of Ortvale Klde (Fig. 9.1), experienced palaeontological study in the 1970s and 1980s. The results indicate the dominance of cave bear (*Ursus spelaeus*) at the base of the sequence (Layer 2) and Bos/Bison and red deer at the top of the sequence (Layer 1) (Lioubin and Barychnikov, 1984; Adler and Tushabramishvili, 2004: Table 7) (Table 9.4). These data are interpreted as evidence that cave bears and humans utilized the site intermittently, with the remains of the former resulting from deaths during hibernation (Layer 2) or perhaps human predation. In either event, competing use such as this may help explain the relatively ephemeral archaeological signature in Layer 2. The faunal assemblage from Layer 1 is most likely the result of active hunting and is accompanied by a lithic assemblage of large retouched points (see Adler and Tushabramishvili, 2004). This pattern of fluctuating site use is thought to reflect changes in faunal exploitation patterns between the two layers, although to date, systematic zooarchaeological analyses have not been conducted and the faunal assemblage is no longer available for reanalysis. Lithic and faunal data point to the intermittent (Layer 2) and task-specific (Layer 1) use of this cave, an interpretation largely consistent with that offered by previous researchers (e.g., Liubin, 1977; Tushabramishvili, 1984; Liubin, 1989; Tushabramishvili, 1994). In this respect Djruchula Cave does not represent a central habitation site, but rather a specialized hunting camp. Regrettably, the available faunal data are not of suitable quality to allow detailed consideration of seasonal foraging behaviors or diet breadth.

The faunal assemblage from Bronze Cave, one among a series of stratified LMP sites located near the village of Tsutskhvati, approximately 35 km southwest of Ortvale Klde (Fig. 9.1), contains diverse taxa, but is dominated by steppe bison (80.0% by NISP for all layers), followed by lower frequencies of Caucasian tur (10.4% by NISP for all layers) (Tushabramishvili, 1978; Adler and Tushabramishvili, 2004) (Table 9.4). This assemblage is notable because it contains a relatively high frequency of carnivore remains in each layer (3–20%), specifically cave bear (Table 9.4), however this species does not dominate the assemblage as it does in Layer 2 at Djruchula Cave (Tushabramishvili, 1978; A. Vekua, 2001, personal commmunication; Adler and Tushabramishvili, 2004: Table 9). The available lithic and faunal data suggest that Bronze Cave served periodically as a campsite but that occupations were intermittent and often ephemeral in comparison to those documented at Ortvale Klde (Adler and Tushabramishvili, 2004: Table 10). Again, the available faunal data from Bronze Cave do not allow us to consider the issue of seasonality in any meaningful way, and increased diet breadth is only weakly suggested (Table 9.4).

Dzudzuana Cave, a neighboring site contemporaneous with and archaeologically identical to the EUP of Ortvale Klde, represents a valuable comparative case. Studies by Bar-Oz et al. (2004b, 2007), employing identical research methods and protocols to those used at Ortvale Klde, show a greater reliance on steppe bison and aurochs in all layers relative to that witnessed at Ortvale Klde (Table 9.4). Direct comparisons are hampered slightly since the taphonomic histories of the two sites are not identical, with the assemblage from Ortvale Klde spanning the full range of bone densities while that from Dzudzuana Cave is dominated by dense bones – mainly shaft fragments and teeth (Bar-Oz et al., 2007). We find that the density-mediated bias identified in Dzudzuana Cave is partially caused by in situ attrition resulting from a combination of post-depositional forces (weathering, trampling, and physical erosion). However, this observation is not likely to account for the apparent increase in steppe bison, as durable skeletal elements such as teeth are abundant among the remains of both steppe bison and Caucasian tur. While the demographic data for Caucasian tur suggests that Ortvale Klde was occupied during late fall–early spring, the demographic pattern of bison and aurochs at Dzudzuana Cave, and in particular the presence of newborn Bos/Bison, suggest that most individuals were procured during the summer and possibly early fall (Bar-Oz et al., 2007). While still relatively narrow in terms of taxonomic diversity, the data in Table 9.4 suggest a possible increase in diet breadth compared to Ortvale Klde or seasonal fluctuations in prey availability.

Unfortunately, it is difficult to construct a regional understanding of seasonal site use and species exploitation patterns based on the aforementioned data (Adler and Tushabramishvili, 2004: Table 11). Only two sites, Ortvale Klde and Dzudzuana Cave, have been studied with these specific questions in mind, while the others, Djruchula and Bronze caves, experienced excavation and analysis long before the advent of zooarchaeological and taphonomic techniques. So the question remains whether the differences observed between these sites reflect differences in seasonal hunting patterns, site use, environment, relative availability of taxa on the landscape and thus encounter rate, diet breadth, butchery and processing behaviors, taphonomy, or methods of analysis. Without the re-excavation of key sites and the discovery of new ones, this issue cannot be resolved. While it is not possible at present to assess how representative Ortvale Klde might be of regional Neanderthal or modern human hunting behaviors, the site does serve as a sensitive behavioral gauge of temporal trends in resource acquisition at a particular point on the landscape. These data demonstrate continuity between the LMP and EUP in resource knowledge and acquisition, specifically with respect to seasonal prey choice, culling practices, and niche preference.

Conclusions

Relying primarily on zooarchaeological and taphonomic analyses we document the hunting, butchering, and processing behaviors of the LMP and EUP inhabitants of Ortvale Klde and show that LMP groups targeted prime-age adult Caucasian tur and transported entire, or near entire carcasses back to the site for processing and consumption, behaviors attested also to for modern humans elsewhere across Eurasia. Consideration of Caucasian tur life history characteristics suggests that hunting activities in and around Ortvale Klde were structured according to this species' specific migratory behaviors, which made males, females, and young locally abundant from late fall to early spring. These data allow us to build a case for the intensive, seasonal use of the site, whereby Ortvale Klde operated as a locus from which hunting forays could be launched and to which hunted prey could be returned for processing and consumption.

Where direct comparisons were possible we found no significant economic differences between the LMP and EUP. This is not surprising if we accept that Neanderthals and modern humans were top predators in their chosen environments and that, given this position in the predator-prey hierarchy, each population exploited the same niches in a similar manner, especially in the arena of medium–large game procurement. Within the southern Caucasus, where populations are largely circumscribed by their surrounding landscape (greater and lesser Caucasus mountains to the north and south, respectively; Black and Caspian seas to the west and east, respectively), it is difficult for two top predators armed with equally effective methods of acquiring ecological knowledge and the same medium–large prey to occupy the same ecological niche for long. Following the principle of competitive exclusion (see Pianka, 1988; Begon et al., 1996), we contend that if these two populations encountered one another there was little chance for long-term coexistence; two populations in direct competition for identical limited resources cannot remain sympatric, especially if one population possesses or develops a major technological, biological, demographic, or social advantage.

In this paper we document the exploitation of the same limited resources by both groups. We have also explored the differential existence of technological, biological, demographic, or social advantages in a series of recent publications (Bar-Oz and Adler, 2005; Adler et al., 2006a, b; Bar-Yosef et al., 2006), but lacking human fossil material and reliable data on demography – a perennial shortcoming of the Palaeolithic record – we have only been able to conduct tests based on technology and mobility. Regarding the former we find clear rifts in material culture across the Middle–Upper Palaeolithic "transition" represented by the abrupt and unanticipated appearance of sophisticated projectile technology (backed microliths and bone points) in the EUP, and the synchronous (within the precision of radiocarbon) disappearance of typical LMP assemblages of scrapers and Levallois technology on the southern side of the mountains (e.g., Ortvale Klde, Adler et al., 2006a, b) and para-Micoquian, non-Levallois assemblages on the northern side of the mountains (e.g., Adler et al., 2006b; Cleghorn, 2006; Mezmaiskaya Cave, Golovanova and Doronichev, 2003). In the present context it is unclear what advantage(s) this new technology offered but it may have served to increase foraging efficiency (e.g., increased kill vs. encounter rate) as well as the distance between predators and prey (Shea, 2006). Technological and tactical modifications such as these should correlate with profound alterations in hunting efficiency and thus individual and group fitness, but the faunal data from Ortvale Klde do not document any significant difference in patterns of prey exploitation or diet breadth between those with and those without such technology. Instead cultural (stylistic) needs that serve to build and maintain social relationships between immediate and extended group members may be linked closely to the development of elaborate projectile technology. The foraging data from Ortvale Klde indicate that, at least seasonally, both Neanderthal and modern human technological repertoires were equally well suited to the task of medium–large game acquisition.

Regarding mobility, the EUP occupants of Ortvale Klde routinely exploited obsidian sources located in excess of 100 km to the south, a behavior for which there is no parallel among local Neanderthals (Adler et al., 2006a). These patterns of raw material exploitation are generally representative of both populations, with modern humans routinely exploiting relatively large territories and establishing extended social networks, while Neanderthals foraged in relatively small territories and likely interacted with far fewer individuals (Adler et al., 2006a). Given the available data, we argue that Neanderthal-modern human coexistence in the southern Caucasus, if it occurred, was probably very short, and the major behavioral edge that would have allowed modern human populations to grow at the expense of the Neanderthals was a cultural one, namely their ability to establish larger extended social networks (see Gamble, 1999; Whallon, 2006) and exploit larger territories, behaviors that in and of themselves likely elevated individual and group fitness, affording a considerable competitive advantage (see Adler et al., 2006a).

Archaeological, taphonomic and chronometric data from Ortvale Klde suggest that the arrival of modern humans likely signaled the rapid – and by all accounts permanent – cessation of local Neanderthal traditions within the southern Caucasus. Modern humans may have gained further advantage by exploiting a wider array of resources (Stiner et al., 2000; Hockett and Haws, 2005), although such data are currently lacking at Ortvale Klde and only weakly suggested at Dzudzuana Cave. This may be due to the early state of zooarchaeological and taphonomic research in the region as much as the seasonally structured economic record currently available for study. We find no evidence that Neanderthal populations attempted to broaden their resource base prior to extinction or that local densities of Caucasian tur were in decline. From a purely ecological perspective we argue that the "transition" from the LMP to the EUP in the southern Caucasus represents the replacement of one top predator (Neanderthals) by another (modern humans), with little discernable shift in the predator-prey hierarchy, the medium–large game hunted, or their seasons of exploitation. In terms of resource and niche preference, we suggest that local Neanderthal and modern human populations, if they coexisted, were sympatric to the point of exclusion.

Acknowledgment. We are grateful to our colleagues at the Georgia State Museum: D. Lordkipanidze, A. Vekua, N. Tushabramishvili, and T. Meshveliani for their collaboration and assistance during our many years of research in the Republic of Georgia. We recognize the Wenner-Gren Foundation for Anthropological Research (Grants 6881 and 7059), the L. S. B. Leakey Foundation, American School of Prehistoric Research at Harvard University, the Rothschild Post-Doctoral Foundation, and the Davis Center for Russian Studies at Harvard University for their generous financial support. Finally, we wish to thank J.-J. Hublin and M. Richards for their invitation to present our research at the Max Planck Institute for Evolutionary Anthropology, Department of Human Evolution, and for their exceptional hospitality.

References

Adler, D.S., 2002. Late Middle Palaeolithic Patterns of Lithic Reduction, Mobility, and Land Use in the Southern Caucasus. Ph.D. Dissertation, Harvard University.

Adler, D.S., Tushabramishvili, N., 2004. Middle Palaeolithic Patterns of Settlement and Subsistence in the Southern Caucasus. In: Conard, N.J. (Ed.), Middle Palaeolithic Settlement Dynamics. Publications in Prehistory, Kerns: Tübingen, pp. 91–132.

Adler, D.S., Bar-Oz, G., Belfer-Cohen, A., Bar-Yosef, O., 2006a. Ahead of the game: Middle and Upper Palaeolithic Hunting Practices in the Southern Caucasus. Current Anthropology 47(1), 89–118.

Adler, D.S., Belfer-Cohen, A., Bar-Yosef, O., 2006b. Between a Rock and a Hard Place: Neanderthal-Modern Human Interactions in the Southern Caucasus. In: Conard, N.J. (Ed.), Neanderthals and Modern Humans Meet. Publications in Prehistory, Kerns: Tübingen, pp. 165–187.

Adler, D. S., Bar-Yosef, O., Belfer-Cohen, A, Tushabramishvili, N., Boaretto, E., Mercier, N., Valladas, H., Rink, W. J. 2008. Dating the Demise: Neanderthal Extinction and the Establishment of Modern Humans in the Southern Caucasus. Setting the Record Straight: Toward a Systematic Chronological Understanding of the Middle to Upper Paleolithic Boundary in Eurasia, D. S. Adler and O. Jöris, editors. Journal of Human Evolution 55(5), 817–833.

Aiello, L.C., Wheeler, P., 2003. Neandertal Thermo-Regulation and the Glacial Climate. In: van Andel, T.H., Davies, W. (Eds.), Neanderthals and Modern Humans in Europe During the Last Glaciation. MacDonald Institute for Archaeological Research: Cambridge, pp. 147–156.

Bar-Oz, G., Adler, D.S., 2005. Taphonomic History of the Middle and Upper Palaeolithic Faunal Assemblages from Ortvale Klde, Georgian Republic. Journal of Taphonomy 3(4), 185–211.

Bar-Oz, G., Munro, N.D., 2007. Gazelle Bone Marrow Yields and Epipalaeolithic Animal Exploitation Strategies in the Southern Levant. Journal of Archaeological Science 34, 946–956.

Bar-Oz, G., Dayan, T., Weinstein-Evron, M., Kaufman, D., 2004a. The Natufian Economy at el-Wad Terrace with Special Reference to Gazelle Exploitation Patterns. Journal of Archaeological Science 31, 217–231.

Bar-Oz, G., Adler, D.S., Vekua, A., Meshveliani, T., Tushabramishvili, N., Belfer-Cohen, A., Bar-Yosef, O., 2004b. Faunal Exploitation Patterns Along the Southern Slopes of the Caucasus During the Late Middle and Early Upper Palaeolithic. In: Mondini, M., Muñoz, S., Wickler, S. (Eds.), Colonisation, Migration, and Marginal Areas. A Zooarchaeological Approach. Oxbow Books: Oxford, pp. 46–54.

Bar-Oz, G., Belfer-Cohen, A., Mesheviliani, T., Djakeli, N., Bar-Yosef, O., 2007. Taphonomy and Zooarchaeology of the Upper Palaeolithic Cave of Dzudzuana, Republic of Georgian. International Journal of Osteoarchaeology 17, 1–21.

Bar-Yosef, O., 2004. Eat What is There: Hunting and Gathering in the World of Neanderthals and Their Neighbours. International Journal of Osteoarchaeology 14, 333–342.

Bar-Yosef, O., Belfer-Cohen, A., Adler, D.S., 2006. The Implications of the Middle-Upper Paleolithic Chronological Boundary in the Caucasus to Eurasian Prehistory. Anthropologie XLIV/1, 49–60.

Bar-Yosef, O., Vandermeersch, B., Arensburg, B., Belfer-Cohen, A., Goldberg, P., Laville, H., Meignen, L., Rak, Y., Speth, J.D., Tchernov, E., Tillier, A.M., Weiner, S., 1992. The Excavations in Kebara Cave, Mt. Carmel. Current Anthropology 33(5), 497–550.

Baryshnikov, G., Hoffecker, J.F., 1994. Mousterian Hunters of the NW Caucasus: Preliminary Results of Recent Investigations. Journal of Field Archaeology 21(1), 1–14.

Baryshnikov, G., Hoffecker, J.F., Burgess, R.L., 1996. Palaeontology and Zooarchaeology of Mezmaiskaya Cave (Northwestern Caucasus, Russia). Journal of Archaeological Science 23, 313–335.

Begon, M., Harper, J.L., Townsend, C.R., 1996. Ecology. Blackwell: Oxford.

Bergmann, C., 1847. Über die verheltnisse der warmerkonomie de thiere zu iherer grosse. Gottingen Studien, 3, 595–708.

Bird, D.W., O'Connell, J.F., 2006. Behavioral Ecology and Archaeology. Journal of Archaeological Research 14, 143–188.

Blumenschine, R.J., Selvaggio, M., 1988. Percussion Marks on Bone Surface as a New Diagnostic of Human Behavior. Nature, 333, 763–765.

Bocherens, H., Drucker, D., 2006. Dietary Competition Between Neanderthals and Modern Humans: Insights from Stable Isotopes. In: Conard, N.J. (Ed.), Neanderthals and Modern Humans Meet. Publications in Prehistory, Kerns: Tübingen, pp. 129–143.

Bocherens, H., Billiou, D., Mariotti, A., Patou-Mathis, M., Otte, M., Bonjean, D., Toussaint, M., 1999. Palaeoenvironmental and Palaeodietary Implications of Isotopic Biogeochemistry of Last Interglacial Neanderthal and Mammal Bones in Scladina Cave (Belgium). Journal of Archaeological Science 26, 599–607.

Bocherens, H., Drucker, D., Billiou, D., Patou-Mathis, M., Vandermeersch, B., 2005. Isotopic Evidence for Diet and Subsistence Pattern of the Saint-Césaire I Neanderthal: Review and Use of a Multi-source Mixing Model. Journal of Human Evolution 49, 71–87.

Brown, M., Burton, R. (Eds.), 1974. *Funk & Wagnalls Wildlife Encyclopedia*, vol. 10. Funk & Wagnalls: New York.

Bunn, H.T., 1993. Bone Assemblages at Base Camps: a Further Consideration of Carcass Transport and Bone Destruction by the Hadza. In: Hudson, J. (Ed.), From Bones to Behavior: Ethnoarchaeological and Experimental Contributions to the Interpretation of Faunal Remains. Occasional Paper No. 21. Southern Illinois University: Illinois, pp. 156–168.

Butzer, K., 1982. *Archaeology as Human Ecology: Method and Theory for a Contextual Approach*. Cambridge University Press: Cambridge.

Cleghorn, N.E., 2006. A Zooarchaeological Perspective on the Middle to Upper Paleolithic Transition at Mezmaiskaya Cave, The Northwestern Caucasus, Russia. Ph.D. Dissertation, Stony Brook University.

Davis, S.J.M., 1981. The Effect of Temperature Change and Domestication on the Body Size of Late Pleistocene to Holocene Mammals of Israel. Paleobiology 7, 101–114.

Domínguez-Rodrigo, M., 2002. Hunting and Scavenging by Early Humans: the State of the Debate. Journal of World Prehistory, 16, 1–54.

Driesch, A. von den., 1976. *A Guide to a Measurement of Animal Bones from Archaeological Sites*. Peabody Museum Bulletins 1. Peabody Museum of Archaeology and Ethnology: Cambridge.

Finlayson, C., 2004. *Neanderthals and Modern Humans: An Ecological and Evolutionary Perspective*. Cambridge University Press: Cambridge.

Gamble, C., 1999. *The Palaeolithic Societies of Europe*. Cambridge University Press: Cambridge.

Gaudzinski, S., Roebroeks, W., 2000. Adults Only. Reindeer Hunting at the Middle Palaeolithic Site Salzgitter Lebenstedt, Northern Germany. Journal of Human Evolution 38, 497–521.

Gaudzinski, S., Roebroeks, W., 2003. Profile Analysis at Salzgitter-Lebenstedt. A Reply to Munson & Marean. Journal of Human Evolution 44, 275–281

Golovanova, L.V., Doronichev, V.B., 2003. The Middle Paleolithic of the Caucasus. Journal of World Prehistory 17, 71–140.

Grayson, D.K., Delpech, F., 2002. Specialized Early Upper Palaeolithic Hunters in Southwestern France? Journal of Archaeological Science 29, 1439–1449.

Grayson, D.K., Delpech, F., 2003. Ungulates and the Middle-to-Upper Paleolithic Transition at Grotte XVI (Dordogne, France). Journal of Archaeological Science 30, 1633–1648.

Grayson, D.K., Delpech, F., 2006. Was There Increasing Dietary Specialization Across the Middle-To-Upper Paleolithic Transition in France? In: Conard, N.J. (Ed.), Neanderthals and Modern Humans Meet. Publications in Prehistory, Kerns: Tübingen, pp. 377–417.

Henshilwood, C.S., Marean, C.W., 2003. The Origin of Modern Human Behavior: Critique of the Models and Their Test Implications. Current Anthropology 44, 627–651.

Heptner, V.G., Nasimovich, A.A., Bannikov, A.G., 1988. *Mammals of the Soviet Union, vol. I: Artiodactyla and Perissodactyla*. Smithsonian Institute Libraries and National Science Foundation: Washington, DC.

Heptner, V.G., Nasimovich, A.A., Bannikov, A.G., 1989. *Mammals of the Soviet Union, vol. 1: Ungulates*. E.J. Brill: Leiden.

Hockett, B., Haws, J.A., 2005. Nutritional Ecology and the Human Demography of Neandertal Extinction. Quaternary International 137, 21–34.

Hoffecker, J.F., 1999. Neanderthals and Modern Humans in Eastern Europe. Evolutionary Anthropology 7, 129–141.

Hoffecker, J.F., 2002. *Desolate Landscapes: Ice-Age Settlement in Eastern Europe*. Rutgers University Press: New Brunswick.

Hoffecker, J.F., Cleghorn, N., 2000. Mousterian Hunting Patterns in the Northwestern Caucasus and the Ecology of the Neanderthals. International Journal of Osteoarchaeology 10, 368–378.

Hoffecker, J.F., Baryshnikov, G., Potapova, O., 1991. Vertebrate Remains from the Mousterian Site of Il'skaya I (Northern Caucasus, U.S.S.R.): New Analysis and Interpretations. Journal of Archaeological Science 18, 113–147.

Jaubert, J., Lorblanchet, M., Laville, H., Slott-Moller, R., Turq, A., Brugal, J.-P., 1990. *Les Chasseurs d'Aurochs de La Borde*. Documents d'Archéologique Français, No. 27, Editions de la Maison des Sciences de L'Homme: Paris.

Kelly, R.L., 1995. *The Foraging Spectrum: Diversity in Hunter-Gatherer Lifeways*. Smithsonian Institution Press: Washington, DC.

Kelly, R.L., 2003. Colonization of New Land by Hunter-Gatherers: Expectations and Implications Based on Ethnographic Data. In: Rockman, M., Steele, J. (Eds.), Colonization of Unfamiliar

Landscapes: The Archaeology of Adaptation. Routledge: London, pp. 44–58.

Klein, R.G., 1976. The Mammalian Fauna of the Klasies River Mouth Sites, Southern Cape Province, South Africa. South African Archaeological Bulletin 31, 75–96.

Klein, R.G., Cruz-Uribe, K., 1984. *The Analysis of Animal Bones from Archaeological Sites*. University Chicago Press: Chicago.

Klein, R.G., Wolf, C., Freeman, L.G., Allwarden, K., 1981. The Use of Dental Crown Heights for Constructing Age Profiles of Red Deer and Similar Species in Archaeological Samples. Journal of Archaeological Science 8, 1–31.

Kuhn, S.L., 1995. *Mousterian Lithic Technology: An Ecological Approach*. Princeton University Press: Princeton.

Kuhn, S.L., Stiner, M.C., 2006. What's a Mother to Do? The Division of Labor Among Neandertals and Modern Humans in Eurasia. Current Anthropology 47, 953–980.

Kurten, B., 1965. Canivora of the Palestine Caves. Acta Zoologica Fennica, 107, 1–74.

Lay, D.M., 1967. A Study of the Mammals of Iran. Fieldiana: Zoology 54, 3–282.

Levine, M.A., 1983. Mortality Models and the Interpretation of Horse Population Structure. In: Bailey, G. (Ed.), Hunter-Gatherer Economy in Prehistory: A European Perspective. Cambridge University Press: Cambridge, pp. 23–46.

Lioubin, V.P., Barychnikov, G.F., 1984. L'activité de Chasse des Plus Anciens Habitants du Caucase (Acheuléen, Moustérien). L'Anthropologie 88(2), 221–229.

Liubin, V.P., 1977. *Mustierskie kul'turi Kavkaza (Mousterian Cultures of the Caucasus)*. Nauka: Leningrad.

Liubin, V.P., 1989. Paleolit Kavkaza (Paleolithic of the Caucasus). In: Boriskovski, P.I. (Ed.), Paleolit Kavkaza i Sredney Azii (The Palaeolithic of the Caucasus and Northern Asia). Nauka: Leningrad, pp. 9–144.

Liubin, V.P., 1998. La Grotte Mousterienne Barakaevskaïa (Nord Caucase). L'Anthropologie 102, 67–90.

Lupo, K.D., 1995. Hadza Bone Assemblages and Hyena Attrition: An Ethnographic Example of the Influence of Cooking and Mode of Discard on the Intensity of Scavenger Ravaging. Journal of Anthropological Archaeology 14, 288–314.

Marean, C.W., Kim, S.Y., 1998. Mousterian Large-Mammal Remains from Kobeh Cave: Behavioral Implications for Neandertals and Early Modern Humans. Current Anthropology 39, S79–S113.

Marean, C.W., Spencer, L.M., 1991. Impact of Carnivore Ravaging on Zooarchaeological Measures of Element Abundant. American Antiquity, 56, 645–658.

Marean, C.W., Spencer, L.M., Blumenschine, R.J., Capaldo, S., 1992. Captive Hyena Bone Choice and Destruction, the Schlepp Effect, and Olduvai Archaeofaunas. Journal of Archaeological Science, 19, 101–121.

Marks, A.E., Chabai, V.P., 2001. Constructing Middle Palaeolithic Settlement Systems in Crimea. In: Conard, N.J. (Ed.), Settlement Dynamics of the Middle Palaeolithic and Middle Stone Age. Kerns: Tübingen, pp. 179–204.

Marks, A.E., Chabai, V.P., 2006. Stasis and Change During the Crimean Middle Paleolithic. In: Hovers, E., Kuhn, S.L. (Eds.), Transitions Before the Transition: Evolution and Stability in the Middle Paleolithic and Middle Stone Age. Springer: New York, pp. 121–135.

Meltzer, D.J., 2003. Lessons in Landscape Learning. In: Rockman, M., Steele, J. (Eds.), Colonization of Unfamiliar Landscapes: The Archaeology of Adaptation. Routledge: London, pp. 222–241.

Meshveliani, T., Bar-Yosef, O., Belfer-Cohen, A., Djakeli, N., Kraus, A., Lordkipanidze, D., Tvalchrelidze, M., Vekua, A., 1999. Excavations at Dzudzuana Cave, Western Georgia (1996–1998): Preliminary Results. Préhistoire Europeenne 15, 79–86.

Meshveliani, T., Bar-Yosef, O., Belfer-Cohen, A., 2004. The Upper Paleolithic in Western Georgia. In: Brantingham, P.J., Kuhn, S.L., Kerry, K.W. (Eds.), The Early Upper Paleolithic East of the Danube. University of California Press: Berkeley, pp. 129–143.

Metcalfe, D., Jones, K.T., 1988. A Reconsideration of Animal Body Part Utility Indices. American Antiquity 53, 486–504.

Miracle, P., 2005. Late Mousterian Subsistence and Cave Use in Dalmatia: The Zooarchaeology of Mujina Pećina, Croatia. International Journal of Osteoarchaeology 15, 84–105.

Munro, N.D., Bar-Oz, G., 2005. Gazelle Bone Fat Processing in the Levantine Epipalaeolithic. Journal of Archaeological Science, 32, 223–239.

Munson, P.J., Marean, C.W., 2003. Adults Only? A Reconsideration of Middle Palaeolithic 'Prime Dominated' Reindeer Hunting at Salzgitter Lebenstedt. Journal of Human evolution, 44, 263–273.

Nasimovich, A.A., 1949. Ocherk ekologii zapadnokavkazskogo tura (Outline of Ecology of Western Caucasian tur). Trudy Kavkazskogo Gosudarstvennogo Zapovednika 3.

Nowak, R.M. (Ed.), 1991. Walker's Mammals of the World, vol. 2. The Johns Hopkins University Press: Baltimore.

O'Connell, J.F., 2006. How Did Modern Humans Displace Neanderthals? Insights from Hunter-Gatherer Ethnography and Archaeology. In: Conard, N.J. (Ed.), Neanderthals and Modern Humans Meet. Publications in Prehistory, Kerns: Tübingen, pp. 43–64.

Parker, S.P. (Ed.), 1990. Grzimek's Encyclopedia of Mammals, vol. 5. McGraw-Hill: New York.

Pianka, E.R., 1988. Evolutionary Ecology. 4th ed. Harper & Row: New York.

Pike-Tay, A., Valdés, V.C., de Quirós, F.B., 1999. Seasonal Variations of the Middle–Upper Paleolithic Transition at El Castillo, Cueva Morin and El Pendo (Cantabria, Spain). Journal of Human Evolution 36, 283–317.

Richards, M.P., Pettitt, P.B., Trinkaus, E., Smith, F.H., Paunovi´c, M., Karavani´c, I., 2000. Neanderthal Diet at Vindija and Neanderthal Predation: The Evidence from Stable Isotopes. Proceedings of the National Academy of Science, USA 97, 7663–7666.

Richards, M.P., Pettitt, P.B., Stiner, M.C., Trinkaus, E., 2001. Stable Isotope Evidence for Increasing Dietary Breadth in the European Mid-Upper Paleolithic. Proceedings of the National Academy of Sciences, USA 98, 6528–6532.

Roberts, T.J., 1977. The Mammals of Pakistan. Ernest Benn: London.

Shea, J.J., 1998. Neandertal and Early Modern Human Behavioral Variability. Current Anthropology 39, S45–S78.

Shea, J. J., 2003. Neandertals, Competition, and the Origin of Modern Human Behavior in the Levant. Evolutionary Anthropology 12, 173–187.

Shea, J. J., 2006. The Origins of Lithic Projectile Point Technology: Evidence from Africa, the Levant, and Europe. Journal of Archaeological Science, 33, 823–846.

Shennan, S., 2002. Archaeology and Evolutionary Ecology. World Archaeology 34, 1–5.

Speth, J. D., 1983. Bison Kills and Bone Counts: Decision Making by Ancient Hunters. University Chicago Press, Chicago.

Speth, J.D., 1987. Early Hominid Subsistence Strategies in Seasonal Habitats. Journal of Archaeological Science 14, 13–29.

Speth, J.D., 1989. Early Hominid Hunting and Scavenging: the Role of Meat as an Energy Source. Journal of Human Evolution 18, 329–343.

Speth, J.D., 1990. Seasonality, Resource Stress, and Food Sharing in So-Called 'Egalitarian' Foraging Societies. Journal of Anthropological Archaeology 9, 148–188.

Speth, J.D., Clark, J.L., 2006. Hunting and Overhunting in the Levantine Late Middle Palaeolithic. Before Farming 3, 1–42.

Speth, J.D., Spielmann, K., 1983. Energy Source, Protein Metabolism and Hunter-Gatherer Subsistence Strategies. Journal of Anthropological Archaeology 2, 1–31.

Speth, J.D., Tchernov, E., 1998. The Role of Hunting and Scavenging in Neanderthal Procurement Strategies: New Evidence from Kebara Cave (Israel). In: Akazawa, T., Aoki, K., Bar-Yosef, O. (Eds.), Neandertals and Modern Humans in Western Asia. Plenum Press: New York, pp. 223–240.

Steele, T.E., 2004. Variation in Mortality Profiles of Red Deer (Cervus elaphus) in Middle Palaeolithic Assemblages from Western Europe. International Journal of Osteoarchaeology, 14, 307–320.

Stewart, J.R., 2004. Neanderthal-Modern Human Competition? A Comparison Between the Mammals Associated with Middle and Upper Palaeolithic Industries in Europe during OIS 3. International Journal of Osteoarchaeology 14, 178–189.

Stewart J.R., Markova, A., Musil, R., van Kolfschoten, T., 2003. Neanderthals as Part of the Broader Late Pleistocene Megafaunal Extinctions? In: Van Andel, T., Davies, S.W. (Eds.), Neanderthals and Modern Humans in the European Landscape During the Last Glaciation, 60,000 to 20,000 years ago: Archaeological Results of the Stage Three Project McDonald Institute Monograph Series: Cambridge, pp. 221–231.

Stiner, M.C., 1990. The Use of Mortality Patterns in Archaeological Studies of Hominid Predatory Adaptations. Journal of Anthropological Archaeology 9, 305–351.

Stiner, M., 1994. Honor Among Thieves: A Zooarchaeological Study of Neandertal Ecology. Princeton University Press: Princeton.

Stiner, M.C., 2002. Carnivory, Coevolution, and the Geographic Spread of the Genus Homo. Journal of Archaeological Research 10, 1–63.

Stiner M.C., 2005. The Faunas of Hayonim Cave (Israel): A 200,000-Year Record of Paleolithic Diet, Demography and Society. Peabody Museum of Archaeology and Ethnology: Cambridge.

Stiner, M.C., Weiner, S., Bar-Yosef, O., Kuhn, S.L., 1995. Differential Burning, Fragmentation, and Preservation of Archaeological Bone. Journal of Archaeological Science 22, 223–237.

Stiner, M.C., Munro, N.D., Surovell, T.A., 2000. The Tortoise and the Hare: Small-Game Use, the Broad-Spectrum Revolution, and Paleolithic Demography. Current Anthropology 41, 39–73.

Tushabramishvili, D.M. (Ed.), 1978. Arkheologicheskie panyatniki Tsutskhvatskogo mnogoetajnogo peshernogo kompleska (Archaeological Sites of the Tsutskhvati Cave Complex). Metsniereba: Tbilisi.

Tushabramishvili, D.M., 1984. Paleolit Gruzii (Palaeolithic of Georgia). Vestnik – Gosudarstvennogo Muzeia Gruzii (Newsletter of the Georgian State Museum) 37B, 5–27.

Tushabramishvili, N., 1994. Srednii Paleolit Zapadnoi gruzii i etapy ego perekhoda k verkhnemu paleolitu (po materialam Ortvala-Klde) (The Middle Palaeolithic of Western Georgia and the Stages of its Transition to the Upper Palaeolithic Based on Material from Ortvale Klde). Ph. D. Dissertation, University of Tbilisi.

Tushabramishvili, N., Lordkipanidze, D., Vekua, A., Tvalcherlidze, M., Muskhelishvili, A., Adler, D.S., 1999. The Middle Palaeolithic Rockshelter of Ortvale Klde, Imereti Region, The Georgian Republic. Préhistoire Europeenne 15, 65–77.

Vereshchagin, N.K., 1967. The Mammals of the Caucasus: A History of the Evolution of the Fauna. Israel Program for Scientific Translations: Jerusalem.

Villa, P., Mahieu, E., 1991. Breakage Patterns of Human Long Bones. Journal of Human Evolution, 21, 27–48.

Wang, Z.-L., Zhang, J.-G., Liu, X.-M., 2000. Competition and Coexistence in Spatially Subdivided Habitats. Journal of Theoretical Biology 205, 631–639.

Whallon, R., 2006. Social Networks and Information: Non-"Utilitatian" Mobility Among Hunter-Gatherers. Journal of Anthropological Archaeology 25, 259–270.

Winterhalder, B., Smith, E.A., 2000. Analyzing Adaptive Strategies: Human Behavioral Ecology at Twenty-Five. Evolutionary Anthropology 9, 51–72.

Wobst, M., 1978. The Archaeo-Ethnology of Hunter-Gatherers, or, the Tyranny of the Ethnographic Record in Archaeology. American Antiquity 43, 303–309.

Zharkov, I. V. 1940. Zimnii uchet kopytnykh progonom v Kavkazskom zapovednike (Winter counting of ungulates by driving them in Caucasian preserve). Nauchnye-Metody Zapovednika Glavnogo Upravleniya Po Zapovednikam 6.

10. Epipaleolithic Subsistence Intensification in the Southern Levant: The Faunal Evidence

Natalie Munro
University of Connecticut
Department of Anthropology
Unit 2176, 354 Mansfield Road
Storrs, CT, USA
Natalie.Munro@uconn.edu

Keywords Epipaleolithic • subsistence intensification • Southwest Asia • foraging efficiency • zooarchaeology • southern Levant

Abstract Subsistence intensification – the extraction of increased amounts of energy from a given area at the expense of foraging efficiency – figures prominently in discussions of the Epipaleolithic period (ca. 21,000–11,500 cal BP). Despite their paramount status, intensification trends are not often subjected to rigorous testing using multiple archaeological data sets. This study aims to fill this gap by synthesizing relevant zooarchaeological data from Epipaleolithic sites in Southwest Asia. Intensification is examined not only at the level of the animal community, but also at the taxonomic, and individual carcass levels. Multiple lines of faunal evidence including prey age profiles, large and small game abundance, species diversity, and the fragmentation of animal bone support an intensification trend across the Epipaleolithic period that culminates in the "transition" to agriculture. The pathway toward intensification was not always gradual – variation in Southwest Asian hunting practices indicates that there were some stops and starts along the way. The intensification trend reported here corresponds with other long-term processes, including increased site use intensity and growing human populations, but crosscuts major climatic events.

Introduction

The intensification of hunter-gatherer subsistence strategies is commonly associated with the Epipaleolithic, the period immediately preceding the "transition" to agriculture in much of the Old World. It is generally assumed that intensification is caused by population packing – imbalances between human populations and resources lead to the extraction of more resources per unit area at a greater cost to the forager. Epipaleolithic intensification is thus often linked to population pressure models, which have long figured prominently in explanations for agricultural origins (Boserup, 1965; Binford, 1968; Flannery, 1969; Cohen, 1977; Bar-Yosef and Belfer-Cohen, 1989, Henry, 1989; Bar-Yoset and Meadow, 1991; Bar-Yosef and Meadow, 1995; Bar-Oz, 2004; Munro, 2004b; Davis, 2005).

J.-J. Hublin and M.P. Richards (eds.), *The Evolution of Hominin Diets: Integrating Approaches to the Study of Palaeolithic Subsistence*, 141–155.

Epipaleolithic intensification has most often been investigated by searching for evidence of Flannery's (1969) Broad Spectrum Revolution (BSR). The BSR is described as an expansion of human diets to include new small game species, and ultimately results in reduced foraging efficiency. Documentation of the BSR has focused primarily on recording prey diversity, which has not always been sufficiently sensitive to expose changes in foraging efficiency (Edwards, 1989; Neeley and Clark, 1993; Bar-Oz et al., 1999). This has led some researchers to conclude that the BSR and demographic sources of population imbalance were unimportant in the "transition" to agriculture. More recently, however, studies compiling multiple zooarchaeological measures of human foraging efficiency have pinpointed periods of human demographic expansion and clearly intensified resource use in the Epipaleolithic (Stiner et al., 1999, 2000; Stiner, 2001, Munro, 2003, 2004b; Bar-Oz, 2004; Stiner 2005; Munro and Bar-Oz, 2005). These studies differ from earlier attempts to characterize intensification by combining a number of complementary zooarchaeological measures, including the relative abundance of prey types exploited by humans, prey mortality profiles, and carcass butchering strategies. Below, a synthesis of multiple lines of zooarchaeological evidence for Epipaleolithic subsistence intensification in Southwest Asia, in particular the southern Levant, is presented at community, population, and individual animal scales.

Here, intensification is defined as an increase in the amount of energy extracted from a given environment per unit time. Intensification is thus synonymous with a reduction in foraging efficiency – more energy is invested for each unit of energy returned. Intensification is a common solution to imbalances between human population size and resource availability (carrying capacity). Resource imbalances can occur as the result of human demographic growth or as a change in the availability of high-ranked prey in response to environmental change or localized predation (Lyman, 2003). Excellent ways to resolve imbalances include adding new, but less cost-effective resources to the diet, and intensifying the exploitation of already hunted species. The latter strategy can be executed by targeting previously unexploited age groups or intensifying the extraction of nutrients from individual animal carcasses. These solutions are effective, but because they return less energy for the amount invested in comparison to strategies focusing on higher-ranked taxa, age groups, or body-parts, they result in intensification.

The intensification of animal resources can take place on a number of ecological scales, including the community, the population, and the individual animal scale. Classically, detection of the broad spectrum revolution has occurred at the community scale by documenting the diversity (the number) of hunted prey species through time (Edwards, 1989; Neeley and Clark, 1993; Bar-Oz et al., 1999). In this case, an increase in species diversity is assumed to represent intensification as dietary breadth expands. Diversity measures can provide good gauges of intensification, but are less valuable if they do not also consider the relative abundances of each prey group and the differential costs and benefits of prey taxa in the diet. Relative abundance indices provide instructive alternatives to diversity analyses by comparing the frequency of high- to low-ranked game groups to detect evidence for declining foraging efficiency (Winterhalder, 1981; Szuter and Bayham, 1989; Broughton, 1994; Madsen and Schmitt, 1998; Stiner et al., 1999, 2000; Butler, 2000; Cannon, 2000; Stiner, 2001; Munro 2003, 2004b). Relative rankings are assigned according to the cost-benefits of each prey species, and are most often based on prey body size (Broughton, 1994) and cost of capture (Stiner et al., 1999, 2000). Prey relative abundance indices require that compared taxa have differential cost/benefits, and thus an increase in the relative representation of low- in relation to high-ranked prey indicates intensification – a decrease in foraging efficiency.

At the scale of a single animal population, the degree of intensification can be measured as the relative proportion of young to adult animals of a single prey species. Mortality profiles can be successful indicators of intensification for two primary reasons. First, because prey mortality increases with heightened hunting pressure, human hunting releases a prey population from the constraints of carrying capacity and thus encourages a prey population to grow (Elder, 1965; Caughley, 1977; Koike and Ohtaishi, 1985; Stiner, 1994). Growing populations have higher proportions of juvenile animals, which will in turn be hunted and deposited in the archaeological record. The ratio of juvenile to adult animals in a hunted population will thus be elevated in comparison to non-hunted populations. Second, like in the community scale analyses mentioned previously, intensification should result in an increased abundance of low- in relation to high-ranked animals or, in this case, in the abundance of smaller juvenile individuals that yield less meat and fat, in comparison to the significantly larger prime-aged adults (Stiner, 1994). Intensified hunting of a given animal population should thus be expressed as an increase in the proportion of young individuals in the hunted assemblage. In animal populations that grow throughout their lives, such as tortoises, a decrease in the average age of the population will be depicted as a reduction in average body-size (Klein, 1999; Stiner et al., 1999, 2000).

Finally, intensification can be investigated at the level of the individual animal by examining how aggressively humans processed carcasses to extract edible products such as meat, marrow, and grease. Edible products are differentially distributed across animal carcasses and some cost more to extract than others (Binford, 1978; Lupo and Schmitt, 1997; Saint-Germaine, 1997; Munro and Bar-Oz, 2005). Thus, substantial variation in the costs and benefits of the extraction of different products and body parts exists. Carcass processing intensity can be measured first by determining what products (meat, marrow, grease) were regularly extracted from animal carcasses (Munro and Bar-Oz, 2005). Grease, for example requires more effort to extract than bone marrow, which itself requires more energy to process than most cuts of meat. Second, carcass processing intensity can be measured by the size and quality of the meat and fat packages that humans regularly butchered and consumed. Low intensity processing strategies should thus be represented by the focused removal of prime energy packages. This will be expressed zooarchaeologically by the presence of high-utility body-parts, associated cutmarks, and routine breakage of only the richest marrow-bearing bones. In contrast, intensive processing strategies will be indicated by the presence of both high- and low-utility body-parts, and the fragmentation of all marrow-bearing bones, including those with small marrow stores.

In combination, analyses from each of the three ecological scales have the power to provide a robust picture of dietary intensification. Here, I present faunal evidence from each scale to establish a broad picture of Epipaleolithic subsistence intensification in Southwest Asia. I focus primarily on the southern Levant because currently, it is home to the richest Epipaleolithic record in the region, and large comparative faunal databases exist. The Epipaleolithic of the southern Levant can be divided into three primary cultural periods (Table 10.1): the Kebaran (21,500–17,000 cal BP); the Geometric Kebaran (17,000–14,500 cal BP); and the Natufian which is commonly subdivided into Early (14,500–13,000 cal BP) and Late (13,000–11,500 cal BP) phases. Below, I combine my own data (Munro, 2001, 2004b) with comparable datasets published by Stiner (2005) and Bar-Oz (2004) from 11 discrete Epipaleolithic phases from nine archaeological sites (Fig. 10.1; Table 10.1). All sites are located in northern Israel and occupy either the Mediterranean zone or the ecotone between the Mediterranean hills and the coastal plain. The archaeological components are organized from oldest to youngest to depict changes in human subsistence use through time.

TABLE 10.1. Site names, cultural phases and time ranges of southern Levantine assemblages discussed herein.

Site	Cultural period	Time range (cal BP)	Reference
El-wad Terrace	Late Natufian	13,000–11,500	Bar-Oz, 2004; Bar-Oz et al., 2004
Hilazon Tachtit	Late Natufian	13,000–11,500	Munro, 2001, 2004b
Hayonim Terrace	Late Natufian	13,000–11,500	Munro, 2001, 2004b
Hayonim Cave	Late Natufian	13,000–11,500	Munro, 2001, 2004b
Hayonim Cave	Early Natufian	14,500–13,000	Munro,2001, 2004b
El-wad Cave	Early Natufian	14,500–13,000	Rabinovich, 1998; Munro, 2001, 2004b
Hefzibah 7–18	Geometric Kebaran	17,000–14,500	Bar-Oz and Dayan, 2003; Bar-Oz, 2004
Neve David	Geometric Kebaran	17,000–14,500	Bar-Oz et al., 1999; Bar-Oz, 2004
Nahal Hadera V	Kebaran	21,000–17,000	Bar-Oz and Dayan, 2002; Bar-Oz, 2004
Hayonim Cave	Kebaran	21,500–17,000	Stiner, 2005
Meged Rockshelter	Kebaran	21,500–17,000	Stiner, 2005

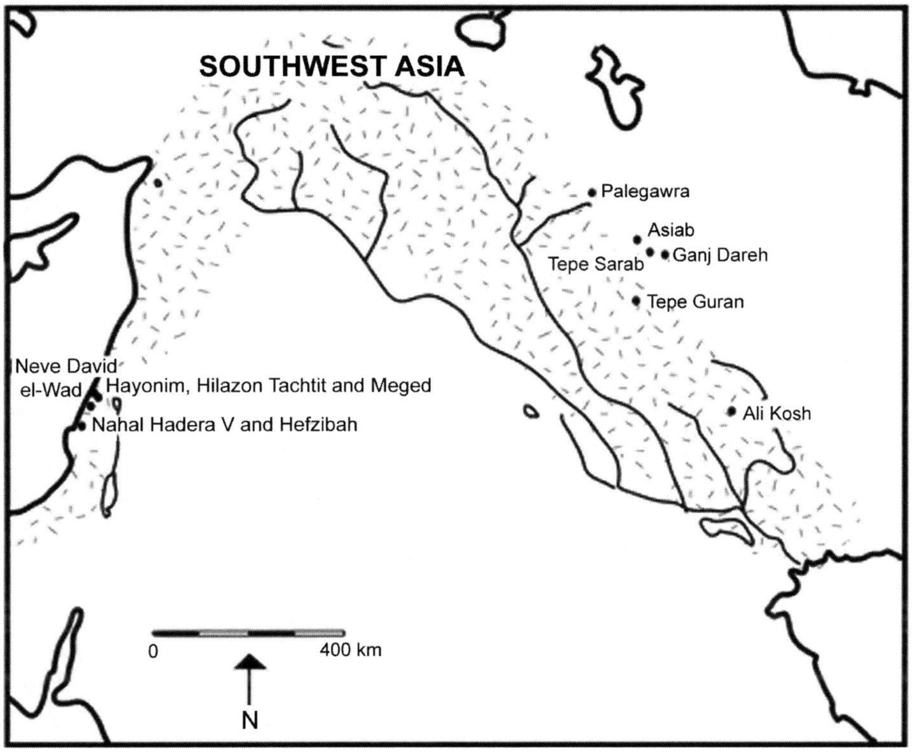

FIG. 10.1. Map of the Southwest Asian sites discussed in the text.

Methods

Epipaleolithic intensification will be investigated using classic archaeozoological methods including: (a) relative taxonomic abundance indices; (b) mortality profiles; (c) average body-size data; and (d) bone fragmentation data.

The relative abundance analyses are comprised of three indices that compare the frequencies of taxa that are differentially ranked according to their body-size and escape strategy. A decline in the abundance of high-ranked game in relation to lower-ranked taxa indicates a decline in foraging efficiency and thus, intensification. First, an ungulate index compares the large and medium ungulates to the smaller-bodied (and thus lower-ranked) small ungulates. The large ungulates include wild cattle (*Bos primigenius*) and red deer-sized (*Cervus elaphus*) animals, while the medium ungulates are dominated by Persian fallow deer (*Dama mesopotamica*), but also include small frequencies of wild boar (*Sus scrofa*) and bezoar goat (*Capra aegagrus*). Finally the small ungulate category is nearly exclusively comprised of mountain gazelles (*Gazella gazella*), although roe deer (*Capreolus capreolus*) also make an occasional appearance. Second, a small game index compares the abundance of lower-ranked small game animals to the higher-ranked ungulates. The small game category is dominated by three species – Mediterranean spur-thighed tortoise (*Testudo graeca*), cape hare (*Lepus capensis*), and chukar partridge (*Alectoris chukar*). Other avian species, most commonly raptors and waterfowl, also make occasional appearances. Finally, a fast small game index compares the abundance of more difficult to catch fast-moving small game taxa, such as hare and partridge, to the gatherable and thus higher-ranked tortoise.

In combination, the three indices comparing different kinds of taxa provide a more comprehensive picture of local and regional intensification in the southern Levant than the use of a single index. The ungulate index compares large-bodied taxa that range over substantial territories, and thus provide good evidence for region-scale hunting pressure. In contrast, the comparisons of small game groups that have small home ranges and often live at high densities close to human habitation sites, provide a more local picture of intensification that most likely reflects the intensity of site occupation, rather than regional game depression (Hames, 1980; Szuter and Bayham, 1989; Tchernov, 1993; Stiner and Munro, 2002).

Intensification is investigated on a population scale by tracking the percentage of juvenile individuals in the populations of the two most common hunted species – the gazelle and the tortoise. In gazelle, the first phalanx fuses between 5–8 months of age (Davis, 1980), before the animal has reached reproductive maturity. Thus, the percentage of unfused first phalanges (based on MNE) represents the proportion of gazelle fawns in the assemblage. Unlike gazelles and other mammals, tortoises grow throughout their lives, and thus changes in the average age of individuals in a population are reflected in their average body size. The average body size of the tortoise population from each of the studied assemblages is estimated using the average breadth of the narrowest point on the humeral shaft. The humerus is one of the primary weight-bearing bones in the body and therefore provides a good proxy for overall body-size (G. Hartman, 2006, personal communication). The tortoise humerus is a small element, but its distinctive hourglass shape provides a natural constriction that can easily be measured. The narrow length of this constriction makes it easy to pinpoint, and thus the measurements are highly replicable.

Finally, carcass processing intensity is evaluated by considering the completeness indices of gazelle first and second phalanges. Gazelle first and second phalanges contain marrow, but in quantities that are substantially smaller than those provided by the six major limb bones (Phalanx 1 = 0.17 g; Phalanx 2 = 0.03 g average in seven sampled animals; Bar-Oz and Munro, 2007). It is likely that the large gap between these two groups of elements represents a point of diminishing returns, beyond which humans were unlikely to invest in the cracking of bones for marrow unless necessary (Bar-Oz and Munro, 2007).

The percentage of broken first and second phalanges should reflect how willing humans were to invest in even small packages of bone marrow. Breakage, however, may also occur in response to other pre- and post-depositional forces. Fortunately, because foot bones are small, compact, and relatively dense in relation to other body-parts, they are less likely to break in response to most other pre- and post-depositional processes (Marean, 1991). The percent completeness of compact foot bones that have no nutritional value (astragalus and third phalanx) in five of the Epipaleolithic sites of interest (Hefzibah, Nahal Hadera V, Early and Late Natufian Hayonim Cave and el-Wad Terrace) emphasizes this point. In these assemblages, the astragalus and third phalanx contain no marrow and have relatively high completeness ratios (81.6% for the astragalus and 74.6% for the third phalanx on average; Munro and Bar-Oz, 2005). Because of their low nutritional content and high bone mineral density, the breakage of foot bones that do not contain bone marrow (astragalus and third phalanx) is most likely caused by post-depositional rather than pre-depositional processes (Marean, 1991). Indeed, of the five Epipaleolithic assemblages, the lowest completeness values originate in open air sites where they were subject to greater disturbance by post-depositional fluvial action, weathering, and other factors (Munro and Bar-Oz, 2005). In summary, the completeness values for the astragalus and the third phalanx indicate that some degree of toe bone breakage occurs post-depositionally and is unrelated to marrow extraction. Nevertheless, breakage of the marrow-bearing first and second phalanges above and beyond these figures undoubtedly reflects human agents.

Temporal change in the intensification indices described above are depicted below in a series of graphs that present the results from relevant sites in chronological order from bottom to top. The time scale is not continuous because the occupation of some sites overlap in time, and the duration of site occupation, the rate of accumulation of material, and the amount of time between site occupations varies within and between sites. Nevertheless, these graphs provide good general indicators of the nature of subsistence change throughout the southern Levantine Epipaleolithic.

Results

Indices of Relative Taxonomic Abundance

Ungulate Index

The ungulate index indicates a clear, gradual increase in the abundance of small ungulates at the expense of medium and large ungulates across the Epipaleolithic periods (Figs. 10.2 and 10.3). A Spearman's rank-order correlation coefficient indicates a statistically significant increase in the abundance of small ungulates from the Kebaran through the Late Natufian periods (Figs. 10.2 and 10.3; Spearman's rho = 0.887, $p < .0001$, $n = 11$). Throughout the sequence, large ungulates make

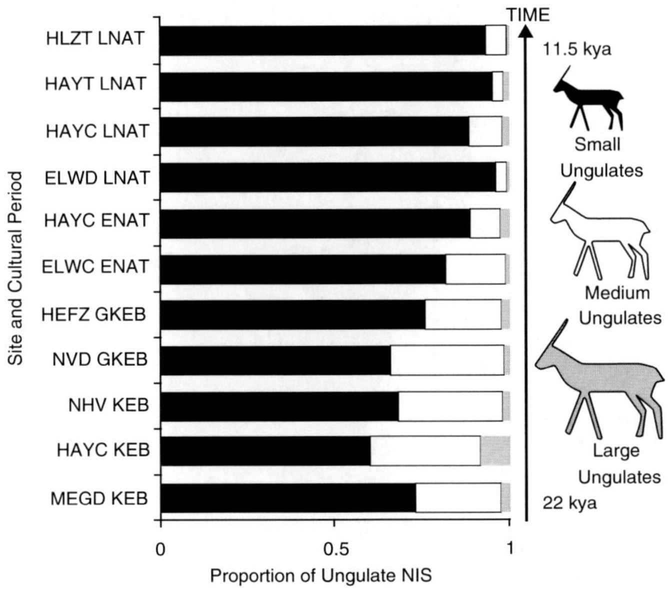

FIG. 10.2. Ungulate Index – Relative abundance of small, medium and large ungulates in Epipaleolithic assemblages from the southern Levant. Sites are arranged in chronological order from bottom to top. Cultural period abbreviations are as follows: KEB – Kebaran; GKEB – Geometric Kebaran; ENAT – Early Natufian; and LNAT – Late Natufian. Site abbreviations are as follows: MEGD – Meged Rockshelter; HAYC – Hayonim Cave; NHV – Nahal Hadera V; NVD – Neve David; HEFZ – Hefzibah; ELWC – el-Wad Cave; ELWD – el-Wad Terrace; HAYT – Hayonim Terrace; and HLZT – Hilazon Tacthit.

only incidental appearances, and thus the bulk of this shift is in the relative abundance of medium and small ungulates. Patterned variation in the relative abundance of gazelle and fallow deer during the Paleolithic and Epipaleolithic was first observed by Dorothy Bate in the 1930s. Bate (Garrod and Bate, 1937) argued that the fluctuations in *Gazella:Dama* abundance reflected climatic changes and associated shifts in the distribution of forested and open habitats. Fallow deer favor forested, closed environments; therefore their numbers are expected to blossom during wet and warm periods, and contract during cool and dry periods. In contrast, gazelle favor open grassland habitats which expand during cool and dry climatic episodes. The majority of the Epipaleolithic coincides with a warming trend that begins after the Late Glacial Maximum (ca. 18,000 BP), and continues until the onset of the Younger Dryas, a major drying and cooling event contemporaneous with the Late Natufian phase (Baruch and Bottema, 1991; Bar-Matthews et al., 1999; Frumkin et al., 1999). In Fig. 10.2, fallow deer (medium ungulates) decline in abundance in relation to gazelles (small ungulates) throughout the Epipaleolithic sequence. Thus the pattern shown in the Epipaleolithic faunas is the opposite of what is expected if climate is driving ungulate abundances. While fluctuations in *Gazella Dama* abundance may be linked to climatic change in the preceding Paleolithic periods, this explanation does not hold up for the Epipaleolithic. Instead, the Epipaleolithic trend is unidirectional in nature and shows

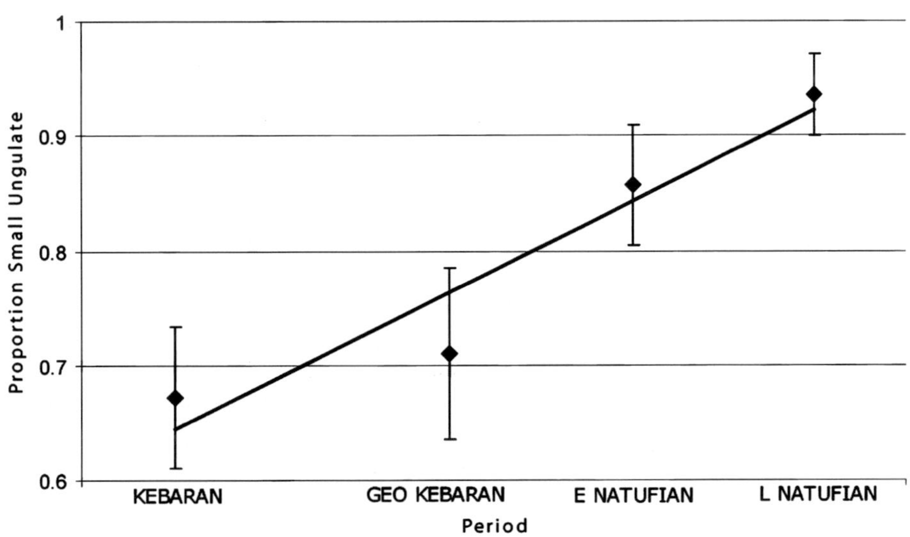

FIG. 10.3. The average abundance of small ungulates in the Kebaran, Geometric Kebaran, and Early and Late Natufian periods. The bars indicate one standard deviation. The trend toward increasing small ungulate abundance is statistically significant (Spearman's rho = 0.887, $p < .0001$, $n = 11$).

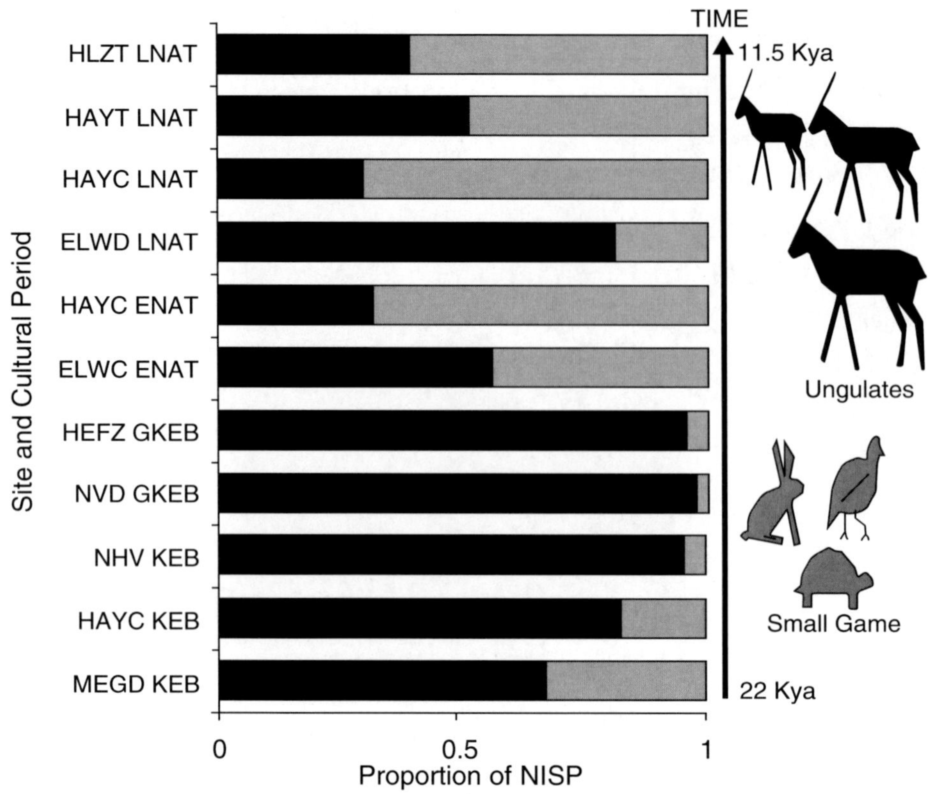

FIG. 10.4. Small game index – Relative abundance of ungulates and small game in Epipaleolithic assemblages from the southern Levant. Sites are arranged in chronological order from bottom to top. Cultural period abbreviations are as follows: KEB – Kebaran; GKEB – Geometric Kebaran; ENAT – Early Natufian; LNAT – Late Natufian. Site abbreviations are as follows: MEGD – Meged Rockshelter; HAYC – Hayonim Cave; NHV – Nahal Hadera V; NVD – Neve David; HEFZ – Hefzibah; ELWC – el-Wad Cave; ELWD – el-Wad Terrace; HAYT – Hayonim Terrace; and HLZT – Hilazon Tacthit.

no major interruptions, despite the pronounced climatic changes of the Younger Dryas.

The gradual decline in large and medium ungulate abundance despite favorable environmental conditions supports an interpretation of large game depression in the Epipaleolithic, especially during the Natufian period. The dramatic increase in gazelle abundance during the Natufian has classically been described as the beginning of a specialized gazelle economy (Davis, 1982, 1983; Cope, 1991; Tchernov, 1991, 1993). The results presented here demonstrate that the focus on gazelle is not the result of human selectivity, but a response to changing resource availability. Human-derived resource depression is argued to have depleted large and medium game assemblages, producing a landscape dominated by gazelle and other small game species. Fallow deer and larger animals were still collected when encountered, but clearly encounter rates declined substantially throughout the duration of this period. The shift in investment from the capture of higher-ranked large ungulates to lower-ranked gazelle reflects gradual intensification across the Epipaleolithic period.

Small Game Index

Despite some variability, there is a substantial increase in the proportion of low-ranked small game in relation to higher-ranked ungulates between the early Epipaleolithic and Natufian periods (Fig. 10.4). This florescence in small game abundance is undoubtedly the phenomenon that Flannery (1969) observed when he identified the broad spectrum revolution, although he did not quantify it at the time. Because the small game species that increase in abundance in the Natufian were also being captured in smaller numbers in earlier Epipaleolithic periods, this dietary expansion can not be detected using classic taxonomic diversity indices (Edwards, 1989; Neeley and Clark, 1993; Bar-Oz et al., 1999). Nevertheless, evidence for the BSR is abundant using other indicators of expanding dietary breadth including relative abundance indices, age profiles, and carcass processing strategies (Munro, 2003, 2004b). Like the ungulate index, the small game index depicts a decline in foraging efficiency and thus an intensification in human foraging strategies. In this case, however, the shift begins in the Natufian period rather than in the early Epipaleolithic and is likely partially linked to contemporaneous increases in site use intensity (see below).

Fast Small Game Index

Further evidence for intensification is apparent within the small game fraction itself (Fig. 10.5). The small game component is comprised of animals of similar body size. In this

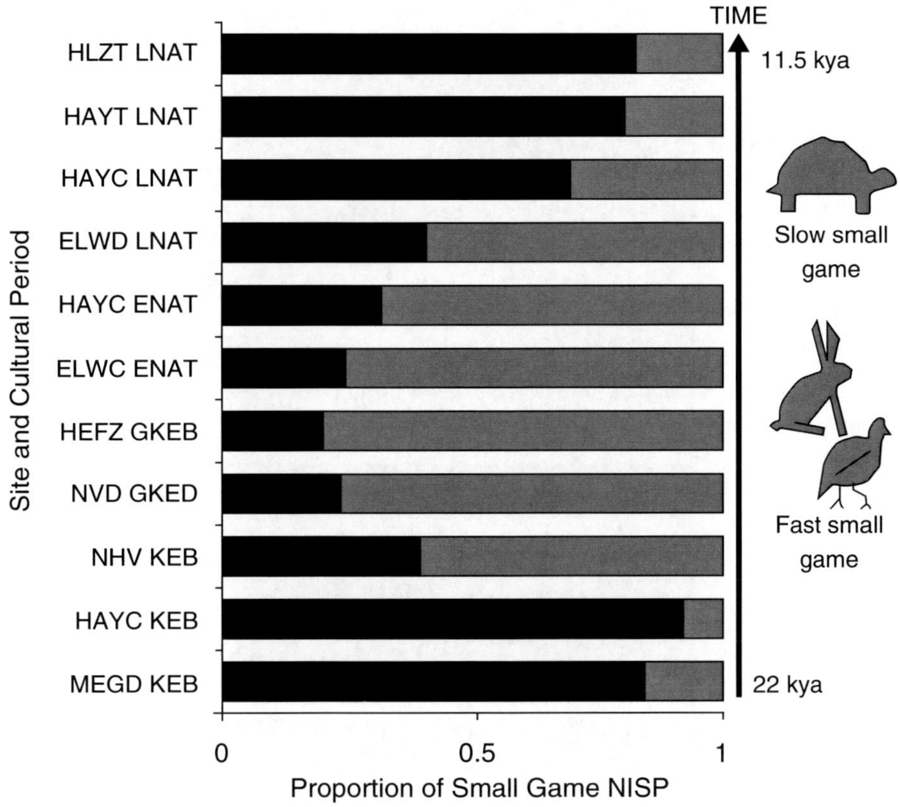

FIG. 10.5. Fast small game index – Relative abundance of fast to slow small game in Epipaleolithic assemblages from the southern Levant. Sites are arranged in chronological order from bottom to top. Cultural period abbreviations are as follows: KEB – Kebaran; GKEB – Geometric Kebaran; ENAT – Early Natufian; LNAT – Late Natufian. Site abbreviations are as follows: MEGD – Meged Rockshelter; HAYC – Hayonim Cave; NHV – Nahal Hadera V; NVD – Neve David; HEFZ – Hefzibah; ELWC – el-Wad Cave; ELWD – el-Wad Terrace; HAYT – Hayonim Terrace; and HLZT – Hilazon Tacthit.

case, relative cost effectiveness is differentiated by cost of capture, which varies according to escape strategy. Slow small game, namely tortoises, are highly ranked because they can be hunted more efficiently than fast small game species like hares and partridges. Although, fast small game were undoubtedly captured using special technologies (i.e., bow and arrow, traps, and nets), these technologies incur high production, maintenance, and acquisition costs. Although they may ultimately defray the cost of capture, operational costs are still substantially higher than the collection of tortoises using bare hands (Bailey and Aunger, 1989; Churchill, 1993; Lupo and Schmitt, 2002). Although hunting innovations enabled small shifts in small game acquisition early in the Epipaleolithic, they cannot account for the dramatic reversion in fast small game abundance in the Late Natufian when these technologies were clearly in use. Nor can they explain why technologies that existed in the early Epipaleolithic were not used in earnest until at least 7,000 years later in the Early Natufian period.

Figure 10.5 depicts an initial increase in the abundance of fast small game starting in the Late Kebaran period. Unlike the preceding two indices, the path of this index is not unidirectional. Although initially there is a clear trend toward intensification, the direction reverses and prey abundances return to Kebaran-like proportions in the Late Natufian phase. The ratio of slow to fast small game likely indicates something quite different than the preceding two indices. Because small game tends to be captured in local environments surrounding archaeological sites, it provides a more sensitive monitor of site use intensity (i.e., the degree of sedentism) than of region-wide resource depression. Changes in the fast small game index are thus argued to reflect an increase in sedentization during the Geometric Kebaran and Early Natufian periods, followed by a return to less permanent settlements and increased mobility in the Late Natufian, immediately preceding the origins of agriculture (Munro, 2004b). This interpretation is supported by numerous other archaeological indicators, in particular architectural investment, site size, diversity of material culture, and human burial patterns (Garrod and Bate, 1937; Belfer-Cohen, 1991; Valla et al., 1991; Bar-Yosef, 1996; Goring-Morris and Belfer-Cohen, 1998; Valla, 1998; Belfer-Cohen and Bar-Yosef, 2000; Grosman, 2003; Munro, 2003, 2004b).

Together, the relative abundance indices provide a nuanced picture of subsistence intensification throughout the Epipaleolithic period. All three indices indicate considerable

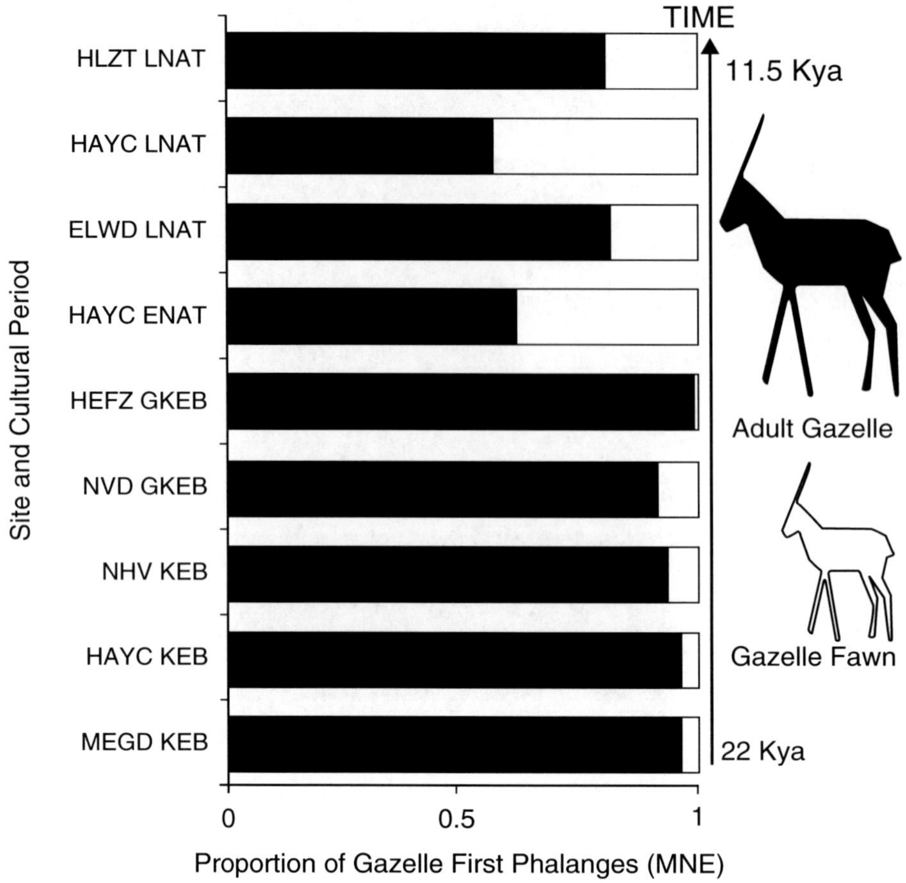

FIG. 10.6. Relative abundance of gazelle fawns and adults in Epipaleolithic assemblages from the southern Levant. Sites are arranged in chronological order from bottom to top. Cultural period abbreviations are as follows: KEB – Kebaran; GKEB – Geometric Kebaran; ENAT – Early Natufian; and LNAT – Late Natufian. Site abbreviations are as follows: MEGD – Meged Rockshelter; HAYC – Hayonim Cave; NHV – Nahal Hadera V; NVD – Neve David; HEFZ – Hefzibah; ELWD – el-Wad Terrace; and HLZT – Hilazon Tacthit.

resource intensification both at the local and regional scale, especially between the Geometric Kebaran and Early Natufian periods. Although intensive region-wide subsistence strategies continue into the Late Natufian, they are accompanied by a drop in site occupation intensity.

Mortality Profiles

Gazelle Mortality Profiles

The proportion of gazelle fawns in hunted Epipaleolithic assemblages increases considerably – from less than 10% in the Geometric Kebaran and Kebaran periods to between 20% and 40% in the Early Natufian (Fig. 10.6). The hunting of high proportions of fawns is sustained (up to 45%) throughout the Late Natufian (up to 45%). The increased hunting of gazelle fawns, which are substantially smaller than adults, again indicates an intensified gazelle hunting regime. This trend corresponds exactly with the increase in the abundance of small game, confirming that humans expanded dietary breadth at this critical juncture by adding smaller, lower-ranked resources to the diet. By the Late Natufian, humans were actively hunting all available gazelle age classes. It is therefore not surprising that intensive gazelle exploitation was quickly followed by the adoption of domestic goat and sheep in the mid Pre-Pottery Neolithic B (Helmer, 1989; Legge, 1996; Bar-Yosef, 2000; Munro, 2004a). Because gazelles are not well-suited for domestication (Clutton-Brock, 1999; Diamond, 1999), the adoption of other small ungulate species more amenable to increased production through human selection was the only direction left to go for further extraction of ungulate products from Mediterranean environments.

Tortoise Body-Size Data

Although a clear decline in tortoise body-size exists from the Middle to the Upper Paleolithic to the Epipaleolithic period in the southern Levant (Stiner et al., 1999, 2000; Stiner, 2005), the diminution trend does not continue across the Epipaleolithic sequence (Fig. 10.7). All Epipaleolithic tortoise populations are smaller on average than Paleolithic populations, but seemingly random fluctuations in average body size occur within the Epipaleolithic sequence itself. The only obvious pattern that springs from the body-size

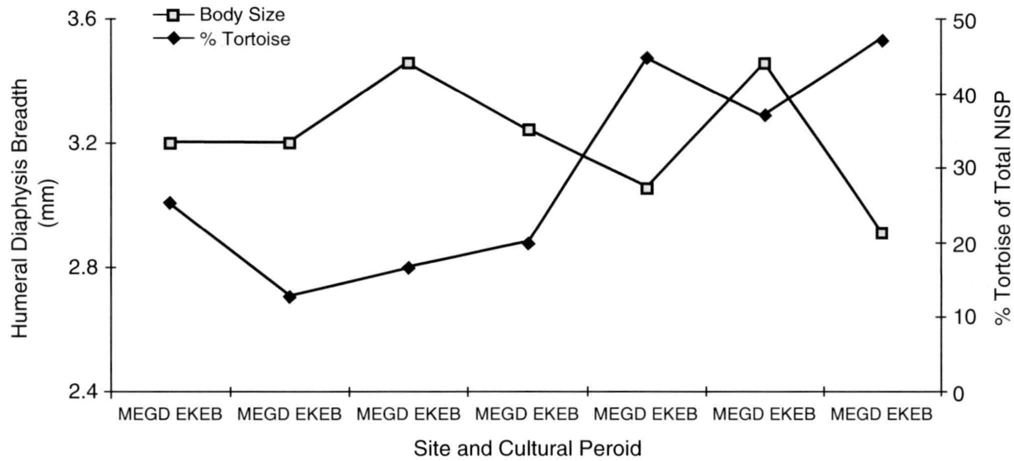

FIG. 10.7. Relative abundance of tortoises and average tortoise body-size based on the narrowest breadth of tortoise humerus shafts in Epipaleolithic assemblages from the southern Levant. Cultural period abbreviations are as follows: KEB – Kebaran; GKEB – Geometric Kebaran; ENAT – Early Natufian; and LNAT – Late Natufian. Site abbreviations are as follows: MEGD – Meged Rockshelter; HAYC – Hayonim Cave; NHV – Nahal Hadera V; HAYT – Hayonim Terrace; and HLZT – Hilazon Tacthit.

data is an inverse relationship between average body-size and the proportion of tortoises in each assemblage. In other words, when the body size of the tortoises increases, there is a decrease in the abundance of tortoises in the assemblage and vice versa. This trend is not directly proportional, and likely reflects localized hunting pressure on tortoise populations. As hunting pressure, and hence the mortality, of a tortoise population in a given area grew, average body size became correspondingly smaller. This is the combined result of intentional human selection for the largest individuals in the population and the fact that smaller individuals did not have the opportunity to reach full body size before they were predated.

Interestingly, the fluctuations in tortoise body size do not correspond to the u-shaped pattern indicated by the fast small game index as might be expected. A decline in tortoise body size indicates increased hunting pressure, but in the case of Late Natufian populations, it was not sufficiently severe to have pushed humans to further broaden their diets by adding new resources. Thus, in the Epipaleolithic case, tortoise body size seems to be a better measure of local variation in hunting pressures than of overall foraging efficiency.

Gazelle Carcass Processing

Completeness values for gazelle first and second phalanges are available only for a subsample of the Epipaleolithic assemblages (Fig. 10.8a, b). The results inspire three noteworthy observations. First, the results vary tremendously among sites and do not follow a temporal trend. Second, in all cases the first and second phalanges from open air sites (Nahal Hadera V, Hefzibah, el-Wad Terrace, Hayonim Terrace) have lower completeness indices than those from cave sites (Hayonim Cave and Hilazon Tachtit). Third, in all cases the first phalanx, which contains considerably more marrow than the second phalanx, is breached more often. These observations lead to two conclusions regarding the cause of breakage of the gazelle phalanges. First, as previously mentioned, the degree of breakage of astragali and third phalanges which do not contain marrow and the high rates of breakage of first and second phalanges at open air versus cave sites indicate that postdepositional processes are responsible for some damage. Second, lower completeness values for first than second phalanges despite similar structural densities indicate that humans preferentially broke gazelle toes with higher marrow content. Nevertheless, the fact that on average 40.2% of first phalanges and 62.3% of second phalanges are complete, while gazelle marrow-bearing long bone shafts are complete less than 5% of the time (Munro, 2004b; Munro and Bar-Oz, 2005) indicates that there is indeed a point of diminishing returns in the extraction of gazelle bone marrow. The fact that humans still bothered to invest in the extraction of very small marrow stores even half of the time indicates an intensive marrow-extraction strategy that stopped just short of exhausting gazelle marrow supplies (Bar-Oz and Munro, 2007).

That long bone breakage was largely determined by marrow extraction is supported by additional data from select Epipaleolithic assemblages. At Nahal Hadera V, Hefzibah, Hayonim Cave, and el-Wad Terrace bone breakage (measured using NISP/MNE ratios) is significantly correlated with marrow content in gazelle long bones (Munro and Bar-Oz 2005; Bar-Oz and Munro, 2007). Furthermore, the relationship between bone fragmentation and marrow content is maintained across taxa and age groups in both the Early and Late Natufian layers at Hayonim Cave. In both Natufian layers, gazelle bones are substantially more fragmented than hare and partridge bones, which contain little or no marrow. The fragmentation of adult gazelle long bones from the same site is significantly greater than of juvenile gazelles, which use their long bone cavities more for red blood cell production than fat

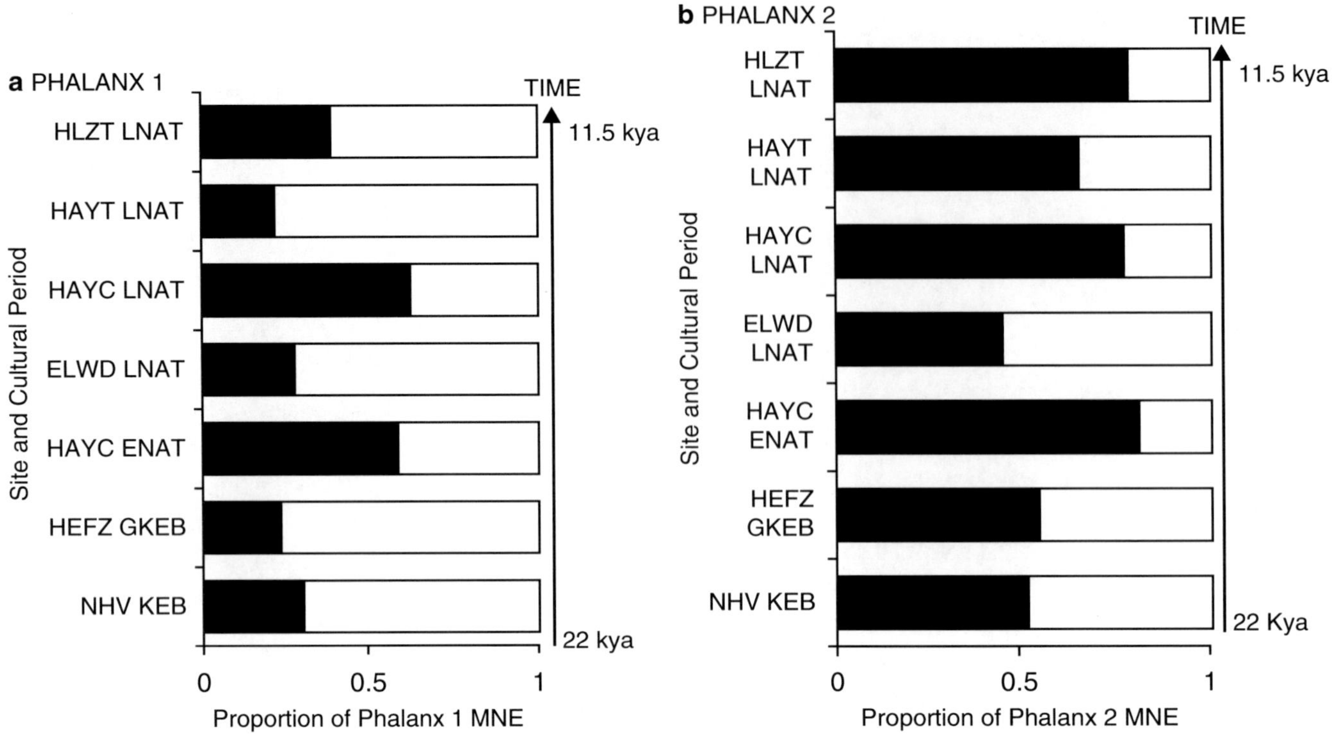

FIG. 10.8. Proportion of breached and unbreached gazelle **(a)** first phalanges and **(b)** second phalanges in Epipaleolithic assemblages from the southern Levant. Black bars indicate complete phalanges, white bars indicate broken phalanges. Cultural period abbreviations are as follows: KEB – Kebaran; GKEB – Geometric Kebaran; ENAT – Early Natufian; and LNAT – Late Natufian. Site abbreviations are as follows: NHV – Nahal Hadera V; HEFZ – Hefzibah; HAYC – Hayonim Cave; ELWD – el-Wad Terrace; HAYT – Hayonim Terrace; and HLZT – Hilazon Tacthit.

storage. More intensive fragmentation of hare (little marrow) than partridge bones (no marrow), despite similar body-size, indicates that humans still bothered to access small marrow stores at least half of the time, again attesting to an intensive marrow exploitation strategy (Bar-Oz and Munro, 2004; Munro, 2004b).

Other Parts of Southwest Asia

Although comparative data from other regions of Southwest Asia is rapidly increasing, the details required to replicate the analyses presented here are not currently available. One study (Munro, 2004a) on the small game faunas from sites dating to the agricultural "transition" in the Zagros region provides some comparative data, at least for the small game index. Unfortunately, this data provides only a small taste of how Epipaleolithic intensification trends are expressed in neighboring regions of Southwest Asia (Fig. 10.9).

The small game index presented in Fig. 10.9 represents a different time scale than that of the southern Levantine series. While, the oldest two sites (Palegawra, Asiab) are Late Epipaleolithic in age, the remaining sites (Ganj Dareh, Ali Gosh, Tepe Guran, and Tepe Sarab) date to Aceramic and Ceramic phases of the Early Neolithic (Fig. 10.1; Table 10.2). Nevertheless, the data show a similar pattern to that depicted in the southern Levant, though on a more restricted scale. Epipaleolithic assemblages in the Zagros region are represented by rich collections of small game fauna which, like in the southern Levant, are comprised of partridge, hare, tortoises, and an additional species, the freshwater turtle, *Maurmys caspica*. Although the Epipaleolithic is represented by only two data points in Fig. 10.9, the comparison with Neolithic assemblages highlights the magnitude of the florescence in small game abundance. By the onset of the Aceramic Neolithic, small game animals begin to drop rapidly from human-collected assemblages. They are replaced by small ungulates, in this case sheep and goat, in the early stages of animal management and domestication. The innovation of herd management relaxed the need for the plethora of subsistence intensification strategies innovated in the Epipaleolithic. Herd animals quickly filled the top slot in the prey rankings, and because of their great productivity, and hence increased availability, eventually eliminated the need for broad diets. Thus, soon after the appearance of managed sheep and goat, small game take a sudden drop from their elevated position during the Epipaleolithic to become a periodic and likely opportunistic dietary supplement.

Although it provides only a single line of evidence, the small game index from the Zagros region suggests that at least some of the intensification strategies employed in the

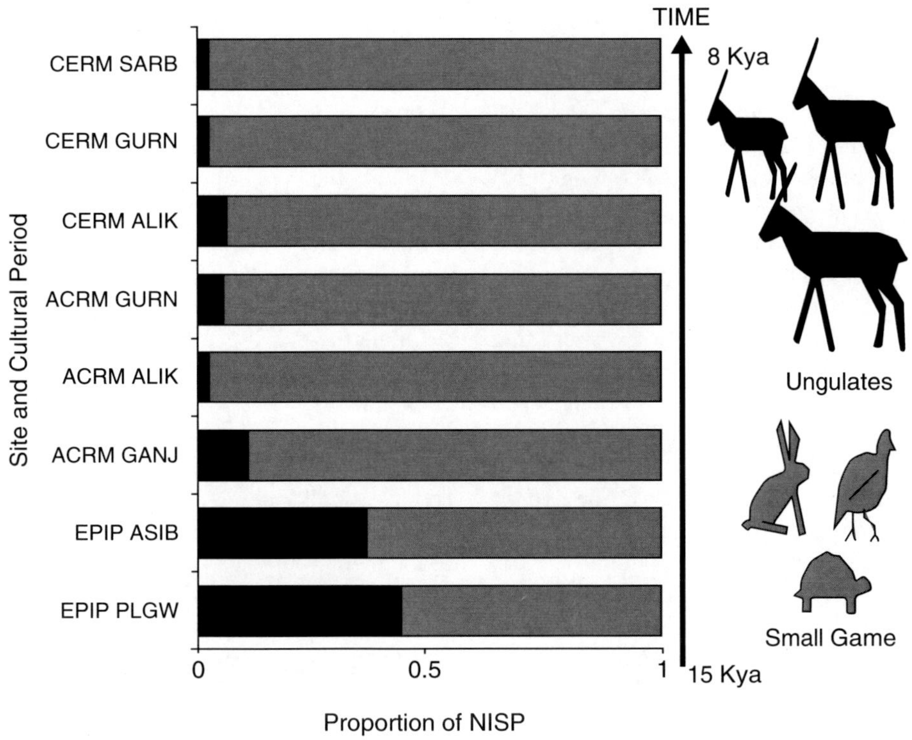

FIG. 10.9. Relative abundance of ungulates and small game in Epipaleolithic and Early Neolithic assemblages from the Zagros region. Sites are arranged in chronological order from bottom to top. Cultural period abbreviations are as follows: EPIP – Epipaleolthic; ACRM – Aceramic Neolithic; CERM – Ceramic Neolithic. Site abbreviations are as follows: PLGW – Palegawra; ASIB – Asiab; GANJ – Ganj Dareh; ALIK – Ali Kosh; GURN – Tepe Guran; SARB – Tepe Sarab.

TABLE 10.2. Site names, cultural phases and time ranges of the Zagros assemblages discussed herein.

Site	Cultural period	Time range (cal BP)	Reference
Sarab	Ceramic Neolithic	9,000–8,000	Bökönyi, 1977; Munro, 2004a
Guran	Ceramic Neolithic	9,000–8,000	Flannery, n.d.; Munro, 2004a
Ali kosh	Ceramic Neolithic	9,000	Hole et al., 1969; Munro, 2004a
Guran	Aceramic Neolithic	9,200–9,000	Flannery, n.d.; Munro, 2004a
Ali kosh	Aceramic Neolithic	9,400–9,000	Hole et al., 1969; Munro, 2004a
Ganj Dareh	Aceramic Neolithic	9,900–9,700	Hesse, 1978; Munro, 2004a
Asiab	Epipaleolithic	10,700	Bökönyi, 1977; Munro, 2004a
Palegawra	Epipaleolithic	15,000–11,500	Turnbull and Reed, 1974; Munro, 2004a

southern Levant were practiced in other parts of the Fertile Crescent. Active research programs and zooarchaeological research investigating the "transition" to agriculture, especially in southeastern Anatolia, will hopefully soon fill out the picture of Epipaleolithic intensification in the broader Southwest Asian region in general.

Discussion

Together, several lines of evidence provide a robust picture for subsistence intensification during the Epipaleolithic of the southern Levant. Some aspects of the intensification process occurred steadily over the course of the Epipaleolithic, while others make a pronounced jump between the Geometric Kebaran and the Natufian periods. Gradual intensification across the Epipaleolithic is supported by both the ungulate and small game indices. A steady reduction in the abundance of high-ranked game and a simultaneous increase in the numbers of small ungulates and small game animals mark a classic expansion in dietary breadth and decreased foraging efficiency, beginning in the Geometric Kebaran and continuing into the Natufian periods.

Together, the small game index and gazelle age profiles evidence a prominent shift in the nature of resource extraction between the Geometric Kebaran and the Natufian periods.

Because early Epipaleolithic foragers had already depressed populations of the most highly ranked animals such as fallow deer and red deer, the Natufians first intensified game extraction by boosting their exploitation of gazelle – the most abundant ungulate on the landscape by this time. Gazelle intensification is marked first by an expansion in the demographic spectrum (i.e., the range of age groups) included in the hunted population. In particular, this includes the unprecedented hunting of fawns that are substantially smaller than adults. Fawns not only make their first appearance in the Natufian period – very few fawns were consumed in the preceding Paleolithic and Epipaleolithic (Bar-Oz, 2004; J. Speth, 2004, personal communication; Stiner, 2005) – they also appear in considerable numbers. In addition, the Natufians consumed greater percentages of older juvenile gazelles (6–18 months of age) than preceding foragers (Munro, 2004b). The exploitation of older juvenile gazelles increases gradually throughout the Epipaleolithic, again peaking in the Natufian period. Finally, gazelle utilization was also intensified by maximizing the use of individual animal carcasses. Virtually all marrow-bearing gazelle long bones were breached to remove bone fat, and even small marrow supplies from gazelle first phalanges were extracted at least half of the time.

In addition to intensifying gazelle exploitation, the Natufians increased their extraction of animal biomass from their environments in relation to preceding Epipaleolithic and Paleolithic periods. This move is indicated by the addition of progressively more costly species to the hunting repertoire – most importantly hares, partridges, and waterfowl, which require the use of special technologies for their capture. Together, the adoption of these new hunting and processing strategies contributed to an overall increase in the total amount of energy that the Natufians extracted from their environments per unit area, albeit at a greater total expense.

The dual nature of the Epipaleolithic intensification trend (i.e., gradual intensification throughout much of the Epipaleolithic and a punctuated leap in the Early Natufian) indicated by the faunal data is supported by the archaeological record. The Epipaleolithic record is characterized by a series of major developments, including the expansion of human populations, increased site permanence, and documented resource intensification evidenced by an increased abundance of groundstone tools, sickle blades, small grain grass seeds, and small animals (Henry, 1989; Bar-Yosef and Belfer-Cohen, 1989, 1991; Tchernov, 1993; Bar-Yosef and Meadow, 1995; Stiner et al., 1999, 2000; Belfer-Cohen and Bar-Yosef, 2000; Munro, 2004a, b; Weiss et al., 2004). These processes intensify over the course of the Epipaleolithic, but the most dramatic changes coincide with the onset of the Early Natufian. The Early Natufian is characterized by the establishment of a number of large settlement sites in the Mediterranean zone, which are typically classified as "basecamps" due to their more permanent nature, size, and diverse material culture (Bar-Yosef and Belfer-Cohen, 1989, 1991). These sites are on a scale of magnitude larger than those occupied in previous periods, and reflect both a shift in the organization of human populations into more aggregated, permanent settlements, and an increase in human population size in the Mediterranean region in general.

For the most part, the data presented here depict unidirectional changes, indicating a one-way intensification trend through time. Despite major climatic changes during the Epipaleolithic, this trend in resource intensification crosscuts major climatic and environmental events. There is, however, one anomaly in the otherwise unidirectional trend. In the Late Natufian, there is a reversion in the ratios of fast to slow small game to more Kebaran-like proportions. This reversion suggests that despite continued intensive hunting at the regional scale, as attested by the ungulate and small game indices and the gazelle age data, there was a change in site use intensity. The fast small game index suggests that people settled down into increasingly permanent villages and had greater impacts on local environments starting in the Geometric Kebaran and continuing through the Early Natufian, but that they reverted to a more mobile settlement pattern in the Late Natufian phase. This shift is supported by numerous lines of archaeological evidence (see above) and corresponds to the onset of the Younger Dryas. The Younger Dryas had a drying and cooling effect on the southern Levant and likely altered the distribution of resources, in particular wild grasses whose distribution was undoubtedly a major determinant in the location and duration of occupation of local villages (Bar-Yosef, 1996). Increased mobility will alleviate human pressure on the environments surrounding habitation sites as fewer local resources are harvested and longer recovery periods occur between exploitation episodes. Pressure on large game resources, however, would not have been alleviated as human populations continued to range over wide territories more on par with those of the large game themselves. This shift in occupation intensity explains why the dramatic Late Natufian shift in the fast small game index is the only major change in the otherwise unidirectional trend toward intensification.

In the Late Natufian phase, humans continued to intensively exploit animals on a regional scale, but there was a brief respite when pressures on local resources lessened. This respite, however, occurred only in localized patches and on small animal resources within the Natufian landscape. Although the leap in intensification at the regional level that occurs with the onset of the Early Natufian continues into the Late Natufian phase, it does not strengthen, but instead levels off. Thus, the intensification trend that ultimately culminates in the origins of agriculture does not actually increase in the period immediately prior to the "transition." The Late Natufians chose mobility, rather than animal management and agriculture, as an immediate solution to the environmental restrictions imposed by the Younger Dryas. Agriculture did not become a viable option until conditions became warmer, wetter, and more stable in the early Holocene (Richerson et al., 2001). Thus, although resource stress or population pressure and corresponding intensification trends were crucial ingredients in the "transition" to agriculture, they proved to be only one of several necessary conditions. The faunal data indicate that the favorable environment provided by a more stable Holocene climate was a second essential factor.

By the Natufian period, the faunal record indicates that there was little place to go to further intensify hunting strategies. Humans were already hunting small, costly species, and the full spectrum of gazelle age groups, and were squeezing the nutrients out of gazelle carcasses. The natural next step along this intensification trajectory was the management of herd animals. Unlike gazelles, sheep and goat are well disposed for this task (Clutton-Brock, 1999; Diamond, 1999), and by ca. 10,000 cal BP managed sheep and goat populations had entered the southern Levant from the north (Helmer, 1989; Legge, 1996; Bar-Yosef, 2000). The role of intensive hunting in the process of animal domestication and the "transition" to agriculture remains poorly understood, and is thus the subject of more detailed investigation under the auspices of the recently launched gazelle project. The gazelle project investigates the impacts of intensive of human hunting on Epipaleolithic and Early Neolithic gazelle population structure, body-size and allometry, and how these impacts may have contributed to the domestication process.

Conclusion

Multiple lines of evidence provide a robust picture of faunal resource intensification during the Epipaleolithic periods in the southern Levant and likely also in other parts of Southwest Asia. Despite the growing clarity of the Epipaleolithic intensification process in the southern Levant, the fine details required for comparative analyses are only beginning to surface in neighboring regions. Nevertheless, emerging evidence suggests that similar intensification processes occurred elsewhere in Southwest Asia and served as important precursors to animal domestication (Rosenberg et al., 1998; Ervynck et al., 2002; Munro, 2004a). The origin of agriculture represents the natural next step in the intensification trend that pushed natural resources, in particular gazelle, to their limits by the Natufian period. Agriculture, however, was not adopted until a few thousand years after this point, indicating that several ingredients – including resource pressure and stable climatic conditions – had to coalesce to enable the "transition" to agriculture.

Acknowledgments. A special thank you goes to Mary Stiner, Guy Bar-Oz and Mindy Zeder whose contribution to this paper goes well beyond the published reports from which I extracted comparative data. More importantly, previous collaborative work with these individuals has lead to many of the ideas and interpretations represented herein. Thanks to Jean-Jacques Hublin and Mike Richards for the invitation to join the fine conference in Leipzig at which this paper was presented and to Alison Cleveland, Silke Streiber and Diana Carstens for the impeccable organization. Thanks also to Guy Bar-Oz for commenting on an earlier draft of this paper. This research was funded in part by the University of Connecticut Research Foundation, a Smithsonian Institution Postdoctoral Fellowship and a National Science Foundation dissertation improvement grant (SBR-9815083).

References

Bailey, R.C., Aunger, R. Jr., 1989. Net hunters vs. archers: variation in women's subsistence strategies in the Ituri Forest. Human Ecology 17, 273–297.

Bar-Matthews, M., Ayalon, A., Kaufman, A., Wasserburg, G.J., 1999. The Eastern Mediterranean paleoclimate as a reflection of regional events: Soreq Cave, Israel. Earth and Planetary Science Letters 166, 85–95.

Bar-Oz, G., 2004. Epipalaeolithic Subsistence Strategies in the Levant: A Zooarchaeological Perspective. American School of Prehistoric Research (ASPR) Monograph Series, Brill Academic, Boston.

Bar-Oz, G., Dayan, T., 2002. "After twenty years": a taphonomic re-evaluation of Nahal Hadera V, an Epipalaeolithic site on the Israeli coastal plain. Journal of Archaeological Science 29, 145–156.

Bar-Oz, G., Dayan, T., 2003. Testing the use of multivariate inter-site taphonomic comparisons: the faunal analysis of Hefzibah in its Epipalaeolithic cultural context. Journal of Archaeological Science 30, 885–900.

Bar-Oz, G., Munro, N.D., 2004. Beyond cautionary tales: a multivariate taphonomic approach for resolving equifinality in zooarchaeological studies. Journal of Taphonomy 2(4), 201–220.

Bar-Oz, G., Munro, N.D., 2007. Gazelle bone marrow yields and Epipalaeolithic carcass exploitation strategies in the southern Levant. Journal of Archaeological Science 34, 946–956

Bar-Oz, G., Dayan, T., Kaufman, D., 1999. The Epipalaeolithic faunal sequence in Israel: a view from Neve David. Journal of Archaeological Science 26, 67–82.

Bar-Oz, G., Dayan, T., Weinstein-Evron, M., Kaufman, D., 2004. The Natufian economy at el-Wad Terrace with special reference to gazelle exploitation patterns. Journal of Archaeological Science 31, 217–231.

Baruch, U., Bottema, S., 1991. Palynological evidence for climatic changes in the Levant ca. 17,000–9,000 B.P. In: Bar-Yosef, O., Valla, F.R. (Eds.), The Natufian Culture in the Levant. International Monographs in Prehistory, Ann Arbor, pp. 11–20.

Bar-Yosef, O., 1996. The impact of Late Pleistocene – Early Holocene climatic changes on humans in Southwest Asia. In: Straus, L.G., Eriksen, B.V., Erlandson, J.M., Yesner, D.Y. (Eds.), Humans at the End of the Ice Age: The Archaeology of the Pleistocene-Holocene Transition. Plenum Press, New York, pp. 61–76.

Bar-Yosef, O., 2000. The context of animal domestication in southwestern Asia. In: Mashkour, M. (Ed.), Archaeozoology of the Near East IVa. ARC, Groningen, pp. 186–195.

Bar-Yosef, O., Belfer-Cohen, A., 1989. The origins of sedentism and farming communities in the Levant. Journal of World Prehistory 3, 447–498.

Bar-Yosef, O., Belfer-Cohen, A., 1991. From sedentary hunter-gatherers to territorial farmers in the Levant. In: Gregg, S.A. (Ed.), Between Bands and States. Center for Archaeological Investigations, Occasional Paper No. 9. Southern Illinois University, Carbondale, pp. 181–202.

Bar-Yosef, O., Meadow, R.H., 1995. The origins of agriculture in the Near East. In: Price, T.D., Gebauer, A.B. (Eds.), Last Hunters-First Farmers: New Perspectives on the Prehistoric Transition to Agriculture. School of American Research Press, Santa Fe, pp. 39–94.

Belfer-Cohen, A., 1991. The Natufian in the Levant. Annual Review of Anthropology 20, 167–186.

Belfer-Cohen, A., Bar-Yosef, O., 2000. Early sedentism in the Near East: a bumpy road to village life. In: Kuijt, I. (Ed.), Life in Neolithic Farming Communities: Social Organization, Identity and Differentiation. Plenum Publishers, New York, pp. 19–37.

Binford, L.R., 1968. Post-Pleistocene adaptations. In: Binford, S.R., Binford, L.R. (Eds.), New Perspectives in Archaeology. Aldine, Chicago, pp. 313–341.

Binford, L.R., 1978. Nunamiut Ethnoarchaeology. Academic Press, New York.

Bökönyi, S., 1977. Animal Remains Form the Kermanshah Valley, Iran. BAR International Series S34. British Archaeological Reports, Oxford.

Boserup, E., 1965. The Conditions of Agricultural Growth. Aldine, Chicago.

Broughton, J.M., 1994. Declines in mammalian foraging efficiency during the Late Holocene, San Francisco Bay, California. Journal of Anthropological Archaeology 13, 371–401.

Butler, V.B., 2000. Resource depression on the Northwest Coast of North America. Antiquity 74, 649–661.

Cannon, M.D., 2000. Large mammal relative abundance in Pithouse and Pueblo Period archaeofaunas from southwestern New Mexico: resource depression among the Mimbres-Mogollan? Journal of Anthropological Archaeology 19, 317–347.

Caughley, G., 1977. Analysis of Vertebrate Populations. Wiley, New York.

Churchill, S.E., 1993. Weapon technology, prey size selection and hunting methods in modern hunter-gatherers: implications for hunting in the Palaeolithic and Mesolithic. In: Peterkin, G.L., Bricker, H.M., Mellars, P.A. (Eds.), Hunting and Animal Exploitation in the Later Palaeolithic and Mesolithic of Eurasia. Archeological Papers of the American Anthropological Association 1993, 11–24.

Clutton-Brock, J., 1999. A Natural History of the Domesticated Animals. Cambridge University Press, Cambridge.

Cohen, M.N., 1977. The Food Crisis in Prehistory: Overpopulation and the Origins of Agriculture. Yale University Press, New Haven.

Cope, C., 1991. Gazelle hunting strategies in the southern Levant. Bar-Yosef, O., Valla, F.R. (Eds.), The Natufian Culture in the Levant. International Monographs in Prehistory, Ann Arbor, pp. 341–358.

Davis, S.J., 1980. A note on the dental and skeletal ontogeny of Gazella. Israel Journal of Zoology 29, 129–134.

Davis, S.J., 1982. Climatic change and the advent of domestication: the successions of ruminant artiodactyls in late Pleistocene-Holocene in the Israel region. Paléorient 8, 5–15.

Davis, S.J., 1983. The age profiles of gazelles predated by ancient man in Israel: possible evidence for a shift from seasonality to sedentism in the Natufian. Paléorient 9, 55–62.

Davis, S.J., 2005. Why domesticate food animals? Some zoo-archaeological evidence from the Levant. Journal of Archaeological Science 32, 1408–1416.

Diamond, J., 1999. Guns, Germs and Steel: The Fates of Human Societies. W.W. Norton, New York.

Edwards, P.C., 1989. Revising the broad spectrum revolution: its role in the origins of Southwest Asian food production. Antiquity 63, 225–246.

Elder, W.H., 1965. Primeval deer hunting pressures revealed by remains from American Indian middens. Journal of Wildlife Management 29, 366–371.

Ervynck, A., Dobney, K., Hongo, H., Meadow, R., 2002. Born free!: new evidence for the status of pigs from Çayönü Tepesi, Eastern Anatolia. Paléorient 27, 47–73.

Flannery, K.V., 1969. Origins and ecological effects of early domestication in Iran and the Near East. In: Ucko, P.J., Dimbleby, G.W. (Eds.), The Domestication and Exploitation of Plants and Animals. Aldine, Chicago, pp. 73–100.

Flannery, K., n.d. Hunting and early animal domestication at Tepe Guran. Unpublished manuscript on file in the Archaeobiology Lab, Smithsonian Institution, National Museum of Natural History, Washington.

Frumkin, A., Ford, D.C., Schwarcz, H.P., 1999. Continental paleoclimatic record of the last 170,000 years in Jerusalem. Quaternary Research 51, 317–327.

Garrod, D.A., Bate, D.M., 1937. The Stone Age of Mount Carmel: Excavations at the Wady El-Mughara, Volume 1. Clarendon Press, Oxford.

Goring-Morris, N., Belfer-Cohen, A., 1998. The articulation of cultural processes and Late Quaternaryenvironmental changes in Cisjordan. Paléorient 23, 107–119.

Grosman, L., 2003. Preserving cultural traditions in a period of instability: the Late Natufian of the hilly Mediterranean zone. Current Anthropology 44, 571–580.

Hames, R.B., 1980. Game depletion and hunting zone rotation among the Ye'kwana and Yanomamo of Amazonas, Venezuela. In: Hames, R.B. (Ed.), Studies in Hunting and Fishing in the Neotropics. Working Papers on South American Indians, No. 2. Bennigton College, Vermont, pp. 31–66.

Helmer, D., 1989. Le développement de la domestication au Proche-Orient de 9500 à 7500 BP: les nouvelles données d'el Kowm et de Ras Shamra. Paléorient 15, 111–121.

Henry, D.O., 1989. From Foraging to Agriculture, The Levant at the End of the Ice Age. University of Pennsylvania Press, Philadelphia.

Hesse, B., 1978. Evidence for Husbandry from the Early Neolithic Site of Ganj Dareh in Western Iran. Ph.D. dissertation, Columbia University.

Hole, F., Flannery, K., Neely, J., 1969. Prehistory and Human Ecology of the Deh Luran Plain: An Early Village Sequence from Khuzistan, Iran. Memoir 1, Museum of Anthropology, University of Michigan, Ann Arbor.

Klein, R.G., 1999. The Human Career: Human Biological and Cultural Origins, 2nd edition. University of Chicago Press, Chicago.

Koike, H., Ohtaishi, N., 1985. Prehistoric hunting pressure estimated by the age composition of excavated sika deer (Cervus nippon) using the annual layer of tooth cement. Journal of Archaeological Science 12, 443–456.

Legge, A., 1996. The beginning of caprine domestication in southwest Asia. In: Hillman, G. (Ed.), The Origins and Spread of Agriculture and Pastoralism in Eurasia. University College London Press, London, pp. 238–262.

Lupo, K.D., Schmitt, D.N., 1997. Experiments in bone boiling: nutritional returns and archaeological reflections. Anthropozoologica 25/26, 137–144.

Lupo, K.D., Schmitt, D.N., 2002. Upper Paleolithic net-hunting, small prey exploitation, and women's work effort: a view from the ethnographic and ethnoarchaeological record of the Congo Basin. Journal of Archaeological Method and Theory 9, 147–179.

Lyman, R.L., 2003. Pinniped behavior, foraging theory, and the depression of metapopulations and nondepression of a local population on the southern Northwest Coast of North America. Journal of Anthropological Archaeology 22, 376–388.

Madsen, D.B., Schmitt, D.N., 1998. Mass collecting and the diet breadth model: a Great Basin example. Journal of Archaeological Science 25, 445–455.

Marean, C.W., 1991. Measuring the postdepositional destruction of bone in archaeological assemblages. Journal of Archaeological Science 18, 677–694.

Munro, N.D., 2001. A Prelude to Agriculture: Game Use and Occupation Intensity During the Natufian Period in the Levant. Ph.D. dissertation, University of Arizona.

Munro, N.D., 2003. Small game, the Younger Dyras, and the transition to agriculture in the southern Levant. Mitteilungen der Gesellschaft für Urgeschichte 12, 47–71.

Munro, N.D., 2004a. Small game indicators of human foraging efficiency and early herd management at the transition to agriculture in Southwest Asia. In: Brugal, J.-P., Desse, J. (Eds.), Petits Animaux et Sociétés Humaines, du Complément Alimentaire aux Resources Utilitaires. CÉPAM, CNRS, Sophia Antipolis, pp. 515–531.

Munro, N.D., 2004b. Zooarchaeological measures of hunting pressure and occupation intensity in the Natufian: implications for agricultural origins. Current Anthropology 45, S5–S33.

Munro, N.D., Bar-Oz, G., 2005. Gazelle bone fat processing in the Levantine Epipalaeolithic. Journal of Archaeological Science 32, 223–239.

Neeley, M.P., Clark, G.A., 1993. The human food niche in the Levant over the past 150,000 years. In: Peterkin, G.L., Bricker, H., Mellars, P. (Eds.), Hunting and Animal Exploitation in the Later Palaeolithic and Mesolithic of Eurasia. Archaeological Papers of the American Anthropological Association 4, 221–240.

Rabinovich, R., 1998. Taphonomical aspects of the recent excavations at El-Wad, Appendix III. In: Weinstein Evron, M., (Ed.), Early Natufian El-Wad Revisited. ERAUL 77. Études et Recherches Archéologiques de l'Université de Liège, Liège, pp. 199–224.

Richerson, P.J., Boyd, R., Bettinger, R.L., 2001. Was agriculture impossible during the Pleistocene but mandatory during the Holocene? A Climate Change Hypothesis. American Antiquity 66, 387–411.

Rosenberg, M., Nesbitt, R., Redding, R.W., Peasnall, B.L., 1998. Hallan Çemi, pig husbandry, and post Pleistocene adaptations along the Taurus-Zagros arc (Turkey). Paléorient 24, 25–41.

Saint-Germaine, C., 1997. The production of bone broth: a study in nutritional exploitation. Anthropozoologica 25/26, 153–156.

Stiner, M.C., 1994. Honor Among Thieves: A Zooarchaeological Study of Neandertal Ecology. Princeton University Press, Princeton.

Stiner, M.C., 2001. Thirty years on the "Broad Spectrum Revolution" and Paleolithic demography. Proceedings of the National Academy of Science of the United States 98, 6993–6996.

Stiner, M.C., 2005. The Faunas of Hayonim Cave, Israel: A 200,000-Year Record of Paleolithic Diet, Demography, and Society. American School of Prehistoric Research Bulletin 48. Peabody Museum, Cambridge.

Stiner, M.C., Munro, N.D., 2002. Approaches to prehistoric diet breadth, demography, and prey ranking systems in time and space. Journal of Archaeological Method and Theory 9, 175–208.

Stiner, M.C., Munro, N.D., Surovell, T.A., Tchernov, E., Bar-Yosef, O., 1999. Paleolithic population growth pulses evidenced by small animal exploitation. Science 283, 190–194.

Stiner, M.C., Munro, N.D., Surovell, T.A., 2000. The tortoise and the hare: small game use, the Broad Spectrum Revolution, and Paleolithic demography. Current Anthropology 41, 39–73.

Szuter, C.R., Bayham, F.E., 1989. Sedentism and prehistoric animal procurement among desert horticulturalists of the North American Southwest. In: Kent, S. (Ed.), Farmers as Hunters: The Implications of Sedentism. Cambridge University Press, Cambridge, pp. 80–95.

Tchernov, E., 1991. Biological evidence for human sedentism in southwest Asia during the Natufian. In: Bar Yosef, O., Valla, F.R. (Eds.), The Natufian Culture in the Levant. International Monographs in Prehistory, Ann Arbor, pp. 315–340.

Tchernov, E., 1993. The impact of sedentism on animal exploitation in the southern Levant. In: Buitenhuis, H., Clason, A.T. (Eds.), Archaeozoology of the Near East. Universal Book Service, Leiden, pp. 10–26.

Turnbull, P., Reed, C., 1974. The fauna from the terminal Pleistocene of Palegawra Cave, a Zarzian occupation site in northeastern Iraq. Fieldiana Anthropology 63, 81–146.

Valla, F.R., 1998. Natufian seasonality: a guess. In: Rocek, T.R., Bar-Yosef, O. (Eds.), Seasonality and Sedentism Archaeological Perspectives from Old and New World Sites. Peabody Museum, Cambridge, pp. 93–108.

Valla, F.R., Le Mort, F., Plisson, H., 1991. Les fouilles en cours sur la Terrase d'Hayonim. In: Bar-Yosef, O., Valla, F.R. (Eds.), The Natufian Culture in the Levant. International Monographs in Prehistory, Ann Arbor, pp. 93–110.

Weiss, E.,Wetterstrom, W., Nadel, D., Bar-Yosef, O., 2004. The Broad Spectrum revisited: evidence from plant remains. Proceedings of the National Academy of Sciences of the United States 101, 9551–9555.

Winterhalder, B., 1981. Foraging strategies in the boreal forest: an analysis of Cree hunting and gathering. In: Winterhalder, B., Smith, E.A. (Eds.), Hunter-Gatherer Foraging Strategies. University of Chicago Press, Chicago, pp. 66–98.

11. Paleolithic Diet and the Division of Labor in Mediterranean Eurasia

Mary C. Stiner
Department of Anthropology
P.O. Box 210030
University of Arizona
Tucson, AZ, 85721-0030, USA
mstiner@email.arizona.edu

and

Steven L. Kuhn
Department of Anthropology
P.O. Box 210030
University of Arizona
Tucson, AZ, 85721-0030, USA
skuhn@email.arizona.edu

Keywords Division of labor • hunter-gatherers • Middle Paleolithic • Upper Paleolithic • hunting • zooarchaeology • demography • prey selection • diet breadth • resource intensification

Abstract Hunter-gatherers of the recent era vary in many aspects of culture, yet they display great uniformity in their tendency to divide labor along the lines of gender and age. We argue on the basis of zooarchaeological, technological, and demographic evidence that the complementary economic roles of men and women so typical of ethnographically documented hunter-gatherers did not appear in Eurasia until the beginning of the Upper Paleolithic. The rich archaeological record of Middle Paleolithic cultures in Eurasia suggests, by contrast, that earlier hominins (Neandertals, among others) pursued narrowly focused economies, with women's activities more closely aligned to those of men with respect to schedules and territory use patterns. Hoofed animals were the principal source of meat for virtually all Middle and Upper Paleolithic foragers, but Upper Paleolithic people supplemented diets to large game with a broader spectrum of small animals, leading to considerable expansion in dietary breadth. Parallel trends are apparent in the technological record. Evidence of skill-intensive, time-consuming craft work that normally supports the food quest among recent forager economies also emerged in the early Upper Paleolithic, including indications of dry hide scraping based on lithic micro-wear evidence and widespread use of bone tools suitable for working hide, plant fibers or both. The comparatively narrow reliance on large game animals during the Middle Paleolithic for meat would have constrained the demographic potential of these endemic populations. More broadly based economies, as indicated both by the faunal record and the increasing complexity of foraging and related technologies, appeared earliest in the eastern Mediterranean region and spread (with modification) to the north and west. The behavioral changes associated with the Upper Paleolithic record signal a wider range of economic and technological roles in forager societies, and these changes in adaptation may have provided the expanding *Homo sapiens* populations with a demographic advantage over other hominins in Eurasia. Middle Paleolithic human reproductive units probably were not robust at the micropopulation scale, and localized extinctions were likely to have been common. The demographic robustness of the Upper Paleolithic systems may be explained by the rise of new, diversified strategies for evening-out or sharing risk. When and where Middle and Upper Paleolithic populations first came into contact, the marginal advantages provided by collaborative economies meant that replacement of the Middle Paleolithic groups was only a matter of time.

J.-J. Hublin and M.P. Richards (eds.), *The Evolution of Hominin Diets: Integrating Approaches to the Study of Palaeolithic Subsistence*, 157–169.

Introduction

Any mention of the Middle and Upper Paleolithic also raises the topic of the biological and cultural "transition" in which *Homo sapiens* supplanted endemic members of the genus in Asia and Europe. An essentially African origin is widely recognized for anatomically modern human populations, and this continent is *the* (or at least the main) source of early behaviorally modern humans as well (Klein, 1999; Klein and Edgar, 2002). Eurasia would have posed unequal challenges for endemic and invading hominin populations because of its great environmental and climatic diversity. It is for this reason that Eurasia presents a natural laboratory for comparing some of the inferable limits of Middle and Upper Paleolithic adaptations during the Late Pleistocene. The Mediterranean Basin is particularly interesting for inter-period comparisons of human diets on account of its exceptional wealth of species and general community stability during the Late Pleistocene (Tchernov, 1998a, b).

Within Eurasia, the Upper Paleolithic period certainly earns its reputation as a revolution in human behavior. The appearance of the Upper Paleolithic, or the broader complex of behavioral characters sometimes called "modern human behavior," occurs abruptly in most Eurasian stratigraphic sequences. In contrast, elements of "modern human behavior" appear earlier and coalesce more gradually in some African Middle-Late Stone Age sequences (McBrearty and Brooks, 2000; Henshilwood and Marean 2003; Henshilwood et al., 2004). The contrasting pace of cultural change among regions and between continents is taken as evidence of an invasive expansion of behaviorally modern human populations into Eurasia at the expense of pre-existing hominins that possessed Middle Paleolithic culture.

Zooarchaeological evidence likewise testifies to important economic differences between Middle and Upper Paleolithic populations. The earliest demonstrable impacts of humans on biotic communities in Eurasia associate with early Upper Paleolithic hunter-gatherers in the eastern Mediterranean Basin roughly 45,000 years ago (Tchernov, 1992). Concomitant with the spread of anatomically modern *Homo sapiens* into Eurasia, we see the evolution of novel technological and social mechanisms for buffering or redistributing environmental risk. These behavioral developments coincide with permanent changes in human demographic potentials and the carrying capacities of environments.

Of central importance to social and economic change is the manner in which labor was divided within social groups. Human labor divided by gender and age is universal to recent small-scale societies and thus is characteristic of the modern human species. We propose that labor allocation was structured differently between the Middle and Upper Paleolithic periods. Though male and female foraging agendas differ within many higher vertebrates, this aspect of modern human behavior can be distinguished from examples of niche separation within other animals mainly in that human cultural adaptations *combine* complementary foraging roles for individuals of different ages and sexes through regular, pervasive sharing (Kuhn and Stiner, 2006). Divided, collaborative labor among recent humans is thought to confer very significant benefits in terms of foraging efficiency.

Although the so-called "division of labor" is an essential feature of the recent human condition, the circumstances in which this tendency evolved are poorly known. Criteria are needed to recognize changes in the organization of labor, and these criteria must have correlates in material culture and the dietary record if they are to serve archaeological studies. Such criteria can be developed from cross-cultural patterns in recent forager systems. This use of ethnographic information is very different from simple analogy: we are not looking for matches between present and past societies, but instead are using generalized cross-cultural patterns of recent forager systems to isolate anomalies in extinct culture systems. The anomalies must then be explained independently of these referents.

Zooarchaeologists and technologists have collected vast amounts of data on Paleolithic subsistence, far too much to review in this short essay. It is useful in light of this fact to consider how these findings might address some of the bigger questions of human behavioral evolution during the Late Pleistocene. As one small step towards this end, we propose that the typical patterns of labor division documented by ethnographic studies today emerged relatively late in human evolutionary history. With respect to Eurasia, the archaeological record of Middle Paleolithic humans (including the Neandertals, the most recent of the "non-modern" hominins) exhibits less evidence for the array of distinct economic roles typically fulfilled by women and older children in recent hunter-gatherer groups than do the records of the Upper Paleolithic. We propose that Middle Paleolithic males, females, and juveniles all participated in a narrower range of economic activities that centered on obtaining large terrestrial game. This is not to say that every person performed identical activities or that they ate no plants or small animals; the availability of personnel to participate in some way in communal large game hunting was paramount, encouraging group members to remain in fairly close proximity to one another so as not to miss hunting opportunities. Such groups would exhibit less inter-individual variation in land use if compared to recent foragers.

We also argue that collaborative economic systems are more likely to have *first* evolved in the tropics or subtropics, where biotic diversity and evolutionary opportunities for diet diversification are greatest. This is not a matter of environmental determinism of social roles, but rather a product of widely repeated selection for dietary diversification as a function of low latitude and high biotic diversity. At some point in the past, cooperative economic systems may have given Upper Paleolithic humans a demographic advantage over Middle Paleolithic groups and their contemporaries, facilitating the rapid expansion of Upper Paleolithic culture throughout Eurasia.

The first part of this essay concerns observations about the division of labor by age and gender among historically and ethnographically documented foraging peoples. The second part reviews Paleolithic evidence for differentiated, complementary economic roles, or the lack thereof, before the appearance of modern humans in Eurasia, based partly on data from the Mediterranean Basin. Meat use is emphasized in this essay for reasons of visibility in the archaeological record, but plant use is also addressed to the extent possible. The final part of the chapter speculates on where this unique, nearly universal human pattern might have originated, and how it contributed to the geographic spread and evolutionary success of Upper Paleolithic populations.

The Division of Labor in Recent Hunter-Gatherers

The axiomatic "division of subsistence labor" in recent foraging peoples holds that men tend to hunt large animals, and women and children tend to focus more of their efforts on gathering plants, capturing smaller animals or both. Ethnographic accounts also teach us that the boundaries between these broad economic roles are highly permeable, and that there are many individual departures from the general pattern. It is clear from some ethnographic studies, for example, that women possess the knowledge and skills needed for hunting large game, and that boys without mothers may learn to sew weatherproof clothing in regions where it is needed (Jenness, 1922; Landes, 1938; Briggs, 1970; Bailey and Aunger, 1989; Bailey, 1991; Endicott, 1999). It remains true, however, that in most circumstances individuals are prepared to cross roles mainly under duress or as temporary, rare arrangements. In addition, children at times forage for themselves (e.g., Laughlin, 1969; Watanabe, 1969), but they seldom can manage without supplementary food from parents or relatives (Blurton Jones et al., 1989, 1997; Bird and Bliege Bird, 2000; Kaplan et al., 2000; Bliege Bird and Bird, 2002; Walker et al., 2002).

Anthropologists offer at least three non-competing explanations for divided, collaborative labor among recent hunter-gatherers (Panter-Brick, 2002; Shennan, 2002). First, men and women may have different agendas that relate to their roles in childcare and the certainty of their genetic relationships to offspring (Hawkes and Bliege Bird, 2002). Second, avoidance of the more dangerous subsistence pursuits by women and children protects the reproductive core of population from undue risk, exposing "expendable" males more of the time. Third, the demands of childcare often cause women to favor activities that can be interrupted with minimum cost and entail relatively limited mobility (Kelly, 1995).

Certain other economic tendencies are also important to understanding the division of labor among recent foragers. These tendencies are of great interest in that they may provide archaeological criteria for identifying changes in socioeconomic patterns in Paleolithic records:

(a) Recent forager systems are very responsive to variations in the physical environment. It is for this reason as much as any other that large game, and meat in general, is most important to human diets in higher latitude terrestrial environments of the world, whereas gathered vegetable foods and small game tend to be more important at low latitudes (Keeley, 1988; Kelly, 1995; Binford, 2001; Marlowe, 2005). Following global patterns of biotic diversity, the range of variation in diet composition is also greater in the tropics and sub-tropics.

(b) Virtually all foragers eat fruits and greens if they are available. In circumstances where people depend on plants as major calorie sources, however, some combination of seeds, nuts, and tubers almost always serves as a staple (Keeley, 1988). Energy-rich seeds and nuts (and certain tubers) can be very abundant locally, but they tend to be time-consuming to collect and process (Kelly, 1995).

Staple plant resources present an extreme contrast to large game animals with respect to prevailing economic currencies (Table 11.1). Large animals generally yield high returns per unit foraging time (kJ per hour), but are unpredictable resources. Seeds and nuts give much lower net yields per increment time (kJ per kilogram acquired), but they have potentially high yields with respect to the volume obtained and the area of land utilized. Note in Table 11.1 that the returns from large and small game animals are similar to one another in terms of kJ/kg but not in terms of kJ per hour, and that all meat sources have lower returns per volume than those from intensively processed seeds and nuts.

Splitting tasks between the two broad food-getting agendas could result in a more efficient and less risky economic system, especially if the major food types have patchy and non-congruent distributions in the environment. By sharing resources at central places, a divided yet collaborative system spreads risk among individuals while getting the most from high-risk and low-risk resources in a single territory. Such a system requires, however, that the schedules, ranging patterns, and technology used by individuals differ on a regular basis and that at least some of the products of foraging are widely shared.

(c) In high-latitude environments where the options for gathering plants or small game hunting are quite limited, women and even children tend to assume responsibility for non-subsistence tasks that support the food quest, including collection of water and fuel, transport and construction of housing, and skill-intensive and time-consuming manufacture of tools, shelter, and clothing (Osgood, 1940; Balikci, 1970; Halperin, 1980; Waguespack, 2005). As important as some of these activities can be to the food quest, the workloads and daily and seasonal schedules of activity may differ greatly from those required to capture large animals.

Because many of these aspects of recent hunter-gatherer behavior are expressed in technology, it sometimes is possible to identify archaeological signatures of alternative female roles in environmental circumstances where most of the food came

TABLE 11.1. Net energy yields of various food classes consumed by recent foragers, organized in terms of yield per hour (kJ/h) as opposed to yield by unit weight (kJ/kg) (From Kuhn and Stiner, 2006).

		kJ/h		kJ/kg		
	N cases	Mean	Minimum	Maximum	Mean	SD
Large game	4	63,398[a]	36,000	75,115	6,980[b]	1,383
Small mammals	14	16,034[a]	1,672	56,317	6,980[b]	1,383
Reptiles	3	15,850[a]	17,556	12,435	4,489[b]	715
Birds	3	4,472[a]	961	8,255	–	–
Roots and tubers	14	6,120[a]	418	26,133	2,926[c]	1,680
Roots and tubers	13	1,882[e]	1,045	2,300	3,136[e]	2,338
Seeds and nuts	34	3,520[a]	380	18,538	13,188[c]	9,334
Seeds and nuts	9	6,508[d]	1,203	24,933	13,188[c]	9,334
Seeds and nuts	6	–	–	–	19,372[e]	6,250
Foliage	–	–	–	–	1,250[c]	819
Foliage	3	–	–	–	1,534[e]	186
Fruits	–	–	–	–	2,403[c]	1,463

[a]Data from Kelly (1995, Table 3.3).
[b]Data from Hawkes et al. (1982); Hurtado and Hill (1987).
[c]Data from Pennington (1989).
[d]Data from Wright (1994, Table 11.2).
[e]Data from Wiessner (P. Wiessner, 2004, personal communication).

from hunting large animals – tools for craftwork – and where plant seeds and nuts served as staples – milling and grinding tools.

Division of Labor in the Paleolithic

Is there archaeological evidence for divided labor before modern humans. If so, how does it compare to known hunter-gatherer systems? Middle Paleolithic (Neandertals) and Upper Paleolithic humans were quasi-contemporaries in Eurasia, yet behaviorally distinct based on technological and other evidence. The Neandertals were distributed from northern Europe to the southern Mediterranean basin and well into Asia. The great expanse of this Old World geography permits systematic examination of how subsistence and technology varied across a range of environments and latitudes from 250,000 to 30,000 years ago. Neandertals and other Middle Paleolithic hominins were very intelligent, judging from the sizes of their brains, but Middle Paleolithic hunting implements were comparatively simple (Kuhn and Stiner, 2001), and modes of aesthetic expression were non-existent or are not preserved, though mineral pigments were used for purposes yet to be determined (Mellars, 1996; Klein, 1999).

Zooarchaeological Comparisons

A range of archaeofaunal studies indicate that Middle Paleolithic humans maintained exceptionally narrow foraging regimens, even in the Mediterranean Basin, with 95–99% of the animal foods procured by weight coming from large game (Fig. 11.1). Most surprisingly, Middle Paleolithic folk behaved this way across a wide range of environments, and adjustments to variation in latitude and biotic diversity were quite limited. In the southern part of their ranges, Middle Paleolithic humans supplemented their meat intake with easily collected (gatherable) small prey animals – tortoises, marine shellfish, ostrich eggs, and large lizards. This makes sense in terms of optimal foraging models: the low handling costs of these small prey animals make up for their small size, bringing net yields closer to those of large game. It is strange, however, that small animals with higher handling costs were avoided in nearly all circumstances. Whatever flexibility existed in Middle Paleolithic foraging systems, it seldom extended to animals or plants with high capture or processing costs; this stands in stark contrast with the behavior of many recent hunter-gatherers, and even Upper Paleolithic hunter-gatherers.

Evidence for large game hunting, whether in the form of prey biomass comparisons, mortality patterns, or food transport patterns, fails to differentiate much between the economies of Middle and Upper Paleolithic societies in Eurasia, or even between the late Lower Paleolithic and the Middle Paleolithic (e.g., Stiner, 1994; Grayson and Delpech, 2003; Gaudziaski, 2005; Stiner, 2005; Alder et al., 2006). One great difference between the predatory economics of these societies concerns the strategies for filling the gaps in large game availability. Upper Paleolithic foragers had more versatile methods of meat and plant food supplementation, which they used to even-out lows in the availability of highly ranked foods. This translates to permanent changes in dietary breadth, greater the inherent flexibility and internal diversity in foraging regimens, and, in some cases, increased environmental carrying capacity for some groups during the Upper Paleolithic period.

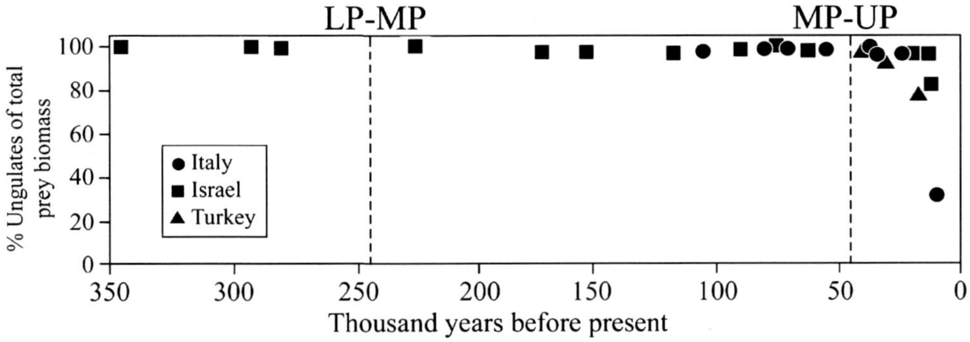

FIG. 11.1 Percentage of total prey biomass represented by ungulate prey in the assemblages of each Mediterranean faunal series. LP–MP refers to the Lower to Middle Paleolithic cultural transition; MP–UP the Middle to Upper Paleolithic transition.

Diet-breadth models suggest that the breadth of or variety within forager diets depends, among other things, upon the availability of high quality, high-yield foods (Stephens and Krebs, 1986: 17–24). Narrow diets, in which low-quality prey are usually ignored, are practical only if the chance of finding more profitable prey types is high. If the encounter rates with preferred prey types decline, humans should and generally do broaden their diets by taking more lower-yield types. Dietary diversification is especially likely to occur when and where foragers put excessive pressure on preferred (i.e., highly ranked) resources, thereby forcing them into decline, and a reduction in the predator population may result (e.g., Broughton, 1997; Nagaoka, 2002). Some variation in diet breadth is reversible and to be expected within the confines of a given cultural adaptation ("adjustment") (Kelly, 1995). However, certain trends occur in forager diets over the long term, and these changes appear to represent evolutionary transitions in human adaptations (Stiner et al. 2000; Tchernov, 1998a, b).

A major, early shift in the breadth of the meat diet coincides with the Middle to Upper Paleolithic culture boundary in southern Eurasia (Stiner, 2001). Evidence for this "transition" is widespread in the Mediterranean area, based on the relative exploitation of small quick animals, such as birds and lagomorphs, relative to slow-moving collectable small animals such as tortoises and shellfish. Although small animals may have served mainly as back-up resources, the highly conditional nature of small game use in the Paleolithic, along with great differences in the reproductive ecology of the species commonly hunted, reveal important changes in human ecology and demography. Small animals differ tremendously in their capture requirements, and some species – such as Mediterranean tortoises (*Testudo*) – are very sensitive to over-hunting and therefore represent the proverbial "canaries in the coal mine" for studying shifting human predator–prey interactions (Stiner, 2001). The relative emphasis that humans placed on small prey types as grouped by predator defense traits – slow-moving or "sessile" animals, fast-running hares and rabbits, and quick flying game birds – implies significant reorganization in foraging economics in the Mediterranean region. It is striking that Middle Paleolithic foragers seldom

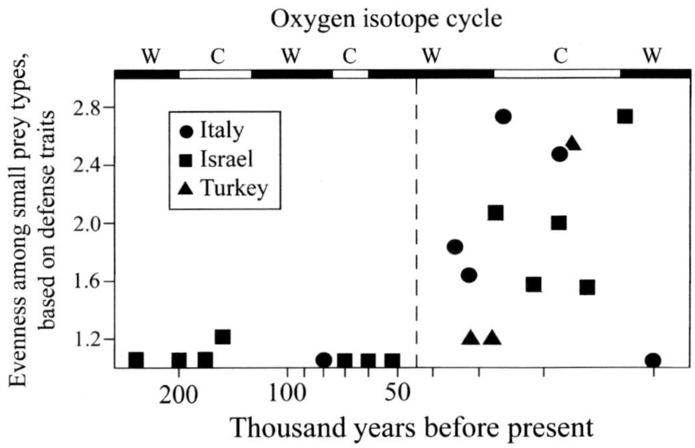

FIG. 11.2. Comparison of the degree of evenness across three small game categories in Paleolithic faunal assemblages, based on prey defense mechanisms (slow game, quick running terrestrial mammals and quick flying birds) (3 = most even, 1 = least even). Symbols are for assemblages from Italy (circle), Israel (square), and Turkey (triangle). Time is expressed on a logged scale, as are oxygen isotope climate cycles; (C) cold stage (W) warm stage (From Stiner, 2001).

pursued small prey unless the animals could be collected with little effort. The situation changed abruptly around 45,000–50,000 years ago in the eastern end of the Mediterranean basin. Though archaeologists may differ in their proposed explanations, the pattern spread or emerged repeatedly in adjacent areas over the remainder of the Upper and Epi-Paleolithic (consult, for example, various authors in Brugal and Desse, 2004).

An index of "evenness" (the Inverse of Simpson's Index; Simpson, 1949) in the small prey types collected by foragers summarizes the abrupt expansion in Paleolithic diets in Mediterranean contexts (Fig. 11.2). Much of the expansion in diet took place during a phase of climate cooling (Oxygen Isotope Stage 2; following Martinson et al., 1987). Had it occurred only in conjunction with global warming (OIS 3), the trend would be difficult to distinguish from natural shifts in the natural diversity and structure of humans' food supply. Instead, the evidence points to a categorical change in how humans interacted with small animal populations around the time of the Middle-Upper Paleolithic cultural "transition". As noted above, the burgeoning importance of small, quick prey in Upper Paleolithic diets is also detectable, albeit later, in the northern interior of Europe as well as in the warmer, arid lands to the south. Not every Upper Paleolithic forager group made use of costly small prey, but many did, in contrast to a nearly uniform lack of this behavior in the Middle Paleolithic in the same regions.

Differences in prey species productivity are a key to interpreting the economic trends for Paleolithic demography. An important quality of small prey animals that reproduce quickly is their greater potential reliability as a food source. Warm-blooded small animals, mainly partridges, hares, and rabbits, mature in 1 year or less, and their populations rebound easily from heavy hunting Predator–prey simulation modeling (Stiner et al., 2000). results show major differences in the scale at which humans could possibly hope to depend on tortoises, hares, and partridge-like birds for meat. Other things being equal, hare populations can support up to seven times greater off-take of adults and sub-adults by predators than tortoises can support, and partridges can support up to ten times greater off-take than tortoises. This means that humans' reliance on tortoises is only sustainable if human population densities are very low. Human's reliance on partridges and hares is sustainable in both low- and high-density conditions.

It is odd that Middle Paleolithic foragers in the Mediterranean region focused on slow-growing prey types so consistently, to the extent they pursued small animals at all. Where tortoises were an important food source in the Levant, there is no evidence for over-harvesting of the tortoises (i.e., no reduction in the mean body size of individuals or skewed age structures) until the very end of the Middle Paleolithic. At the threshold of the Middle-Upper Paleolithic cultural "transition", when fast-reproducing but difficult to capture small animals were added to human diets in significant numbers, the mean sizes of tortoises declined and unnatural skewing is evidenced in the age (size) structure of the harvested animals (Stiner, 2005: 139–147). Taken together, these observations imply that human populations of this particular region had first exceeded the availability or potential of high-ranked, high-return resources to support them as early as 50,000 years ago (but see Speth and Clark, 2006). The zooarchaeological evidence testifies to further demographic growth in the Mediterranean basin over the remainder of the Late Pleistocene, accelerating particularly 15,000 years ago (Binford, 1968; Flannery, 1969; Bar-Yosef, 1981; Keeley, 1988).

Technological Comparisons

Recent hunter-gatherers supplement their meat intake from large game with a variety of small animals and plant foods. Seed and nut hull fragments are found in a few Lower and Middle Paleolithic sites (e.g., Barton et al., 1999; Goren-Inbar et al., 2002; Madella et al., 2002), but there are no indications of stockpiled nuts or seeds. Of course, preservation of organic remains is rare for reasons of sediment chemistry and great time depth, but the kinds of large durable artifacts needed to grind or crush seeds and nuts are also absent, or exceedingly rare, in the Middle Paleolithic toolkit. Another potential complication is that many Middle Paleolithic groups occupied relatively cool, high-latitude environments, where a dietary bias towards large game is to be expected in recent foragers as well. Here, ethnographic experience would predict that females and other non-hunting members living at high latitudes would have taken on the role of technology specialists. The ethnographic record fails to predict Middle Paleolithic's low level of technological elaboration.

European Neandertals almost certainly wore simple skin garments of some kind. However, the types of artifacts commonly used by recent hunter-gatherers to manufacture tailored, weather-resistant clothing – bone needles and awls – did not become a regular part of the archaeological record until the Upper Paleolithic. Edge damage on stone tools from meat cutting and wet/fresh hide is relatively common in Middle Paleolithic assemblages, but evidence of working dry hide (cured leather) in the form of micro-wear traces on stone artifacts is comparatively scarce (e.g., Beyries, 1987; Anderson-Gerfaud, 1990; Lemorini, 2000; Martínez-Molina, 2005). Taken together, the archaeological evidence suggests that Middle Paleolithic females and juveniles did not undertake a suite of economic roles equal in diversity to that within recent hunter-gatherer groups living in similar environments.

A different side of the technological record concerns innovations in the tools and techniques for increasing the digestibility of plant seeds and squeezing more nutrition out of animal carcasses. Food processing grew much more complex in the late Upper Paleolithic (<30,000 years ago) and especially during the Epi-Paleolithic. Significant increases in carcass processing efficiency were gained by the practice of bone grease rendering. Spongy bone parts were broken into small pieces, put in water, and the mixture was heated by adding fire-warmed stones. In this way pure fats could be skimmed off as they rose to the surface and the fat stored for months in this purified form. Grease rendering is evidenced in a variety of late Upper Paleolithic sites (ca. 26,000–14,000 years ago) in Portugal, France, Germany, and the Czech Republic by the combined presence of worn anvil stones, distinctive patterns of bone fragmentation, and thick litter of fire-cracked rocks (Audouze, 1987; Weniger, 1987; West, 1997; Stiner, 2003). Though heat-in-liquid techniques are labor-intensive, ethnoarchaeological studies show that people can raise the protein and fat yields per carcass well beyond what is possible from simpler extraction techniques (Binford, 1978; Lupo and Schmitt, 1997). This hard work can be worthwhile provided that it does not interfere with getting another carcass, or if the natural availability of large prey is limited. Only simpler "cold-marrow extraction" techniques were practiced during the Lower and Middle Paleolithic periods, which focused on the most concentrated marrow reserves in large medullary cavities (e.g., Bunn et al., 1980; Potts, 1984; Stiner, 1994; Speth and Clark, 2006).

Women's Work in the Middle Paleolithic

The marked skeletal and muscular robusticity of the Neandertals indicates high levels of activity among both males and females (Trinkaus, 1983, 1986). Recent experimental work suggests that activity levels during childhood development were high as well (Lieberman et al., 2001). Thus there is little reason to think that females were simply reproductive vessels.

Some paleoanthropologists have proposed that Middle Paleolithic women and their children were economically independent of mature males (Binford 1984; Soffer, 1994), perhaps consuming small resources in the field rather than carrying them to base camps. This hypothesis would predict a class of archaeological evidence – field consumption stations – that currently is not known for the Middle Paleolithic, in contrast to some Holocene archaeological records (e.g., Great Basin, Elston and Zeanah, 2002) and some recent aboriginal populations of Australia. The archaeological evidence instead points to a third hypothesis in which women, children, and men all participated actively in the exploitation of large animals. This third model assumes that the archaeological record *is* representative of Middle Paleolithic diets, and women and juveniles somehow participated in large game hunting more extensively, more consistently, and more directly than is generally seen among recent foragers. This is not to deny the importance of large game hunting among many recent foragers. Nor should we assume that social and economic roles in Middle Paleolithic societies were identical among age groups or by gender. Rather, it is the narrowness of the large game focus during the Middle Paleolithic and its associated search requirements that constitute an important difference in behavior and in labor organization in particular.

Hunting large animals is a rough and dangerous business, all the more so if the hunters were equipped with thrusting spears to be used at close range (Churchill, 1993; Shea, 1997). Like other social carnivores, humans can gain advantages over prey if some members of the hunting party act as artificial surrounds or funnels for directing the movement of quarry toward the killers. Evidence for healed fractures is common on Neandertal skeletons, although the sex distribution for these injuries is unclear (Berger and Trinkaus, 1995). Individual roles in hunts can vary according to several categories of risk, from direct physical contact to the more generalized challenges of frequent moves and foraging on rough or unfamiliar ground. We can be sure that those individuals who came in closest contact with large prey generally incurred the greatest risks. Other individuals could take the more circumspect roles of beating the bushes, processing carcasses, and carrying meat and still can be essential to the pay-off of a hunt. A vast zooarchaeological literature shows us that large amounts of ungulate meat and bone often were carried back to base camps during the Middle Paleolithic. Body parts of prey were processed and apparently shared at these camps, so the close proximity of group members was not simply about keeping hungry hunters honest. Still, the fates of Middle Paleolithic women and children would have been very closely allied to male hunting with respect to activity schedules and ranging patterns.

Top-level carnivores can only exist at very low population densities in terrestrial environments for at least two reasons. First, a heavy dependence on large game for food energy implies an ecological position high in the trophic pyramid where entropy effects are extreme. Second, large game animals can yield high average return rates, but these resources are unpredictable as staple food sources (Hawkes, 1996; Bliege Bird, 1999; Wrangham et al., 1999; Kaplan et al., 2000). In humans, the high day-to-day variance in protein and fats available to children and pregnant or lactating women will limit the reproductive potential of a highly carnivorous population. While the long existence of Middle Paleolithic lifeways across the Old World indicates that adult females in these societies enjoyed reasonable levels of reproductive success, women's fertility would have remained very low due to the unpredictable diet and the necessity of women's cooperation and ready proximity for hunting operations. Middle Paleolithic populations seldom attained large sizes and were subject to frequent crashes (Semino et al., 2000; Pennington, 2001; Richerson et al., 2001; Boone, 2002; Shennan, 2002).

To summarize thus far, Middle Paleolithic society must have been constrained by the combined influences of high risks of injury all around, frequent residential moves, and economies based on a high quality but unpredictable nutrient supply subject to boom and bust cycles. The demographic consequence of everyone participating closely in the exploitation of large game helps to explain the limited demographic potential for Middle Paleolithic populations.

Labor Allocation and Population Competition

Where did the modern pattern of divided labor first evolve, and how might it have contributed to the evolutionary success of Upper Paleolithic *Homo [sapiens] sapiens*? Upper Paleolithic humans were big game hunters, as were the Middle Paleolithic people before them, but Upper Paleolithic groups supplemented their diets in more versatile ways, particularly where the diversity of animal and plant species was naturally great. In the Jordan and Nile River valleys, a variety of small quick animals were added to the diets quite early on. Even fish, seeds, and tubers, late additions to human diets elsewhere in the Old World, were exploited along the rivers and lakes of the Great Rift System that links Africa and Asia (Stewart, 1989; Wright, 1994; Weiss et al., 2004) during the Last Glacial Maximum 23,000–20,000 years ago.

The Upper Paleolithic also presents widespread evidence for the manufacture of elaborate clothing and shelters. Craft toolkits were, as one might expect from ethnographic experience, more elaborate and abundant in sites of the higher latitudes (e.g., Berke, 1984; Soffer et al., 1998; Owen, 2005), but they occur in sites throughout Eurasia from the Upper Paleolithic onward (Kuhn and Stiner, 2001). Micro-wear evidence from stone tools documents similar increases in the complexity of craftsmanship; edge damage on Upper Paleolithic stone tools from hide and particularly leather (both wet and dry) preparation is common (e.g., Vaughn, 1985; Donahue, 1988), even in the Initial Upper Paleolithic phase in southern Turkey (Martínez-Molina, 2005). The behavioral contrast in labor organization between the Middle and Upper Paleolithic therefore is suggested *both* in the realm of subsistence and the domain of technological support.

Here lies a critical point: the two dimensions of variability in recent forager behavior – technology and dietary breadth – must be considered simultaneously across cases and environments in order to appreciate a fundamental socio-economic distinction between Middle and Upper Paleolithic populations. The technologies of recent foragers are most complex at high latitudes, and thus we can expect the strongest archaeological evidence for divided, collaborative labor in colder environments to be expressed in material culture. The diets and foraging activities of individuals are more varied in lower latitude environments (Kelly, 1995), where ecosystems also harbor greater natural biodiversity, and the division of labor is most vividly expressed by the great breadth of individual and group diets. Both of these dimensions of forager behavior testify to the inherent flexibility of modern human responses to environmental variation. Together, the two dimensions expose coarse but useful generalizations about how recent foragers tend to solve the problems of environmental risk and variable food supplies.

While these observations belabor the obvious from an ethnographic perspective, such adjustments to environmental variation are not typical of Middle Paleolithic populations. Diversification during the Upper Paleolithic may be emphasized more in subsistence or in technology from one region to another, but either (or both) were clearly within their capacity. This behavioral contrast in the technological and dietary dimensions of culture is significant for the niche evolution, environmental carrying capacity, and changes in the internal organization of Paleolithic societies. It also seems that Upper Paleolithic societies differed from Middle Paleolithic societies in Eurasia due to a wider range of economic and social roles overall.

We do not believe that stereotypical patterns of divided labor were an inevitable evolutionary development. Rather, the modern pattern of divided cooperative labor by age and gender could have been an historical accident, stemming in part from the tropical and sub-tropical environments where *Homo sapiens* first evolved. Cooperative economies organized around complementary subsistence roles are more likely to develop spontaneously and repeatedly in low latitude regions for at least three reasons: (1) all classes of food resource other than large game are more diverse and abundant in low latitude ecosystems, and some are available for a greater part of the year; (2) plant resources such as tubers may be especially abundant, and early examples of intensive plant processing first appear in archaeological records at low latitudes; (3) resources that children can collect for themselves are more diverse and abundant, giving children and their mothers greater options for economic independence. Under these conditions individuals have more options in their foraging agendas, and strategies are free to diverge into a greater range of roles.

The African Middle Stone Age is not as extensively documented as the Paleolithic record in Eurasia, but there are hints of a greater range of subsistence and technological roles in some areas, based on fish remains and bone harpoons (Yellen et al., 1995) and scattered reports of early grinding tools suitable for processing seeds or nuts (McBrearty and Brooks, 2000). The eastern Mediterranean Basin (a.k.a. the Levant), situated at the northern end of the rift, is host to what are arguably the earliest Upper Paleolithic culture complexes in Eurasia (Bar-Yosef, 2000). Some, though not all, of these early Upper Paleolithic assemblages also manifest early evidence for dietary expansion (Stiner, 2001; Stiner et al., 2002).

Many of the new foods added to human diets in the Upper Paleolithic associate with distinct foraging substrates, are costly to capture or process, and differ from large game hunting in the energetic currencies most relevant to their exploitation. Specifically, time spent searching and capturing prey is a central consideration in the case of large hoofed animals; the absolute volume acquired and access rights may be much more important in the case scattered patches of edible plants or small animals (Table 11.1; see also Stephens and Krebs, 1986: 7–9 on currency assumptions in foraging models). The benefits of niche separation within human groups are likely to increase as the diet broadens over the long run, due to less overlap or symmetry in the schedules and locations in which various foods can be obtained (reviewed by Pianka, 1988: 254). A population of diverse specialists (our hypothetical UP population) might actually out-compete a population of generalists (MP population) in which all individuals hold more similar roles, even if average individual foraging efficiency is lower within the Upper Paleolithic population (Horan et al., 2005). This kind of within-population diversification would be especially advantageous in environments where a variety of key resources occur at disparate locations or times (MacArthur and Levins, 1964, 1967; Pianka, 1988: 254), or if distinct mechanical strategies are required to obtain them efficiently. Though we propose that the advantages of divided, cooperative labor first arose in the tropics or subtropics, such economic systems would have conferred different but equally valuable advantages in patchy and seasonally variable temperate environments. In cooler regions, individual roles in producing and maintaining high-quality clothing and shelter may substitute for diversifying foraging roles.

McBrearty and Brooks' (2000) important summary article documents precocious developments in technology and art on the African continent. Also very significant in our view is the uneven distribution of these phenomena in time and space, as if the features of interest in Middle Stone Age cultures genuinely came and went many times. The same may be said for the archaeological record of the Near East (Bar-Yosef, 2000). No single continent of origin is necessary to this hypothesis about the origins of the modern division of labor, even if this is how it turned out. Instead the pattern was one of many and often isolated experiments across diverse subtropical habitats. Low latitude ecosystems provide consistently rich opportunities for dietary diversification, should natural selection favor this behavior for any reason. Problems of dating notwithstanding, the geography of early developments in "modern human behavior" – art, stone-tipped weapons, bone tools are often cited – includes much of Middle Stone Age Africa and adjacent areas of western Asia (e.g., Kuhn et al., 2001; Vanhaeren et al., 2006) and possibly also eastern Europe.

The spread of "collaborative economies" would have stemmed from their demographic consequences. As the modern humans entered new environments with novel food supplies and physical challenges, the basic (tropical) system would have continued to change without losing its diverse, collaborative character, because of the competitive advantage of greater efficiency (Winterhalder and Goland, 1993). In patchy and seasonally variable temperate environments, clothing and shelter technology became as or more important than diversifying foraging roles.

There is little reason to believe that the Middle-Upper Paleolithic cultural "transition" marks the beginning of human "inventiveness." The value of innovations for labor collaboration was set by the unprecedented juxtaposition of two economically distinct populations. The demographic edge of Upper Paleolithic foragers over neighboring Middle Paleolithic populations was probably quite subtle, yet more than enough to make a difference over a few thousand years (Zubrow, 1989).

Most narratives of human evolutionary history are orthogenetic in that they portray change as a simple progression of physical or cultural forms. Yet the history of hominin diets after roughly 500,000 years ago is not marked by progressive increases in their basic aptitude as hunters of large animals. Rather, there seems to have been an evolutionary trade-off between tightly bound cooperation among group members in surrounding prey and killing them with simple weapons at close range, and an increasing incidence of solo or small-party hunting, wherein advantages were gained from the greater efficiency of high-investment weapons and traps. The rising importance of the latter system, even in combination with the first, effectively freed some individuals to engage at times in quite different and spatially remote foraging activities. This trade-off represents over the long term a shift in the value of individual forager's time and a relaxation of the requirement that individuals stay in nearly continuous proximity to one another. With these kinds of changes in hunting, we also see greater or more intensive exploitation of plants and the generation and maintenance of complex material culture in support of the food quest.

On the Problem of Neandertal Extinction

How might one behaviorally defined "population" have driven another to extinction or largely replaced it, if the differences in demographic potential were marginal? The Neandertals, the most recent of Middle Paleolithic hominins, thrived for more than 100,000 years and then went extinct soon after 30,000 years ago (Hublin et al., 1995; Mellars, 1996). How could they have persisted for such a long time and then suddenly become so fragile?

Competing populations need not be terribly different from one another in order for one to absorb or supplant the other. Competitive exclusion in its original conception is said to occur only if two populations come into competition and cannot achieve an evolutionary stable coexistence (Pianka, 1988: 221–222). If both populations were small at the outset and existed well below environmental carrying capacity, each could at first grow exponentially, limited mainly by their respective reproductive potentials. Later, as the habitat(s) fills in, the growth rate of each population would decelerate. Populations are unlikely to have identical rates of increase, competitive abilities or carrying capacities, and a threshold will be reached when one population suddenly gains an incremental advantage. While one population stops growing, the other's rate of increase is still positive and eventually inhibits the first population. In this way, the second population will eventually exclude the first.

In the concept of a "rugged fitness landscape" (Wright, 1932; Palmer, 1991), within which high points represent adaptive configurations of relatively greater fitness and low points areas of reduced fitness, selection tends to drive populations toward the optimal peak that is closest to the population's starting point. This occurs even though the rugged fitness landscape includes many fitness peaks of varying heights, representing local sub-optima, separated by many "valleys." Historical contingency plays a critical part in this process: having ascended one fitness peak, it is very difficult for that population to shift to another, even if a distant peak provides greater maximum fitness. Radical peak-shifting is seldom rewarded under these circumstances, because crossing deep valleys necessarily involves a major reduction in fitness. Only severe environmental or demographic perturbations may dislodge the subject population from the sub-optimal fitness peak it currently occupies, metaphorically clearing the way to a higher peak now visible on the horizon and accessible.

Neandertals, though long successful in Eurasia, were ascending a local fitness peak that may not have been the highest in the total fitness landscape. While they were alone in Eurasia, existing at low densities, there would have been no benefit to diversified subsistence. Indeed there may not have been room in much of their natural range for experimentation with lower-return resources. The low latitudes have, by contrast, probably always supported the densest or largest hominin populations and also presented the most consistent incentives for dietary diversification due to their inherently higher biodiversity. Diversified economies in these regions could represent a high fitness peak relative to all others. Here, human populations would likely undergo repeated episodes of expansion, and possess certain social and economic pre-adaptations as part of their evolutionary legacies.

We suggest the demographic expansion of early Upper Paleolithic populations caused perturbations in the fitness landscape that had not existed previously in Eurasia. The more flexible foraging and technological systems of Upper Paleolithic populations provided greater micro-population stability, which allowed them to supplant indigenous Middle Paleolithic populations. It seems that Upper Paleolithic groups were also exceptionally good at holding on to any habitat gained, apparently by reducing the probability of population crashes and effectively hugging the ceiling of environmental carrying capacity. These qualities allowed Upper Paleolithic populations to expand rapidly through Eurasia after 45,000 years ago.

In this way of thinking, the group-level economic advantages of Upper Paleolithic populations were a byproduct of a gender-based division of labor and food sharing. The "inflexibility" of Middle Paleolithic culture was a product of the success and stability of that adaptation, rather than being a question of lower intelligence. In a few areas at least, late Neanderthals did develop or adopt some features of the "modern behavioral repertoire" (d'Errico et al., 1998, 2003). However, estimates of the rates of long-term demographic increase before 50,000 years ago are remarkably low, a conclusion that also is supported by the zooarchaeological evidence.

There seems to have been a lack of economic incentives for large-brained, mobile Middle Paleolithic hunters to squeeze more out of traditional food supplies, and little if any long-term selection for greater foraging efficiency. This implies that, like other organisms – but unlike recent humans – Middle Paleolithic hominins responded to population-resource imbalances almost exclusively through localized depopulation, rather than by increasing food yields by intensifying resource extraction or diet diversification. More difficult to explain than Middle Paleolithic conservatism is how Upper Paleolithic populations managed to exist at higher densities in the same range of environments. An important clue as to how higher population density could become a permanent condition lies in persistent diversification during the Upper Paleolithic period onward and a net lowering of humans position in regional food webs.

Acknowledgments. We are very grateful to the organizers of the hominid diet conference, J.-J. Hublin, M. Richards, the Max Planck Institute Department of Human Evolution staff, and the lively group of participants, for providing an exciting opportunity and forum in which to present and refine this paper. We also thank three anonymous reviewers for their insightful comments, criticisms, and suggestions on the manuscript.

References

Adler, D.S., Bar-Oz, G., Belfer-Cohen, A., Bar-Yosef, O., 2006. Ahead of the game – Middle and Upper Paleolithic hunting behaviors in the southern Caucasus. Current Anthropology 47, 89–118.

Anderson-Gerfaud, P., 1990. Aspects of behavior in the Middle Paleolithic: functional analysis of stone tools from southwest France. In: Mellars, P. (Ed.), The Emergence of Modern Humans: An Archaeological Perspective. University Press, Edinburgh, pp. 389–418.

Audouze, F., 1987. The Paris Basin in Magdalenian times. In: Soffer, O. (Ed.), The Pleistocene Old World: Regional Perspectives. Plenum, New York, pp. 183–200.

Bailey, R.C., 1991. *The Behavioral Ecology of Efe Pygmy Men in the Ituri Forest, Zaire.* Museum of Anthropology, University of Michigan, Ann Arbor, MI.

Bailey, R.C., Aunger, A., 1989. Net hunters vs. archers: variations in women's subsistence strategies in the Ituri Forest. Human Ecology 17 (3), 273–297.

Balikci, A., 1970. *The Netsilik Eskimo.* Natural History Press, Garden City, NY.

Barton, R.N.E., Currant, A., Fernandez-Jalvo, Y., Finlayson, C., Goldberg, P., Macphail, R., Pettit, P., Stringer, C., 1999. Gibraltar Neanderthals and results of recent excavations in Gorham's, Vanguard, and Ibex caves. Antiquity 73, 13–24.

Bar-Yosef, O., 1981. The Epi-Palaeolithic complexes in the southern Levant. In: Cauvin, J., Sanlaville, P. (Eds.), Préhistoire du Levant. Éditions du C.N.R.S, Paris, pp. 389–408.

Bar-Yosef, O., 2000. The Middle and early Upper Paleolithic in Southwest Asia and neighboring regions. In: Bar-Yosef, O., Pilbeam, D. (Eds.), The Geography of Neandertals and Modern Humans in Europe and the Greater Mediterranean. Peabody Museum Bulletin 8. Peabody Museum Press, Harvard University, Cambridge, MA, pp. 107–156.

Berger, T., Trinkaus, E., 1995. Patterns of trauma among the Neanderthals. Journal of Archeological Science 22, 841–852.

Berke, H., 1984. The distributions of bones from large mammals at Petersfels. In: Berke, H., Hahn, J., Kind, C.-J. (Eds.), Upper Palaeolithic Settlement Patterns in Europe. Institut fur Urgeschichte der Universitat Tübingen, Verlag Archaeologica Venatoria, Tübingen, pp. 103–108

Beyries, S., 1987. *Variabilité de l'Industrie Lithique as Mousterien. Approche Functionelle sur queleues Gisments Fraçaises.* International Series S328. British Archaeological Reports, Oxford.

Binford, L.R., 1968. Post-Pleistocene adaptations. In: Binford, S.R., Binford, L.R. (Eds.), New Perspectives in Archaeology. Aldine, Chicago, IL, pp. 313–341.

Binford, L.R., 1978. *Nunamiut Ethnoarchaeology.* Academic Press, New York.

Binford, L.R., 1984. *Faunal Remains from Klasies River Mouth.* Academic Press, Orlando, FL.

Binford, L.R., 2001. *Constructing Frames of Reference.* University of California Press, Berkeley, CA.

Bird, D.W., Bliege Bird, R., 2000. The ethnoarchaeology of juvenile foragers: shellfishing strategies among the Meriam children. Journal of Anthropological Archaeology 19, 461–476.

Bliege Bird, R., 1999. Cooperation and conflict: the behavioral ecology of the sexual division of labor. Evolutionary Anthropology 8, 65–75.

Bliege Bird, R., Bird, D., 2002. Constraints of knowing or constraints of growing? Fishing and collecting by the children of Mer. Human Nature 13, 239–267.

Blurton Jones, N., Hawkes, K., O'Connell, J.F., 1989. Studying the costs of children in two foraging societies: implications for schedules of reproduction. In: Standon, V., Foley, R. (Eds.), Comparative Socieoecology of Mammals and Man. Blackwell, London, pp. 365–390.

Blurton Jones, N., Hawkes, K., O'Connell, J.F., 1997. Why do Hadza children forage? In: Seigel, N., Weisfield, G.E., Weisfield, C.C. (Eds.), Uniting Psychology and Biology: Integrative Perspectives on Human Development. American Psychological Association, Washington, DC, pp. 279–313.

Boone, J., 2002. Subsistence strategies and early human population history: an evolutionary ecological perspective. World Archaeology 34, 6–25.

Briggs, J.L., 1970. *Never in Anger: Portrait of an Eskimo Family*. Harvard University Press, Cambridge, MA.

Broughton, J., 1997. Widening diet breadth, declining foraging efficiency, and prehistoric harvest pressure: ichthyofaunal evidence from the Emeryville Shellmound, California. Antiquity 71, 845–862.

Brugal, J.-P., Desse, J. (Eds.), 2004. Petits Animaux et Sociétiés Humaines, du Complément Alimentaire aux Resources Utilitaires, XXIVe rencontres internationales d'archéologie et d'histoire d'Antibes, 546 pages.

Bunn, H.T., Harris, J., Isaac, G., Kaufulu, Z., Kroll, E., Schick, K., Toth, N., Behrensmeyer, A.K., 1980. FxJj50, an early Pleistocene site in northern Kenya. World Archaeology 12, 109–136.

Churchill, S.E., 1993. Weapon technology, prey size selection and hunting methods in modern hunter-gatherers: implications for hunting in the Palaeolithic and Mesolithic. In: Peterkin, G.L., Bricker, H.M., Mellars, P. (Eds.), Hunting and Animal Exploitation in the Later Palaeolithic and Mesolithic of Eurasia. Archeological Papers of the American Anthropological Association No. 4, Washington, DC, pp. 11–24.

d'Errico, F., Zilhão, J., Baffier, D., Julien, M., Pelegrin, J., 1998. Neanderthal acculturation in western Europe? A critical review of the evidence and its interpretation. Current Anthropology 39, S1–S44.

d'Errico, F., Julien, M., Liolios, D., Vanhaeren, M., Baffier, D., 2003. Many awls in our argument. Bone tool manufacture and use in the Châtelperronian and Aurignacian levels of the Grotte du Renne at Arcy-sur-Cure. In: Zilhão, J., d'Errico, F. (Eds.), The Chronology of the Aurignacian and of the Transitional Technocomplexes. Dating, Stratigraphies, Cultural Implications. Trabalhos de Arqueologia 33. Instituto Português de Arqueologia, Lisbon, pp. 247–270.

Donahue, R., 1988. Microwear analysis and site function of Paglicci Cave, level 4a. World Archaeology 19, 357–375.

Elston, R.G., Zeanah, D.W., 2002. Thinking outside the box: a new perspective on diet breadth and sexual division of labor in the Pre-archaic Great Basin. World Archaeology 34 (1), 103–130.

Endicott, K.L., 1999. Gender relations in hunter-gatherer societies. In: Lee, R.B., Daly, R. (Eds.), The Cambridge Encyclopedia of Hunter-Gatherers. Cambridge University Press, Cambridge, pp. 411–418.

Flannery, K.V., 1969. Origins and ecological effects of early domestication in Iran and the Near East. In: Ucko, P.J., Dimbleby, G.W. (Eds.), The Domestication and Exploitation of Plants and Animals. Aldine, Chicago, IL, pp. 73–100.

Gaudzinski, S., 2005. Monospecific or species-dominated faunal assemblages during the Middle Paleolithic in Europe. In: Hovers, E., Kuhn, S. (Eds.), Transitions Before the Transition: Evolution and Stability in the Middle Paleolithic and Middle Stone Age. Springer, New York, pp. 137–149.

Goren-Inbar, N., Sharon, G., Melamed, Y., Kislev, M., 2002. Nuts, nut cracking, and pitted stones at Gesher Benot Ya'aqov, Israel. Proceedings of the National Academy of Sciences of the USA 99 (4), 2455–2460.

Grayson, D.K., Delpech, F., 2003. Ungulates and the Middle-to-Upper Paleolithic transition at Grotte XVI (Dordogne, France). Journal of Archaeological Science 30, 1633–1648.

Halperin, R.H., 1980. Ecology and mode of production: seasonal variation and the division of labor by sex among hunter-gatherers. Journal of Anthropological Research 36, 379–399.

Hawkes, K., 1996. Foraging differences between men and women: behavioral ecology of the sexual division of labor. In: Steele, J., Shennan, S. (Eds.), The Archaeology of Human Ancestry: Power, Sex, and Tradition. Routledge, London, pp. 283–305.

Hawkes, K., Bliege Bird, R., 2002. Showing off, handicap signaling, and the evolution of men's work. Evolutionary Anthropology 11, 58–67.

Hawkes, K., Hill, K., O'Connell, J., 1982. Why hunters gather: optimal foraging and the Ache of eastern Paraguay. American Ethnologist 9, 378–398.

Henshilwood, C.S., Marean, C.W., 2003. The origin of modern human behavior: critique of the models and their test implications. Current Anthropology 44, 627–652.

Henshilwood, C., d'Errico, F., Vanhaeren, M., van Niekerk, K., Jacobs, Z., 2004. Middle Stone Age shell beads from South Africa. Science 304 (5669), 404.

Horan, R.D., Bulte, E., Shogren, J., 2005. How trade saved humanity from biological exclusion: an economic theory of Neanderthal extinction. Journal of Economic Behavior and Organization 58, 1–29.

Hublin, J.-J., Barroso, C., Medina, P., Fontugne, M., Reyss, J., 1995. The Mousterian site of Zafarraya (Andalucia, Spain): dating and implications on the paleolithic peopling processes of Western Europe. Comptes rendus de l'Académie des sciences. Série 2, Sciences de la terre et des planets 321, 931–937.

Hurtado, A.M., Hill, K., 1987. Early dry season subsistence ecology of Cuiva (Hiwi) foragers of Venezuela. Human Ecology 15, 163–187.

Jenness, D., 1922. *The Life of the Copper Eskimo. Reports of the Canadian Arctic Expedition 1913–1918, N. 12*. F.A. Acland, Ottawa.

Kaplan, H., Hill, K., Lancaster, J., Hurtado, A.M., 2000. A theory of human life history evolution: diet, intelligence, and longevity. Evolutionary Anthropology 9, 156–185.

Keeley, L.H., 1988. Hunter-gatherer economic complexity and population pressure. Journal of Anthropological Archaeology 7, 373–411.

Kelly, R.L., 1995. *The Foraging Spectrum*. Smithsonian Institution Press, Washington, DC.

Klein, R.G., 1999. *The Human Career: Human Biological and Cultural Origins*. (2nd edition). University of Chicago Press, Chicago, IL.

Klein, R.G., Edgar, B., 2002. *The Dawn of Human Culture*. Wiley, New York.

Kuhn, S.L., Stiner, M.C., 2001. The antiquity of hunter-gatherers. In: Panter-Brick, C., Layton, R.H., Rowley-Conwy, P.A. (Eds.), Hunter-Gatherers: Interdisciplinary Perspectives. Cambridge University Press, Cambridge, pp. 99–142.

Kuhn, S.L., Stiner, M.C., 2006. What's a mother to do? A hypothesis about the division of labor among Neanderthals and modern humans in Eurasia. Current Anthropology 47, 953–980.

Kuhn, S.L., Stiner, M.C., Reese, D.S., Güleç, E., 2001. Ornaments in the earliest Upper Paleolithic: new results from the Levant. Proceedings of the National Academy of Science 98 (13), 7641–7646.

Landes, R., 1938. *The Ojibwa Woman*. Norton, New York.

Laughlin, W., 1969. The demography of hunters: an Eskimo example. In: Lee, R.B., Devore, I. (Eds.), Man the Hunter. Aldine, Chicago, IL, pp. 241–243.

Lemorini, C., 2000. *Reconnaître de Tactiques d'Exploitation du Milieu au Palélithique Moyen. La Contribution de l'Analyse Fonctionelle. Étude Fonctionelle des Indsutries Lithiques au Grotta Breuil (Latium, Italie) et de La Colombette (Bonnieux, Valcluse, France).* International Series 858. British Archaeological Reports, Oxford.

Lieberman, D.E., Devlin, M., Pearson, O.M., 2001. Articular surface area responses to mechanical loading: effects of exercise, age and skeletal location. American Journal of Physical Anthropology 116, 266–277.

Lupo, K.D., Schmitt, D.N., 1997. Experiments in bone boiling: nutritional returns and archaeological reflections. Anthropozoologica 25–26, 137–144.

Macarthur, R.H., Levins, R., 1964. Competition, habitat selection, and character displacement in a patchy environment. Proceedings of the National Academy of Sciences of the USA 51, 1207–1210.

Macarthur, R.H., Levins, R., 1967. The limiting similarity, convergence, and divergence of coexisting species. American Naturalist 101, 377–385.

Madella, M., Hovers, E., Jones, M.K., Goldberg, P., Goren, Y., 2002. The exploitation of plant resources by Neanderthals in Amud Cave (Israel): the evidence from Phytolith studies. Journal of Archaeological Science 29, 703–719.

Marlowe, F.W., 2005. Hunter-gatherers and human evolution. Evolutionary Anthropology 14, 54–67.

Martínez Molina, K., 2005. Análisis Functional de Industrias Líticas del Pliestoceno Superior. El Paleolítico Medio del Abric Romaní (Capellades, Barcelona) y del Paleolítico Superior de Üçağızlı (Hatay, Turquía) y del Molí del Salt (Vimbodí, Tarragona). Cambios en los Patrones Funcionales entre el Paleolítico medio y el Superior. Ph.D. Dissertation, Faculty of Letters, Universitat Rovira I Virgili.

Martinson, D.G., Pisias, N.G., Hays, J.D., Imbrie, J., Moore, T.C., Shackleton, N.J., 1987. Age dating and the orbital theory of the ice ages: development of a high-resolution 0 to 300,000-year chronostratigraphy. Quaternary Research 27, 1–29.

McBrearty, S., Brooks, A., 2000. The revolution that wasn't: a new interpretation of the origin of modern human behavior. Journal of Human Evolution 39, 453–563.

Mellars, P., 1996. *The Neanderthal Legacy: An Archaeological Perspective from Western Europe*. Princeton University Press, Princeton, NJ.

Nagaoka, L., 2002. The effects of resource depression on foraging efficiency, diet breadth, and patch use in southern New Zealand. Journal of Anthropological Archaeology 21, 419–442.

Osgood, C., 1940. *Ingalik Material Culture*. Yale University Publications in Anthropology, No. 22. Yale University Press, New Haven, CT.

Owen, L., 2005. *Distorting the Past: Gender and Division of Labor in the European Upper Paleolithic*. Kerns, Tübingen.

Palmer, R., 1991. Optimization on rugged fitness landscapes. In: Perelson, E., Kaufman, S. (Eds.), Molecular Evolution on Rugged Fitness Landscapes. Addison, Redwood City, CA, pp. 3–25.

Panter-Brick, C., 2002. Sexual division of labor: energetic and evolutionary scenarios. American Journal of Human Biology 14 (5), 627–640.

Pennington, L., 1989. *Bowes and Church's Food Values of Portions Commonly Used*. Harper & Row, New York.

Pennington, R., 2001. Hunter-gatherer demography. In: Panter-Brick, C., Layton, R.H., Rowley-Conwy, P.E. (Eds.), Hunter-Gatherers: Interdisciplinary Perspectives. Cambridge University Press, Cambridge, pp. 170–204.

Pianka, E.R., 1988. *Evolutionary Ecology* (4th edition). Harper & Row, New York.

Potts, R., 1984. Home bases and early hominids. Scientific American 72, 338–347.

Richerson, P., Boyd, R., Bettinger, R., 2001. Was agriculture impossible during the Pleistocene but mandatory during the Holocene? A climate change hypothesis. American Antiquity 66, 387–412.

Semino, O., Passarino, J., Oefner, P.J., Lin, A.A., Arbuzova, S., Beckman, LE., De Benedictis, G., Francalacci, P., Kouvatsi, A., Umborska, S., Marcikiae, M., Mika, A., Mika, B., Primorac, D., Santachiara-Benerecetti, A.S., Cavalii-Sforza, L.L., Underhill, P.A., 2000. The genetic legacy of Paleolithic *Homo sapiens* in extant Europeans: a Y-chromosome perspective. Science 210, 1155–1159.

Shea, J., 1997. Middle Paleolithic spear point technology. In: Knecht, H. (Ed.), Projectile Technology. Plenum, New York, pp. 79–106.

Shennan, S., 2002. *Genes, Memes and Human History. Darwinian Archaeology and Cultural Evolution*. Thames & Hudson, London.

Simpson, E.H., 1949. Measurement of diversity. Nature 163, 688.

Soffer, O., 1994. Ancestral lifeways in Eurasia – the Middle and Upper Paleolithic records. In: Nitecki, M.H., Nitecki, D.V. (Eds.), Origins of Anatomically Modern Humans. Plenum, New York, pp. 101–119.

Soffer, O., Adovasio, J.M., Hyland, D.C., Klima, B., Svoboda, J., 1998. Perishable technologies and the genesis of the Eastern Gravettian. Anthropologie 36 (1–2), 43–68.

Speth, J.D., Clark, J.L., 2006. Hunting and overhunting in the Levantine Late Middle Palaeolithic. Before Farming 3, 1–42.

Stephens, D.W., Krebs, J.R., 1986. *Foraging Theory*. Princeton University Press, Princeton, NJ.

Stewart, K.M., 1989. *Fishing Sites of North and East Africa in the Late Pleistocene and Holocene: Environmental Change and Human Adaptation*. International Series 521. British Archaeological Reports, Oxford.

Stiner, M.C., 1994. *Honor Among Thieves: A Zooarchaeological Study of Neandertal Ecology*. Princeton University Press, Princeton, NJ.

Stiner, M.C., 2001. Thirty years on the "Broad Spectrum Revolution" and Paleolithic demography. Proceedings of the National Academy of Sciences of the USA 98 (13), 6993–6996.

Stiner, M.C., 2003. Zooarchaeological evidence for resource intensification in Algarve, southern Portugal. Promontoria 1 (1), 27–61.

Stiner, M.C., 2005. *The Faunas of Hayonim Cave (Israel): A 200,000-Year Record of Paleolithic Diet, Demography & Society*. American School of Prehistoric Research Monographs. Peabody Museum Press, Harvard University, Cambridge, MA.

Stiner, M., Kuhn, S., 2006. Changes in the "connectedness" and resilience of Paleolithic societies in Mediterranean ecosystems. Human Ecology 34, 693–712.

Stiner, M.C., Munro, N.D., Surovell, T.A., 2000. The tortoise and the hare: small game use, the Broad Spectrum Revolution, and Paleolithic demography. Current Anthropology 41, 39–73.

Stiner, M.C., Pehlevan, C., Sagir, M., Özer, I., 2002. Zooarchaeological studies at Üçağizlı Cave: preliminary results on Paleolithic subsistence and shell ornaments. Araşterma Sonuçları Toplantısı, Ankara 17, 29–36.

Tchernov, E., 1992. Evolution of complexities, exploitatation of the biosphere and zooarchaeology. Archaeozoologica V (1), 9–42.

Tchernov, E., 1998a. Are Late Pleistocene environmental factors, faunal changes and cultural transformations causally connected? The case of the southern Levant. Paléorient 23 (2), 209–228.

Tchernov, E., 1998b. The faunal sequence of the southeast Asian Middle Paleolithic in relation to hominid dispersal events. In: Akazawa, T., Aoki, K., Bar-Yosef, O. (Eds.), Neanderthals and Modern Humans in West Asia. Plenum Press, New York, pp. 77–90.

Trinkaus, E., 1983. Neandertal postcrania and the adaptive shift to modern humans. In: Trinkaus, E. (Ed.), The Mousterian Legacy. International Series 164, British Archaeological Reports, Oxford, pp. 165–200.

Trinkaus, E., 1986. The Neandertals and modern human origins. Annual Review of Anthropology 15, 193–218.

Vanhaeren, M., d'Errico, F., Stringer, C., James, S.L., Todd, J.A., Mienis, H.K., 2006. Middle Paleolithic Shell Beads in Israel and Algeria. Science 312, 1785–1788

Vaughn, P.C., 1985. *Use-Wear Analysis of Flakes Stone Tools*. University of Arizona Press, Tucson, AZ.

Waguespack, N.M., 2005. The organization of male and female labor in foraging societies: implications for early Paleoindian archaeology. American Anthropologist 107 (4), 666–676.

Walker, R., Hill, K., Kaplan, H., McMillan, G., 2002. Age-dependency in hunting ability among the Ache of eastern Paraguay. Journal of Human Evolution 42, 639–657.

Watanabe, H., 1969. Subsistence and ecology of northern food gatherers with special reference to the Ainu. In: Lee, R., Devore, I. (Eds.), Man the Hunter. Aldine, Chicago, IL, pp. 69–77.

Weiss, E., Wetterstrom, W., Nadel, D. Bar-Yosef, O., 2004. The broad spectrum revisited: evidence from plant remains. Proceedings of the National Academy of Sciences of the USA 101 (26), 9551–9555.

Weniger, G.-C., 1987. Magdalenian settlement pattern and subsistence in Central Europe: the southwestern and central German cases. In: Soffer, O. (Ed.), The Pleistocene Old World: Regional Perspectives. Plenum, New York, pp. 201–215.

West, D., 1997. *Hunting Strategies in Central Europe During the Last Glacial Maximum*. International Series 672. British Archaeological Reports, Oxford.

Wiessner, P., 2004. Owners of the future: calories, cash and self-sufficiency in the Nyae Nyae area between 1996 and 2003. Visual Anthropology Review 19 (1–2), 149–159.

Winterhalder, B., Goland, C., 1993. On population, foraging efficiency, and plant domestication. Current Anthropology 34, 710–715.

Wrangham, R., Jones, J.H., Layden, G., Pilbeam, D., Conklin-Brittain, N., 1999. The raw and the stolen: cooking and the ecology of human origins. Current Anthropology 40, 567–594.

Wright, K., 1994. Ground-stone tools and hunter-gatherer subsistence in southwest Asia: implications for the transition to farming. American Antiquity 59, 238–263.

Wright, S., 1932. The roles of mutation, inbreeding, crossbreeding and selection in evolution. Proceedings of the 6th International Congress on Genetics 1, 356–366.

Yellen, J.E., Brooks, A.S., Cornelissen, E., Klein, R.G., Mehlman, M., Stewart, K., 1995. A middle Stone Age worked bone industry from Katanda, upper Semliki River valley (Kivu), Zaire. Science 268, 553–556.

Zubrow, E., 1989. The demographic modelling of Neandthal extinction. In: Mellars, P., Stringer, C.B. (Eds.), The Human Revolution. Princeton University Press, Princeton, NJ, pp. 212–231.

12. Moving North: Archaeobotanical Evidence for Plant Diet in Middle and Upper Paleolithic Europe

Martin Jones
Department of Archaeology
University of Cambridge
Cambridge CB2 3DZ, UK
mkj12@cam.ac.uk

Keywords Archaeobotany • plant foods

Abstract This paper reviews the evidence for Middle and Upper Paleolithic plant foods in Europe and neighboring regions. Up until now, most research into the prehistory of plant foods has been conducted on Neolithic and post-Neolithic communities. In recent decades, that has been extended to explore preludes to agriculture and to connect contemporary ethnobotany with the recent archaeological record. The review conducted here shifts focus to the problems of Paleolithic human ecology itself and the challenge of acquiring sufficient plant foods in the novel environments that the first generations of humans crossed and colonized.

Introduction: Archaeobotanical Issues for a Carnivorous Diet

The search for early food plants has largely focused upon the most recent 10,000 years, and on the origins and development of agriculture. It has often been argued that the "transition" to agriculture was marked by a substantial shift in the human food web towards plants, and towards seed and tuber crops in particular. The notion of the Paleolithic meat-eater has been broadly corroborated by isotopic studies (Bocherens, 2009; Richards, 2009), and has left archaeobotany a rather undeveloped field within Paleolithic studies (Mason and Hather, 2002). However, even in the context of a human ecology largely concerned with meat, there are certain significant issues about the Paleolithic plant diet that need to be raised in order to make sense of the expansion of Paleolithic communities across the northerly latitudes of Europe. Central among these are the metabolic need to moderate the high nitrogen levels of lean meat and the role that plants may play in achieving that moderation. Arising from this is the increasing challenge of acquiring sufficient plant food to achieve that in northerly ecosystems, and an increasing dependence upon multi-stage processing to render available plant tissue edible. These two issues form the basis of the following speculative exploration of northward expansion in the Middle and Upper Paleolithic, and its implications for understanding the small but growing archaeobotanical database.

J.-J. Hublin and M.P. Richards (eds.), *The Evolution of Hominin Diets: Integrating Approaches to the Study of Palaeolithic Subsistence,* 171–180.

Lowering the Nitrogen Load

The isotopic evidence from Middle and Upper Paleolithic human skeletons recurrently indicate individuals who derived the majority of their protein from animal sources (Bocherens, 2009; Richards, 2009). That much can be confidently inferred from the isotopic data, although its precision does not really allow the size of that majority to be closely ascertained. If it were reasonably high, as the balance of broader palaeoecological evidence might suggest, then very many of the nutritional requirements would be met by consuming sizeable quantities of fresh meat and associated tissues. The dietary needs of the non-meat component are not so much related to quality as to bulk, and an important function of the bulk is to dilute the nitrogen load within the diet, bringing it below toxic levels. Among living communities, that dilution is effected by consuming in reasonable quantities some combination of fat and plant foods.

In a recent survey of hunter-gather diets, Cordain et al. (2000) explored the dietary balance of a number of extant societies. These include Eskimo communities in which the animal product component of diet reaches 96% and 99%, and which depend upon a high consumption of fat to dilute their dietary nitrogen. That fat is most easily accessible as blubber from marine mammals, but the inland Nunamiut Eskimo gather a great deal of fat from the caribou.

Moving to the temperate zone hunter-gatherer communities in the survey of Cordain et al. (2000), there is a significant jump in levels of plant consumption. In place of the diet of 1–4% plant consumption that has been possible with the high fat content of the cold-climate animals preyed upon by North American Eskimo, they record 21% plant foods among the

Onge of the Andaman Islands, 22% in the Paraguayan Ache, 23% in the Arnhem Land Aborigines, and 25% in both the Australian Anbarra and the Venezuelan Hiwi. The bones of these hunter-gatherers would carry a "meaty" isotopic signal similar to that of Paleolithic Europeans, but a fifth to a quarter of their diet was plant based. We clearly need to entertain the possibility that these temperate zone ethnographic parallels may well be a truer guide to the emphasis upon plant foods in their diet than the Eskimo. Or perhaps more likely, in different times in different places in Paleolithic Europe, dietary nitrogen dilution was effected by varying combinations of fat and plant tissue, and in many cases, the availability of either or both may have constrained the rate and extent of northerly expansion.

Preparing Plants for Consumption

The skill in eating animals largely relates to their capture while live; once caught, killed, and split open, much of their innards may be consumed without further treatment. The reverse is true with plants; they may be relatively easy to "catch," but they frequently require a fair bit of preparation, unpacking, grinding, pounding, pasting, fermenting, heating and so on, before they are digestible. These multiple step processing sequences, often involving radical changes of state between solid, mush, dough, liquid, etc. bear many of the features associated with cognitive "modernity," i.e., the same mental facility that allows the construction of multi-component multi-stage artifact production. As we look back in time to the Middle Paleolithic and beyond, there is a debate as to how far back in time the modern cognition that might be associated with multi-step plant processing sequences can be taken (Mellars, 1996; Mellars and Gibson, 1996; Mithen, 1996).

The Evidence: Looking Back and Looking Forward

As emphasised at the outset, archaeobotanical effort is still drawn predominantly to the question of agriculture, and Paleolithic archaeobotany remains in its infancy. Even among a series of excellent pioneering studies, much of the existing data is concerned with exploring either hunter-gatherers that are essentially coeval with Holocene farmers, or a "prelude" to agriculture, in the context of its emergence among Late Pleistocene and early Holocene, Epipaleolithic and Mesolithic communities. Notable among these are studies in the Upper Euphrates Valley and surrounding regions of Anatolia (Van Zeist and Bakker-Heeres, 1984; Hillman et al., 1989; Hillman, 1996; Willcox, 2002; Martinoli, 2004; Martinoli and Jacomet, 2004), studies of a few key cave sites in the Aegean and Albania (Hansen, 1991, 1999), and studies on the North European coastal Mesolithic (Kubiak Martens, 1999, 2002; Perry, 2002; Robinson and Harild, 2002). Among them, they have allowed us to "look back" beyond agriculture to a Broad Spectrum Revolution in which an immense and regionally very variable range of plant foods formed a key constituent of that broad spectrum of foods. Also worthy of note is the range of North American archaeobotanical research, often conducted in the context of historical ethnobotany, and to a great extent exploring wild plant use over the last 4,000 years, with a few studies relating to the earlier Holocene (Bernstein, 2002; Deal and Butt, 2002; Lepofsky, 2002; Peacock, 2002; Wolgemuth, 2002).

If we switch from looking back beyond agriculture, to looking forward to the problems faced by a meat-eating African genus as it spread northwards into the seasonal and

TABLE 12.1. Key to sites mentioned in the text.

Site name	Country	Cultural association	Reference
Abu Hureyra	Syria	Epipalaeolithic	Hillman et al., 1989; Moore et al., 2000
Amud	Israel	Mousterian	Madella et al., 2002
Buka Island	Solomon islands	28 kya	Loy et al., 1992
Dederiyeh Cave	Syria	Mousterian	Hillman, 2004
Dolní Věstonice	Czech Republic	Upper Palaeolithic	Mason et al., 1994; Beresford-Jones, 2006
Doura Cave	Syria	Mousterian	Hillman G.C., personal communication
Franchthi Cave	Greece	Mousterian	Hansen, 1991
Gesher Benot Ya'aquov	Israel	Acheulian	Goren-Inbar et al., 2002
Goreham's and Vanguard Caves	Gibralter	Mousterian	Gale and Carruthers, 2000
Hallam Cemi	Anatolia	PPNA	Savard et al., 2006
Halsskov	Denmark	Mesolithic	Kubiak-Martens, 2002
Jerf el Ahmar	Syria	PPNA	Willcox, 2002
Kebara Cave	Israel	Mousterian	Lev et al., 2005
Konispol Cave	Albania	Upper Palaeolithic	Hansen, 1999
Laacher See	Germany	Upper Palaeolithic	Baales et al., 2002
Niah Cave	Borneo	Upper Palaeolithic	Barton, 2005
Ohalo II	Israel	Upper Palaeolithic	Kislev et al., 1992; Weiss et al., 2004, 2005
Öküzini	Anatolia	Epipalaeolithic	Martinoli, 2002, 2004; Martinoli and Jacomet, 2004
Santa Maira	Spain	Upper Palaeolithic	Aura et al., 2005
Tybrind Vig	Denmark	Mesolithic	Kubiak-Martens, 1999
Wadi Kubbaniya	Egypt	Epipalaeolithic	Hillman, 1989

sometimes sparse environments of Europe, the available data is considerably thinner. Across Asia, there are now a few archaeobotanical studies extending our knowledge of plant use by anatomically modern humans back 20,000 years and beyond, most notably Ohalo II on the Sea of Galilee (Weiss et al., 2004, 2005), and Niah Cave in Borneo (Barton, 2005). The former site has yielded 90,000 plant remains of around 150 species of seed, fruit, and nut, and the latter the starch granules and parenchyma from a range of monocot tubers and stems (Table 12.2), including aroids, yams, and sago palms. As yet, the plant evidence from Europe itself is much sparser, but is currently being enhanced by an intensive flotation program at a series of Gravettian sites (25–30 kya) in the Moravian Gate in the Czech Republic (Mason et al., 1994; Beresford-Jones, 2006). For earlier hominins, the archaeobotanical data is sparser still, with a scattering of Mousterian examples across Europe and the East Mediterranean, the richest by far being the Kebara Cave in Israel (Lev et al., 2005). Israel is also home to an important Acheulian plant assemblage, at Gesher Benot Ya'aquov (Goren Inbar et al., 2002).

In the remainder of this paper, I shall draw on such evidence as has been collected in the development of an argument about the ecological problems faced by a community of meat-eating *Homo* expanding northwards into Europe in their acquisition of plant foods.

An Ancestral Plant-Gathering Behavior

There are now a number of detailed studies of plant use by our closest living relatives among the great apes, which provide us with a baseline of behavior that we can be reasonably confident was shared by all great apes, including hominins. As indicated above, there is some debate about the point in evolutionary history at which we can add "modern" human cognition to that baseline (Mellars, 1996; Mellars and Gibson, 1996; Mithen, 1996). I would suggest there are two key components to that baseline in relation to the challenges of plant gathering in an expansive *Homo* populations: they are an extensive "*ecological intelligence*" and a facility for "*timely dextrous unpacking*."

TABLE 12.2. Summary of monocot taxa mentioned in the text.

Family	Genus + sp.	Common name	Principal range
Alismataceae	*Alisma* spp.	Water plantain	G(trop + temp)
Alliaceae	*Allium*	Leeks, onions	G(N temp)
Araceae	*Colocasia esculenta*	Taro	As + O(trop)
Araceae	*Xanthosoma* spp.	Cocoyam, tannier	Am(trop)
Asparagaceae	*Asparagus officinalis*	Asparagus	E
Cyperaceae	*Cyperus esculentus*	Rush nut	G(N)
Cyperaceae	*Cyperus papyrus*	Papyrus	Af
Cyperaceae		Sedges	G
Dioscoraceae	*Dioscorea schimperiana*	Kipsigis	Af
Dioscoraceae	*Dioscorea* spp.	Yams	Af + As + O
Juncaceae	*Juncus* spp.	Rushes	G
Marantaceae	*Maranta arundinacea*	Arrowroot	Am
Marantaceae		Arrowroots	G
Musaceae	*Ensete* spp.	Ensete	Af + As
Palmae	*Metroxylon sagu*	Sago palms	As(SE)
Palmae		Palms	G
Poaceae	*Aristida* spp.	Spidergrass	G
Poaceae	*Bambusa*	Bamboo	As
Poaceae	*Hordeum bulbosum*	Bulbous barley	E + Af(med)
Poaceae	*Panicum maximum*	Guinea grass	Af
Poaceae	*Phragmites*	Reeds	G
Poaceae	*Saccharum*	Sugar cane	As
Poaceae		Grasses	G
Geographical key			
Global	G		
Asia	As		
America	Am		
Africa	Af		
Europe	E		
Oceania (including Australasia)	O		
Mediterranean regions	(med)		
Temperate regions	(temp)		
Tropical regions	(trop)		
Arid regions	(arid)		
Compass points	(N, S, E, W)		

Among our close relatives, the chimpanzees, that extensive ecological intelligence is well attested in comparisons between the plant diets of chimpanzee populations in the Gombe and Mahale reserves. The facility for learning and species recognition is evidently extensive, and has allowed adaptation within a single species to the floristic differences between woodlands in a single region of Africa (Nishida et al., 1983; Malensky et al., 1994). That facility would be tested to the extreme by a movement out of Africa through European latitudes; the sheer scale of the taxonomic shift among plants is considerable, and probably beyond the cognitive flexibility of most great apes, which nonetheless exhibit, to varying degrees, an extensive ecological intelligence in relation to more regionally constrained environments.

"*Timely dextrous unpacking*" I use to refer to the use of prehensile digits, jaws, and sometimes a piece of wood or stone held with the digits to unpeel, remove spines, hard lignin coats, or layers of soil in order to reach a food source. The "timely" element is really an extension of ecological intelligence, and relates to an awareness that at certain stages in the growth cycle, a plant food source is less woody or toxic, and generally more edible. Thus, poisonous fruits can be culled when slightly unripe, and hardened stems "unpacked" when soft and immature. Timely dextrous unpacking becomes an important facility as soft, edible parts are more and more seeds enclosed within hardened shells, stem pith within woody coats, and tubers buried underground. It is also a facility which we share with a number of great apes, including the element of "tool"-use. Chimpanzees have been observed cracking nuts with hard wood anvils and stone hammers, the latter being a requirement to break open the hard shells of such taxa as *Panda oleosa* (McGrew, 1992; Boesch and Boesch-Acherman, 2000; Mercader et al., 2002).

Moving North

In equatorial latitudes, great apes spend a great deal of time gathering soft plant tissues, leaves, blossoms, fruits, and pith, as we might presume early hominins did. As *Homo* extended its northerly range, much of that plant resource became heavily restricted as a consequence of three factors. First, the soft tissue production became increasingly seasonal, available for just part of the year. Second, the total biomass productivity progressively fell. Third, and as an evolutionary consequence of these first two factors, whatever soft tissue was seasonally available, particularly in storage organs upon which the plant itself depended for survival, was increasingly well defended from predators such as ourselves, by toxins, spines and other barriers to digestion. Looking to the plant diet of our closest living relatives in equatorial Africa, and reflecting upon what of a great ape natural history might be transferable in such a northerly expansion is clearly a speculative exercise, but there are some clear candidates.

Monocot Stems

Chimpanzees and bonobos consume a variety of monocot stems, unpacking the pith from *Dioscorea schimperana* in the yam family, and *Panicum maximum* in the grass family, for example (Wrangham, 1975). At Wamba, the hard outer husks are stripped off the slender sprouts of at least ten species of Marantaceae, revealing the edible white leaf roll within (Kano, 1992). Like the yams and grasses, the Marantaceae also provide underground stems as a human food source, notably *Maranta arundinacea*, or arrowroot. Indeed, edible stems that have at some stage served as human food in different parts of the world are widely distributed across the 50 or so families of monocotyledous plants. What part of the actual part of the stem varies between species, seasons, and cultural groups? They range from the tips of above ground shoots (e.g., bamboo, asparagus, ensete), the sugar-rich rising sap (e.g., sugar cane, palm, reeds), pith or stem pulp (e.g., sago, ensete, leeks, and other *Alliums*), tuberous corms at ground level (e.g., taro, cocoyam, tanier, water plantain), underground stems or 'rhizomes' (arrowroots, rushes, spidergrass) and swollen into underground bulbils and tubers (e.g., yams, rush nuts, bulbous barley).

Monocots have a number of characteristic features which distinguish them from the dicots (Table 12.3) that make up the remaining flowering plant families. One critical feature relates to the thickening and strengthening of the stem. In dicots this is achieved through laying down considerable quantities of lignin to produce a woody stem to which hominin dentition is poorly suited. Monocot stems are structured differently, with a far more restricted deployment of lignin. As a consequence, the often softer pith has more potential as a hominin food source. A second feature is the parallel arrangement of veins upon the leaf. Humans, like other great apes, have repeatedly discovered that if a parallel veined leaf is followed back to the stem, and the stem followed inwards or downwards away from the toughest parts, there is a good chance that some accessible energy food will be found, and a relatively low chance that serious toxins will be encountered. In other words, the enclosed or buried food source carries a clear visual "flag" above ground.

If we imagine an expansion northwards from equatorial Africa, while the actual species encountered repeatedly changed, parallel-veined leaves could still be followed inwards and downwards to find energy-rich food. As seasonality increased and biological productivity decreased, the emphasis grew upon the downward direction of search, under the surface of water and beneath the soil surface, for stems swollen into storage organs, and these buried monocot stems, in particular the many species of yam, have remained central to human diet in low latitudes. Another interesting group of submerged monocot stems are the starchy rhizomes of *Cyperus papyrus*, which some have argued could explain the C4 plant isotopic signal found in the skeletal remains of some australopithecines (Sponheimer and Lee-Thorpe, 1999; Lee-Thorpe et al., 2000; Sponheimer and Lee-Thorpe, 2003; Van der Merwe et al., 2003).

Many of the earliest records of Old World food plants are of monocot stems, often submerged or underground. These include the sedge and rush tubers from Wadi Kubbaniya at 19 kya (Hillman, 1989), and the starch granules and

TABLE 12.3. Summary of dicot taxa mentioned in the text.

Family	Genus + sp.	Common name	Principal range
Anacardiaceae	*Pistacia*	Pistache, Pistacio	E + Af(med) + As(SW)
Apiaceae		Umbells	G
Asteracae		Composites	G
Boraginaceae	*Anchusa* spp.		E + Af + As
Boraginaceae	*Alkanna tinctoria*	Alkanet	E + Af(med)
Boraginaceae	*Lithospermum officinale*	Gromwell	E
Cannabaceae	*Celtis* spp.	Hackberry	G
Convolvulaceae	*Ipomoea batatas*	Sweet potato	Am(trop)
Ericaceae		Heather family	G(temp)
Euphorbiaceae	*Manihot esculenta*	Manioc	Am(S)
Fabaceae	*Acacia*	Thorntree, wattle	G(trop + arid)
Fabaceae	*Lens*	Lentil	As(SW)
Fabaceae	*Pisum*	Pea	As(SW) + Af(NE)
Fabaceae	*Vicia ervilia*	Bitter vetch	E + Af(med) + As(SW)
Fabaceae	*Vicia/ Lathyrus* spp.	Vetches	G
Fagaceae	*Quercus* spp.	Acorns	G
Moraceae	*Ficus*	Fig	G
Oleaceae	*Olea*	Olive	E + As + O
Pandaceae	*Panda oleosa*		Af
Rhamnaceae	*Ziziphus* spp.	Jujube	G
Rosaceae	*Crataegus* spp.	Haw	G(N)
Rosaceae	*Malus* spp.	Wild apple	G
Rosaceae	*Mespilus germanica*	Medlar	As(SW)
Rosaceae	*Prunus dulcis*	Almonds	As(SW)
Rosaceae	*Prunus padus*	Bird cherry	E(N) + As(N)
Rosaceae	*Prunus spinosa*	Sloe	E + As(W) + Af(N)
Rosaceae	*Pyrus* spp.	Wild pear	E + As(W)
Rosaceae	*Rosa* spp.	Rosehip	G
Rosaceae	*Rubus caesius*	Dewberry	G(temp)
Rosaceae	*Rubus fruticosus*	Blackberry	G
Rosaceae	*Rubus idaeus*	Raspberry	G(temp + N)

Geographical key	
Global	G
Asia	As
America	Am
Africa	Af
Europe	E
Oceania (including Australasia)	O
Mediterranean regions	(med)
Temperate regions	(temp)
Tropical regions	(trop)
Arid regions	(arid)
Compass points	(N, S, E, W)

phytocrystals of aroids, yams, and sago from various South East Asian sites between 20–30 kya (Loy et al., 1992; Barton, 2005; Paz, 2005).

Legume Pods

Among the 184 plant foods that Wrangham (1975) recorded from the chimpanzee diet at the Tanzanian reserve of Gombe, only 14 taxa are consumed as seeds. Ten of those 14 seed plants are legumes, pointing to another easily transferable element of plant food knowledge. While leguminous plants may vary from small herbs to substantial trees, their fruiting pods have much structurally in common, and remain visually recognisable from one leguminous taxon to another. Legumes as a family also have a vast geographical and ecological range. It is hard to imagine anywhere along the route of hominin dispersal where at least one legume or another might not be found. Even the very driest landscapes may sustain some species of *Acacia*. Ripe legume seeds are often hard, and may be toxic, but nevertheless in the unripe state may often be safely and effectively ingested.

The richest food plant assemblage of Mousterian date, at Kebara Cave in Israel, is dominated by a legume seeds of a range of species, the form of some of which might suggest collection while underripe (Lev et al., 2005). Towards the end of the Paleolithic, legume finds are scattered across Europe, for example the pea and bitter vetch at Öküzini, Turkey; lentil at Konispol cave, Albania; and vetches and other legumes at

Santa Maira, Alacant, Spain (Baales et al., 2002). Another rich example of pre-agricultural legume foragers comes from Hallam Cemi in Turkish Anatolia (Savard et al., 2006).

New Environments and New Plant Strategies

Monocot stems and legume pods may have provided a significant mass of plant foods during the expansion into the lower latitudes of Eurasia, where a vast array of yams and legumes have emerged in the modern human food web as domesticated plants. Moving further northwards still, these kinds of resources diminish significantly, both in diversity and in biomass availability. The quest for plant foods will have presented an increasing challenge.

The cooler northern vegetation stands would have been characterised by a range of open biomes including "arctic steppe" (Cwynar and Ritchie, 1980; Zazula et al., 2003) and closed vegetations characterised by woody dicots and coniferous trees. Woody dicots are reasonably rich in edible nuts, kernels, and fruits and, in certain families, edible roots and tubers. As mentioned above, the lower the biological productivity, the greater evolutionary pressure to protect seeds and storage organs from predation, and so it is generally true that, especially as they move northwards, human feeders are presented with a more complex "landscape of toxicity" by dicots than tends to be the case with monocots, particularly in the context of the seeds and tubers upon which the plant itself relies to cross the non-growing season. The kind of transferable ecological knowledge that allowed feeders to move from one monocot to another in more southerly biomes is not directly transferable to the dicots in northerly biomes. Experimenting with new plants in a new landscape is chemically less predictable in these taxa. An interesting ethnographic case has been recorded among the Inuit of Alaska and Siberia, who hunt for roots that mice collect and assemble in their nests. A detailed botanical knowledge is required to use this resource as the mice also store roots that are toxic to humans, but not to mice (Jones, 1983; cited in Owen, 2002).

I would conjecture that moving around these northern landscapes required a different style of employment of ecological intelligence, one that perhaps involved a taxonomy of plant types based on multiple features. Such a taxonomy would need to be both hierarchical, recognizing particular "plant families," such as the Asteraceae and Apiaceae as being broadly useful, and also specific, distinguishing several palatable species from their toxic close relatives. I would also conjecture that an increased encounter with plant toxicity also facilitated an exploration of plants for uses other than food, for example as medicines, mind-altering substances, and hunting toxins. Whatever the changes that accompanied the necessary adaptation to dicot use in the more northerly latitudes of Eurasia, that dicot adaptation sustained as the northerly expansion continued southwards through America. While the lower latitude tubers of the Old World are monocots, their most important New World counterparts, manioc, sweet potato, and potato, are dicots, incidentally each from a family of plants that includes a number of toxic and psychoactive relatives.

The most conspicuous markers of dicot usage in the archaeobotanical record are the kernels of tree fruits. In the existing Old World archaeobotanical record, the most notable of these is a trio of acorns, pistacia nuts and almonds, which recurs from one of the very earliest archaeobotanical records in Israel. All three are found at the Acheulian site of Gesher Benot Ya'aquov (Goren-Inbar et al., 2002). Acorns and pistacia are found again at the Mousterian site of Kebara Cave (Lev et al., 2005), and all three have been recovered from Upper Paleolithic Ohalo II (Weiss et al., 2004, 2005). Towards the end of the Paleolithic, these three are found in other regions around the Mediterranean, for example pistacia at Abu Hureyra (Hillman et al., 1989), acorn, pistacia, and almond at Öküzini, Turkey, and acorn and pistacia together at Öküzini, Turkey, Konispol Cave, Albania, and Santa Maira, Alacant, Spain (Hansen, 1999; Martinoli, 2004; Aura et al., 2005).

Dicots also supply a series of key subsidiary nutrients, the most familiar of which is vitamin C, present in a range of fresh plant tissue, and particularly fresh leaves and fruit. Although only seasonally available, a small number of plant families have fruits that are easily recognizable, and one or other species of two families in particular, the Rosaceae and the Ericaceae, could have been consumed by human migrants all the way to the Arctic Circle. While macro remains of Ericaceae (especially *Vaccinium*) are recurrent features of North American sites (e.g., Bernstein, 2002; Deal and Butt, 2002; Lepofsky, 2002) this family is not well attested in Old World Paleolithic macrofossils, reflecting the paucity of archaeobotanical data in northern regions. However, there are several archaeobotanical records of Old World Paleolithic use of Rosaceae fruits. Haw fruits have been recovered from Ohalo II in Israel and Konispol cave Albania, wild pear and sloe at Abu Hureyra in Syria, and wild pear at Öküzini, Turkey (Hillman et al., 1989; Hansen, 1999; Martinoli, 2004; Weiss et al., 2004). A series of German sites have yielded bird cherry, blackberry, dewberry, and raspberry, and Santa Maira, Alacant in Spain has produced evidence of sloe, rosehip, sorbus, and wild apple (Baales et al., 2002; Aura et al., 2005). In addition to this wide range of Paleolithic Rosaceae fruits, hackberry has been recovered at Dederiyeh Cave and Abu Hureyra in Syria, grape at Santa Maira, grape and olive from Ohalo II and Öküzini, at the latter site with fig. Medlar has been found at Abu Hureyra, ziziphus at Ohalo II, and juniper at Santa Maira. Paleolithic communities encountered a very wide range of fruits, but only during certain seasons (Hillman et al., 1989; Martinoli, 2004; Weiss et al., 2004; Aura et al., 2005).

Another important group of nutrients is the essential fatty acid group (EFAs), critical to effective brain growth (Mostofsky et al., 2001). If nursing hominin mothers had fairly open access to the kill (and it is a matter of conjecture whether or not they did) they could well have acquired enough EFAs for the brain growth of their young by consuming generous quantities of meat and fat. It is also worth considering whether this was at least supplemented by plant-based EFAs. The Paleolithic archaeobotanical record has a number of candidates, particularly members of the Boraginaceae family whose seeds are rich in EFAs. Franchthi cave yielded a range of such seeds, of anchusa, alkanet, and gromwell (Hansen, 1991), and Boraginaceae also recur at a number of Eastern Mediterranean sites. Most intriguing are the Mousterian Boraginaceae seeds from the Mousterian levels of Douara Cave, which appeared to be smashed open (Akazawa, 1987; Hillman G.C., personal communication).

The Boreal Zone

Boreal woodland is perhaps the most challenging of northerly biomes for hominins and many other mammals. It is nonetheless evident from much pollen and charcoal data that many Middle and Upper Paleolithic communities would have been most familiar with stands of coniferous trees (Van Andel and Davies, 2003). While a certain amount of digestible tissue can be derived from the inner bark and the kernels of some of its trees, the vast bulk of biomass within coniferous woodland offers far less to the human grazer than other woodland types. True, it supports a wealth of fungi, but these are more significant as sources of flavour, toxins, and mind-altering substances rather than calorific bulk. The best place for a human plant gatherer in boreal woodland is in bands of woodland edge vegetation, particularly as they approached the water's edge, where the age old resource of monocot stems may be found, alongside a range of aquatic tubers. While the Paleolithic archaeobotanical record for the boreal zone itself is currently too thin to elucidate, aquatic tubers do feature in more southerly sites. Water lily is recorded from both Acheulian Gesher Benot Ya'aquov, and Upper Paleolithic Wadi Kubbaniya (Hillman, 1989; Goren-Inbar et al., 2002). The tubers of *Trapa natans* (one of the taxa referred to as water chestnut) also appear at Gesher Benot Ya'aquov, and moving forward in time to the Mesolithic, clearly become a recurrent resource further north in Europe, as do water lily tubers, found, for example at well-studied Mesolithic sites in Denmark (Kubiak-Martens, 1999, 2002).

Long Processing Sequences

One of the key issues introduced at the beginning of this paper was the relationship between long processing sequences and human cognitive evolution. Two discussions of cognition are relevant in this respect. One is the idea of a mental "blueprint" that allows a multiple step processing sequence, such as is seen in Levallois stone tool technology (Mellars, 1996; Mellars and Gibson, 1996). Most of the plant acquisition strategies alluded to above entail the ancient facility of "timely dextrous unpacking" and to some extent this can be compared with the mechanical "unpacking" of nodules to create Lower Paleolithic stone tools. The augmentation of timely dextrous unpacking with multiple step processing would have opened up enormous possibilities with toxic food sources. Such procession also had bearing upon two groups of plants that may have been of significance for much of the Paleolithic – legumes and monocot stems – but upon a novel usage of these plants. Multiple step processing would open up access to what has today become the major food source of the human species, their hard, dry seeds.

The components of multiple step processing may include pounding, grinding, roasting, wetting, and boiling, fermenting, and germinating. They may involve the construction of an oven, a leaching device, sieve, or container, and also involve applying the right treatment in the right order. A number of processes gathered under the term "cooking" involve multiple step procedures; however, they fall along a continuum, and not all forms of "external digestion" (Aiello and Wheeler, 1995) are multi-step. The exploitation of natural fermentation and rotting could simply be seen as just one facet of an extensive ecological intelligence, and some forms of roasting may entail no more than a one-step exposure to the campfire. The earliest instances of "cooked food" may indeed relate to foraging in the aftermath of natural fires. Such targeted exploitation of natural occurrences lie at one end of a spectrum of "external digestion" that reaches through to the kind of intricate culinary aesthetics and "cuisine" that is arguably found at the 12,000 year old "kitchen" at Jerf-el Ahmar (Willcox, 2002). Thus there may not necessarily be a contradiction between an argument for a history of external digestion that spans the full history of *Homo* (Wrangham et al., 1999), and one that envisages elements of cooking and food preparation relating to subsequent episodes of the cognitive evolution of the genus. A further feature of multi-step plant processing with implications for cognition is the radical changes of state in which a living organism (plant source) can be transformed into a powder, a paste or a liquid, transformations that imply what Mithen (1996) describes as a fluid interaction between ecological and technical domains within a generalized multipurpose intelligence.

The clearest archaeobotanical corollaries of multi-step plant processing are acorns and grass seeds, neither of which are especially digestible without some form of pre-treatment. At the 23,000 year old site of Ohalo II in Israel, there is clear evidence of modern human exploitation of acorns and grass seeds (Piperno et al., 2004; Weiss et al., 2004, 2005). While it would be valuable to have earlier evidence of plant food from anatomically modern humans, we can infer, also by reference to both fish consumption (Brooks, 2007) and to lithic technologies, that long processing sequences have been characteristic of the plant food quest as well.

The Middle Paleolithic is less well attested. There is much evidence of external digestion, primarily in the form of built hearths at Mousterian sites, some with associated cracked and burnt food (Gale and Carruthers, 2000), but the evidence of grass seed eating is slim, and rests upon one sample of phytoliths, derived from grass seed heads, within Mousterian hearth levels at Amud Cave in Israel (Madella et al., 2002).

Conclusions

The spread of Middle and Upper Paleolithic communities into northerly regions presented considerable challenges in relation to the gathering of plant food. These challenges related to the drop in biological productivity, the increasing seasonality of biomes, and the radical taxonomic shifts between ecosystems. A great deal, however, could be achieved by drawing on the combination of features widespread among great apes, bringing a facility for timely dextrous unpacking together with an extensive ecological intelligence. That knowledge would need to be transferable between species, but at a quite basic level of transferability, for example, from one parallel-veined leaf to another, and from one legume pod to another. Such a broad level transferability of ecological knowledge would enable a significant capacity for plant food acquisition across the lower latitudes of Eurasia.

A movement further northwards would have reduced the available biomass, especially of monocot stems. I have speculated that a sustainable expansion may have depended on two developments in dietary behavior. The first was a shift in emphasis from stems to ripened seeds, in which both legumes and monocots, particularly grasses, continued to play an important role. The second was a shift in emphasis from monocots, to the more lignified, and sometimes toxic, landscape of dicotyledonous seeds, fruits, roots and tubers. Consumption of seeds and kernels regularly involved some multi-step sequence of processing. Consumption of geographically variant species of dicots may have required more than a broad transferability of ecological knowledge. I have suggested some multi-factor categorization of recognized plant families, rather than just the direct recognition of a pod or a parallel veined leaf. Both these developments thus have implications for cognitive evolution.

When humans encountered the Boreal Zone and extensive stands of coniferous woodland, then the quest for plant food was at its most challenging, and it is conceivable that the expansion of humans into this zone was constrained less by the availability of meat, and more by the combined availability, through the full cycle of seasons, of fats and plant food.

This paper has developed an argument that is of necessity fairly speculative. There is much need for an enhancement of the archaeobotanical database at various latitudes, in the context of understanding the broad shape of the food web through isotopic analysis, and the likely fat availability through the course of the year through zooarchaeological analyses. The strategy for acquiring the first component of this database has to be the application of flotation methods on a scale that has become customary on Neolithic and later sites, but is still rare on Paleolithic excavations. Even quite sizeable fragments of charred seed, root, and tuber remain effectively invisible during the process of manual excavation, and only come to light when flotation fractions are scanned under the microscope. There are furthermore certain styles of trowelling and sediment cleaning, widely practiced in Paleolithic fieldwork, which are highly destructive of plant evidence contained within the sediment. Catering for the collection of macrofossil evidence entails the incorporation of flotation and sieving as a sustained and integral element of the excavation strategy.

Such programs have occasionally been put in place, notably at Ohalo II, and more recently at sites of the Moravian Gate in the Czech Republic. These two Upper Paleolithic projects reveal two complementary aspects of the novel plant use that may have played a key role in enabling a northerly expansion, Ohalo II elucidating the use of hard and toxic seeds, and the Moravian Gate sites the use of dicot tubers and roots. However, these are but two sites, and we shall need to build up a much richer database of Paleolithic archaeobotany before we can understand the roles of plants in evolving Paleolithic diets.

References

Aiello, L.C., and Wheeler, P., 1995. The expensive tissue hypothesis. Current Anthropology 34, 184–193.

Akazawa, T., 1987. The ecology of the Middle Paleolithic occupation at Douara Cave, Syria. In: T. Akazawa and Y. Sakaguchi (Eds.), Paleolithic Site of Douara Cave and Palaeogeography of Palmyra Basin in Syria. Part 4 1984 Excavations. University of Tokyo Museum bulletin 29, pp. 155–166.

Aura, J.E., Carrión, Y., Estrelles, E., Peres Jorda G., 2005. Plant economy of hunter-gatherer groups at the end of the last Ice Age: plant macroremains from the cave of Santa Maira (Alacant, Spain) ca. 12000–9000 B.P. Vegetation History and Archaeobotany 14(4), 542–550.

Baales, M., Jöris, O., Street, M., Bittmann, F., Weninger, B., Wiethold, J., 2002. Impact of late glacial eruption of the Laacher See volcanao, central Rhineland, Germany. Quaternary Research 58, 273–288.

Barton, N., 2005. The case for rainforest foragers: the starch record at Niah Cave, Sarawak. Asian Perspectives 44, 1 56–72.

Beresford-Jones, D., 2006. Preliminary Appraisal of the Archaeobotanical Data from the 2005 Dolni Věstonice Field Season. Department of Archaeology, University of Cambridge, Cambridge, unpublished report.

Bernstein, D.J., 2002. Patterns in the prehistoric use of non-agrarian botanical resources in the Long Island and Block Island Sound region of eastern North America. In: S.L.R Mason and J.G. Hather (Eds.), Hunter-Gatherer Archaeobotany: Perspectives from the Northern Temperate Zone. Institute of Archaeology Press, London, pp. 74–83.

Bocherens, H., 2009. In: J.-J. Hublin and M.P. Richards (Eds.), The Evolution of Hominid Diets: Integrating Approaches to the Study of Paleolithic Subsistence. Springer, Dordrecht, 9.

Boesch, C., Boesch-Acherman, H., 2000. The Chimpanzees of Tai Forest. Oxford University Press, Oxford.

Brooks, A., 2009. In: J.-J. Hublin and M.P. Richards (Eds.), The Evolution of Hominid Diets: Integrating Approahces to the Study of Paleolithic Subsistence. Springer, Dordrecht.

Cordain, L., Miller, J.B., Eaton, S.B., Mann, N., Holt, S.H.A., Speth, J.D., 2000. Plant-animal subsistence ratios and macronutrient energy estimations in worldwide hunter-gatherer diets. American Journal of Clinical Nutrition 71/3, 682–692.

Cwynar, L.C., Ritchie, J.C., 1980. Arctic steppe-tundra: a Yukon perspective. Science 20/8, 1375–1377.

Deal, M., Butt, A., 2002. The great want: current research in Beothuk palaeoethnobotany. In: S.L.R Mason and J.G. Hather (Eds.), Hunter-Gatherer Archaeobotany: Perspectives from the Northern Temperate Zone. Institute of Archaeology Press, London, pp. 15–27.

Gale, R., Carruthers, W., 2000. Charcoal and charred seed remains from Middle Paleolithic levels at Goreham's and Vanguard Caves. In: C.B. Stringer, R.N.E. Barton and J.C. Finlayson (Eds.), Neanderthals on the Edge: Papers From a Conference Marking the 150th Anniversary of the Forbes' Quarry Discovery, Gibralter. Oxbow books, Oxford, pp. 207–210.

Goren-Inbar, N., Sharon, G., Kislev, M., 2002. Nuts, nut cracking, and pitted stones at Gesher Benot Ya'aquov, Israel. Proceedings of the National Academy of Science 99, 2455–2460.

Hansen, J.M., 1991. The Palaeoethnobotany of Franchthi Cave. Excavations at Franchthi Cave, Greece, 7. Indiana University Press, Bloomington.

Hansen, J.M., 1999. Konispol Cave plant remains. In: P.P. Betancourt, V. Karageorghis, R. Laffineur and W.D. Niemeier (Eds.), Meletemata. Studies in Aegean Archaeology Presented to Malcolm H. Wiener as He Enters his 65th year. Aegeum 20. Liège. pp. 333–338.

Hillman, G.C., 1989. Late Paleolithic plant foods from Wadi Kubbaniya in Upper Egypt: dietary diversity, infant weaning, and seasonality in a riverine environment. In: D.R. Harris and G.C. Hillman (Eds.), Foraging and Farming, the Evolution of Plant Exploitation. Unwin Hyman, London, pp. 207–239.

Hillman, G.C., 1996. Late Pleistocene changes in wild plant foods available to the hunter-gatherers of the northern Fertile Crescent: possible preludes to cereal cultivation. In: D. Harris (Ed.), The Origins and Spread of Agriculture and Pastoralism in Eurasia. University College Press, London, pp. 159–203.

Hillman, G.C., 2004. The rise and fall of human dietary diversity: an overview of archaeobotanical evidence from Western Eurasia, and of experiments with some of the key food plants. The International Society of Ethnobiology – Ninth International Congress – abstract.

Hillman, G.C., Colledge, S.M., Harris, D.R., 1989. Plant food economy during the Epipaleolithic Period at Tell Abu Hureyra, Syria: dietary diversity, seasonality and modes of exploitation. In: D.R. Harris and G.C. Hillman (Eds.), Foraging and Farming, the Evolution of Plant Exploitation. Unwin Hyman, London, pp. 240–268.

Jones, A., 1983. Nauriat Nigiñaqtuat: Plants that We Eat. Maniilaq Association, Kotzebue, Alaska.

Kano, T., 1992. The Last Ape: Pygmy Chimpanzee Behavior and Ecology. Stanford University Press, Palo Alto.

Kislev, M.E., Nadel, D., Carmi, I., 1992. Epipaleolithic (19,000) cereal and fruit diet at Ohalo II, Sea of Galilee, Israel. Review of Palaeobotany and Palynology 73, 161–166.

Kubiak-Martens, L., 1999. The plant food component of the diet at the late Mesolithic (Ertebølle) settlement at Tybrind Vig, Denmark. Vegetation History and Archaeobotany 8, 117–127.

Kubiak-Martens, L., 2002. New evidence for the use of root foods in pre-agrarian subsistence recovered from the late Mesolithic site at Halsskov, Denmark. Vegetation History and Archaeobotany 11, 23–31.

Lee-Thorpe, J.A., Thackeray, F.J., van der Merwe, N.J., 2000. The hunters and the hunted revisited. Journal of Human Evolution 39, 565–576.

Lepofsky, D., 2002. Plants and pithouses: archaeobotany and site formation processes at the Keatley Creek village site. In: S.L.R Mason and J.G. Hather (Eds.), Hunter-Gatherer Archaeobotany: Perspectives from the Northern Temperate Zone. Institute of Archaeology, London, pp. 62–73.

Lev, E., Kislev, M.E., Bar-Yosef, O., 2005. Mousterian Vegetal Food in Kebara Cave, Mt Carmel. Journal of Archaeological Science 32, 475–484.

Loy, T.H., Spriggs, M., Wickler, S., 1992. Direct evidence for the human use of plants 28,000 years ago: starch residues on stone artifacts from the northern Solomons. Antiquity 66, 898–912.

Madella, M., Jones, M.K., Goldberg, P., Goren, Y., Hovers, E., 2002. The exploitation of plant resources by Neandertals in Amud Cave (Israel): the evidence from phytolith studies. Journal of Archaeological Science 29, 703–719.

Malensky, R.K., Kuroda, S., Vineberg, E.O., Wrangham, R.W., 1994. The significance of terrestrial herbaceous foods for bonobos, chimpanzees and gorillas. In: R.W. Wrangham, W.C. McGrew, F.B. de Waal and P.G. Heltne (Eds.), Chimpanzee Cultures. Harvard University Press, Cambridge, pp. 59–75.

Martinoli, D., 2002. Les macrorestes botaniques de la grotte d'Öküzini. In: I. Yalçinkaya, M. Otte, J. Kozlowski and O. Bar-Yosef (Eds.), La grotte d'Öküzini: Evolution du Paléolithique final du sud-ouest de l'Anatolie, Etudes et Recherches Archéologiques de l'Université de Liège ERAUL 96, 91–94.

Martinoli, D., 2004. Food plant use, temporal changes and site seasonality at EpiPaleolithic Öküzini and Karain B Caves, Southwest Anatolia, Turkey. Paléorient 30/2, 61–80.

Martinoli, D., Jacomet, S., 2004. Identifying endocarp remains and exploring their use at EpiPaleolithic Öküzini in Southwest Anatolia, Turkey. Vegetation History and Archaeobotany 13(1), 45–54.

Mason, S.L.R., Hather, J.G. (Eds.), 2002. Hunter-Gatherer Archaeobotany: Perspectives from the Northern Temperate Zone. Institute of Archaeology, London.

Mason, S.L.R., Hather, J.G., Hillman, G.C., 1994. Preliminary investigation of the plant macro-remains from Dolní Věstonice II, and its implications for the role of plant foods in Paleolithic and Mesolithic Europe. Antiquity 68, 48–57.

McGrew, W., 1992. Chimpanzee Material Culture. Cambridge University Press, Cambridge.

Mellars, P.A., 1996. The Neanderthal Legacy: An Archaeological Perspective from Western Europe. Princeton University Press, Princeton.

Mellars, P.A., Gibson, K., 1996. Modelling the Early Human Mind: Archaeological and Psychological Perspectives on the Evolution of Human Intelligence. McDonald Institute for Archaeological Research Monographs, Cambridge.

Mercader, J., Panger, M., Boesch, C., 2002. Excavation of a chimpanzee stone tool site in the African rainforest. Science 296 (May 24), 1452–1455.

Mithen, S., 1996. The Prehistory of the Mind: The Cognitive Origins of Art and Science. Thames & Hudson, London.

Moore, A.M.T., Hillman, G.C., Legge, A.J., 2000. Village on the Euphrates. Oxford University Press, New York.

Mostofsky, D. I., Yehuda, S., Salem, N., 2001. Fatty Acids: Physiological and Behavioral Functions. Humana Press, Totowa.

Nishida, T., Wrangham, R.W., Goodall, J., Uchara, S., 1983. Local differences in plant feeding habits of chimpanzees between Mahale Mountains and Gombe National Park, Tanzania. Journal of Human Evolution 12, 467–489.

Owen, L.R., 2002. Reed tents and straw baskets? Plant resources during the Magdalenian of Southwest Germany. In: S.L.R Mason and J.G. Hather (Eds.), Hunter-Gatherer Archaeobotany: Perspectives from the Northern Temperate Zone. Institute of Archaeology, London, pp. 156–174.

Paz, V., 2005. Rock Shelters, Caves, and Archaeobotany in Island Southeast Asia. Asian Perspectives 44/1, 107–118.

Peacock, S.L., 2002. Perusing the pits: the evidence for prehistoric root resource processing on the Canadian plateau. In: S.L.R Mason and J.G. Hather (Eds.), Hunter-Gatherer Archaeobotany: Perspectives from the Northern Temperate Zone. Institute of Archaeology, London, pp. 44–61.

Perry, D., 2002. Preliminary results of an archaeobotanical analysis of Mesolithic sites in the Veenkoloniën province of Groningen, Netherlands. In: S.L.R Mason and J.G. Hather (Eds.), Hunter-Gatherer Archaeobotany: Perspectives from the Northern Temperate Zone. Institute of Archaeology, London, pp. 108–116.

Piperno, D.R., Weiss, E., Holst, I., Nadel, D., 2004. Processing of wild cereal grains in the Upper Paleolithic revealed by starch grain analysis. Nature 430, 670–673.

Richards, M.P., 2009. In: J.-J. Hublin and Richards, M.P. (Eds.), The Evolution of Hominid Diets: Integrating Approaches to the Study of Paleolithic Subsistence. Springer, Dordrecht.

Robinson, D.E., Harild, J.A., 2002. Archaeobotany of an early Ertebølle (Late Mesolithic) site at Halsskov, Zealand, Denmark. In: S.L.R Mason and J.G. Hather (Eds.), Hunter-Gatherer Archaeobotany: Perspectives from the Northern Temperate Zone. Institute of Archaeology, London, pp. 84–95.

Savard, M., Nesbit, M., Jones, M.K., 2006. The role of wild grasses in subsistence and sedentism: new evidence from the northern Fertile Crescent. World Archaeology 38/2, 179–196.

Sponheimer, M., Lee-Thorp, J.A., 1999. Reconstructing the diet of the early hominid *Australopithecus africanus* using 13C/12C analysis. Science 283, 368–370.

Sponheimer, M., Lee-Thorp, J.A., 2003. Differential resource utilization by extant great apes and Australopithecines: towards solving the C4 conundrum. Comparative Biochemistry and Physiology A 136, 27–34.

Van Andel, T., Davies, W. (Eds.), 2003. Neanderthals and Modern Humans in the European Landscape Duration the Last Glaciation. McDonald Institute Monographs, Cambridge.

Van der Merwe, N.J., Thackeray, J.F., Lee-Thorp, J.A., Luyt, J., 2003. The carbon isotope ecology and diet of Australopithecus africanus at Sterkfontein, South Africa. Journal of Human Evolution 44, 581–597.

Van Zeist, W., Bakker-Heeres, J.A.A., 1984. Archaeobotanical studies in the Levant 3: Late Paleolithic Mureybit. Palaeohistoria 26, 171–199.

Weiss, E., Wetterstrom, W., Nadel, D., Bar-Yosef, O., 2004. The broad spectrum revisited: evidence from plant remains. Proceedings of the National Academy of Science 101(26), 9551–9555.

Weiss, E., Kislev, M.E., Simchoni, O., Nadel, D., 2005. Small-grained wild grasses as staple food at the 23,000 year-old site of Ohalo II, Israel. Economic Botany 588, 125–134.

Willcox, G., 2002. Charred plant remains from a 10th millennium B.P. kitchen at Jerf el Ahmar. Vegetation History and Archaeobotany 11, 55–60.

Wolgemuth, E., 2002. Late prehistoric plant resource intensification in the eastern San Francisco Bay area: plant remains from ALA-42 and ALA-555, Pleasanton, California. In: S.L.R Mason and J.G. Hather (Eds.), Hunter-Gatherer Archaeobotany: Perspectives from the Northern Temperate Zone. Institute of Archaeology Press, London, pp. 28–43.

Wrangham, R.W., 1975. The Behavioral Ecology of Chimpanzees in Gombe National Park, Tanzania. Ph.D. Dissertation, University of Cambridge, Cambridge.

Wrangham, R.W., Jones, J.H., Laden, G., Pilbeam, D., Conklin-Brittain, N.L., 1999. The raw and the stolen: cooking and the ecology of human origins. Current Anthropology 40, 567–594.

Zazula, G.D., Froese, D.G., Schweger, C.E., Mathewes, R.W., Beaudoin, A.B., Telka, A.M., Harington, C.R., Westgate, J.A., 2003. Ice-age steppe vegetation in east Beringia. Nature 423, 603.

13. Diet in Early Hominin Species: A Paleoenvironmental Perspective

Zeresenay Alemseged
California Academy of Sciences
55 Concourse Drive
San Francisco, CA, 94118 USA
zeray@calacademy.org

and

René Bobe
Department of Anthropology
The University of Georgia
250 A Baldwin Hall, Jackson St.
Athens, GA 30602, USA
renebobe@uga.edu

Keywords Pliocene • early hominin diet • Shungura-Omo

Abstract Unraveling the dietary adaptations of early hominin species has become a very important question in the paleobiological studies of extinct taxa. For long, dentognathic morphological studies have been used to infer diet. More recently, new systematic and quantitative approaches have been developed to explore this issue. In the present study, a paleoenvironmental approach is used to infer possible dietary adaptation of two hominin genera: *Paranthropus* and *Homo*. Our results show that both genera likely depended on all types of available resources in diverse types of habitats but differed on how they exploited fallback resources. *Homo* exploited resources that were found in more wooded conditions, probably meat, and *Paranthropus* fell back on those common in open environments such as abrasive and hard diet. We think that there is no compelling evidence for the claim that *Paranthropus* was more specialized than *Homo*, at least as far as habitat preference is concerned. Studies that posited that the cause for the demise of *Paranthropus* was that it was too specialized and was not able to compete with the more versatile hominin species of *Homo* warrant revision as it is clear from our study and that of others that the two species fell back on different resources during scarcity, and thus were not in direct competition when resources were inadequate.

J.-J. Hublin and M.P. Richards (eds.), *The Evolution of Hominin Diets: Integrating Approaches to the Study of Palaeolithic Subsistence,* 181–188.

Introduction

For an animal, diet means survival. This simple fact has important implications for our understanding of a species' paleobiology. Comprehending the dietary adaptation of a species entails having substantial insight into the selection forces and mechanisms that are responsible for shaping the morphology and behavior of that species. Specifically, questions pertaining to a range of paleobiological aspects of any extinct taxon cannot be fully understood without some knowledge about what the species ate. Nevertheless, reconstructing the diet of fossil species remains an elusive and sometimes speculative endeavor.

Diet can be understood in two main ways: in terms of food texture (brittle, soft, etc.) or nutrition (high vs. low quality). The amount and quality of data available on these two characteristics of diet for any extinct species is highly dependent on the quality of the fossil evidence, the methods used, and our ability to interpret the data using actualistic models. While doing this is difficult, it becomes even more problematic when the considered taxa are fossil hominins, particularly the earliest ones. The difficulties emanate from several facts: (a) fossil remains of early hominins are rare; (b) many approaches used to study dietary adaptation are destructive and thus cannot be applied to fossil specimens; (c) some methods are not applicable to earlier species where organic matter does not preserve; and (d) using modern analogues to understand diets in fossil species is problematic. Despite these limitations, considerable efforts and diverse approaches have been put to addressing dietary issues in the earliest species of our family. Below, after a brief historical background of the study of diet in early hominins, we summarize some of the commonly used methods and highlight the main results obtained.

Historical Background

It is fair to say that Raymond Dart was the first to write in a documented manner about diet in early hominins with his discovery of the Taung Child and subsequent naming of the species *Australopithecus africanus* (Dart, 1925, 1949). He imagined this species as a mighty hunter practicing an "osteodontokeratic culture" (Dart, 1949). Subsequently, however, the very detailed, and now classic, taphonomic studies by Brain showed that *Au. africanus* was the hunted rather than the hunter (Brain, 1981, 1993). In the 1950s, John Robinson developed what came to be known as the dietary hypothesis (Robinson, 1954, 1964). According to this idea, there were fundamentally two kinds of hominins in the Plio-Pleistocene. One was the "robust" australopithecine (called *Paranthropus*) specialized for herbivory, and the other was the "gracile" australopithecine, *Australopithecus*, with an omnivore/carnivore diet. The latter, with its moderately enlarged canines, evenly proportioned tooth row, and smaller masticatory muscles was contrasted to the more robust species from Kromdraai and Swartkrans. *Paranthropus robustus*, with its small anterior teeth, very large premolars and molars with thick enamel, and heavy cheek bones seemed to Robinson to represent a vegetarian. *Australopithecus* was likely to be an omnivore and more closely related to later humans.

Based on gross dental morphologies and inferred dietary adaptation, and departing from previous interpretations that recognized several species (Broom and Robinson, 1949, 1952), Robinson argued that the ape-men from South Africa could be put into only two genera, *Paranthropus* and *Australopithecus* each containing a single species. According to Robinson, the former became extinct while the latter evolved into *Homo*. This changed considerably our understanding of early hominin adaptation and taxonomy, and in the following years Robinson's ideas were embraced by many. In subsequent years, the recovery of fossil hominins increased rapidly, and almost every important discovery was published with an interpretation of the species' dietary adaptation derived from dental and bone morphology and inferred muscle functions. The dubbing of *Australopithecus boisei* with the name "Nutcracker" when Louis Leakey discovered OH-5 is among the many good examples of this (Leakey, 1959).

Today there is an appreciation that dietary reconstructions involve not just understanding what a species ate, but a fundamental aspect of adaptation that has far reaching implications for taxonomy, phylogeny, and behavior. In describing *Kenyanthropus platyops*, a new genus and species from the Lomekwi area in Northern Kenya, Leakey et al. (2001) wrote, "…a 3.5 Myr-old cranium, showing a unique combination of derived facial and primitive neurocranial features, is assigned to a new genus of hominin. These findings point to an early *diet-driven adaptive radiation*, and provide new insight on the association of hominin craniodental features, and have implications for our understanding of Plio-Pleistocene hominin phylogeny." More recently, commenting on the origin of the genus *Australopithecus* based on newly discovered fossils from the site of Asa Issie in the Middle Awash, Ethiopia, White et al. (2006) pointed out that "Whatever the geometry of early hominid phylogeny, diagnostic megadontia and related dentognathic morphology of *Australopithecus* herald its appearance at or before 4.2 Myr ago. Its masticatory apparatus appears better adapted to a more heavily chewed diet of tough and abrasive items than that of *Ardipithecus*. These phenotypic signals indicate an adaptive shift towards the exploitation of tougher and more abrasive food resources."

The studies mentioned above and conclusions drawn therein underscore the need for, and the importance of, understanding the dietary adaptation of a species, as it is used to address major issues in behavior, locomotion, phylogeny, and even speciation events. If this is the case then the critical question is how much do we know about the diet of early hominin species? And what kinds of tools do we have to address these questions? Here we summarize various approaches as they are applied to dietary reconstructions in hominin species.

Systematic Approaches to Diet

Over the past decades considerable efforts have been put toward the reconstruction of primate diet using systematic and quantifiable approaches in addition to the traditional studies of external dental and cranial gross morphology. One of these approaches attempts to infer and quantify muscle mass from gross bone morphology (e.g., Rak, 1978; Hylander, 1979; Picq, 1983; Rak, 1983; Sakka, 1984). Mandibular biomechanics indicate that the deep corpus and high ramus in *Au. afarensis* and *Au. africanus* reflect resistance to mastication stress, whereas *Au. anamensis* has a configuration that is intermediate between that of the apes and later hominins, showing the possible dietary shift discussed above (White et al., 2006). Tooth shape and size are also widely used for similar purposes (e.g., Luke and Lucas, 1983; Lucas et al., 1985). Small incisors relative to body size, as in gorillas, are thought to indicate a diet that requires little preparation with the anterior teeth. In early hominins there is a trend of increasing molar size in the probable lineage *Ardipithecus ramidus – Australopithecus anamensis – Australopithecus afarensis – Paranthropus*, most likely pointing to increasingly tough and abrasive diets. Enamel thickness is widely used to interpret early hominin diet but quantifying it is not as clear-cut as it is often described (Kay, 1981; Dumont, 1995). Still, enamel thickness has played an important role in interpreting diet and taxonomy of hominins (e.g., White et al., 1994). Most hominins are described as having thick enamel, but thin enameled taxa such as *Ardipithecus* raise questions regarding the validity of using enamel thickness as a strong taxonomic character (Senut et al., 2001). The "shearing quotient" approach, which measures the relative shear potential of molar teeth, was pioneered by Richard Kay to understand the diets of different primates (Kay, 1984; Meldrum and Kay, 1997). Various studies show that, in general, Pliocene hominins have lower shearing quotients than extant apes. This would indicate that, in general, hominins were feeding on hard and brittle foods rather than on tough and pliant items such as stems and meat. Microwear analysis was also successfully applied to some hominins (Grine, 1981; Kay and Grine, 1988). Folivorous species have long narrow scratches on their molar teeth and frugivores have more pits on those surfaces. Among the latter, hard object feeders have more pits than soft fruit eaters. Compared to *Paranthropus*, *Au. africanus* was found to have longer scratches and fewer pits, and like *Au. afarensis* it was likely feeding on fruits and some abrasive material. In addition, isotope studies showed that *Au. africanus* ate not only fruits and leaves, but also large quantities of carbon-13-enriched foods such as grasses and sedges, or animals that ate these plants, or both (Sponheimer and Lee-Thorp, 1999). The results suggest that early hominins regularly exploited relatively open environments such as woodlands or grasslands for food. They may also suggest that hominins consumed high-quality animal foods before the development of stone tools and the origin of the genus *Homo*.

Finally, a method known as the "dental topography" approach, developed by Peter Ungar, quantifies slopes and reliefs of the tooth and is shedding fresh light on the details of hominin diet (Ungar, 2004). This work shows that among primates, high-crowned teeth that interlock between the upper and lower jaws are adapted to shearing foods during folivory and insectivory. Frugivorous primates or hard-object feeding animals tend to have flatter teeth (without high crests) adapted for crushing or grinding. On this basis, examination of the occlusal topography of teeth provides information about the diets of primates, including hominins. Ungar applied his method to *Gorilla gorilla gorilla*, *Pan troglodytes troglodytes*, *Australopithecus afarensis* and *Homo* to explore dietary adaptations. Among his conclusions he notes that gorillas and chimps overlap considerably in their diet. However, during times of scarcity gorillas fall back on tougher, more fibrous foods (such as leaves and stems) than those eaten by chimpanzees. Differences in occlusal morphology between *P. t. troglodytes* and *G. g. gorilla* thus reflect differences in fallback resources rather than preferred foods. Ungar suggests that the differences between *Au. afarensis* and early *Homo* are similar in magnitude to those between gorillas and chimpanzees, and that *Au. afarensis* would have relied on hard, brittle fallback foods, whereas early *Homo* would have relied on resources that included meat (Ungar, 2004). Thus, these various approaches to the study of hominin (and primate) diets are yielding encouraging results, but we believe that dietary reconstructions are best understood within the paleoenvironmental context of the species in question.

The Paleoenvironmental Perspective

An understanding of a species' paleoenvironmental context allows an assessment of its reconstructed diet in relation to inferred resources in the landscape. Paleoenvironments can be reconstructed in different ways. Hominins are often assumed to have lived in the type of environment inferred from the most common taxa that make up the fossil assemblage, and sedimentological or other contextual evidence is often used to evaluate this assumption. Yet, if taphonomical issues are not addressed in detail, conclusions on the type of environment of the species can only be tentative and preliminary (Brain, 1981; Behrensmeyer, 2006). Also, such approaches usually result in static and site-specific reconstructions that do not (or cannot) take spatial and temporal variations into account. Using this approach, WoldeGabriel et al. (1994) suggested, for example, that in the Aramis site of Ethiopia *Ardipithecus ramidus* lived and died in a wooded setting, with an associated fauna characterized by abundant colobine monkeys and tragelaphine bovids but scarce equids and proboscideans. In addition to sedimentological and paleobotanical evidence, these faunal abundances indicate a prevalence of wooded habitats associated with the Aramis hominins. However, the paleoenvironment of the same species at As Duma in the Gona site is reconstructed as a moderate-rainfall woodland and woodland/grassland (Semaw et al., 2005). *Australopithecus anamensis* is suggested to have occupied diverse environments. At Allia Bay in the Turkana Basin of Kenya, *Au. anamensis* is associated with a mixed faunal assemblage sampling aquatic, forest, grassland, and bushland habitats. At Kanapoi, the fauna points to a mixed environment including dry, possibly open, wooded or bushland conditions with wide gallery forest in the vicinity (Leakey et al., 1995, 1998; Wynn, 2000), whereas, at the site of Asa Issie in the Middle Awash, *Au. anamensis* was a regular occupant of a wooded biome (White et al., 2006). At Hadar, the species *Au. afarensis* is in general found in habitats containing water and trees (Reed, 1997), and in the adjacent site of Dikika, the vertebrate fauna indicates the presence of a woodland-grassland landscape close to water and with frequent flooding (Alemseged et al., 2005; Wynn et al., 2006). Thus, it should be underlined that specimens of *Au. afarensis* have been encountered in diverse contexts.

The reconstructed environments of early hominins are often portrayed as mosaic habitats (Wynn, 2004). There is much evidence in support of this idea, and it is likely that populations of extinct hominins were exploiting diverse habitats with changing ecological conditions. Various studies have shown that hominin paleoenvironments can be studied within a dynamic context, with ecological variables changing temporally and spatially, and with hominin species adapting to complex and dynamic settings (e.g., Behrensmeyer et al., 1997; Potts, 1998; Bobe et al., 2002; Alemseged, 2003; Bonnefille et al., 2004; Wynn, 2004). A key question here is whether hominins were using some of the habitats available to them preferentially over other habitats. If we can determine hominin habitat preferences, then we can assess how environmental changes or fluctuations over time would have affected resource availability. This would allow us to study the role of environmental change as a potential motor of selection and evolution in early hominins. In this study we focus on two examples to approach the question of habitat preference in extinct species.

Material and Method

The Shungura Formation in the Lower Omo Valley of southwest Ethiopia is well-known for yielding thousands of mammalian fossils including hominins from temporally well-constrained fluvio-lacustrine sediments. Details of the geology, stratigraphic context, and paleontology on this formation have been widely published over the past 3 decades (e.g., Coppens et al., 1976; de Heinzelin and Haesaerts, 1983; Bobe and Eck, 2001). With over 50,000 fossil vertebrates deriving from about 1,500 chronologically well-constrained localities, the 800 m thick Shungura Formation spans from about 3.6–1.0 Ma and covers an area of ~180 km^2, offering a rare opportunity to examine vertical and lateral variation of hominin paleoenvironmental and paleoecological dynamics. Numerous paleoenvironmental studies of the Shungura Formation have been undertaken focusing on faunal evidence (e.g., Coppens, 1978), particularly with taxa that are environmentally sensitive and show habitat specificity (e.g., bovids), with special attention drawn to one or more of the four hominin species encountered there (e.g., Alemseged, 2003). As a result, the general environmental context of these hominins and associated taxa in the Shungura Formation is understood with some degree of confidence. Nonetheless, questions pertaining to habitat preference of different species of hominins have been addressed only to a very limited extent. In the present study we use data from the Shungura Formation to look at habitat preferences of two bovid species and two hominin species. We examine the geographic distribution of the bovid species *Menelikia lyrocera* and *Kobus sigmoidalis* (tribe Reduncini) across many localities to determine if these taxa had preferential occupation of particular areas within the Shungura paleolandscape. For the hominin genera *Paranthropus* and *Homo*, we test for significant associations with other, more common mammalian taxa whose environmental and dietary adaptations are reasonably well understood. Thus, if hominin and other mammalian taxa are frequently found together (significantly associated) across several localities, then we assume that these taxa may have had similar habitat preferences. This environmental context would then help us infer dietary adaptations of early hominins.

Bovids

Bovid tribes are known to be environmentally specialized with known habitat preferences (e.g., Vrba, 1980). For example, Alcelaphini and Antilopini include species that are cursorial and usually occupy open and dry environments, whereas Tragelaphini tend to occupy more wooded environments and Reduncini require grasslands close to permanent water (Kingdon, 1982a, b; Greenacre and Vrba, 1984). While this is generally true, it is clear that each species within a given bovid tribe has its own habitat preferences and ecological requirements. Field observations show that there can be subtle adaptational differences among species of the same tribe (Dorst and Dandelot, 1976).

Here we examine the geographic distribution of two fossil species of the tribe Reduncini, *Menelikia lyrocera* and *Kobus sigmoidalis*, across the Shungura paleolandscape. We consider 71 localities that yielded at least 100 fossil mammals each from both the French and American collections. The American collections derive primarily from the northern exposures of the Shungura Formation, whereas the French collections derive from more southern exposures. Table 13.1 presents the localities

TABLE 13.1. Abundance of two reduncine species, *Menelikia lyrocera* and *Kobus sigmoidalis* in 71 Shungura, Omo localities.

Locality	*Menelikia lyrocera*	*Kobus sigmoidalis*
L 338	0	0
L 1	0	0
L 398	0	0
L 7	14	37
L 627	4	4
L 32	3	0
L 28	4	1
L 2	0	0
L 628	0	0
L 345	0	0
L 238	0	1
L 16	5	32
L 25	4	1
L 51	0	0
L 626	1	2
L 668	0	0
L 67	8	35
L 146	0	2
L 26	2	3
L 9	0	1
L 27	0	0
L 52	3	7
L 5	0	0
L 193	0	0
L 465	2	1
L 82	0	0
L 17	1	0
L 78	0	0
L 47	0	0
Omo 47	39	0
Omo 18	3	1
Omo 75	50	27
Omo 323	81	0
Omo 29	8	6
Omo 28	0	0
Omo 75N	19	1
Omo 57	0	0

(continued)

TABLE 13.1. (continued)

Locality	*Menelikia lyrocera*	*Kobus sigmoidalis*
Omo 75S	2	2
Omo 76	0	0
OSH-1	0	0
Omo 40	0	0
Omo 74	2	0
Omo 50	9	0
Omo 310	66	0
Omo 71	0	0
Omo 84	1	0
Omo 100	0	0
Omo 141	0	4
Omo 217	0	0
Omo 58	0	2
Omo 195	0	3
Omo 78	0	0
Omo 81	0	0
Omo 75S	2	2
O3-2	0	0
Omo 75i	1	4
Omo 132	0	0
Omo 277	2	0
O 270	0	0
Omo 127	0	0
OVE 0	0	0
O1-B	6	6
Omo 41	0	0
Omo 227	2	0
Omo 3-0	0	0
Omo 269	0	0
Omo 48	1	0
Omo 103	0	4
Omo 136	0	2
Omo 254	1	0
Omo 33	0	1

considered and the number of fossils recovered belonging to each species.

It is apparent from Table 13.1 that *M. lyrocera* is more common in the French localities (in the south) whereas *K. sigmoidalis* is more common in the American localities (in the north). This indicates that the relative abundance of these two species indeed varies across localities, and that the two species were not evenly distributed in the north and south of the Shungura area. In order to test the significance of these differences we used chi square test among the two collections and the difference in relative abundance between them is significant ($X^2 = 159$, $P < 0.001$).

These two closely related species derive from the same time period and from similar sedimentary contexts, yet show significantly different abundances between the northern and southern exposures. Explanations for this pattern may be sought in taphonomic or collection biases, but it has been shown elsewhere that different collection methods could not have introduced these differences (Alemseged et al., 2007). There is no documented cause, nor is there any reason to think that either of the research teams would have been biased for or against any specific bovid species. Still, the two samples come from geographically proximate but distinct areas: the American team having worked in the north and the French in the south of the Shungura exposures. The abundance differences between these two bovids suggest that there may have been environmental differences between the northern and southern areas during this time. The work of Spencer (1997) on dietary adaptation of Plio-Pleistocene bovids sheds important light on this issue. After reconstructing diet in many living and extinct bovid species, she concluded that "... the feeding apparatus of *M. lyrocera* was unlike any modern African bovid ... whereas *K. sigmoidalis* was a grass feeder in edaphic grasslands." Thus the two species had different diets. We suggest, therefore, that the uneven distribution of these two species in the north and south of the Shungura areas was caused by preferences for a specific type of resources. Further work is required, in addition to that of Spencer, to tease out the type of diet that *Menelikia lyrocera* and *Kobus sigmoidalis* had in the Plio-Pleistocene. Ecomorphological approaches using postcranial elements might help explore this important question pertaining to dietary adaptation and habitat preferences.

Hominins

The relative abundance approach used above cannot be applied to hominins because they are rare elements in the fossil record. Our approach is therefore to use "proxy taxa" of known environmental or dietary adaptation to assess hominin taxonomic associations and habitat preferences. Analysis of broad associations of hominin species with other mammals in the Shungura Formation have been reported elsewhere (Geraads and Coppens, 1995; Alemseged, 2003; Bobe and Behrensmeyer, 2004). Using paleontological localities as the unit of analysis, Bobe and Behrensmeyer (2004) demonstrated that hominin taxa from the Shungura Formation showed significant associations with non-hominin species. Specimens of *Australopithecus* sp. were significantly associated with the suid species *Kolpochoerus limnetes*, presumed to have lived in wooded environments (Cooke, 1976). *Paranthropus aethiopicus* was significantly associated with *Papio* sp., assumed to have been a semi-terrestrial and omnivorous primate. This might imply that *Australopithecus* sp. (gracile) and *Paranthropus aethiopicus* occupied different habitats in the late Pliocene Omo valley. This conclusion is however, tempered by the fragmentary nature of the fossil record of these two species. After around 2.4 Ma, there is a better record of *Paranthropus* and *Homo* even though the fossil material is fragmentary and composed mainly of teeth. In the present study we consider the interval from 2.4 to 2.3 Ma which includes 55 localities with samples of at least ten specimens (Bobe and Behrensmeyer, 2004). For these localities we calculated the Ochiai association index to evaluate the strength of the association of *Homo* and *Paranthropus* with other mammal species, and the chi-square test to establish the significance of the associations. Ochiai index (OI) is a

TABLE 13.2. Association of *Homo* and *Paranthropus* with taxa that are encountered in the Shungura Omo localities.

	Homo		Paranthropus	
	Chi-square	Ochiai index	Chi-square	Ochiai index
Tragelaphus nakuae	2.53	0.39	2.48	0.55
Tragelaphus gaudryi	**7.51**	**0.48**	0.73	0.44
Aepyceros	1.88	0.37	0.12	0.50
Alcelaphini	4.77	0.43	0.03	0.36
Antilopini	0.00	0.00	0.00	0.11
Bovini	4.32	0.42	0.37	0.41
Kobus sigmoidalis	0.00	0.24	1.21	0.29
Menelikia	4.03	0.40	0.46	0.36
Notochoerus scotti	1.02	0.32	0.07	0.39
Kolpochoerus limnetes	2.25	0.37	1.06	0.32
Metridiochoerus andrewsi	4.61	0.41	4.78	0.50
Colobinae	2.24	0.33	0.25	0.32
Procolobus	0.00	0.00	2.26	0.00
Papio sp.	**9.61**	**0.51**	2.35	0.44
Theropithecus oswaldi	0.12	0.15	**6.97**	**0.47**
Theropithecus brumpti	0.21	0.14	0.32	0.27
Paranthropus	**9.61**	**0.51**		
Homo			**9.61**	**0.51**

FIG. 13.1. Graphic representation of the Ochiai association indices showing how much *Homo* and *Paranthropus* are associated with each other and with other taxa. These results show that *Homo* is primarily and significantly associated with *Paranthropus* and the omnivorous *Papio* sp. When this is not the case the second preferential association of *Homo* is with *Tragelaphus gaudryi*, an extinct kudu which probably lived in more wooded environment. Likewise, Paranthropus is mostly encountered in localities where the genus *Homo* is found. But when this is not the case *Paranthropus* is significantly associated with *Theropithecus oswaldi*, an extinct open environment dwelling, and hard object feeding species of cercopithecid.

measure of the strength of association between species pairs. It measures the ratio of the number of joint occurrences of two species to the total occurrences of each of the species separately at localities. This index ranges from 0 (indicating no association between the species) to 1 (indicating maximum association). Of special interest here are results obtained on the significant associations of the genus *Paranthropus* and *Homo* with taxa whose habitat preference and dietary adaptations are known (Bobe and Behrensmeyer, 2004). In the present paper however, we go further in interpreting the results in light of dietary reconstructions of the two genera that come from independent approaches.

These results show that *Homo* is primarily and significantly associated with *Paranthropus* and the *Papio* sp. (Table 13.2, Fig. 13.1). The second preferential association of *Homo* is with *Tragelaphus gaudryi*, an extinct kudu, which probably lived in wooded environment. *Paranthropus* is frequently encountered in localities where the genus *Homo* is found, but *Paranthropus* is also significantly associated with *Theropithecus oswaldi*, an extinct grazing cercopithecid that probably inhabited relatively open environments. These results show that both *Homo* and *Paranthropus*, like *Papio* sp., were primarily opportunistically exploiting all types of available resources in diverse types of habitats. The strong association of the two hominin genera with each other and with *Papio* sp. supports this claim. *Homo* and *Paranthropus* associations however, differ considerably when they are not found together. *Homo* is significantly but less strongly associated with the woodland living *Tragelaphus gaudryi* whereas *Paranthropus* associates with *Theropithecus oswaldi*. This pattern might indicate that both hominin genera were similar in depending on diverse types of habitat and diet but differed on how they exploited fallback resources. In the late Pliocene of the lower Omo Valley, *Homo* exploited fallback resources that were found in wooded conditions and *Paranthropus* fell back on those common in more open environments where *T. oswaldi* was frequently found. We suggest that early *Homo* fell back on animal resources that were found in the wooded habitats, whereas *Paranthropus* preferred hard and abrasive objects similar to those exploited by *T. oswaldi*. These conclusions are in accord with interpretation based the gross dental morphology of these taxa, with *Homo* teeth better adapted for shearing while megadont *Paranthropus* was more specialized for grinding. Our interpretations are consistent with those of Ungar (2004), who noted based on dental topography that *Homo* probably fell back on meat while *Paranthropus* did so on abrasive and hard foods. This study shows that dietary adaptation analyses based on associations can be used as a tool for insights into the patterns of distribution of resources that were available to our ancestors. Furthermore, it suggests that hominin species likely occupied similar environments (not necessarily the same geographic location), but at times used different resources, probably as fallback foods during lean periods. Habitat preferences have also been shown in non-hominin taxa such as *Menelikia lyrocera* and *Kobus sigmoidalis*, but the specific dietary preferences of these bovids and how they exploited fallback resources remain to be demonstrated.

The findings on both bovids and hominins, we think, are evidence for the presence of habitat preferences in early hominin species and other taxa. Our study shows that *Paranthropus* was not necessarily specialized for a given type of environment as has been suggested. Instead, *Paranthropus* and early *Homo* relied on different fallback resources during times of scarcity. We concur with Wood and Strait (2004) in that there is no compelling evidence to claim that *Paranthropus* was more specialized (stenotopic) than *Homo*, at least as far as habitat preference is concerned. It has been suggested in the past that one of the causes of the demise for *Paranthropus* was that it was too specialized and was not able to compete with the more versatile hominin species of *Homo*. We think that this important conclusion deserves re-evaluation as it is clear from our association study that the two species fell back on different resources during times of scarcity, thus perhaps were not in direct competition when resources were scarce.

References

Alemseged, Z., 2003. An integrated approach to taphonomy and faunal change in Shungura Formation (Ethiopia) and its implication for hominid evolution. Journal of Human Evolution 44, 461–478.

Alemseged, Z., Wynn, J.G., Kimbel, W.H., Reed, D., Geraads, D., Bobe, R., 2005. A new hominin from the Basal Member of the Hadar Formation, Dikika, Ethiopia, and its geological context. Journal of Human Evolution 49, 499–514.

Alemseged, Z., Bobe, R., Geraads, D., 2007. Comparability of fossil data and its significance for the interpretation of hominin environments: a case study in the lower Omo valley, Ethiopia. In: Bobe, R., Alemseged, Z., Behrensmeyer, A.K. (Eds.), Hominin Environments in the East African Pliocene: An Assessment of the Faunal Evidence. Springer, Dordrecht, pp. 159–181.

Behrensmeyer, A.K., 2006. Climate change and human evolution. Science 311, 476–478.

Behrensmeyer, A.K., Todd, N.E., Potts, R., Mcbrinn, G.E., 1997. Late Pliocene faunal turnover in the Turkana Basin, Kenya and Ethiopia. Science 278, 1589–1594.

Bobe, R., Behrensmeyer, A.K., Chapman, R.E., 2002. Faunal change, environmental variability and late Pliocene hominin evolution. Journal of Human Evolution 42(4), 475–497.

Bobe, R., Behrensmeyer, A.K., 2004. The expansion of grassland ecosystems in Africa in relation to mammalian evolution and the origin of the genus *Homo*. Palaeogeography, Palaeoclimatology, Palaeoecology 207, 399–420.

Bobe, R., Eck, G.G., 2001. Responses of African bovids to Pliocene climatic change. Paleobiology 27 (suppl. to No. 2). Paleobiology Memoirs 2, 1–47.

Bonnefille, R., Potts, R., Chalie, F., Jolly, D., Peyron, O., 2004. High-resolution vegetation and climate change associated with Pliocene *Australopithecus afarensis*. Proceedings of the National Academy of Sciences USA 101(33), 12125–12129.

Brain, C.K., 1981. The Hunters or the Hunted? An Introduction to African Cave Taphonomy. University of Chicago Press, Chicago.

Brain, C.K., 1993. A taphonomic overview of the Swartkrans fossil assemblage. In: Brain, C.K. (Ed.), Swartkrans: A Cave's Chronicle of Early Man. Transvaal Museum Monograph, Pretoria, pp. 257–264.

Broom, R., Robinson, J.T., 1949. A new type of fossil man (from Swartkrans). Nature 164, 322–323.

Broom, R., Robinson, J.T., 1952. Swartkrans Ape-Man *Paranthropus crassidens*. Transvaal Museum Monograph, Pretoria.

Cooke, H.B.S., 1976. Suidae from Plio-Pleistocene strata of the Rudolf basin. In: Coppens, Y., Howell, F.C., Isaac, G.L., Leakey, R.E. (Eds.), Earliest Man and Environments in the Lake Rudolf Basin. University of Chicago Press, Chicago, pp. 251–263.

Coppens, Y., 1978. Evolution of the hominids and their environment during the Plio-Pleistocene in the lower Omo Valley, Ethiopia. In: Bishop, W.W. (Ed.), Geological Background to Fossil Man. Scottish Academic Press, Edinburgh, pp. 499–506.

Coppens, Y., Howell, F.C., Isaac, G.L., Leakey, R.E. (Eds.), 1976. Earliest Man and Environments in the Lake Rudolf Basin: Stratigraphy, Paleoecology, and Evolution. University of Chicago Press, Chicago.

Dart, R., 1925. *Australopithecus africanus*: the man-ape of South Africa. Nature 115, 195–199.

Dart, R., 1949. The predatory implemental technique of *Australopithecus*. American Journal of Physical Anthropology 7, 1–38.

Dorst, J., Dandelot, P., 1976. Guide des grands mammifères d'Afrique. Delachaux & Niestle, Paris, pp. 187–279.

Dumont, E.R., 1995. Enamel thickness and dietary adaptation among extant primates and chiropterans. Journal of Mammalogy 76(4), 1127–1136.

Geraads, D., Coppens, Y., 1995. Evolution des faunes de mammifères dans le Plio-Pléistocène de la basse vallée de l'Omo (Ethiopie): apports de l'analyse factorielle. Comptes rendus de l'Académie des sciences Paris, IIa, 320, 625–637.

Greenacre, M.J., Vrba, E.S., 1984. Graphical display and interpretation of antelope census data in African wildlife areas, using correspondence analysis. Ecology 65(3), 984–997.

Grine, F.E., 1981. Trophic differences between 'gracile' and 'robus' australopithecines: a scanning electron microscope analysis of occlusal events. South African Journal of Science 77, 203–230.

de Heinzelin, J., Haesaerts, P., 1983. The Shungura Formation. In: de Heinzelin, J. (Ed.), The Omo Group. Musée Royale de l'Afrique Central, Tervuren, pp. 25–127.

Hylander, W.L., 1979. The functional significance of primate mandibular form. Journal of Morphology 160(2), 223–239.

Kay, R.F., 1981. The nut-crackers – a new theory of the adaptations of the Ramapithecinae. American Journal of Physical Anthropology 55, 141–151.

Kay, R.F., 1984. On the use of anatomical features to infer foraging behavior in extinct primates. In: Rodman, P.S., Cant, J.G.H. (Eds.), Adaptations for Foraging in Non-human Primates. Columbia University Press, New York, pp. 21–53.

Kay, R., Grine, F.E., 1988. Tooth morphology, wear, and diet in *Australopithecus* and *Paranthropus*. In: Grine, F.E. (Ed.), The Evolutionary History of the Robust Australopithecines. Aldine de Gruyter, New York, pp. 427–447.

Kingdon, J., 1982a. East African Mammals: An Atlas of Evolution in Africa: Volume III Part C: Bovids. Academic Press, London.

Kingdon, J., 1982b. East African Mammals: An Atlas of Evolution in Africa: Volume III Part D: Bovids. University of Chicago Press, Chicago.

Leakey, L.S.B., 1959. A new fossil skull from Olduvai. Nature 184, 491–493.

Leakey, M.G., Feibel, C.S., McDougall, I., Walker, A., 1995. New four-million-year-old hominid species from Kanapoi and Allia Bay, Kenya. Nature 376, 565–571.

Leakey, M.G., Feibel, C.S., McDougall, I., Ward, C.V., Walker, A., 1998. New specimens and confirmation of an early age for *Australopithecus anamensis*. Nature 393, 62–66.

Leakey, M.G., Spoor, F., Brown, F.H., Gathogo, P.N., Kiarie, C., Leakey, L.N., McDougall, I., 2001. New hominin genus from eastern Africa shows diverse middle Pliocene lineages. Nature 410, 433–440.

Lucas, P.W., Corlett, R.T., Luke, D.A., 1985. Plio-Pleistocene hominid diets: an approach combining masticatory and ecological analysis. Journal of Human Evolution 14(2), 187–202.

Luke, D.A., Lucas, P.W., 1983. The significance of cusps. Journal of Oral Rehabilitation 10(3), 197–206.

Meldrum, D.J., Kay, R.F., 1997. *Nuciruptor rubricae*, a new pitheciin seed predator from the Miocene of Colombia. American Journal of Physical Anthropology 102, 407–427.

Picq, P., 1983. L'articulation temporomandibulaire des hominides fossiles: anatomie comparee, biomecanique, evolution, biometrie. Ph.D. dissertation, l'Universite' Pierre et Marie Curie, Paris.

Potts, R., 1998. Variability selection in hominid evolution. Evolutionary Anthropology 7, 81–96.

Rak, Y., 1978. The functional significance of the squamosal suture in *Australopithecus boisei*. American Journal of Physical Anthropology 49(1), 71–78.

Rak, Y., 1983. The Australopithecine Face. Academic Press, New York.

Reed, K.E., 1997. Early hominid evolution and ecological change through the African Plio-Pleistocene. Journal of Human Evolution 32, 289–322.

Robinson, J.T., 1954. The genera and species of the Australopithecinae. American Journal of Physical Anthropology 12, 181–200.

Robinson, J.T., 1964. Adaptive radiation in the australopithecines and the origin of man. In: Howell, F.C., Bourlière, F. (Eds.), African Ecology and Human Evolution. Methuen, London, pp. 385–416.

Sakka, M., 1984. Cranial morphology and masticatory adaptations. In: Chivers, D.J., Wood, B.A., Bilsborogh, A. (Eds.), Food Acquisition and Processing in Primates. Plenum Press, New York, pp. 415–427.

Semaw, S., Simpson, S.W., Quade, J., Renne, P.R., Butler, R.F., McIntosh, W.C., Levin, N., Dominguez-Rodrigo, M., Rogers, M.J., 2005. Early Pliocene hominids from Gona, Ethiopia. Nature 433(7023), 301–305.

Senut, B., Pickford, M., Gommery, D., Mein, P., Cheboi, K., Coppens, Y., 2001. First hominid from the Miocene (Lukeino Formation, Kenya). Comptes rendus de l'Académie des sciences Paris 332, 137–144.

Spencer, L.M., 1997. Dietary adaptations of Plio-Pleistocene Bovidae: implications for hominid habitat use. Journal of Human Evolution 32(2–3), 201–228.

Sponheimer, M., Lee-Thorp, J.A., 1999. Isotopic evidence for the diet of an early hominid, *Australopithecus africanus*. Science 283, 368–370.

Ungar, P., 2004. Dental topography and diets of *Australopithecus afarensis* and early *Homo*. Journal of Human Evolution 46(5), 605–622.

Vrba, E.S., 1980. The significance of bovid remains as indicators of environment and predation patterns. In: Behrensmeyer, A.K., Hill, A. (Eds.), Fossils in the Making: Vertebrate Taphonomy and Paleoecology. University of Chicago Press, Chicago, pp. 247–271.

White, T.D., Suwa, G., Asfaw, B., 1994. *Australopithecus ramidus*, a new species of early hominid from Aramis, Ethiopia. Nature 371, 306–312.

White, T.D., WoldeGabriel, G., Asfaw, B., Ambrose, S., Beyene, Y., Bernor, R.L., Boisserie, J.-R., Currie, B., Gilbert, H., Haile-Selassie, Y., Hart, W.K., Hlusko, L.J., Howell, F.C., Kono, R.T., Lehmann, T., Louchart, A., Lovejoy, C.O., Renne, P.R., Saegusa, H., Vrba, E.S., Wesselman, H., Suwa, G., 2006. Asa Issie, Aramis and the origin of *Australopithecus*. Nature 440(7086), 883–889.

WoldeGabriel, G., White, T.D., Suwa, G., Renne, P., de Heinzelin, J., Hart, W.K., Heiken, G., 1994. Ecological and temporal placement of early Pliocene hominids at Aramis, Ethiopia. Nature 371, 330–333.

Wood, B.A., Strait, D., 2004. Patterns of resource use in early *Homo* and *Paranthropus*. Journal of Human Evolution 46, 119–162.

Wynn, J.G., 2000. Paleosols, stable carbon isotopes, and paleoenvironmental interpretation of Kanapoi, Northern Kenya. Journal of Human Evolution 39(4), 411–432.

Wynn, J.G., 2004. Influence of Plio-Pleistocene aridification on human evolution: evidence from paleosols of the Turkana Basin, Kenya. American Journal of Physical Anthropology 123(2), 106–118.

Wynn, J.G., Alemseged, Z., Bobe, R., Geraads, D., Reed, D., Roman, D.C., 2006. Geological and palaeontological context of a Pliocene juvenile hominin at Dikika, Ethiopia. Nature, 443, 278–281.

14. The Impact of Projectile Weaponry on Late Pleistocene Hominin Evolution

John J. Shea
Anthropology Department
Stony Brook University
Stony Brook, NY 11794–4364 USA
John.Shea@sunysb.edu

Keywords Hunting • Middle Stone Age • Middle Paleolithic • stone tools • tip cross-section area • Africa • Levant • Europe

Abstract: Projectile weaponry is a key component of all recent human subsistence strategies, but its origins and antiquity remain poorly understood. Cross-sectional area variation among North American arrowheads and spearthrower dart tips is used as a criterion for evaluating hypotheses about possible stone projectile points from Eurasian Middle Paleolithic and African Middle Stone Age contexts. Analysis of pointed artifacts from Africa, the Levant, and Europe suggest projectile technology emerged first in Africa, around 50–100 Ka. Projectile technology probably reflects part of a broader pattern of ecological diversification and subsistence intensification among early *Homo sapiens* populations.

Introduction

The use of projectile weaponry as an aid to subsistence is a human cultural universal. Nearly every ethnographic, historic, and archaeologically known human population over the last 50,000 years used projectile weapons (i.e., low-mass/high velocity weapon systems like the spearthrower-and-dart, the bow and arrow, or more recently firearms). Those few groups who seem to have lacked projectile weaponry, such as eighteenth-nineteenth century Tasmanian Aborigines, are descended from populations who possessed projectile technology. Projectile weaponry also appears to be uniquely derived among *Homo sapiens* populations. The humeri and scapulae of Neandertals and earlier hominins do not feature the characteristic patterns of bone remodeling that form in response to the torsional forces involved in habitual throwing (Churchill, 2002). Understanding the origins of projectile technology is important because versatile projectile technology is so clearly implicated in changes in Pleistocene hominin diets. For early hominins, meat and fat from large animals were prizes hard-won from carnivore competitors. Today, *Homo sapiens* is the apex predator in every ecosystem we inhabit.

The origins of projectile technology are controversial. Most archaeologists agree that clear and convincing evidence for projectile technology is present in the Eurasian early Upper Paleolithic and the African Later Stone Age. There is also a broad consensus against the hypothesis that Lower Paleolithic stone tools, such as handaxes and spheroids, were specialized projectile weapons. Wooden spears from European Middle Pleistocene contexts at Clacton (England), Lehringen, and Schöningen (both in Germany) are regarded by some researchers as possible javelins ("würfspeere") (Movius, 1950; Theime, 2000). However, the dimensions of these wooden artifacts' tips are much larger than those of ethnographic throwing spears (Oakley et al., 1977; Shea, 2006). Even if they are accepted as throwing spears on the basis of replicative experiments (Rieder, 2003), their effective range (that is to say, their users' "lethal radius") is likely to have been smaller than that of ethnographic projectile weaponry.

The focal periods of interest for research on the origins of projectile weaponry are the Eurasian Middle Paleolithic (MP) and African Middle Stone Age (MSA) (referred to here collectively as MP/MSA). Both periods span approximately 250–40 Ka. They both also feature numerous pointed stone artifacts that have been proposed as possible projectile weapons (Fig. 14.1). Bone artifacts from both MP and MSA contexts have been proposed as possible weapon armatures too, but taphonomic biases against the preservation of such artifacts in tropical contexts make it difficult to compare the evidence for bone tools between Africa and Eurasia (Henshilwood and Marean, 2003).

There has long been a division in archaeological views about these hypothetical projectile weapon armatures (Shea, 1997). European researchers have generally regarded claims of European MP projectile weapon armatures with skepticism (Holdaway, 1989; Anderson-Gerfaud, 1990; Debénath and Dibble, 1994; Plisson and Béyries, 1998; Kuhn and Stiner, 2001). The most commonly-proposed candidates for European MP projectile points are Levallois points,

J.-J. Hublin and M.P. Richards (eds.), *The Evolution of Hominin Diets: Integrating Approaches to the Study of Palaeolithic Subsistence*, 189–199.

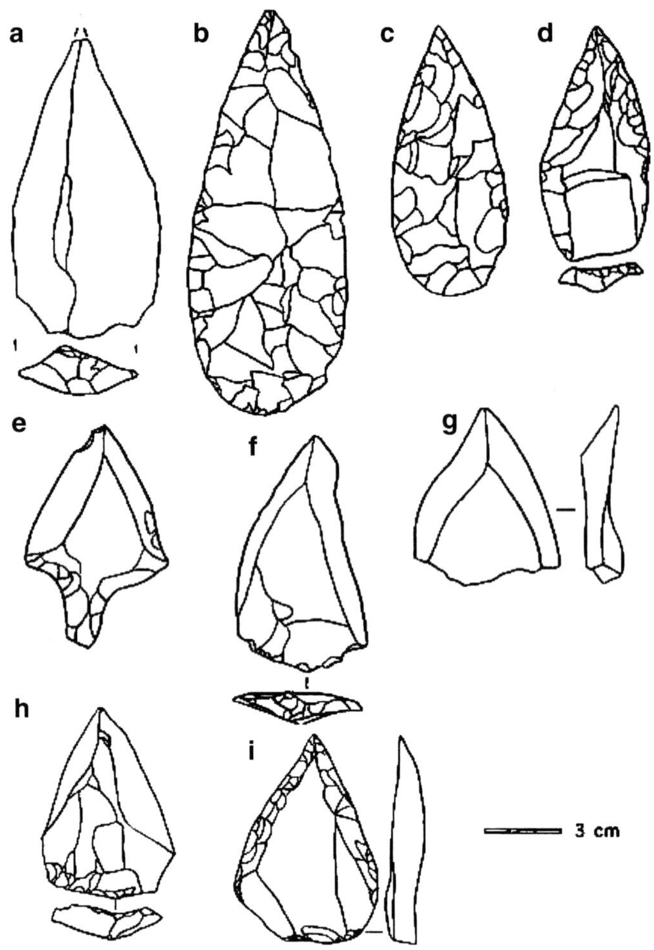

FIG. 14.1. Examples of Middle Paleolithic and Middle Stone Age point-types discussed in this study: **a. Triangular** Flake (Klasies River Mouth Cave, South Africa), **b**. Still Bay point (Skildergat, South Africa), **c**. Bifacial point (Porc Epic Cave, Ethiopia), **d**. Unifacial point (Porc Epic Cave, South Africa), **e**. Aterian Tanged point (Beni Abbas, Morocco), **f**. Levallois point (Qafzeh Cave, Israel), **g**. Levallois point (Kebara Cave, Israel), **h**. Levallois point (Houppeville, France), **i**. Mousterian point (Combe Grenal, France).

Mousterian points, and foliate bifaces (the latter mainly from Central and Eastern Europe). Such points are usually described as being too large to have been projectile points, lacking obvious modification for hafting, and not showing characteristic impact breakage and allied wear traces. Nor do they display the patterned chronological and regional variation seen among stone projectile points of recent human hunter-gatherers. Moreover, Mousterian and Levallois points are generally rare among European MP assemblages. An additional subjective factor underlying European prehistorians' views about MP projectile weapons is a longstanding tradition of viewing Neandertals' behavioral capacities as significantly inferior to those of recent *Homo sapiens* (Trinkaus and Shipman, 1993).

Historically, Africanist researchers have been more receptive to hypotheses about MSA projectile point technology. Many African MSA points in question are light, thin, and narrow, heavily retouched, and often feature basal modification congruent with hafting. They further display patterned chronological and regional variation (Leakey, 1960; Clark, 1988; Brooks et al., 2005). Pointed artifacts are relatively common among MSA assemblages, particularly those dating to <100 Ka (McBrearty and Brooks, 2000). The willingness of Africanists to credit MSA humans with projectile technology, too, may have a subjective dimension. The bow and arrow remains in widespread use in rural sub-Saharan Africa, and there is a long history in Africanist research of projecting characteristics of the ethnographic present into the remote past (Lane, 2005). Recent affirmations of *Homo sapiens*' vast antiquity in Africa by fossil and genetic evidence have also energized the search for "advanced" technology in MSA and earlier contexts (McBrearty and Brooks, 2000).

These contrasting regional perspectives are clear in the debate about the possible use of Levallois points as spear points in the East Mediterranean Levant. At an early stage in his doctoral research, this author (originally trained as an Africanist) identified Levallois points and allied tool types from Kebara, Tabun, Qafzeh, and Hayonim caves (all in Israel) and Tor Faraj rockshelter (Jordan) as possible spear points on the basis of wear patterns and their size, shape, and ecogeographically-patterned variation (Shea, 1988, 1997, 1998). Other Levantine prehistorians, primarily ones trained in European prehistory, viewed these interpretations skeptically (Debénath and Dibble, 1994; Geneste, 1998; Marks, 1998; Plisson and Béyries, 1998). The "debate" seems to have subsided somewhat after the discovery of a Levallois point embedded in the vertebra of an equid at Umm el Tlel (Syria) (Boëda et al., 1999), but the question of whether these tools were spear points or projectile points remains unresolved.

Because finds of stone points embedded in animal bones are rare, and thus unlikely to answer questions about *variability* in artifact function, there clearly needs to be a more objective standard against which to judge hypotheses about MP/MSA projectile weaponry. This paper describes a method for testing hypotheses about prehistoric stone projectile points developed in North American research and applies it to evidence from Middle Paleolithic contexts in Africa, Southwest Asia, and Europe. Stone-tipped projectile weapons whose ballistically-significant dimensions are congruent with design for use as projectile points appear among stone tool assemblages from Africa prior to 50 Kya. Plausible stone projectile points appear in the Near East and in Europe after the Middle/Upper Paleolithic "transition," ca. 35–45 Kya (Shea, 2006). Possible reasons for the chronological priority of projectile weapon use in Africa are explored, as are the larger effects of projectile weaponry on evolutionary changes in hominin diet.

Materials and Methods

The problem of identifying projectile weapon systems from characteristics of stone armatures has a long history in American archaeology. The Americas appear to have been colonized before 13 Kya by populations already equipped with stone and bone tipped projectile weaponry. As they dispersed through the New World, and as that world was transformed by Holocene climate change, American populations fine-tuned their hunting equipment in response to novel ecological circumstances (Waters and Stafford, 2007). Consequently, North and South American archaeological records feature an extraordinary range of possible projectile points. Most of these artifacts are bifacially shaped and thinned with some modification for hafting, such as notches, a stem, basal edge-abrasion, and/or a distal concavity. Recognizing that change and variability in hunting weaponry may track larger changes in ecological adaptation, New World archaeologists have explored a wide range of approaches for inferring projectile point function (Hughes, 1998). These efforts have been aided by museum collections of hafted stone points of known function from ethnographic and recent archaeological contexts.

There are significant obstacles to using many of these New World approaches to test hypotheses about MP and MSA projectile points. Microwear and residue analyses have obvious value for identifying point function, but taphonomic/preservation biases limit the number of contexts from which such archaeological evidence can be gathered. Tip shape and mass have obvious functional significance, but to use them in MP/MSA contexts ignores the possible effects of secondary use and recycling on point morphology. Ethnoarchaeological studies and numerous refitting-based analyses of archaeological lithic assemblages suggest that elements of curated "personal gear," such as projectile points, were probably recycled for other tool functions and modified accordingly (Binford, 1979; Flenniken and Raymond, 1986). Neck width (the minimum distance between the notches present on the side, corners, or base of many New World projectile points) is demonstrably an effective variable for discriminating stone arrowheads and dart tips from each other (Thomas, 1978; Shott, 1997). Other than North African Aterian tanged points, few hypothetical MP and MSA stone projectile points feature such distinctive modifications for hafting.

Hughes' (1998) survey of various ballistically-significant dimensions of projectile point design identified tip cross-sectional area (TCSA) as a potentially useful variable for identifying functionally-significant variation among prehistoric weapon armatures. TCSA is estimated by the formula ([*0.5* × *maximum width*] × *thickness*). TCSA effectively discriminates between hafted arrowheads (both ethnographic and recent archaeological) and spearthrown dart tips in museum collections (Shea, 2006). TCSA requires neither basal modification for hafting nor a complete tip on archaeological points being measured. It has a further advantage in that the measurements needed to calculate TCSA are among those routinely recorded by Paleolithic archaeologists in the course of their typological and technological analyses of Middle and Upper Paleolithic artifacts. Thus, it is possible to obtain large samples of point measurements whose statistical characteristics provide insights into the origins of projectile weaponry.

TABLE 14.1. Tip Cross Sectional Area values (mm^2) for hafted projectile points in the control samples.

Samples	*n*	Mean	SD	Source
Arrowheads	118	33	20	Thomas, 1978
Dart tips	40	58	18	Thomas, 1978; Shott, 1997

Results

One way to seek evidence for differences in the timing of MP/MSA stone projectile point use is to compare statistical variation of TCSA values among archaeological point samples to the TCSA values of projectile points of known function. Measurements of hafted ethnographic and recent archaeological arrowheads and dart tips from museum collections published by Thomas (1978) and Shott (1997) form a comparative database (Table 14.1). Thomas' and Shott's dart tip samples do not differ significantly from one another in terms of TCSA values, and they are pooled here for comparative purposes. TCSA values for the Thomas arrowhead and Thomas/Shott dart tip samples differ from each other at a high level of statistical significance ($t = 7.56$, $p < .01$). If design for projectile point use was a factor in the production of MP/MSA points, those points ought to exhibit TCSA values approximating those of stone projectile points from recent contexts. If the points were designed for non-projectile uses, such as tips for thrusting spears or knives, their values ought to significantly exceed those of the Thomas/Shott projectile point "control" samples.

Table 14.2 summarizes TCSA variation in 20 samples of MP and MSA point measurements from sites in Africa, Southwest Asia, and Europe. Some of these samples are the same ones previously published by the author (Shea, 2006), but here they have been parsed of the long statistical "tail" of TCSA values above 200 mm^2 that each sample exhibited. Some samples of Mousterian points from western Asia and Europe have been omitted, either because the raw data needed to calculate descriptive statistics were unavailable or because of small sample size. New data on Mousterian points have been added from two European sites, Orgnac and Payre (both in France).

Africa

African MSA assemblages feature a wide range of stone tools variously described as "points." Samples of these points are available from the sites of Klasies River Mouth, Blombosch

TABLE 14.2. African MSA and Eurasian MP Archaeological Point Samples.

Site and level	Point types	*n*	Mean	SD	*t*-statistic versus arrowheads	*t*-statistic versus dart tips	Source
Africa							
Klasies River Mouth MSA 1	Points/triangular flakes	58	138	37	20.47	14.37	1
Klasies River Mouth MSA 2 Lower	Points/triangular flakes	328	139	36	39.22	23.43	1
Klasies River Mouth MSA 2 Upper	Points/triangular flakes	213	131	39	30.24	18.80	1
Various South African Sites (see text)	Still Bay points	168	108	39	21.50	12.23	2
Blombos Cave Level M1	Still Bay points	190	100	53	15.89	8.86	3
Porc Epic Cave Levels A–K	Bifacial points	87	92	39	13.23	6.89	4
Porc Epic Cave Levels A–K	Unifacial points	302	87	37	19.36	8.20	4
Various North African sites (see text)	Aterian tanged points	121	114	50	17.14	10.28	5
SW Asia/Levant							
Rosh Ein Mor	Levallois points	91	107	40	16.23	9.72	6
Tabun Levels B–D	Levallois points	63	133	38	19.67	13.56	7
Skhul Level B	Levallois points	24	126	44	9.97	3.28	7
Kebara Units IX–XII	Levallois points	244	110	44	23.07	13.11	8
Tor Faraj Level C	Levallois points	129	101	44	16.97	9.12	9
Yabrud 1, Levels 2–9	Levallois points	79	135	42	20.31	14.10	10
Europe							
Orgnac 3 Level 1	Mousterian points	35	146	33	−19.29	−14.06	11
Orgnac 3 Level 6	Mousterian points	9	134	54	−6.18	−4.70	11
Payre Level D	Mousterian points	122	113	43	−18.79	−11.54	11
Payre Level Fa	Mousterian points	30	131	42	−12.47	−8.93	11
Payre Level Ga	Mousterian points	99	129	43	−20.34	−13.66	11

This table only includes data for points with TCSA values less than 200 mm^2. All *t*-statistics are significant at the level of $p < .01$.
1. S. Wurz (2003), unpublished data, 2. T. Minichillo (2004), unpublished data, 3. M. Soressi (2005), unpublished data, 4. D. Pleurdeau (2003), unpublished data, 5. T. Tillet (1992) and Harvard collections, measured by author, 6. A. Marks, unpublished data, 7. Harvard collections, measured by author, 8. L. Meignen, unpublished data, 9. D. Hery (1995), unpublished data, 10. R. Solecki and R. Solecki (1994), unpublished data, 11. M.-H. Moncel (2006), unpublished data.

Sands, Cape Hangklip, Dale Rose Parlour, Hollow Rock Shelter, Kleinjongensfontein, Peers Cave, Blombos Cave, Porc Epic Cave, Bir el-Ater, Aoulef, Asriouel, Azrag, and Izouzaden. Together, these point samples span the full length of the African continent and that part of the MSA dating <100 Ka.

Klasies River Mouth Cave (South Africa) contains a long sequence of MSA occupations dating to 60–115 Ka (Singer and Wymer, 1982). Most of the points from Klasies River Mouth are either retouched or unretouched triangular flakes (Wurz, 2002). Data provided by Sara Wurz allowed TCSA values to be calculated for 914 triangular flakes from the MSA I and MSA II Upper and MSA II Lower levels of Klasies River Mouth.

Also from South Africa are two samples of "Still Bay" points, thinned bifaces shaped by invasive thinning retouch over both their dorsal and ventral surfaces. Most dated occurrences of these foliate points are younger than 100 Ka (McBrearty and Brooks, 2000). The first Still Bay point sample consists of 203 Still Bay points from Blombosch Sands, Cape Hangklip, Dale Rose Parlour (aka Trappieskop, Eales' Cave), Hollow Rock Shelter, Kleinjongensfontein, Peers Cave (aka Skildergat), and include miscellaneous finds from the Cape Flats and Fishoek Valley. All points were measured by Minichillo (2005). TCSA values were calculated from his measurements.

Blombos Cave provides a second sample of Still Bay points. Blombos Cave is located on the southern Cape of South Africa. Its MSA occupations are dated to 73–140 Ka (Henshilwood et al., 2001). Most of the 239 Still Bay points are from Blombos Level M1, ca. 73 Ka. TCSA data were calculated from measurements provided by Marie Soressi.

Porc Epic Cave is located in west-central Ethiopia at an altitude of 1,400 m above sea level (Clark et al., 1984). Excavations at Porc Epic in the 1920s, 1930s, and 1970s recovered a rich series of Middle Stone Age assemblages, dating to 61.5–77.5 Ka by obsidian hydration analysis. Measurements necessary to calculate TCSA values were provided by Pleurdeau (2005). The points from Porc Epic have been divided into bifacially-retouched points and unifacial points.

Tanged points from the Aterian industry of Northwest Africa have long been thought to have been possible projectile points (Tixier, 1960; Clark, 1989). Aterian tanged points include both retouched flakes and bifacially-shaped tools. TCSA data have been calculated for 121 Aterian points from five open-air sites, Bir el-Ater (Morocco), Aoulef and Asriouel (Algeria), Azrag (Mauritania), and Izouzaden (Niger). The Bir el-Ater points are in collections of the Peabody Museum at Harvard University. The other point samples are from collections at the University of Grenoble (Université de Grenoble) curated by Thierry Tillet. All these points were measured by the author. The exact ages of these Aterian lithic assemblages are unknown, but most researchers place the Aterian near the end of the MSA, ca. 30–50 Ka (Wendorf and Schild, 1992).

Southwest Asia – The Levant

Middle Paleolithic assemblages from the East Mediterranean Levant date to between 250–45 Ka. Many Levantine MP assemblages feature large numbers of symmetrical Levallois points that have been proposed as possible hunting weapon armatures (Shea, 1988). Levallois point samples are available from the sites of Rosh Ein Mor, Tabun Cave, Skhul Cave, Kebara Cave, Tor Faraj Rockshelter, and Yabrud Shelter 1.

Rosh Ein Mor is an open-air site located in the Central Negev Desert in Israel (Crew, 1976). Middle Paleolithic deposits from Rosh Ein Mor date to 200 Ka (Rink et al., 2003). Anthony Marks provided measurements for 102 Levallois points from Rosh Ein Mor.

Tabun Cave is located on the western escarpment of Mount Carmel overlooking the Israeli Coastal Plain. Garrod's 1929–1934 excavation of this cave revealed deeply-stratified Early and Middle Paleolithic occupations (Garrod and Bate, 1937). Tabun Cave Levels B–D contain Middle Paleolithic assemblages, dated to 256–60 Ka by various radiometric methods (Shea, 2003). Eighty Levallois points from these levels are preserved in the collections of the Peabody Museum at Harvard University and were measured there by the author.

Skhul Cave is a collapsed rockshelter located several hundred meters east of Tabun Cave (McCown, 1937). Skhul Level B contains Middle Paleolithic artifacts in association with the remains of early *Homo sapiens*, in sediments dated to 80–130 Ka by a variety of methods (Shea, 2003). Much of the Skhul Level B assemblage was discarded at the excavation site, but 27 Levallois points from Skhul are preserved in the collections of the Peabody Museum at Harvard University and were measured there by the author.

Kebara Cave is located on southwestern Mount Carmel. The most recent excavations at Kebara recovered Neandertal fossils and Middle Paleolithic assemblages from Units IX–XII, all dating to 47–65 Ka (Bar-Yosef et al., 1992). Liliane Meignen provided measurements for 295 Levallois points from which TCSA values were computed. TCSA values of Levallois point samples from Kebara excavation Units IX–XII do not differ significantly from one another and are therefore combined in this study.

Tor Faraj is a rockshelter located in southern Jordan. Middle Paleolithic occupations in Level C date to 69 Ka (Henry, 2003). Donald Henry provided measurements for 142 Levallois points from Tor Faraj that were used to calculate TCSA values.

Yabrud Rockshelter 1 in western Syria was first excavated by Rust in the 1930s (Rust, 1950) and subsequently by the Soleckis in the 1960s (Solecki and Solecki, 1986). Yabrud 1 contains a long sequence of as-yet-undated Middle and Lower Paleolithic occupations (Solecki and Solecki, 1995). TCSA values for 103 Levallois points from MP assemblages in levels 2–9 were calculated from measurements provided by the Soleckis.

Europe

In Europe, the principal candidates for Lower and Middle Paleolithic projectile armatures are Mousterian points and allied tool forms (pseudo-Levallois points, retouched Levallois points, convergent sidescrapers, etc.). Mousterian points occur in many assemblages, but they are rarely numerous. The sites of Orgnac 3 and Payre provide large samples of Mousterian points.

Orgnac 3 Rockshelter is located in southeastern France (Moncel et al., 2005). Metric data allowing calculation of TCSA values were provided by Marie Hélène Moncel for Mousterian points and allied pointed tool types from levels 1 and 6. The Level 1 assemblage is early Middle Paleolithic, probably in the range of 200–250 Ka, based on a radiopotassium date for the underlying Level 2. The Level 6 assemblage dates to ca. 300 Ka, and is thought to represent a late phase of the Acheulean, because Levallois débitage is absent.

Payre is a rockshelter in southeastern France (Moncel, 2003). Middle Paleolithic occupations of Payre, Levels D, Fa, and Ga, preserve numerous Mousterian points and other retouched pointed tools (i.e., convergent scrapers). ESR dating and biostratigraphic analysis of faunal remains date the Payre assemblages to 100–140 Ka.

Analysis

All the African MSA and Eurasian MP points present a consistent picture. Their TCSA values are significantly larger than those of the dart tip and arrowhead control samples. One might be tempted to reject the hypothesis that projectile points were in use anywhere in MP/MSA times. but there are two arguments against such a conclusion.

First, it is far from certain that these point samples are functionally homogeneous. They were compiled by archaeologists without knowledge about their actual functions. As such, they may include pointed artifacts used for many different purposes. If there are projectile points among these samples, they are most likely among the points with the smallest TCSA values. Table 14.3 presents statistical data on those points in the sample assemblages that have TCSA values of less than 100 mm^2, together with the results of t-tests comparing them to the control dart tip sample. The choice of 100 mm^2 is an arbitrary one, but it is slightly more than two standard deviations above the mean TCSA value of the dart tip control sample. All the samples were significantly larger than the control arrowhead sample. The TCSA values of points from all levels of Klasies River Mouth, the two South African Still Bay point samples, and the several Payre assemblages have been pooled in order to retain samples of sufficient size for statistical

TABLE 14.3. Pooled samples of MP and MSA points with TCSA values less than 100 mm2 compared to the control dart tip sample.

Sample	Point types	*n*	Mean	SD	*t*-statistic	*p*
Control sample values – Arrowheads		118	33	20		
Control sample values – Dart Tips		40	58	18		
Klasies River Mouth MSA I–II	Points/triangular flakes	150	86	19	−8.81	<.01
Blombos and various South African sites.	Still Bay points	90	45	20	3.76	<.01
Porc Epic	Bifacial points	22	49	9	2.67	<.01
Porc Epic	Unifacial points	76	45	9	4.17	<.01
Various North African sites	Aterian tanged points	29	59	13	−0.34	0.74
Rosh Ein Mor	Levallois points	23	58	19	−0.01	0.99
Kebara IX–XII	Levallois points	61	55	14	0.89	0.38
Tor Faraj C	Levallois points	32	48	10	2.89	<.01
Payre D, Fa, Ga	Mousterian points	63	61	15	−0.89	0.38

comparison. Samples with fewer than 20 points have been omitted from further analysis.

Among the <100 mm^2 African MSA point samples, TCSA values for the South African Still Bay points and the unifacial and bifacial points from Porc Epic are significantly smaller than those of the control dart tip sample. This is consistent with a design for use as projectile points. The mean for the Aterian tanged point <100 mm^2 subsample is essentially equal to that for the control dart tip sample, but the t-test results are not statistically significant. There are Aterian points small enough to have been projectile points, but it is possible these points are systematically under-represented in older surface collections. Further analysis of Aterian points from controlled excavations is needed to reduce the potential role of collection bias in shaping our impression of their variability.

Among the TCSA values for the Levantine <100 mm^2 MP point samples, those from Rosh Ein Mor and Kebara do not differ significantly from the dart tip sample. Only Tor Faraj Level C exhibits significantly smaller TCSA values than the dart tip control sample. Although it is possible this reflects projectile point design, other factors may be at work. Refitting and technological analysis of the Tor Faraj Level C assemblage show that their creators were producing vast quantities of Levallois points at this site (Henry, 1995). Indeed, Tor Faraj C features one of the highest ratios of Levallois points to cores of all Levantine MP assemblages (Shea, 1998). Larger numbers of small Levallois points at Tor Faraj may reflect more intense core reduction, rather than specific design of projectile points.

The TCSA values for the European <100 mm^2 Mousterian point sample from Payre are larger on average than those of the dart tip control sample, but the difference is not statistically significant. An independent study of TCSA variation among Middle Paleolithic points from the Middle Paleolithic site of Bouheben (France) (Villa and Lenoir, 2006) has also found such points' values larger than those of ethnographic arrowheads.

Only the African MSA data preserves evidence for stone tools whose ballistically-significant metric dimensions suggest design for use as projectile points.

A second argument supporting a possibly earlier antiquity for projectile technology in Africa involves the lithic "backed pieces" (aka "crescents" and "lunates") found in some sub-Saharan MSA assemblages, most famously the Howiesonspoort variant of the South African MSA (Lombard, 2005b) (Fig. 14.2). These backed pieces have long been thought to have been possible projectile armatures, either points, or more likely barbs set into the sides of spears, darts, arrows, and other weapons (Deacon and Deacon, 1999; Barham, 2002). The maximum thickness of these backed pieces and their shaft imposes a cost in terms of the penetrating power of the weapon to which they were attached. Thick points have to be mounted in thick shafts. One can assess the plausibility of MSA backed pieces having been projectile armatures in much the same way as has been done here for stone points using TCSA. Table 14.4 presents data on the thickness of backed pieces from Howiesonspoort MSA levels of Klasies River Mouth cave dating to ca. 70 Ka, originally published by Wurz (2002). It also presents thickness data for backed pieces from Jebel Sahaba (Sudan) measured by the author. Jebel Sahaba is a Later Paleolithic (ca. 14 Ka) cemetery site from which numerous backed pieces were recovered in such close association with human skeletons (in some cases, embedded in them) that they are reasonably interpreted as projectile armatures (Wendorf, 1968). The backed pieces from MSA Howiesonspoort contexts at Klasies River Mouth are not appreciably thicker on average than various backed pieces from Jebel Sahaba. Thus, there is no reason to reject the hypothesis that at least some MSA backed pieces were projectile armatures. In addition, recent microwear analysis has also identified diagnostic traces of projectile impact on MSA backed pieces from Klasies River Mouth (Lombard, 2005c).

Stone points (retouched and unretouched triangular flakes as well as thinned bifaces), as well as backed pieces thin and narrow enough make them highly-effective projectile weapon armatures, are broadly distributed among South, Central, and East African MSA assemblages. Such artifacts do not comprise a majority of any named artifact-type from any MSA site, but this is undoubtedly a reflection of the fact that

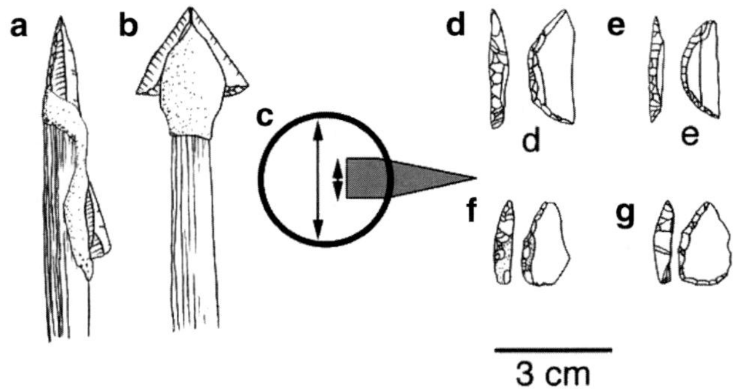

FIG. 14.2. Backed pieces as projectile armatures: **a** & **b**. hypothetical mounting of baked pieces in wooden shafts, based on ethnographic examples, **c**. relationship between thickness of backed pieces and thickness of shaft, **d** & **e**. backed pieces from Klasies River Mouth Cave (South African) Howiesonspoort MSA, **f** & **g**. backed pieces from Late Paleolithic cemetery at Jebel Sahaba (Sudan).

TABLE 14.4. Maximum thickness measurements of backed pieces from African MSA contexts at Klasies River Mouth and Later Stone Age contexts at Jebel Sahaba.

Sample	*n*	Mean	SD	Source
Klasies River Mouth, Howiesonspoort MSA, SW Cave 1a, Backed Pieces	828	4.6	1.2	1
Klasies River Mouth, Howiesonspoort MSA, D-Sample Cave 1a, Backed Pieces	28	4.6	1.1	1
Klasies River Mouth, Howiesonspoort MSA, D-Sample Cave 2, Backed Pieces	74	4.3	1.2	1
Jebel Sahaba, Backed Flakes	16	6.81	1.7	2
Jebel Sahaba Partially-Backed Flakes	6	4.67	1.9	2
Jebel Sahaba-Arched-Backed Flakes	5	6.80	1.3	2
Jebel Sahaba-Crescents	5	7.80	1.9	2
Jebel Sahaba-Triangular Arched-Backed Flakes	12	6.17	1.8	2

1. Wurz (2002: 1011), 2. Jebel Sahaba collections at SMU Dallas measured by Shea.

archaeological typologies for MSA stone tools do not take into account ballistically-significant size variation in making distinctions among named artifact-types. The presence of such plausible projectile points in so many African MSA contexts suggests projectile weaponry may have been in use in Africa long before it appeared in Eurasia. Comparisons of Southwest Asian and European Upper Paleolithic stone point types (Shea, 2006) reveal TCSA values and other evidence consistent with their having been used as dart tips and even arrowheads.

This comparison of recent projectile points with MP and MSA point samples is far from comprehensive or problem-free. The control sample of ethnographic and archaeological point measurements is a small database from which to test hypotheses about projectile point design variation over hundreds of thousands of years. Experimentation with replicas of MP and MSA points (Shea et al., 2001, 2002) may help us refine our criteria for recognizing possible stone projectile points from their ballistically-significant dimensions. Bordes' (1961) recommendation that typologists test hypotheses about Middle Paleolithic spear points by hunting wild boar with them would make for good television, but (as he noted) would probably rather quickly exhaust the supply of typologists.

A second problem concerning the difficulty of obtaining published measurements of European Middle Paleolithic points was one of the "big surprises" of this research project, and (as outlined below) a limiting factor in the conclusions that can be drawn from it. A systematic investigation of microwear and residues on large numbers of hypothetical European Middle Paleolithic points is needed. Even if no evidence for spear point use is found, such "negative results" have potentially probative value for models of prehistoric behavioral variability.

A third seemingly unavoidable problem involves the typological structure of the samples being compared. Although great strides have been made to systematize stone tool typology, it retains a strong subjective component, especially insofar as points are concerned. The statistical characteristics of pointed artifacts from particular archaeological contexts may vary widely depending on the specific criteria used by different analysts to identify "points." Few typologies explicitly specify size thresholds for particular tool types, yet size is a crucial factor in projectile point performance. As shown here, the inclusion of a few very large artifacts in a point sample can potentially obscure the presence of a significant small point component. Future studies of MP and MSA point variation should use additional metric variables, such as mass and symmetry, in selecting samples for comparison and analysis.

Discussion

The results of this study support the hypothesis that projectile weaponry first developed among African *Homo sapiens* populations between 50 and 100 Ka, and that it spread out of Africa along with those populations after 50 Kyr (Shea, 2006). The chronological priority of African projectile technology is supported by other studies of MSA point variation (Brooks et al., 2005; Lombard, 2005a).

The nature of the weapons systems used by Middle Stone Age Africans remains unknown. This study suggests their kinematic and ballistic properties were broadly comparable to those of recent arrows and spearthrower darts. For MSA contexts, some combination of stone-tipped spears and spearthrower darts seems most likely, but one cannot rule out the use of bow and arrow, particularly with respect to "backed pieces." Although the bow is thought to be a recent invention in Eurasia and the Americas, in many cases these estimates are based on untested assumptions about the functions of stone and bone weapon armatures, the antiquity of fortuitously preserved wooden artifacts, or what might charitably be described as archaeological "mythology." Chief among these myths is that projectile weapons fall into some sort of "natural" sequence of evolutionary complexity. Surveys of the ethnographic record indicate complex relationships between the choice of particular weapons systems, modes of use, and prey characteristics (Oswalt, 1976; Churchill, 1993; Ellis, 1997; Yu, 2006). There is no reason to view bow-and-arrow technology as necessarily originating from spearthrower-and-dart technology, or vice versa. It is possible that the use of these weapon systems came and went over the course of the Late Pleistocene, with wide variation in the degree to which regional human populations used one, the other, or both of them.

The suggestion of chronological priority for projectile weaponry in Africa begs the question, "Why Africa?" It is unlikely that there were uniquely-derived evolutionary capacities that predisposed early African *Homo sapiens* populations to invent projectile weaponry. There may be a gene for language (Enard et al., 2002), but even the most die-hard molecular anthropologist would have trouble convincing a sober audience that there was a gene for projectile weapon use. Ambrose (2001) argued for a link between compound technologies (i.e., hafting) and language, but wear trace and residue evidence for hafting is now associated with Neandertals as well as with early *Homo sap*iens (Grünberg, 2002; Lombard, 2005a; Mazza et al., 2006). It is vastly more likely that projectile technology emerged first in Africa and dispersed rapidly among its human populations for contingent evolutionary and environmental reasons.

Potential prey behavior is likely to have influenced early humans' choice of subsistence aids. Heavy, hand-cast, and slow-moving weapons, such as thrusting spears or throwing sticks, lose kinetic energy rapidly in flight. To be effective, they have to be launched in close proximity to their intended targets. With larger prey, such increased proximity would have involved considerable risk of injury. The probability of injury would have been greater still in the cases of large terrestrial African species with whom humans had co-evolved for millions of years. Familiar with hominins and their antics, many of these species (e.g., elephant, hippo, rhino, Cape buffalo) have developed proactive defense strategies that held them in good stead until the invention of high-powered firearms. Faced with such "killer herbivores," African hominins would have had powerful incentives to develop projectile technology at very early stages in their evolution. Indeed, probably the most remarkable thing about the African record is that clear evidence for projectile technology appears so relatively late. As evidence for early hominin carnivory continues to grow stronger, one might reasonably expect evidence for projectile technology to appear in Middle Pleistocene or even Plio-Pleistocene contexts. Perhaps what we are seeing in the MSA is not the origin of projectile technology so much as a strategic shift towards the use of specialized stone and bone armatures with high archaeological "visibility."

Second, projectile technology also has to be understood as a niche-broadening technology, rather than just as a means for killing large game. The arrow or dart that can kill a giraffe can, with minor adjustments, kill a bison, a gazelle, a deer, an ibex, a seal, a rabbit, a bird, or a fish (Churchill, 1993; Yu, 2006). The modern counterparts of the equatorial African habitats associated with Pleistocene human fossils in Africa feature greater plant and animal species diversity than their temperate Eurasian counterparts (Groombridge and Jenkins, 2002). Thus, they also offer greater opportunities and more commonly-recurring incentives for human economic diversification (pursuing a wider range of species) and intensification (putting more effort into procuring the same species) (Kuhn and Stiner, 2006; O'Connell, 2006). Incentives for diversification and intensification are likely to have become increasingly common after 75 Ka, as the accumulation of polar ice pushed the Inter-Tropical Convergence Zone southward, drying out much of northern and equatorial Africa (Goudie, 1996). Numerous "bottlenecks" (reductions of breeding populations) inferred from genetic variation among living humans support a scenario in which African human populations were packed into equatorial woodland refugia around this time (Ambrose, 1998). Geographic circumscription has a disproportionately strong effect on predators, increasing both intra-specific competition and inter-specific competition (MacArthur and Wilson, 1967). Isolation and packing of *Homo sapiens* populations into equatorial refugia may have provided strong incentives for increasing the effectiveness and versatility of projectile weaponry, as well as that of traps, nets, and perhaps plant-harvesting equipment as well. The amelioration of arid conditions would have provided opportunities for these innovations, and for the populations who created them to expand out of equatorial habitats within Africa, north towards the Eurasia, and further on to Australia and the Americas.

The weak evidence for projectile weaponry associated with Neandertals is not easy to explain. It is unlikely to reflect a shortcoming of intellect. Having survived for hundreds of thousands of years in some of the harshest habitats ever occupied by primates, Neandertals were plainly no fools (Speth, 2004). Neandertals and their predecessors were also hunting large, dangerous mammals, and they would have had similar incentives to devise weapons that reduced the risks associated with hunting these animals. The difference between Eurasian Middle Paleolithic and African MSA projectile weapon use, therefore, is not plausibly explicable in terms of big game hunting tactics. On the other hand, if Neandertals and their predecessors lived at low population densities, something one can plausibly infer from analogy with recent patterns of variation in hunter gatherer population density (Binford, 2001), they may have had fewer incentives to adopt versatile hunting weaponry than MSA humans. If this hypothesis is correct, then the apparent retraction of Neandertal settlement into southern Spain, France, Italy, and Greece during MIS 3 (Finlayson and Carrion, 2007) should have increased incentives for niche-broadening projectile weaponry. Some of the most plausible stone projectile armatures associated with Neandertals (or at least those with close morphological analogs among African MSA and Eurasian Early Upper

Paleolithic points) are Chatelperronian points, Uluzzian backed pieces, and allied tool forms. These artifacts found in precisely such late Mousterian and "transitional" contexts in southern Europe.

That isolation by itself did not necessarily elicit projectile weapon development by *Homo sapiens* can be seen in the Levant evidence. The Skhul/Qafzeh humans lived in the Levant during a relatively warm, humid period, 130–80 Kya (McGarry et al., 2004). Deserts were extensive, and sites dating to this period are concentrated near the Mediterranean Coast (Shea, 2003). Yet, there is no stronger evidence for the use of projectile weaponry associated with the Skhul/Qafzeh humans than there is with the later Middle Paleolithic assemblages associated with Neandertals. Sorting out the relative contributions of ecogeographic isolation, the particular qualities of African and Eurasian hominin habitats, and the role of species-specific behavior patterns in the origins of projectile technology remain productive subjects for future research.

Conclusions

There is much that we do not know about the origins and impact of projectile weaponry on the course of human evolution. Criteria for recognizing actual, as opposed to plausible, projectile armatures, including those made of osseous materials, need to be further refined. The samples examined by this study from Africa and the Levant seem sufficient, but there remains much to learn about functions of backed pieces in African MSA contexts. It is unlikely these backed pieces were designed solely for use as projectile armatures. That sort of functional specificity would be highly counterintuitive for toolkit components of mobile hunter-gatherers.

There can be no question that the European Middle Paleolithic samples considered here are an inadequate basis from which to generalize about European Neandertal technology. (This is probably one of the very few instances in which one can complain about the paucity of European archaeological evidence compared to any other part of the Paleolithic world!) Detailed functional studies of Mousterian points, Levallois points, and similar artifacts from Mousterian assemblages are clearly needed. In carrying out such research, great emphasis should be placed on procuring large samples from securely-dated contexts with good data on prevailing environmental conditions. A judgmental sample of few dozen Mousterian points is not a sufficient statistical base from which to generalize about hominin behavioral variability. Learning whether Neandertals used projectile points at some point in their long evolutionary history is considerably less interesting than learning about the particular constellations of circumstances that led them to adopt such subsistence aids.

The main impact of projectile weaponry on the course of Late Pleistocene human evolution seems that it conferred on *Homo sapiens* the capacity to become the apex predator of any ecosystem our ancestors chose to inhabit. Throughout much of the world today, large carnivores and many former human prey species continue to exist at our sufferance. Paired with the capacity to construct a broad ecological niche through the use of nets, traps, fishing gear, and bone-degreasing techniques, projectile technology gave *Homo sapiens* an evolutionarily unique quality of strategic flexibility. The capacity to fine-tune their subsistence strategies to novel ecological circumstances would have made it possible for expanding *Homo sapiens* populations to exploit ecozones within the habitats of rival hominin species that had been temporarily abandoned due to reduced foraging returns. It was almost certainly this strategic quality of ecological flexibility, rather than tactical advantage of projectile weaponry alone, that made our ancestors uniquely formidable competitors in an evolutionary contest whose "second prize" was extinction.

Acknowledgments. I thank the following individuals for sharing their unpublished data on MP and MSA point variation and/or for facilitating access to collections: Ofer Bar-Yosef, Don Henry, Anthony Marks, Liliane Meignen, Marie Hélène Moncel, Tom Minichillo, David Pleurdeau, Ralph Solecki, Rose Solecki, Marie Soressi, Thierry Tillet, Fred Wendorf, and Sarah Wurz. The preparation of this paper profited from discussions with Stan Ambrose, Alison Brooks, Steve Churchill, Steve Kuhn, John Speth, Mary Stiner, and Paola Villa, as well as from comments by three anoymous reviewers. Any errors in the presentation of these data and the interpretations derived from them are my own responsibility.

References

Ambrose, S.H., 1998. Late Pleistocene Human Population Bottlenecks, Volcanic Winter, and the Differentiation of Modern Humans. Journal of Human Evolution 34, 623–651.

Ambrose, S.H., 2001. Paleolithic Technology and Human Evolution. Science 291, 1748–1753.

Anderson-Gerfaud, P.C., 1990. Aspects of Behaviour in the Middle Palaeolithic: Functional Analysis of Stone Tools from Southwest France. In: Mellars, P.A. (Ed.), The Emergence of Modern Humans. Edinburgh University Press, Edinburgh, pp. 389–418.

Barham, L., 2002. Backed Tools in Middle Pleistocene Central Africa and Their Evolutionary Significance. Journal of Human Evolution 43, 585–603.

Bar-Yosef, O., Vandermeersch, B., Arensburg, B., Belfer-Cohen, A., Goldberg, P., Laville, H., Meignen, L., Rak, Y., Speth, J.D., Tchernov, E., Tillier, A.-M., Weiner, S., 1992. The Excavations in Kebara Cave, Mt. Carmel. Current Anthropology 33, 497–550.

Binford, L.R., 1979. Organization and Formation Processes: Looking at Curated Technologies. Journal of Anthropological Research 35, 255–273.

Binford, L.R., 2001. Constructing Frames of Reference: An Analytical Method for Archaeological Theory Building Using Hunter-Gatherer and Environmental Data Sets. University of California Press, Berkeley.

Boëda, E., Geneste, J.-M., Griggo, C., Mercier, N., Muhesen, S., Reyss, J.L., Taha, A.,Valladas, H., 1999. A Levallois Point

Embedded in the Vertebra of a Wild Ass (*Equus africanus*): Hafting, Projectiles and Mousterian Hunting Weapons. Antiquity 73, 394–402.

Bordes, F., 1961. Typologie du Paléolithique Ancien et Moyen. Delmas, Bordeaux.

Brooks, A.S., Yellen, J.E., Nevell, L., Hartman, G., 2005. Projectile technologies of the African MSA: Implications for Modern Human Origins. In: Hovers, E., Kuhn, S.L. (Eds.), Transitions Before the Transition. Plenum/Kluwer, New York, pp. 233–256.

Churchill, S.E., 1993. Weapon Technology, Prey Size Selection, and Hunting Methods in Modern Hunter-Gatherers: Implications for Hunting in the Palaeolithic and Mesolithic. In: Peterkin, G.L., Bricker, H., Mellars, P.A. (Eds.), Hunting and Animal Exploitation in the Later Paleolithic and Mesolithic of Eurasia. American Anthropological Association, Washington, DC, pp. 11–24.

Churchill, S.E., 2002. Of Assegais and Bayonets: Reconstructing Prehistoric Spear Use. Evolutionary Anthropology 11, 185–197.

Clark, J.D., 1988. The Middle Stone Age of East Africa and the Beginnings of Regional Identity. Journal of World Prehistory 2, 235–305.

Clark, J.D., 1989. The Origin and Spread of Modern Humans: A Broad Perspective on the African Evidence. In: Mellars, P.A., Stringer, C. (Eds.), The Human Revolution: Behavioural and Biological Perspectives on the Origins of Modern Humans. Edinburgh University Press, Edinburgh, pp. 565–588.

Clark, J.D., Williamson, K.W., Michels, M.J., Marean, C.A., 1984. A Middle Stone Age Occurrence at Porc Epic Cave, Dire Dawa, Somalia. African Archaeological Review 2, 37–71.

Crew, H., 1976. The Mousterian Site of Rosh Ein Mor. In: Marks, A.E. (Ed.), Prehistory and Paleoenvironments in the Central Negev, Israel, Vol. 1. Southern Methodist University Press, Dallas, pp. 75–112.

Deacon, H.J., Deacon, J., 1999. Human Beginnings in South Africa: Uncovering the Secrets of the Stone Age. Altamira Press, Walnut Creek.

Debénath, A., Dibble, H.L., 1994. Handbook of Paleolithic Typology, Vol. 1: Lower and Middle Paleolithic of Europe. University of Pennsylvania Press, Philadelphia.

Ellis, C.J., 1997. Factors influencing the use of stone projectile tips: An ethnographic perspective. In: Knecht, H. (Ed.), Projectile Technology. Plenum, New York, pp. 37–78.

Enard, W., Przeworski, M., Fisher, S.E., Lai, C.S.L., Wiebe, V., Kitano, T., Monaco, A.P., Pääbo, S., 2002. Molecular evolution of FOXP2, a gene involved in speech and language. Nature 418, 869–872.

Finlayson, C., Carríon, J.S., 2007. Rapid Ecological Turnover and Its Impact on Neanderthal and Other Human Populations. Trends in Ecology and Evolution 22, 213–222.

Flenniken, J.J., Raymond, A.W., 1986. Morphological Projectile Point Typology: Replication, Experimentation and Technological Analysis. American Antiquity 51, 603–614.

Garrod, D.A.E., Bate, D.M.A. (Eds.), 1937. The Stone Age of Mount Carmel, Vol. 1: Excavations in the Wady el-Mughara. Clarendon Press, Oxford.

Geneste, J.M., 1998. Comment on Plisson et S. Beyries "Pointes ou outils triangulaires? Données fonctionelles dans le Moustérien Levantin". Paléorient 24, 20–22.

Goudie, A.S., 1996. Climate: Past and Present. In: Adams, W.M., Goudie, A.S., Orme, A.R. (Eds.), The Physical Geography of Africa. Oxford University Press, Oxford, pp. 34–59.

Groombridge, B., Jenkins, M.D., 2002. World Atlas of Biodiversity: Earth's Living Resources in the 21st Century. University of California Press, Berkeley.

Grünberg, J.M., 2002. Middle Palaeolithic birch-bark pitch. Antiquity 76, 15–16.

Henry, D.O., 1995. The Influence of Mobility Levels on Levallois Point Production, Late Levantine Mousterian, Southern Jordan. In: Dibble, H.L., Bar-Yosef, O. (Eds.), The Definition and Interpretation of Levallois Technology. Prehistory Press, Madison, pp. 185–200.

Henry, D.O. (Ed.), 2003. Neanderthals in the Levant: Behavioral Organization and the Beginnings of Human Modernity. Continuum, New York.

Henshilwood, C.S., Marean, C.W., 2003. The Origin of Modern Human Behavior. Current Anthropology 44, 627–651.

Henshilwood, C.S., Sealey, J.C., Yates, R.C., Cruz-Uribe, K., Goldberg, P., Grine, F.E., Klein, R.G., Poggenpoel, C., Niekirk, K.V., Watts, I., 2001. Blombos Cave, Southern Cape, South Africa: Preliminary Report on the 1992–1999 Excavations of Middle Stone Age Levels. Journal of Archaeological Science 28, 421–448.

Holdaway, S., 1989. Were There Hafted Projectile Points in the Mousterian? Journal of Field Archaeology 16, 79–85.

Hughes, S.S., 1998. Getting to the Point: Evolutionary Change in Prehistoric Weaponry. Journal of Archaeological Method and Theory 5, 345–408.

Kuhn, S.L., Stiner, M.C., 2001. The Antiquity of Hunter-Gatherers. In: Panter-Brick, C., Layton, R.H., Rowley-Conwy, P. (Eds.), Hunter-Gatherers: An Interdisciplinary Perspective. Cambridge University Press, Cambridge, pp. 99–142.

Kuhn, S.L., Stiner, M.C., 2006. What's a Mother to Do: A Hypothesis About the Division of Labor Among Neandertals and Modern Humans. Current Anthropology 47, 953–980.

Lane, P.J., 2005. Barbarous Tribes and Unrewarding Gyrations? The Changing Role of Ethnographic Imagination in African Archaeology. In: Stahl, A.B. (Ed.), African Archaeology: A Critical Introduction. Blackwell, Malden, pp. 24–54.

Leakey, L.S.B., 1960. Adam's Ancestors (Fourth Edition). Harper & Row, New York.

Lombard, M., 2005a. Evidence of hunting and hafting during the Middle Stone Age at Sibudu Cave, KwaZulu-Natal, South Africa: A Multianalytical Approach. Journal of Human Evolution 48, 279–300.

Lombard, M., 2005b. The Howiesonspoort fo South Africa: What We Know, What We Think We Know, What We Need to Know. South African Humanities 17, 33–55.

Lombard, M., 2005c. A Method for Identifying Stone Age Hunting Tools. South African Archaeological Bulletin 60, 115–120.

MacArthur, R.H., Wilson, E.O., 1967. Theory of Island Biogeography. Princeton University Press, Princeton.

Marks, A., 1998. Comment on Plisson et S. Beyries "Pointes ou outils triangulaires? Données fonctionelles dans le Moustérien Levantin". Paléorient 24, 18–20.

Mazza, P.P.A., Martini, F., Sala, B., Magi, M., Colombini, M.P., Giachi, G., Landucci, F., Lemorini, C., Modugno, F., Ribechini, E., 2006. A New Palaeolithic Discovery: Tar-Hafted Stone Tools in a European Mid-Pleistocene Bone-Bearing Bed. Journal of Archaeological Science 33, 1310–1318.

McBrearty, S., Brooks, A.S., 2000. The Revolution That Wasn't: A New Interpretation of the Origin of Modern Human Behavior. Journal of Human Evolution 39, 453–563.

McCown, T.D., 1937. Mugharet es-Skhul. Description and Excavations. In: Garrod, D., Bate, D.M.A. (Eds.), The Stone Age of Mount Carmel, Vol. 1: Excavations in the Wady el-Mughara. Clarendon Press, Oxford, pp. 91–112.

McGarry, S., Bar-Matthews, M., Matthews, A., Vaks, A., Schilman, B., Ayalon, A., 2004. Constraints on Hydrological and Paleotemperature Variations in the Eastern Mediterranean Region in the Last 140 ka Given by the [delta]D Values of Speleothem Fluid Inclusions. Quaternary Science Reviews 23, 919–934.

Minichillo, T., 2005. Middle Stone Age Lithic Study, South Africa: An Examination of Modern Human Origins. Doctoral Dissertation, University of Washington.

Moncel, M.-H., 2003. L'exploitation de l'espace et la mobilité des groupes humains au travers des assemblages lithique à la fin du Pléistocene moyen et au début du Pléistocène supérieur: La moyenne vallée du Rhône entre Drôme et Ardèche, France. British Archaeological Reports International Series, No. S1184, Oxford.

Moncel, M.-H., Moigne, A.-M., Combier, J., 2005. Pre-Neandertal Behaviour During Isotopic Stage 9 and the Beginning of Stage 8. New Data Concerning Fauna and Lithics in the Different Occupation Levels of Orgnac 3 (Ardeche, South-East France): Occupation Types. Journal of Archaeological Science 32, 1283–1301.

Movius, H.L. Jr., 1950. A Wooden Spear of the Third Interglacial Age from Lower Saxony. Southwestern Journal of Anthropology 6, 139–142.

Oakley, K., Andrews, P., Keeley, L.H., Clark, J.D., 1977. A Re-appraisal of the Clacton Spearpoint. Proceedings of the Prehistoric Society 43, 13–30.

O'Connell, J.F., 2006. How Did Modern Humans Displace Neanderthals? Insights from Hunter-Gatherer Ethnography and Archaeology. In: Conard, N. (Ed.), Neanderthals and Modern Humans Meet? Kerns, Tübingen.

Oswalt, W., 1976. An Anthropological Analysis of Food-Getting Technology. Wiley, New York.

Pleurdeau, D., 2005. Human Technical Behavior in the African Middle Stone Age: The Lithic Assemblage of Porc-Epic Cave (Dire-Dawa, Ethiopia). African Archaeological Review 22, 177–197.

Plisson, H., Béyries, S., 1998. Pointes ou outils triangulaires? Données fonctionnelles dans le Moustérien levantin. Paléorient 24, 5–24.

Rieder, H., 2003. Der Große Wurf der frühen Jäger: Nachbau altsteinzeitlicher Speere. Biologie im Unserer Zeit 33, 156–160.

Rink, W.J., Richter, D., Schwarcz, H.P., Marks, A.E., Monigal, K., Kaufman, D., 2003. Age of the Middle Palaeolithic Site of Rosh Ein Mor, Central Negev, Israel: Implications for the Age Range of the Early Levantine Mousterian of the Levantine Corridor. Journal of Archaeological Science 30, 195–204.

Rust, A., 1950. Die Höhlenfunde von Jabrud (Syrien). K. Wacholtz, Neumünster.

Shea, J.J., 1988. Spear Points from the Middle Paleolithic of the Levant. Journal of Field Archaeology 15, 441–450.

Shea, J.J., 1997. Middle Paleolithic Spear Point Technology. In: Knecht, H. (Ed.), Projectile Technology. Plenum, New York, pp. 79–106.

Shea, J.J., 1998. Neandertal and Early Modern Human Behavioral Variability: A Regional-Scale Approach to the Lithic Evidence for Hunting in the Levantine Mousterian. Current Anthropology 39, S45–S78.

Shea, J.J., 2003. The Middle Paleolithic of the East Mediterranean Levant. Journal of World Prehistory 17, 313–394.

Shea, J.J., 2006. The Origins of Lithic Projectile Point Technology: Evidence from Africa, the Levant, and Europe. Journal of Archaeological Science 33, 823–846.

Shea, J.J., Brown, K., Davis, Z., 2002. Controlled Experiments with Middle Paleolithic Spear Points: Levallois Points. In: Mathieu, J.R. (Ed.), Experimental Archaeology: Replicating Past Objects, Behaviors, and Processes. British Archaeological Reports, International Series 1035, Oxford, pp. 55–72.

Shea, J.J., Davis, Z., Brown, K., 2001. Experimental Tests of Middle Paleolithic Spear Points Using a Calibrated Crossbow. Journal of Archaeological Science 28, 807–816.

Shott, M.J., 1997. Stones and Shafts Redux: The Metric Discrimination of Chipped-Stone Dart and Arrow Points. American Antiquity 62, 86–102.

Singer, R., Wymer, J., 1982. The Middle Stone Age at Klasies River Mouth in South Africa. University of Chicago Press, Chicago.

Solecki, R.L., Solecki, R.S., 1986. A Reappraisal of Rust's Cultural Stratigraphy of Yabroud Shelter I. Paléorient 12, 53–60.

Solecki, R.L., Solecki, R.S., 1995. The Mousterian Industries of Yabrud Shelter 1: A Reconsideration. In: Dibble, H.L., Bar-Yosef, O. (Eds.), The Definition and Interpretation of Levallois Technology. Prehistory Press Monographs in World Archaeology No. 23, Madison, pp. 381–398.

Speth, J.D., 2004. News Flash: Negative Evidence Convicts Neanderthals of Gross Mental Incompetence. World Archaeology 36, 519–526.

Theime, H., 2000. Lower Palaeolithic Hunting Spears from Schöningen, Germany – The Oldest Spears in the World. Acta Anthropologica Sinica 19, 140–147.

Thomas, D.H., 1978. Arrowheads and Atlatl Darts: How the Stones Got the Shaft. American Antiquity 43, 461–472.

Tixier, J., 1960. Les Pièces Pédonculées de l'Atérien. Libyca 6/7, 127–158.

Trinkaus, E., Shipman, P., 1993. The Neandertals: Changing the Image of Mankind. Knopf, New York.

Villa, P., Lenoir, M., 2006. Hunting Weapons of the Middle Stone Age and the Middle Palaeolithic: Spear Points from Sibudu, Rose Cottage and Bouheben. Southern African Humanities 18, 89–112.

Waters, M.R., Stafford, T.W. Jr., 2007. Redefining the Age of Clovis: Implications for the Peopling of the Americas. Science 315, 1122–1126.

Wendorf, F., 1968. Site 117: A Nubian Final Paleolithic Graveyard near Jebel Sahaba, Sudan. In: Wendorf, F. (Ed.), The Prehistory of Nubia. Southern Methodist University Press, Dallas, pp. 954–996.

Wendorf, F., Schild, R., 1992. The Middle Paleolithic of North Africa: A Status Report. In: Klees, F., Kuper, R. (Eds.), New Light on the Northeast African Past: Current Prehistoric Research. Heinrich-Barth Institut, Cologne, pp. 40–78.

Wurz, S., 2002. Variability in the Middle Stone Age Lithic Sequence, 115,000–60,000 Years Ago at Klasies River, South Africa. Journal of Archaeological Science 29, 1001–1015.

Yu, P.-L., 2006. From Atlatl to Bow and Arrow: Implicating Projectile Technology in Changing Systems of Hunter-Gatherer Mobility. In: Sellet, F., Greaves, R., Yu, P.-L. (Eds.), Archaeology and Ethnoarchaeology of Mobility. University Press of Florida, Gainesville, pp. 201–220.

15. The Evolution of the Human Capacity for "Killing at a Distance": The Human Fossil Evidence for the Evolution of Projectile Weaponry

Steven E. Churchill
Department of Evolutionary Anthropology
Box 90383
Duke University
Durham, NC 27708-0383 USA
churchy@duke.edu

and

Jill A. Rhodes
Department of Anthropology
Bryn Mawr College
101 N. Merion Avenue
Bryn Mawr, PA 19010 USA
jarhodes@brynmawr.edu

Keywords Paleoanthropology • hominins • functional morphology

Abstract Recent analyses of MSA and Middle Paleolithic points suggest that true long-range projectile weaponry – most likely in the form of spearthrower-delivered darts – evolved in Africa sometime between 90–70ky BP, and was part of the tool kit of modern humans who expanded out of Africa after this time. This possibility has important implications for our understanding of behavior change during the MSA, the evolution of modern human predatory behavior and subsistence strategies, and the nature of the competitive interactions that occurred between modern humans and the archaic humans they encountered on their diaspora from Africa. Research into the origins of projectile weapons can be informed by analyses of the skeletal remains of the prehistoric humans who made and used them, since habitual behavior patterns – especially biomechanically stressful actions like forceful throwing – can be imprinted on the skeleton through both genetic and epigenetic pathways. Previous analyses of humeral diaphyseal geometry in Neandertals and early modern Europeans concluded that habitual, forceful throwing is reflected in the fossil record only after 20ky BP, suggesting a relatively late origin of projectile weaponry. In contrast, recent work on humeral torsion angles in these same groups reveals some evidence to suggest that throwing-based projectile weaponry was commonly used by the earliest modern Europeans. Other aspects of the skeleton, such as scapular glenoid fossa and ulnar supinator crest morphology, might contain a signature of habitual throwing, but have not yet been examined. Here we analyze variation in scapular and ulnar morphology within and between groups of fossil and recent humans relative to the question of the origins of projectile weaponry. Although the results are not clear-cut, the overall pattern of osteological indicators is consistent with the claim that projectile weapons arose in the African later MSA and moved into Europe in the hands of modern humans.

J.-J. Hublin and M.P. Richards (eds.), *The Evolution of Hominin Diets: Integrating Approaches to the Study of Palaeolithic Subsistence,* 201–210.

Introduction

The development of projectile weaponry ranks as one of the key innovations of the Paleolithic and one of the signal events in the evolution of human subsistence systems. With true long-range projectile weapons capable of "killing at a distance," prehistoric foragers were able to broaden the range of prey species exploited, by both reducing the risks associated with hunting dangerous game and by lowering the handling costs of some difficult to capture prey (thus changing the economics of prey choice: see Churchill, 1993; Shott, 1993; Kelly, 1995; Churchill and Morris, 1998). Projectile weapons also allowed hunting in new places (such as open country) and the use of new tactics, further altering subsistence economics and human diet breadth (Churchill, 1993; Marlowe, 2005; O'Connell, 2006). Outside the realm of subsistence, projectiles may have altered the nature of human coalitionary and coercive behavior (Bingham, 2000), and changed the nature of human agonism and interpersonal violence.

The timing of the appearance of true long-range projectile weaponry remains a matter of debate. Wooden spears recovered from ca. 400ky BP water-logged deposits at Schöningen, Germany (Thieme, 1997, 1999) are taken by many as representing the earliest archeological evidence of projectile weaponry. Thieme (1997, 1999) has noted design features in these spears that are similar to modern javelins (namely, greater weight distally and tapering proximally), and experimental studies with replicas have demonstrated effective performance over a 15 m distance (Rieder, 2001, 2003).[1] The form of these spears, in conjunction with their association with the remains of open country prey (predominately horses, some of which may bear lesions from spears), has led to the suggestion that by Middle Pleistocene times humans had released – at least to a degree – the hunting constraints imposed by close-range weapon systems (i.e., hand-held spears). This conclusion has not been uniformly accepted, however. The large size of the Schöningen spears (and similar Middle/Late Pleistocene fossil spears from Lehringen and Clacton: Oakley et al., 1977) relative to ethnohistorically documented throwing and thrusting spears has led some to question their use as true projectiles (Schmitt et al., 2003; Shea, 2006; see also Oakley et al., 1977). An argument could be made that the larger average size of Pleistocene prey animals necessitated larger, heavier spears (to maximize momentum and thus improve penetration ability of a thrown projectile), although we note that this is not how modern human foragers deal with large prey (see Churchill, 1993). More work needs to be done on these Middle Pleistocene spears to better define their capabilities and probable function – at present we can only note that, if indeed they were used as javelins, they were larger and heavier than anything modern humans are known to have used in subsistence hunting – whether thrown or thrust[2].

The question of the use of the Schöningen spears highlights the need to distinguish between hand-deployed throwing spears and true, long-range projectile weapons. Based on historic and ethnographic accounts of recent and extant foraging groups, hand-held spears are sometimes thrust at prey and sometimes thrown (Churchill, 1993, 2002; Kortlandt, 2002). When effective weapon distances (the actual distances over which weapon systems are deployed in hunting, as opposed to the more often reported maximum range of the weapon) are considered it is apparent that even hand-thrown spears are close range weapons (Table 15.1). Note in Table 15.1 that one of the 14 groups in the ethnohistoric literature with reported hand-thrown spear effective ranges is an extreme outlier. The Tiwi of Melville Island (Australia) are reported to have thrown very light spears very long distances (40 m) at very small prey (Basedow, 1913; Goodale, 1971) – when the data are "corrected" by removing this outlier, the average effective distance of hand-thrown spears drops to 5.7 ± 0.9 m. Hand-deployed spears are close-range weapons, and accordingly their use in the ethnographic present is associated with hunting methods that either seek to impede the flight of prey (by somehow disadvantaging them) or to use the element of surprise (by ambushing them) to gain the close access required to deploy the weapon (Churchill, 1993). With the advent of the spearthrower, darts could be deployed over a distance of 40 m or more, allowing hunters to effectively get within weapon range – by using stealth or deception – of open country prey before invoking the flight response of their quarry. Thus the spearthrower, and later the bow, allowed hunters to consistently add approach hunting to their repertoire, and to hunt open country prey in new physiographic settings. At a possible effective distance of 15 m, it seems the Schöningen spears would have been of limited help in approach hunting in open country, but may have reduced the risk factors associated with hunting dangerous prey by other means (such as disadvantaging and ambushing).

TABLE 15.1. Effective weapon distance of various weapon systems (From Churchill, 1993; see text for details).

Weapon system	Effective distance (m)
Thrusting spear	Contact
Hand-thrown spear	7.8 (SE = 2.2, n = 14)
Hand-thrown spear, "corrected"	5.7 (SE = 0.9, n = 13) (range = 2–10)
Spearthrower and dart	39.6 (SE = 5.5, n = 9)
Bow and arrow	25.8 (SE = 2.5, n = 25)

Recent analyses of Middle Paleolithic points suggest that true long-range projectile weapons developed possibly before 77ky BP in Africa (McBrearty and Brooks, 2000; Brooks et al., 2005), or around 50–40ky BP in the Levant (Shea, 2006). In terms of weight and tip angles, points from the later African Middle Stone Age (MSA) – from the sites of ≠Gi (Botswana) and Aduma (Ethiopia) – converge on values for spearthrower darts from North America, while comparably-aged points from the Mousterian of the Near East and Europe do not (Brooks et al., 2005). Shea (2006, 2009), on the other hand, notes that the cross-sectional areas (which are related to projectile point penetrating ability) of MSA points are more similar to those of North American thrusting spears. It is only with points of the terminal Middle Paleolithic or earliest

[1] Similar but unpublished replicative studies by both John Shea and one of us (SEC) have found these spears to have poor flight properties and poor penetration performance beyond a distance of ca. 5 m.

[2] As noted by Villa (2009), Roman soldiers are known to have thrown heavy spears 15 m or more in warfare. While we admit that this lends credence to the notion of heavy spears used in hunting, we also wish to note that the economics of warfare and subsistence hunting are quite different, and the nature of target acquisition, target size and weapon accuracy may vary in important ways (see, for example, Guthrie, 1983; and discussion of shock vs. surgical weapons in Churchill, 1993). In general we agree that there is much to learn from military examples of weapon use (see Kortlandt, 2002 vs.Churchill, 2002), but we also argue that the specifics of direct comparisons be carefully considered before conclusions are reached.

Upper Paleolithic after 50 ky BP, as with Ksar Akil or El Wad points of the Levant, that cross-sectional areas become less variable (as one would expect for true projectile points: see Wilmsen, 1974; Hughes, 1998) and become similar to North American dart point and arrowhead values. Small stone (and bone) points are ubiquitous in early Upper Paleolithic assemblages (e.g., Ksar Akil, El Wad, Font Yves, Font Robert, and Teyjat points), and even occur in at least one initial Upper Paleolithic assemblage (Châtelperron points). Their potential function as projectile armatures would be consistent with the African (Brooks et al., 2005) or Near Eastern (Shea, 2006) origins of projectile weaponry and the movement into Europe of that technology in the hands of modern humans (with their appearance in the Chatelperronian reflecting cultural diffusion following modern human-Neandertal contact). This suggests that projectile weaponry may have been one component of a modern human advantage in competition for resources (Shea, 2003), and that projectiles may have even entered into the nature of the direct interactions between modern and archaic humans in Europe and Asia.

Analyses of the human fossil record, focusing on aspects of upper limb morphology thought to reflect habitual throwing, have come to somewhat conflicting conclusions. Examination of the geometric properties of the humeral shaft of various fossil and recent human groups revealed that early Upper Paleolithic (EUP) Europeans had humeri that were weak to torsional loads (as would be generated during throwing) relative to other foragers (Churchill, 1994; Churchill et al., 1996b). In the right limb, Neandertals and modern humans from the European late Upper Paleolithic (LUP) and Mesolithic and North American Holocene (subarctic maritime hunters and woodland Indians) had mean measures of torsional strength (polar moment of inertia standardized to body size) that were significantly greater than seen in EUP males, who are closer to early agricultural and modern industrialized samples in this regard (Churchill et al., 1996b; and Churchill, unpublished data, 2009). In addition, the EUP males had humeral shafts that shared with those of Neandertals and Levantine Mousterian early modern humans in terms of having greater bending strength in the anteroposterior than mediolateral plane, as might be expected of humeri that, through epigenetic modeling responses, have adapted to the types of bending loads that would be consistent with habitual and forceful use of thrusting spears (Churchill et al., 1996b; Trinkaus and Churchill, 1999; Schmitt et al., 2003). In LUP males (after 20 ky BP), humeral shaft cross-sections are generally rounder, as would be expected of bones adapted to torsional loads (consistent with throwing). Bilateral asymmetry in humeral strength is high in Mousterian and Upper Paleolithic males (and drops to modern levels by the Mesolithic), which would be consistent with throwing in these groups, but which might also be expected as an osteological sequelae of habitual spear thrusting (Schmitt et al., 2003). Churchill et al. (1996b) argued that the humeral strength patterns seen in Paleolithic males were consistent with a late advent of true long range projectile weapons, and that this technology only came into use in Europe around or after 20 ky BP, concordant with the first hard evidence of spearthrowers (Breuil, 1912; Cattelain, 1989). If this is correct, throwing-based projectile weaponry may have been part of the fluorescence of resource extractive technology (which included the appearance, among other things, of fish leisters, hooks and weirs, nets, traps and possibly the bow and arrow: see for example papers in Peterkin et al., 1993; Straus, 1995) associated with subsistence intensification and diversification in the LUP (Churchill et al., 1996b).

A more recent analysis of humeral retroversion angles in fossil and recent humans has produced more equivocal results. Humeral retroversion – the angle formed between the axis of the head and that of the distal articular surface – has been shown to be greater in throwing athletes than non-athletes and greater in the throwing than non-throwing limb of those athletes (Pieper, 1998; Osbahr et al., 2002; Reagan et al., 2002; Crockett et al., 2009). Neandertals have mean values of humeral retroversion that are greater than seen in most modern human samples (Rhodes and Churchill, 2009 n.d.), although this finding appears to be confounded by the anteroposteriorly deeper chests of the Neandertals (see Churchill, 1994). Upper Paleolithic males were not found to have high retroversion angles relative to either throwing athletes or modern comparative samples (Rhodes and Churchill, n.d.), but small fossil sample sizes, possible inter-study differences in measurement methods, and relatively great variance in the fossil samples makes it difficult to interpret these results. Levels of bilateral asymmetry suggest that both EUP and LUP males may have been engaged in regular throwing: LUP males exhibit levels of bilateral asymmetry in excess of that seen in samples of throwing athletes, and EUP males demonstrate asymmetry levels within the range of the professional throwing athletes. However, females in these samples also exhibited relatively high levels of bilateral asymmetry in humeral retroversion, which might suggest habitual throwing as a part of female subsistence practices (perhaps females hunted with spear throwers, or threw stones at smaller prey while foraging). Based on three individuals with sufficiently complete right and left humeri, asymmetry was substantially lower in Neandertals. Overall, the humeral retroversion evidence does not rule out the use of projectile technology during the early part of the Upper Paleolithic (Rhodes and Churchill, 2009), although the overall patterns in the data are somewhat equivocal.

Consideration of muscularity and leverage about the shoulder and elbow also suggests that Neandertals and, to a lesser degree, Levantine Mousterian early modern humans were morphologically less suited to throwing than Upper Paleolithic modern humans (Churchill and Rhodes, 2006). Neandertals, and to a lesser extent Levantine Mousterian modern humans, have relatively small humeral deltoid tuberosities. Based on electromyographic studies of overhand throwing, the anterior, middle and posterior fibers of the deltoid are most active during maximum external humeral rotation, and the anterior and middle fibers are active during the release phase in overhand throwing (Perry and Glousman, 1995). Peak activity of all fibers of the deltoid occurs during the early cocking phase of the throw, when the arm is at 95° abduction (Perry and Glousman, 1995). The relatively small size of the deltoid muscle, as inferred from its attachment site on the humerus, would indicate Neandertals and Levantine early modern humans were comparatively weak in activities involving an elevated arm, such as forceful throwing (see Peterson, 1998). EUP and LUP humans, on the other hand, have relatively large deltoid tuberosities (Churchill and Rhodes, 2006). Finally, Neandertals differ from both Levantine Mousterian and EUP/LUP modern humans in having longer ulnar olecranon processes (Churchill and Rhodes, 2006) and more anteriorly directed ulnar trochlear notches (Churchill et al., 1996a). The former indicates greater leverage of the triceps muscle in Neandertals, which would have increased their ability to forcefully extend the forearm at the elbow, but would have substantially decreased their ability to generate angular acceleration of the hand (as needed to accelerate a projectile for a long distance throw). The shorter and presumably more massive forearms of Neandertals would have likewise limited acceleration of a projectile during the earlier stages of the throw, in which humeral internal rotation is providing most of the acceleration (Tullos and King, 1973). The more anteriorly oriented trochlear notches seen in Neandertal ulnae have been interpreted as reflecting adaptation to habitual loading regimes in which peak loads were incurred with the elbow in partial flexion (Trinkaus and Churchill, 1988), as would occur with the forceful use of thrusting spears (Schmitt et al., 2003). Levantine Mousterian modern humans have trochlear notches more similar to other modern humans (including EUP and LUP humans), but with somewhat of a more anterior orientation (Churchill et al., 1996a).

Thus some of the features reviewed above demark Neandertals from all modern humans (which, if they do indeed reflect the degree to which foragers were engaged in forceful throwing, would suggest that Levantine Mousterian modern humans were using projectile weapons), while others demark the Mousterian foragers (both Neandertal and modern human) from the Upper Paleolithic groups (which would be consistent with an African/Near Eastern origin of projectile weaponry between 100–50 ky BP), while still others distinguish Mousterian and EUP from LUP hunters (suggesting a late origin of true long-range projectile weaponry – at least in Europe – around 20 ky BP).

Here we examine two additional osteological features which may provide information on the amount of habitual throwing engaged in by these groups: scapular glenoid fossa shape and ulnar supinator crest hypertrophy. Neandertals are characterized by glenoid fossae that are narrow relative to their height – reflected in a low glenoid index – which has been interpreted as reflecting a joint that was poorly-adapted to withstand dorsally- and ventrally-directed forces at the shoulder that occur during throwing (Churchill and Trinkaus, 1990). Combined abduction and external rotation of the humerus during late cocking phase results in obligatory posterior translation of the humeral head on the glenoid (Howell et al., 1988), and during a throw joint loads go from 400 N of posterior shear (combined with 1,000 N compression) to 75 N of anterior shear (combined with 400 N compression) (Meister, 2000). Throwing is one of the few motions that engender joint loads in the transverse plane, such that the wide glenoid fossae seen in modern humans have been suggested to reflect an ancestry involving regular throwing (Churchill and Trinkaus, 1990). In contrast to Neandertals, modern humans of the early Upper Paleolithic have scapular glenoid fossae that were relatively wider and not significantly different from recent humans in glenoid index (Churchill and Trinkaus, 1990): however, this finding is based on a very small (n = 5) sample of EUP fossils, and here we examine a larger sample of early modern Europeans. In addition to scapular glenoid morphology, degrees of hypertrophy of the ulnar supinator crest may contain information about habitual throwing behavior. The supinator muscle fires forcefully during the deceleration phase of throwing to counter hyperpronation of the forearm (Gainor et al., 1980; Zarins et al., 1985), and accordingly the origin of its ulnar head (the ulnar supinator crest) has been found to be hypertrophied in habitual spear throwing populations in India and the Near East (Kennedy, 1983, 2004; Peterson, 1998). Thus we also evaluated the relative size of this muscle scar in Mousterian and Upper Paleolithic fossil humans. If true long-range projectile weaponry evolved in Africa or the Near East in the late Middle Paleolithic/MSA, we would expect Mousterian-associated humans (both Neandertals and Levantine early modern humans) to have narrow glenoid fossae and small supinator crests, and Upper Paleolithic humans (both early and late) to have wide glenoid fossae and large supinator crests (assuming that projectile weaponry in the Upper Paleolithic was throwing based).

Materials and Methods

The Neandertal sample is composed of ten males, four females and 15 specimens of indeterminate sex, deriving predominately from OIS 3 but with a geographic range from western Europe to the Near East (Table 15.2). The early modern human fossils were divided into three samples: Mousterian-associated early modern humans from the Near East (four males and two females); early Upper Paleolithic (EUP) Europeans (16 males, seven females and three individuals of indeterminate sex, from Gravettian and Protomagdalenian contexts); and late Upper Paleolithic (LUP) Europeans (14 males, nine females and two individuals of indeterminate sex, from Magdalenian and Epigravettian contexts, as well as terminal Pleistocene/early Holocene cultures). Samples of industrialized recent modern humans – both European- and African-American – were included

TABLE 15.2. List of fossil specimens examined.

Males	Females
Mousterian Neandertals	
Amud 1	La Ferrasie 2
La Chapelle 1	Shanidar 6
La Ferrasie 1	Spy 1
Kebara 2	Tabun C1
Lezetxiki 1[a]	**Sex indeterminate**
Neandertal 1	Krapina 125, 127, 129–133, 179, 181–186
Régourdou 1[b]	Vindija 209
Shanidar 1, 4	
Spy 2	
Mousterian Early modern humans	
Qafzeh 8	Qafzeh 9
Skhul IV, V, VI	Skhul II
Early Upper Paleolithic modern humans	
Abri Pataud 6	Cro-magnon 2
Arene Candide 1	Dolní Věstonice 3[d]
Barma Grande 2, 5	Grotte des Enfants 5
Bausso da Torre 2	Muierii 1
Cro-magnon 1, 3	Nahal Ein Gev 1
Dolní Věstonice 13, 14, 16	Paglicci 25
Grotte des Enfants 4, 6	Předmostí 4, 10[c]
Paviland 1	**Sex indeterminate**
Předmostí 3, 9, 14[c]	Abri Pataud 5
	Isturitz 3
	La Rochette 1
Late Upper Paleolithic modern humans	
Arene Candide 2, 4, 5, 10, 12	Arene Candide 3, 13, 14
Chancelade 1	Bruniquel 24
Gough's Cave 1	Cap Blanc 1
Neussing 2	Farincourt 1
Oberkassel 1	Oberkassel 2
Placard 16[e]	Romito 4
Rochereil 1	Saint Germain La Rivière 4
Romanelli 1	**Sex indeterminate**
Romito 3	Arene Candide 15
Veyrier 1	La Madelaine 1

[a]Originally identified as a female based upon size and gracility (Basabe, 1966); it has been considered male in this analysis based on the high mass estimate derived from the specimen (Ruff et al., 1997; Matiegka 1938; Soficaru et al. 2006)
[b]Sex indeterminate, but considered here as male based on high estimated body mass (Ruff et al., 1997)
[c]All data taken from Matiegka (1938)
[d]Data from Soficaru et al. (2006)
[e]Sex indeterminate; considered here as male based on size, muscularity and robusticity

for comparative purposes. The Euro-American sample derives from the collections of the Maxwell Museum of Anthropology (University of New Mexico) and the Terry collection from the National Museum of Natural History (Smithsonian Institution). The Maxwell Museum collection (40 males, 24 females) is an autopsy sample comprised of individuals from the central Rio Grande region of New Mexico; the Terry collection (15 females) was collected in the 1920s at Washington University Medical School in St. Louis, Missouri. Both samples largely originate from low socioeconomic status groups. The African-American sample is also comprised of individuals from the Terry collection (25 males, 25 females), and similarly represents generally lower socioeconomic status individuals. A sample of western Aleuts (24 males, 20 females) from the collections of the National Museum of Natural History were included for comparison as a highly active sample in which males engaged in both bilateral and unilateral movement patterns (paddling ocean-going vessels and hunting with darts propelled by throwing boards, respectively). The Aleut sample originates from pre- or early post- Russian contact periods (Hrdlička, 1930, 1945).

We used the glenoid index to quantify scapular glenoid fossa shape. The index was constructed from glenoid articular length (GAL) and breadth (GAB), as 100*GAB/GAL. Glenoid articular length was taken as the maximum craniocaudal distance, in a coronal plane, between the outermost edges of the line of attachment of the glenoid labrum, following Churchill and Trinkaus (1990). Glenoid articular breadth was similarly defined as the maximum dorsoventral distance, in a transverse plane, between the outermost edges of the attachment for the labrum. For individuals preserving both right and left glenoid fossae, we averaged the indices for use in the analysis, while we used only the preserved side for individuals with unilateral preservation of the glenoid fossa. Given that asymmetry in upper limb joint size tends to be low even in highly active populations (Trinkaus et al., 1994), combining right and left scapulae is not expected to bias the results. We used the supinator index to quantify the relative hypertrophy of the supinator crest. The index was based on the ratio of supinator crest mediolateral diameter (SCD: defined as the mediolateral diameter of the ulnar shaft plus crest at the level of the maximum lateral projection of the crest) to ulnar shaft mediolateral diameter (UML: mediolateral diameter of the shaft just dorsal of, and excluding, the supinator crest), as 100*SCD/UML. Because the hypertrophy of muscle insertion sites is developmentally plastic (see Kennedy, 1989), and because throwing is a unimanual activity, we expect a substantial degree of asymmetry in supinator crest hypertrophy in throwing populations. Thus we analyzed supinator data from the right and left sides separately.

All statistical analyses were conducted using the software program SPSS version 12.0.1. Between group differences were evaluated by one-way analysis of variance (ANOVA), followed by the Hochberg GT2 post-hoc test subsequent to a significant ANOVA. The Hochberg GT2 test was chosen for its accuracy when sample sizes are unequal and the population variance is equal (Field, 2000). Levene's test for the homogeneity of variance returned non-significant results for both sexes for both variables, and the index values were found to be normally distributed (based on calculation of z-scores to evaluate skewness) with the exception of the glenoid index in the female Aleutian Islanders. Thus the non-parametric Kruskal-Wallis test was employed to test for differences between sexes in glenoid index.

Results

Glenoid indices show substantial variation across samples, both between Neandertals and modern humans and between fossil and recent modern human samples (Table 15.3). In accordance with earlier studies, Neandertals (of both sexes) have low mean values of the glenoid index, reflecting a relatively narrow glenoid fossa. The two Levantine Mousterian modern humans for which glenoid dimensions could be measured (Skhul V and Qafzeh 8) also have narrow glenoid fossae and a mean index value (when both sides are combined) virtually identical to that seen in Neandertal males. However, both of these fossils have imperfectly preserved glenoid fossae and thus estimated index values. If these estimates are reliable, it appears that Mousterian foragers – whether Neandertal or modern human – were characterized by narrow scapular glenoid fossae relative to Upper Paleolithic and recent modern humans. The ANOVAs returned significant results for both sexes, but post-hoc testing (Table 15.4) failed to detect any pair-wise significant difference in the female samples. Among the males, the only significant differences detected were between the Neandertals and the two modern human samples with the relatively widest glenoid fossae: European- and African Americans. Post-hoc testing failed to produce significant results when Neandertals or Levantine Mousterian modern humans were compared with early and late Upper Paleolithic males (most likely due to small sample sizes: when male and female values are pooled the Neandertals are found to differ from both EUP and LUP samples at $p < 0.05$).

Levels of bilateral asymmetry in glenoid index are relatively low (Table 15.5), with median values generally between 2–4%, and not exceeding 6%. This accords well with reported values of size asymmetry (Trinkaus et al., 1994), and supports the contention that glenoid fossa shape, like size, is only moderately epigentically plastic, and can thus probably safely be seen as a reflection of the underlying genetics and selective history of the various groups (note that previous analyses have also shown that there is no significant relationship between glenoid fossa shape and thoracic shape, robusticity or body size within and between groups of Pleistocene and Holocene humans: Churchill, 1994,1996).

Mean values of the supinator index are provided in Table 15.6. Again, Neandertals and Mousterian Levantine modern human males have low supinator index values for the right (presumably throwing) side relative to Upper Paleolithic and recent modern humans. Amongst males, the largest mean right-side values were found in the EUP, LUP and Aleutian Island samples. Given that Aleutian Islanders used spear-throwers ("throwing boards") in hunting, the results may be demarking habitual throwers from non-throwers. Post-hoc testing (Table 15.7) shows the Neandertal males to have significantly smaller supinator indices than EUP, LUP and Aleut males. EUP and Aleutian Islander males also have significantly larger supinator indices than African-American males. Relative supinator size was highly variable across samples in male left-side ulnae and female right- and left-side ulnae: however, no significant differences were found between groups for these values.

Bilateral asymmetry in relative supinator size was generally higher than observed for glenoid index (Table 15.8). Median values above 6% were seen in Neandertal, EUP, Aleut and African-American males, and in Aleut and African-American

TABLE 15.3. Glenoid indices in fossil and recent humans (mean, SD, *n*).

	Right	Left	Sides Combined
Males			
Neandertals	64.2	68.3	66.5
	3.03	3.73	3.28
	5	4	6
Levantine Mousterian	65.6	67.6	66.6
	–	–	1.41
	1	1	2
EUP	71.6	71.0	71.1
	3.28	1.77	2.82
	8	7	9
LUP	72.1	70.2	71.2
	2.58	3.50	3.05
	5	7	9
Euro-Americans	73.5	72.6	73.0
	4.11	3.31	3.17
	40	42	41
Afro-Americans	72.8	72.3	72.5
	3.16	2.32	2.20
	25	25	25
Aleutian islanders	70.9	69.9	70.9
	4.65	3.61	3.83
	24	21	25
Females			
Neandertals	65.0	68.9	66.9
	–	–	2.76
	1	1	2
Levantine Mousterian	–	–	–
EUP	69.6	73.6	71.7
	4.02	4.88	3.16
	4	3	6
LUP	69.2	70.6	70.7
	2.96	2.89	2.76
	3	3	4
Euro-Americans	72.2	71.7	71.9
	4.31	4.48	3.73
	39	41	42
Afro-Americans	72.2	71.4	71.8
	2.99	2.57	2.36
	25	25	25
Aleutian islanders	70.0	68.7	70.0
	4.59	4.45	4.32
	20	18	22
Sex indeterminate			
Neandertals	67.5	69.4	68.4
	2.72	6.44	4.68
	4	4	8
EUP	72.0	–	72.0
	–		–
	1		1

TABLE 15.4. Hochberg GT2 Post-Hoc P-values for between-group differences in glenoid index. (Lower diagonal: males; Upper diagonal: females [italicized]).

	Neandertal	Levantine	EUP	LUP	Euro-Am	Aleut	Afro-Am
Neandertal	–		*0.780*	*0.971*	*0.539*	*0.981*	*0.605*
Levantine	1.000	–	–	–	–	–	–
EUP	0.126	*0.765*	–	*1.000*	*1.000*	*0.993*	*1.000*
LUP	0.432	0.956	1.000	–	*1.000*	*1.000*	*1.000*
Euro-Am	<0.001**	0.107	0.857	0.268	–	*0.436*	*1.000*
Aleut	0.053	0.741	1.000	1.000	0.153	–	*0.705*
Afro-Am	0.001**	0.195	0.992	0.652	1.000	0.693	–

**Significant at $p < 0.01$

TABLE 15.5. Bilateral asymmetry in the Glenoid Fossa Index.

		Percent asymmetry		
Sample	Sex	Median	SD	Maximum
Neandertal	Males	1.99	8.63	16.33
EUP	Males	4.63	2.83	8.63
	Females	5.19	2.79	6.35
LUP	Males	3.40	0.85	3.78
	Females	2.65	3.59	5.18
Euro-Am	Males	3.27	3.54	13.02
	Females	3.42	3.96	15.90
Aleut	Males	3.59	3.52	11.28
	Females	5.96	3.11	11.10
Afro-Am	Males	3.88	2.49	11.01
	Females	2.72	2.76	8.75

TABLE 15.6. Supinator Indices in fossil and recent humans (mean, SD, *n*).

	Right	Left
Males		
Neandertals	68.6	73.4
	9.67	–
	4	1
Levantine Mousterian	66.4	67.2
	–	14.91
	1	2
EUP	81.9	79.7
	7.56	5.45
	7	6
LUP	81.4	82.2
	5.85	6.69
	9	11
Euro-Americans	79.0	79.4
	5.88	8.17
	24	24
Afro-Americans	74.7	76.5
	7.83	7.76
	24	25
Aleutian islanders	80.5	83.4
	7.81	6.58
	24	23
Females		
Neandertals	–	74.6
		–
		1
Levantine Mousterian	92.1	–
	22.05	
	2	
EUP	80.5	84.8
	–	8.76
	1	2
LUP	79.3	79.2
	7.05	4.71
	4	3
Euro-Americans	80.1	77.2
	9.55	8.45
	13	13
Afro-Americans	76.9	78.2
	6.48	8.09
	25	25
Aleutian islanders	85.1	89.9
	9.31	7.83
	13	14
Sex indeterminate		
Neandertals	74.3	–
	–	
	1	

females. The greater median percent asymmetry values reflect the greater epigenetic plasticity of the size of muscle insertion sites, and no doubt reflect behavioral asymmetries in upper limb use intensity and patterns.

Discussion

As with recent work on humeral retroversion angles (Rhodes and Churchill, 2009), the results discussed above are somewhat equivocal but generally support the claim that Mousterian foragers – both Neandertal and modern human – were not habitually engaged in throwing-based projectile weapon use, while EUP and LUP foragers were. Neandertals and Mousterian early modern humans tend to have narrow scapular glenoid fossae compared to EUP and LUP peoples, although when sexes are evaluated individually the differences are not statistically significant (no doubt due to small sample sizes). With the exception of the Levantine Mousterian humans from Skhul and Qafzeh, all modern humans share in having relatively wide glenoid fossae. This may reflect a shared selective history of hunting with long-range projectile weaponry, in which individuals with wider glenoid fossae were more resistant to degenerative joint disease (osteoarthritis) at the shoulder, and thus able to provision dependents over a greater portion of their life span. Assuming the pain and limited shoulder mobility associated with glenoid fossa osteoarthritis was selectively disadvantageous, we would predict relatively

TABLE 15.7. Hochberg GT2 Post-Hoc P-values for between-group differences in right-side supinator index. (Lower diagonal: males; Upper diagonal: females [italicized]).

	Neandertal	Levantine	EUP	LUP	Euro-Am	Aleut	Afro-Am
Neandertal	–	*1.00*	*1.00*	*1.00*	*1.00*	*1.00*	*1.00*
Levantine	1.00	–	*1.00*	*0.589*	*0.498*	*0.959*	*0.168*
EUP	0.008**	1.00	–	*1.00*	*1.00*	*1.00*	*1.00*
LUP	0.023*	1.00	1.000	–	*1.000*	*0.926*	*1.000*
Euro-Am	0.058	1.00	0.863	0.998	–	*0.762*	*0.953*
Aleut	0.015*	1.00	0.997	1.000	1.000	–	*0.065*
Afro-Am	0.718	1.00	0.044*	0.144	0.316	0.039*	–

*Significant at $p < 0.05$
**Significant at $p < 0.01$

TABLE 15.8. Bilateral asymmetry in the Supinator Index.

		Percent asymmetry		
Sample	Sex	Median	SD	Max
Neandertal	Males	6.90	–	–
EUP	Males	7.26	2.84	10.47
	Females	2.41	–	–
LUP	Males	5.34	6.94	17.58
	Females	2.75	2.28	6.09
Euro-Am	Males	4.04	4.77	20.87
	Females	2.44	3.80	10.02
Aleut	Males	7.24	4.95	20.71
	Females	7.84	7.14	22.36
Afro-Am	Males	7.93	5.39	20.43
	Females	8.60	7.28	21.21

rapid evolution of glenoid fossa shape following on the development of (or first real dependence upon) spearthrower technology (we would predict, then, that MSA humans post-dating 70 ky BP – following the arguments of Brooks et al., 2005 – would have relatively wide glenoid fossae).

The morphology of the supinator crest is also consistent with the argument for a late Middle Paleolithic/MSA development of throwing based projectile weaponry. The relative size of the crest in the right limbs of males is consistent with the expectation that Neandertals and Mousterian modern humans were not habitual throwers while EUP and LUP foragers were. Here the statistical results are more compelling, but the overall patterns (when male left limbs and female right and left limbs are considered) are difficult to interpret. As with other developmentally plastic traits (such as humeral retroversion angle: Rhodes and Churchill, 2009), the relative size of the supinator crest may be reflecting behaviors other than throwing that involve forceful supination of the forearm (such as, perhaps, hide scraping or woodworking). The small size of the supinator crest in Neandertals may also be a function of the hypertrophy they exhibit in the other supinator of the forearm, M. biceps brachii. The morphology of Neandertal radial bicipital tuberosities suggest that they were able to use the biceps muscle for supination over a wider range of pronation-supination than can most modern humans (Trinkaus and Churchill, 1988). Perhaps Neandertals relied more heavily on M. biceps brachii than M. supinator for actions requiring forceful supination. This possibly does not alter the conclusion, however, that supinator crest size in Neandertals suggests that they were not habitual throwers, since the extension of the forearm in the final stages of throwing prevents the use of M. biceps brachii as an antagonist to hyperpronation (thus even Neandertals would have had to rely on M. supinator in the deceleration phase of throwing). This conclusion, however, rests upon the assumption that spears would have been thrown in a manner resulting in full extension of the forearm in the follow-through phase of the throw (the best but not the only method of throwing a spear: see Miller, 1960), and also may not apply to smaller, spherical objects (such as stones) which might be thrown without the elbow moving into full extension. At present all that we can say is that the small size of the supinator crests suggests either less involvement in habitual spear throwing or biomechanically different patterns of throwing than seen in modern spear using people.

Conclusions

The human osteological data, when taken in total, suggest that Neandertal and modern human Mousterian foragers were not habitually engaging in forceful throwing, further suggesting that true, long-range projectile weaponry was not a component of Mousterian hunting technology (as argued from the archeological evidence by John Shea and others: see Shea, 2006). The consideration of scapular glenoid fossa shape and ulnar supinator crest hypertrophy strengthens this conclusion. The data are also consistent in suggesting that modern humans of the European Late Upper Paleolithic were habitually engaged in throwing, as clearly indicated by the archeological recovery of projectile weapon components dating to this interval. The data are, frustratingly, a bit more difficult to interpret with respect to throwing by early Upper Paleolithic modern humans. While some aspects of upper limb morphology (such as humeral cross-sectional geometry) suggest that EUP males were not habitually engaged in throwing, the majority of osteological indicators support the suggestion that they were. While issues of sample size and statistical significance still haunt inquiries into the osteological correlates of weapon use in the European later Pleistocene, the majority of indicators studied to date are consistent with the claim that projectile weapons arose in the African later MSA (Brooks et al., 2005) or the Near Eastern terminal Middle Paleolithic (Shea, 2006) and moved into Europe in the hands of modern humans. The two variables studied here – despite issues of sample size and difficult-to-interpret patterns of between-group variation – lend support to this claim.

Acknowledgments. Our sincere thanks go to Jean-Jacques Hublin and Mike Richards for the invitation to participate in the *Evolution of Hominid Diets* symposium. For many interesting and helpful discussions about the evolution of projectile weapon systems, we are grateful to John Shea, Alison Brooks, Paola Villa, and Daniel Richter. We would also like to thank three anonymous reviewers of this paper for their insightful comments, and the many curators across Europe and in North America who allowed us access to fossils and comparative modern human skeletal material.

References

Basabe, J.M., 1966. El húmero premusteriense de Lezetxiki (Guipúzcoa). Munibe 18, 13–31.

Basedow, H., 1913. Notes on the natives of Bathurst Island, north Australia. Journal of the Royal Anthropological Institute of Great Britain 43, 291–323.

Bingham, P.M., 2000. Human evolution and human history: a complete theory. Evolutionary Anthropology 9, 248–257.

Breuil, H., 1912. Les subdivisions du paléolithique supérieur et leur signification. Comptes Rendus de la XIVeme Session, Congres Internationale d'Anthropologie, Archéologie et Préhistoire Genève 5–78.

Brooks, A.S., Yellen, J.E., Nevell, L., Hartman, G., 2005. Projectile technologies of the African MSA: implications for modern human origins. In: Hovers, E., Kuhn, S. (Eds.), Transitions Before the Transition: Evolution and Stability in the Middle Paleolithic and Middle Stone Age. Kluwer, New York, pp. 233–255.

Cattelain, P., 1989. Un crochet de propulseur solutréen de la grotte de Combe- Saunière 1 (Dordogne). Bulletin de la Société Préhistorique Française 86, 213–216.

Churchill, S.E., 1993. Weapon technology, prey size selection, and hunting methods in modern hunter-gatherers: implications for hunting in the Palaeolithic and Mesolithic. In: Peterkin, G.L., Bricker, H.M., Mellars, P.A. (Eds.), Hunting and Animal Exploitation in the Later Palaeolithic and Mesolithic of Europe. Archaeological Papers of the American Anthropological Association 4, 11–24.

Churchill, S.E., 1994. Human upper body evolution in the Eurasian Later Pleistocene. Ph.D. Dissertation, University of New Mexico.

Churchill, S.E., 1996. Particulate versus integrated evolution of the upper body in Late Pleistocene humans: a test of two models. American Journal of Physical Anthropology 100, 559–583.

Churchill, S.E., 2002. Of assegais and bayonets: reconstructing prehistoric spear use. Evolutionary Anthropology 11, 185–186.

Churchill, S.E., Morris, A.G., 1998. Muscle marking morphology and labor intensity in prehistoric Khoisan foragers. International Journal of Osteoarchaeology 8, 390–411.

Churchill, S.E., Pearson, O.M., Grine, F.E., Trinkaus, E., Holliday, T.W., 1996a. Morphological affinities of the proximal ulna from Klasies River Main Site: archaic or modern? Journal of Human Evolution 31, 213–237.

Churchill, S.E., Rhodes, J.A., 2006. How strong were the Neandertals? Leverage and muscularity at the shoulder and elbow in Mousterian foragers. Periodicum Biologorum 108, 457–470.

Churchill, S.E., Trinkaus, E., 1990. Neandertal scapular glenoid morphology. American Journal of Physical Anthropology 83, 147–160.

Churchill, S.E., Weaver, A.H., Niewoehner, W.A., 1996b. Late Pleistocene human technological and subsistence behavior: functional interpretations of upper limb morphology. In: Bietti, A., Grimaldi, S. (Eds.), Reduction Processes ("Chaines Opératoires") in the European Mousterian. Quaternaria Nova 6, 18–51.

Crockett, H.C., Gross, L.B., Wilk, K.E., Schwartz, M.L., Reed, J., O'Mara, J., Reilly, M.T., Dugas, J.R., Meister, K., Lyman, S., Andrews, J.R., 2002. Osseous adaptation and range of motion at the glenohumeral joint in professional baseball pitchers. American Journal of Sports Medicine 30, 20–26.

Field, A., 2000. Discovering Statistics Using SPSS for Windows. Sage Publications, London.

Gainor, B.J., Piotrowski, G., Puhl, J., Allen, W.C., Hagen, R., 1980. The throw: biomechanics and acute injury. American Journal of Sports Medicine 8, 114–118.

Goodale, J.C., 1971. Tiwi Wives: A Study of the Women of Melville Island, North Australia. University of Washington Press, Seattle.

Guthrie, R.D., 1983. Osseous projectile points: biological considerations affecting raw material selection and design among Palaeolithic and Palaeoindian people. In: Clutton-Brock, J., Grigson, C. (Eds.), Animals and Archaeology. Volume 1. Hunters and Their Prey. British Archaeological Reports, Oxford, pp. 273–294.

Howell, S.M., Galinat, B.J., Renzi, A.J., Marone, P.J., 1988. Normal and abnormal mechanics of the glenohumeral joint in the horizontal plane. Journal of Bone and Joint Surgery 70-A, 227–232.

Hrdlička, A., 1930. Anthropological Survey in Alaska. 46th Annual Report of the Bureau of American Ethnology, 1928–1929. Washington DC, pp. 19–374.

Hrdlička, A., 1945. The Aleutian and Commander Islands and Their Inhabitants. Wistar Institute, Philadelphia.

Hughes, S.S., 1998. Getting to the point: evolutionary change in prehistoric weaponry. Journal of Archeological Method and Theory 5, 345–408.

Kelly, R., 1995. The Foraging Spectrum. Smithsonian Institution, Washington DC.

Kennedy, K.A.R., 1983. Morphological variations in ulnar supinator crests and fossae as identifying markers of occupational stress. Journal of Forensic Sciences 28, 871–876.

Kennedy, K.A.R., 1989. Skeletal markers of occupational stress. In: Iscan, M.Y., Kennedy, K.A.R. (Eds.), Reconstruction of Life from the Skeleton. Alan R. Liss, New York, pp. 129–160.

Kennedy, K.A.R., 2004. Slings and arrows of predaceous fortune: Asian evidence of prehistoric spear use. Evolutionary Anthropology 13, 127–131.

Kortlandt, A., 2002. Neanderthal anatomy and the use of spears. Evolutionary Anthropology 11, 183–184.

Marlowe, F.W., 2005. Hunter-gatherers and human evolution. Evolutionary Anthropology 14, 54–67.

Matiegka, J., 1938. Homo predmostensis, Fosilni Clovek z Predmosti na Morave II. Nakladem Ceske Akademie Ved a Umeni, Prague.

McBrearty, S., Brooks, A.S., 2000. The revolution that wasn't: a new interpretation of the origin of modern human behavior. Journal of Human Evolution 39, 453–563.

Meister, K., 2000. Injuries to the shoulder in the throwing athlete. Part one: biomechanics/pathophysiology/classification of injury. American Journal of Sports Medicine 28, 265–275.

Miller, J.E., 1960. Javelin thrower's elbow. Journal of Bone and Joint Surgery 42B, 788–792.

Oakley, K.P., Andrews, P., Keeley, L.H., Clark, J.D., 1977. A reappraisal of the Clacton spearpoint. Proceedings of the Prehistoric Society 43, 13–30.

O'Connell, J.F., 2006. How did modern humans displace Neanderthals? Insights from hunter-gatherer ethnography and archaeology. In: Conard, N.J. (Ed.), Neanderthals and Modern Humans Meet. Kerns, Tubingen, pp. 43–64.

Osbahr, D.C., Cannon, D.L., Speer, K.P., 2002. Retroversion of the humerus in the throwing shoulder of college baseball pitchers. American Journal of Sports Medicine 30, 347–353.

Perry, J., Glousman, R.E., 1995. Biomechanics of throwing. In: Nicholas, J.A., Hershman, E.B., Posner, M.A. (Eds.), The Upper Extremity in Sports Medicine. Mosby, St. Louis, pp. 725–750.

Peterkin, G.L., Bricker, H.M., Mellars, P.A. (Eds.), 1993. Hunting and Animal Exploitation in the Later Palaeolithic and Mesolithic of Europe. Archaeological Papers of the American Anthropological Association 4.

Peterson, J., 1998. The Natufian hunting conundrum: spears, atlatls, or bows? Musculoskeletal and armature evidence. International Journal of Osteoarchaeology 8, 378–389.

Pieper, H.-G., 1998. Humeral torsion in the throwing arm of handball players. American Journal of Sports Medicine 26, 247–253.

Reagan, K.M., Meister, K., Horodyski, M.B., Werner, D.W., Carruthers, C., Wilk, K., 2002. Humeral retroversion and its relationship to glenohumeral rotation in the shoulder of college baseball players. American Journal of Sports Medicine 30, 354–360.

Rhodes, J.A., Churchill, S.E., 2009. Throwing in the Middle and Upper Paleolithic: inferences from an analysis of humeral retroversion. Journal of Human Evolution 56, 1–10.

Rieder, H., 2001. Erprobung der holzspeere von Schöningen (400 000 jahre) und folgerungen daraus. In: Wagner, G.A., Mania, D. (Eds.), Frühe Menschen in Mitteleuropa – Chronologie, Kultur, Umwelt, Aachen. Heidelberg.

Rieder, H., 2003. Nachbau altsteinzeitlicher speere der groâe wurf der frühen jäger. Biologie in unserer Zeit 33, 156–160.

Ruff, C.B., Trinkaus, E., Holliday, T.W., 1997. Body mass and encephalization in Pleistocene Homo. Nature 387, 173–176.

Schmitt, D.O., Churchill, S.E., Hylander, W.L., 2003. Experimental evidence concerning spear use in Neandertals and early modern humans. Journal of Archaeological Science 30, 103–114.

Shea, J.J., 2003. Neandertals, competition, and the origin of modern human behavior in the Levant. Evolutionary Anthropology 12, 173–187.

Shea, J.J., 2006. The origins of lithic projectile point technology: evidence from Africa, the Levant, and Europe. Journal of Archaeological Science 33, 823–846.

Shea, J.J., 2009. The impact of projectile weaponry on Late Pleistocene hominin evolution. In: Hublin, J.-J., Richards, M.P. (Eds.), The Evolution of Hominid Diets: Integrating Approaches to the Study of Paleolithic Subsistence. Springer, Dordrecht, pp. XXX–XXX.

Shott, M.J., 1993. Spears, darts, and arrows: Late Woodland hunting techniques in the upper Ohio valley. American Antiquity 58, 425–443.

Soficaru, A., Dobos, A., Trinkaus, E., 2006. Early modern humans from the Peştera Muierii, Baia de Fier, Romania. Proceedings of the National Academy of Sciences 103, 17196–17201.

Straus, L.G., 1995. The Upper Paleolithic of Europe: an overview. Evolutionary Anthropology 4, 4–16.

Thieme, H., 1997. Lower Palaeolithic hunting spears from Germany. Nature 385, 807–810.

Thieme, H., 1999. Lower Palaeolithic throwing spears and other wooden implements from Schöningen, Germany. In: Ullrich, H. (Ed.), Hominid Evolution: Lifestyles and Strategies. Edition Archaea, Gelsenkirchen/Schwelm, pp. 383–395.

Trinkaus, E., Churchill, S.E., 1988. Neandertal radial tuberosity orientation. American Journal of Physical Anthropology 75, 15–21.

Trinkaus, E., Churchill, S.E., 1999. Diaphyseal cross-sectional geometry of Near Eastern Middle Paleolithic humans: the humerus. Journal of Archaeological Science 26, 173–184.

Trinkaus, E., Churchill, S.E., Ruff, C.B., 1994. Postcranial robusticity in Homo. II: Humeral bilateral asymmetry and bone plasticity. American Journal of Physical Anthropology 93, 1–34.

Tullos, H.S., King, J.W., 1973. Throwing mechanism in sports. Orthopedic Clinics of North America 4, 709–720.

Villa, P., 2009. Lower and Middle Paleolithic faunal exploitation in Europe. In: Hublin, J.-J., Richards, M.P. (Eds.), The Evolution of Hominid Diets: Integrating Approaches to the Study of Paleolithic Subsistence. Springer, Dordrecht, pp. XXX–XXX.

Wilmsen, E.N., 1974. Lindenmeier: A Pleistocene Hunting Society. Harper & Row, New York.

Zarins, B., Andrews, J.R., Carson, W.G.J., 1985. Injuries to the Throwing Arm. W.B. Saunders, Philadelphia.

16. An Energetics Perspective on the Neandertal Record

Katharine Macdonald
Faculty of Archeology
Leiden University, P.O. Box 9515
2300 RA Leiden, The Netherlands
k.macdonald@arch.leidenuniv.nl

Wil Roebroeks
Faculty of Archeology
Leiden University, P.O. Box 9515
2300 RA Leiden, The Netherlands
w.roebroeks@arch.leidenuniv.nl

and

Alexander Verpoorte
Faculty of Archeology
Leiden University, P.O. Box 9515
2300 RA Leiden, The Netherlands
a.verpoorte@arch.leidenuniv.nl

Keywords Neandertals • energetics • cognition • subsistence • use of space

Abstract Our aim in this paper is twofold: first to provide a broad overview of current knowledge of the Neandertal archaeological record, and second, to provide an example of an energetics perspective on this archeological evidence. Neandertals hunted large mammals in their prime in a wide range of environments, and exploited a limited range of food items compared with some anatomically modern humans. In general, they made use of simple, low investment tools, with little change over time and space, and with an emphasis on the production of cutting tools, and invested little in spatial structure. The archaeological record left by Neandertals seems limited when compared to the Upper Paleolithic of Europe.

Explanations for the differences between the archeological record of Neandertals and Upper Paleolithic anatomically modern humans generally highlight supposed cognitive differences. We suggest that it would be worth investigating alternative explanations for these differences, starting from the observation that a number of studies have identified differences in energy use and requirements between the two species. We present an example of an application of an energetics perspective to the Neandertal archeological record, focusing on the use of space. This exercise yields interesting predictions for the archeological record, suggesting that this approach may be useful for explaining differences between the Middle and Upper Paleolithic record in Europe, as well as variation within these periods.

Introduction

In the first part of this paper we review current knowledge of the Neandertal archeological record, providing a background to the detailed studies of fauna and particular aspects of technology presented by other authors in this volume. We briefly discuss Neandertal habitats and questions about limits to their environmental tolerance, the evidence for Neandertal diet and foraging strategies, and trends in lithic and other technology. The Neandertal record is often put into perspective by comparison with the European Upper Paleolithic record, generally produced by anatomically modern humans. Many explanations for contrasts between the two records exist, often highlighting supposed cognitive differences between the two species.

In the second part of the paper we suggest an alternative explanation for some of these differences, starting with biological differences between the two species. Innovative approaches to the fossil record have begun to explore the energy used and required by Neandertals. These studies agree that Neandertal energy requirements are higher than those for anatomically modern humans. We take a closer look at the

J.-J. Hublin and M.P. Richards (eds.), *The Evolution of Hominin Diets: Integrating Approaches to the Study of Palaeolithic Subsistence,* 211–220.

point of departure for some of these estimates, and compare Neandertal energy requirements with those of the modern humans who created the Upper Paleolithic record. Finally, we present a preliminary attempt to explore the implications of high energy requirements for the Neandertal record, focusing on the use of space as an example.

The Neandertal Archeological Record

Time Period and Geographical Scale

Neandertals are by far the best-studied extinct hominins, with a rich fossil record sampling hundreds of individuals. They were the end product of a long evolutionary lineage that colonized Europe as early as about 500,000 years ago (Hublin, 1998; Bischoff et al., 2007), and disappeared from the fossil record by approximately 30,000 years ago. The Neandertal record is heavily biased towards the western part of their former range, western Europe, with the northern, eastern and southern limits of their distribution poorly documented because of imbalances in research intensity (Dennell and Roebroeks, 2005). The juvenile from Teshik-Tash, Uzbekistan, is one of the easternmost ones known, at roughly 1,300 miles from its nearest fossil neighbors, Shanidar in Iraq and 2,000 miles from Kiik-Koba in the Black Sea area. Recently, another occurrence of Neandertals has been discovered roughly 1,200 miles further east of Teshik-Tash (Krause et al., 2007). The southern limit of their distribution is unknown, and may have extended over the whole of Arabia and the Indian subcontinent – until these regions produce the necessary fossil evidence, we simply cannot be certain.

Environments

Many recent studies have focused on characterizing the habitats occupied by Neandertals and their environmental limits (Gamble, 1987; Roebroeks et al., 1992; Roebroeks and Gamble, 1999; van Andel and Davies, 2003; Stewart, 2005). From these, we know that for much of the Neandertal time range, western Eurasia was dominated by a biome that has become known as the Mammothap Steppe (Guthrie, 1984, 1990). Compared to tundra and polar deserts, this was a highly productive habitat that supported a very diverse grazing community, including the mammoth as its characteristic species. This habitat stretched from Cantabria to Alaska, and it is clear that within this monolithic concept one can uncover a great deal of chronological and spatial variation. Ice core studies suggest that apart from the Holocene, instability has dominated the North Atlantic climate over the last 110,000 years, and probably longer (Dansgaard et al., 1993; Alley, 2000).

During glacial maxima, animal populations contracted into more temperate refugia, expanding into previously abandoned territory again when the climate improved. Turq (1999) has suggested that the Aquitaine basin (France) may have functioned as one of the Neandertal refugia in western Europe, predicting that cold periods should see an increase in both the number of sites and in the richness of assemblages there. While most researchers seem to agree on an "ebb-and-flow" model for Neandertal occupation patterns at a regional and a continental scale, the question is where their limits were.

Turning known distribution patterns into inferences on environmental limits has led to considerable debate, for example over Neandertals' ability to cope with climax-interglacial forested environments (Gamble, 1986; Roebroeks et al., 1992; Ashton, 2002). Current evidence strongly suggests that Neandertals were indeed able to cope with such interglacial habitats (Roebroeks and Speleers, 2002). However, until recently all sites attributed to the Eemian came from Central Europe. While some authors suggest that this reflected Neandertal preference for continental climates during this period, Roebroeks and Speleers have argued that the distribution of interglacial sites reflects a combination of geological and research-historical factors (Roebroeks and Speleers, 2002). Current excavations at Caours in northern France are particularly interesting in this context. The fine grained upper part of the Somme river deposits have thus far yielded five archeological levels dating to the Eemian s.s. The sediments have OSL dates which situate them around 125,000 years ago, and the fauna is of full interglacial character, containing *Cervus elaphus*, *Capreolus capreolus*, *Bos primigenius*, *Dama dama*, *Sus scrofa*, and the forest rhino *Stephanorhinus kirchbergensis* (Locht, 2006). Recent controversy also concerns Neandertal ability to deal with very cold climates (Ruff et al., 1993; Holliday, 1997; Aiello and Wheeler, 2003; van Andel and Davies, 2003). While the short, stocky body form of Neandertals is often described as arctic (Ruff et al., 1993) or hyperarctic (Holliday, 1997), Aiello and Wheeler have argued that Neandertal morphology would have had little effect on the minimum temperature at which they could survive. Glacial maxima seem to have led to depopulation in northwest Europe, as there are no finds that can be unambiguously attributed to these phases (Roebroeks and Tuffreau, 1999). In conclusion, Neandertals were present in a wide range of environments, and only the most extreme cold phases of the glacial-interglacial cycles seem to have led to range shifts and retreats into refugia.

Subsistence

In line with the wide variety of habitats documented for Neandertals, prey species varied from reindeer at Salzgitter-Lebenstedt, Germany (Gaudzinski and Roebroeks, 2000) to aurochs at La Borde, France (Jaubert et al., 1990), to forest rhino at the last interglacial site of Taubach, Germany (Bratlund, 1999). From the beginning of the Middle Paleolithic onward, assemblages are frequently dominated by particular species of large herbivores (Gaudzinski, 2009). In addition, a focus on better-quality animals and parts of animals has been documented at a range of locations, including Salzgitter-Lebenstedt and La Borde (Gaudzinski, 2009). Neandertals seem to have focused on high-return resources irrespective of the environmental context (Gaudzinski, 2004). In the Levant, archaeozoological studies indicate that Neandertal hunting activities may even have led to a decline of red deer and aurochs populations (Speth, 2004; Speth and Clark, 2006).

Middle Paleolithic exploitation of small prey has been documented at lower latitudes, largely confined to "gatherable" food items (marine molluscs, tortoises, legless lizards, and ostrich eggs, cf. Stiner, 2002). There is limited evidence for use of relatively fast moving, difficult to catch small game (such as rabbits or game birds). Altogether, evidence for the exploitation of small prey in the Middle Paleolithic has been interpreted as indicating lower Neandertal dietary breadth compared with later humans in this region (Stiner, 2001).

At the same time, isotope studies have been carried out on a small sample of Neandertal fossils from northern Europe. Thus far, these indicate that the bulk of Neandertal dietary protein came from animal sources, strongly suggesting that they were top level carnivores (Richards et al., 2000). Isotope studies suggest that Neandertals and Upper Paleolithic modern humans were eating similar diets in Europe, though some modern humans had broader diets, incorporating inland aquatic resources in particular (Richards et al., 2001; Richards, 2009). However, Middle Paleolithic Neandertal specimens show a highly heterogeneous pattern of tooth microwear, suggesting that they exploited plants or meat on the basis of food availability (Pérez-Pérez et al., 2003).

While the different lines of evidence tell us about different aspects of subsistence – the prey species involved and subsistence strategies, or the dominant type of food in the diet – the evidence from archaeozoological studies of faunal remains uncovered at Neandertal sites, as well as stable isotope studies of their skeletal remains indicate that animal products formed an important part of their diet.

Lithic Technology

As far as lithic technology goes, Neandertals generally made use of simple, low investment tools. The overall emphasis in lithic technology was on the production of cutting edges such as scrapers. Neandertal lithic technology is dominated by prepared core technologies, in particular Levallois, and it has become increasingly clear that blade core technology was a regular component of their lithic skills (e.g. Révillion and Tuffreau, 1994). Middle Paleolithic lithic technology is predominantly focused on the use of local raw materials, even though lithics were occasionally transported over larger distances (Geneste, 1985; Roebroeks et al., 1988; Féblot-Augustins, 1999). The transported part of their toolkit was preferentially made on Levallois blanks (Geneste, 1985; Roebroeks et al., 1988) that yield larger cutting edges for less weight compared to other reduction techniques, such as discoidal ones. In comparison, "exotic" raw materials tend to have a greater importance in the Upper Paleolithic, especially from the Middle Upper Paleolithic (MUP) onward (Féblot-Augustins, 1997).

A recent find of pre-MIS 7 flint flakes with remnants of birch tar at Campitello Quarry, Italy (Mazza et al., 2006), shows that the practice of hafting flint tools is considerably older than the late Middle Paleolithic finds from Königsaue, Germany (Grünberg, 2002) and Umm el Tlell, Syria (Boëda et al., 1996). Middle Paleolithic stone point dimensions, as well as features of Neandertal upper limbs, indicate that Neandertals were not using long-range projectile weaponry such as spear-throwers and bows and arrows (Shea, 2006, Churchill, 2007, 2009; Villa, 2009). The 300,000–400,000 year-old wooden spears discovered at Schöningen have been interpreted as throwing spears (Thieme, 1997; Rieder, 2000), but other workers have suggested that they are better interpreted as thrusting spears (Shea, 2006; Churchill, 2007, 2009).

Use of Space

The European Lower and Middle Paleolithic record contains a large number of open as well as rock-shelter locales, with sometimes excellent preservation of the original spatial layout of find scatters. "Structures evidentes" such as tent rings and stone-lined hearths are rare to non-existent. Though some archeologists have seen the traces of "hut structures" or "dwellings" in the distribution maps of Lower and Middle Paleolithic sites, the interpretation of such "structures latentes" is extremely problematic. All claims of Lower and Middle Paleolithic dwellings have been criticized, and all fail to pass close scrutiny of taphonomic processes (e.g. Terra Amata, Bilzingsleben, Rheindahlen, etc., see Kolen, 1999 for a review). Even the presence of unambiguous wind breaks is rare. The same applies to the use of fire. While traces of the former presence of fires are surprisingly rare before the beginning of the Middle Paleolithic, at about 300,000 years ago, we do know remains of fireplaces from Middle Paleolithic sites. On open air sites these are usually in the form of some patches of burnt flints and/or burnt bones, occasionally associated with some evidence for burned sediments (for example, the fireplaces at Beauvais in northern France (Locht, 2004)). Likewise, in caves, fireplaces usually consist of little more than a shallow pit with a lens of charcoal and ashes, e.g. the ones excavated at Roc de Marsal, France (Sandgathe et al., 2005) or the sequence documented at Kebara cave, Israel (Bar-Yosef et al., 1992; Speth, 2006). This does not mean that camp life was not "organized;" for instance, Speth (2006) has convincingly shown that Neandertals used Kebara cave over long periods of time in a manner consistent with the way modern hunter-gatherers might use such a cave, including periodically cleaning the fireplace and habitation area and disposing of debris in a midden.

Neandertals seem not to have made investments in dwellings and camp facilities. The use-life of Middle Paleolithic locales was short, and the main difference between sites is in terms of the predominance of stone tool production versus tool maintenance.

Current Explanations: Neandertal Cognition

The character of the Middle Paleolithic record is usually put in perspective by comparison with the European Upper Paleolithic record, largely produced by anatomically modern humans (AMH). The archeological record for Neandertal use of space and lithic technology seems poor and limited when compared to the Upper Paleolithic. Within the large range of Neandertals, archeologists have uncovered much less chronological and spatial patterning in lithic reduction strategies and tool forms than in the ensuing Upper Paleolithic. While hut structures or tents are not ubiquitous in the Upper Paleolithic record; the absence of any evidence for pits, tents, huts, or stone-lined hearths in the Middle Paleolithic is striking. Apart from the elements described above, there are the well-known absences of representational "art" objects and of burials with elaborate grave goods that add to the contrast between the two periods, as far as the European record is concerned.

There exists a wide range of explanations for the differences between the archeological record of Neandertals and AMH. For example, some authors have focused on intensification of subsistence strategies as a result of increasing population pressure around the Middle to Upper Paleolithic "transition" (e.g. Stiner, 2001, 2002). However, the main line of explanation is in terms of a difference in cognitive capacities. A well-known example is Richard Klein's suggestion that a *genetic* change at 50,000 years ago boosted the AMH's brain's cognitive powers (Klein, 2000, 2001; Klein and Edgar, 2002). This mutation would have led to the development of many of the characteristics that we tend to associate with modern humans, such as use of symbolic expression and language. In a comparable vein, Tomasello et al. (2005) interpret the archeological record of modern humans with its chronological and spatial differentiation as the result of cumulative natural selection acting on culture, rather than a genotype/phenotype. As ideas, traditions and skills are passed down from one generation to the next, they tend to have novel enhancements made to them, almost every generation "ratchets up" specific items from the "prior" culture, and this ratchet effect is based, in their view, upon the (uniquely human) capacity for joint attention, for "shared intentionality." The – compared to the Upper Paleolithic – puzzling stability of earlier technologies would indicate that earlier hominins were cognitively unfit for "shared intentionality." Many more examples can be presented focusing on cognitive differences, for example Mellars' (2004) emphasis on the emergence of more complex language patterns with AMH.

We suggest that it is worthwhile to investigate alternative explanations for the differences in the archeological record of Neandertals and AMH. Anatomical studies have made enormous progress in understanding Neandertal physical features. Recent studies indicate that Neandertal characteristics also reveal major differences in energetics compared to modern humans. We will argue that this may have major implications for Neandertal behavior.

Neandertal Energetics

Introduction

A number of recent studies have concluded that Neandertals had a higher basal metabolic rate (BMR) than AMH due to a higher body mass and different shape (Sorensen and Leonard, 2001; Steegmann et al., 2002; Aiello and Wheeler, 2003; Churchill, 2009). Based on this high BMR, estimates of total energy expenditure (TEE) of Neandertals are proportionally high. Churchill (2009) estimates that it took, on average, between 3,500 and 5,000 kcal/day to feed an adult Neandertal. These estimates are comparable with those from other studies, falling at the lower end of the range of male and the higher end of female values reported by Sorensen and Leonard (2001), and slightly above those of Steegmann et al. (2002). This can be compared with mean daily energy expenditure of 3,000–4,000 kcal/day for males pursuing traditional foraging practices in the circumpolar regions (Steegmann et al., 2002).

The figures cited above point to a strong contrast in energy requirements between Neandertals and modern humans, and it is this comparison to *extant* modern humans that is often implicitly used when comparing Neandertals and AMH.

In the context of this paper, it is worthwhile to have a closer look at the point of departure for some of the estimates given above, and to compare these Neandertal energy demands with those of a Pleistocene sample of AMH.

Estimates of BMR for Neandertals and Upper Paleolithic Humans

The energy needed to maintain life when at rest scales with both body mass and skin surface area, and can be calculated using a number of different methods. We produced estimates of Neandertal BMR using body mass and stature estimates from the literature (Ruff et al., 1997, 2005; Churchill, 2009) and surface area from Churchill (2009), basing our calculations on European fossils (Neandertals in the Near East have different body proportions and are relatively gracile compared to the European ones (Holliday, 2000)). These calculations show that BMR estimates for Neandertals vary by approximately 200 kcal depending on the equation used (see Table 16.1).

The Middle Upper Paleolithic (30,000–20,000 years ago) provides a suitable fossil sample of Pleistocene AMH; this is not available for the Early Upper Paleolithic, given that earlier human fossils are scarce and are not associated with archaeology. Middle Upper Paleolithic (MUP) humans had larger body size than modern humans by about 10 kg (Ruff et al., 1997).

Table 16.1. Estimated BMR for Neandertals and Middle Upper Paleolithic humans (From Sorensen and Leonard, 2001; Aiello and Wheeler, 2003; Ruff et al., 1997, 2005; Churchill, 2007).

Sample	Body mass (kg)[a]	Stature (cm)[a]	Surface Area (m^2)[b]	BMR (kcal day^{-1})[d]	BMR (kcal day^{-1})[e]	BMR (kcal day^{-1})[f]
European Neandertals	74.5 (*n* = 17)	163.5 (*n* = 13)	1.860 (*n* = 12)	1,780 (*n* = 17)		1,938 (*n* = 12)
European male Neandertals	79.2 (*n* = 10)	164.7 (*n* = 9)	1.926 (*n* = 9)	1,863 (*n* = 10	1,891 (*n* = 10)	2,007 (*n* = 9)
European female Neandertals	67.8 (*n* = 7)	160.75 (*n* = 4)	1.760 (*n* = 3)	1,658 (*n* = 7)	1,493 (*n* = 7)	1,834 (*n* = 3)
MUP	66.6 (*n* = 33)	171.0 (*n* = 6)	1.790[c] (*n* = 6)	1,636 (*n* = 33)		1,865 (*n* = 6)
MUP males	68.5 (*n* = 26)			1,671 (*n* = 26)	1,727 (*n* = 26)	
MUP females	59.6 (*n* = 7)			1,505 (*n* = 7)	1,372 (*n* = 7)	
Modern male hunter gatherer	53.3			1,384	1,495	
Modern female hunter gatherer	45.4			1,227	1,163	

[a] Data for Neandertal (*n* = 17) and MUP body mass (*n* = 33) from Ruff et al., 1997. Ruff et al., 2005 yield a somewhat higher estimate for smaller number of MUP individuals (*n* = 6). Data for Neandertal stature from Churchill, 2007. Data for MUP stature from Ruff et al., 2005. Data for modern hunter-gatherer body mass from Sorensen and Leonard (2001).

[b] Data for Neandertal skin surface area from Churchill (2007).

[c] Skin Surface area = $0.0235 \times \text{Stature}^{0.42246} \times \text{Mass}^{0.5146}$ (Gehan and George, 1970).

[d] BMR (W) = $3.4 \times \text{Mass (kg)}^{0.75}$ (1 W = 20.64 kcal/day) (Kleiber, 1961).

[e] BMR = 15.3 (Mass) + 679 (males); 14.7 (Mass) + 496 (females) (FAO/WHO/UNU, 1985).

[f] BMR = 1,042 × Skin surface area (Winslow and Herrington, 1949).

We calculated estimates of the BMR of Middle Upper Paleolithic modern humans using the different formulas available (see Table 16.1). We found that MUP humans still had lower estimated BMR than for Neandertals. Specifically, the difference in BMR is 9% or more, based on calculations from body mass using two different equations (Kleiber, 1961, FAO/UNESCO/WHO, 1985), and 4% using calculations from skin surface area (Winslow and Herrington, 1949).

Living in cold climates is associated with increased metabolic costs, based on short-term acclimatization and genetic adaptations to cold stress (Leonard et al., 2005). This involves an increase in BMR of about 5% in Siberian populations (Leonard et al., 2005). This is not taken into account in the estimates presented in Table 16.1, but would raise estimates for Neandertals and MUP humans living in cold climates.

Estimates of Total Energy Budget

The total energy budget of an animal includes both the costs necessary for keeping an animal alive on a daily basis and the costs of growth and reproduction (Leonard and Ulijaszek, 2002). The component of the energy budget associated with keeping the animal alive on a daily basis includes the energy needed for basal metabolic processes, heat production, and productive work (Leonard and Ulijaszek, 2002; Churchill, 2009). TEE, defined as the amount of energy that an individual expends over the course of a typical, active day, is a measure of this part of the energy budget (Leonard et al., 2005).

Productive work for Neandertals would have included a wide range of different areas of behavior, of which locomotion while foraging is one, making up 7–14% of total energy expenditure in relatively mobile species (Weaver and Steudel-Numbers, 2005). According to Steudel-Numbers and colleagues (Steudel-Numbers and Tilkens, 2004; Weaver and Steudel-Numbers, 2005), large body size, as well as shorter limb length, would have increased the energetic costs of locomotion for Neandertals relative to MUP humans. While this would not necessarily increase Neandertal TEE, it would mean that locomotion costs took up a higher proportion of the budget relative to other activities.

Below a certain critical temperature, an animal can no longer regulate its core temperature by controlling its thermal conductance and has to increase internal heat production. Although Neandertals have been described as having a cold-adapted body form, Aiello and Wheeler (2003) have shown that they would only have had a modest advantage over AMH in their lower critical and minimum sustainable temperature. At low temperatures, both Neandertals and modern humans would have incurred additional energetic costs.

Daily activity levels for human groups are generally assessed using the PAL index developed by the World Health Organization (FAO/WHO/UNESCO, 1985). The PAL is simply a ratio of total energy expenditure to basal energy expenditure, and provides a measure of the relative amount of energy that a person is expending above their basal needs on a typical day (Leonard et al., 2005). Empirical data from contemporary foragers in the arctic suggests that a PAL of 2–2.5 is plausible for Neandertals (Sorensen and Leonard, 2001; Churchill, 2009).

We calculated TEE for Neandertals and Middle Upper Paleolithic humans from our estimates of BMR, assuming a PAL of 2.5. Using BMR calculated from skin surface area (Churchill, 2009), TEE would have been approximately 5,020 kcal/day for male and 4,590 kcal/day for female Neandertals. Using BMR calculated from body mass, we obtain slightly lower values of 4,450 kcal/day or 4,230 kcal/day for an adult Neandertal (using the Kleiber equation and FAO/WHO/UNESCO equation respectively). TEE for an adult MUP human would have been 4,660 kcal/day (using BMR calculated from skin surface area) or 4,090 kcal/day (using the Kleiber equation). These calculations indicate a difference in total energy requirements of 4% (using BMR calculated from skin surface area) or 9% or more (using the Kleiber or FAO/WHO/UNESCO equation).

The energy invested in growth and reproduction (together making up the production costs) constitutes another unknown in our comparison of Neandertals to Upper Paleolithic humans. In addition to meeting personal needs, hunter-gatherers must feed dependent members of the population, and fuel physical growth as well as reproduction; this amounts to daily requirements well above TEE. In view of the larger Neandertal bodies, female energy requirements for reproduction must have been relatively high compared to Upper Paleolithic humans (Aiello, 2007). In addition, energy requirements for growth must have been higher due to larger body size and possibly rapid growth (Ramirez Rozzi and Bermúdez de Castro, 2004; but see Guatelli-Steinberg et al., 2005). Finally, Neandertal life history seems to have been characterized by high adult mortality (Trinkaus, 1995). Among modern hunter-gatherers, production is low throughout childhood and increases dramatically in adulthood (Kaplan et al., 2000). If adults had a shorter lifespan, and energy requirements were high, this would lead to higher daily energy requirements for provisioners (other factors would include kid's work, group size and decreased provisioning for old people).

There seems to be sufficient evidence to indicate that total Neandertal energy requirements exceeded those of MUP humans by more than 10% because of their body size and shape. Given the possible effects of living in cold climates, and species-specific characteristics of locomotion, reproduction and life history, this is a very conservative estimate.

Implications for Neandertal Use of Space

Introduction

In the Middle Paleolithic record, as described above, evidence for huts, pits, tents, or other structures is absent, and investment in space is limited to transient fireplaces and the occasional windbreak. Unambiguous remains of dwellings or stone-lined hearths are not a common phenomenon in every Upper Paleolithic site in Europe, quite the contrary, but the complete absence of such features in Middle Paleolithic contexts is one of the many striking differences between the European Middle and Upper Paleolithic record from the Aurignacian onward. How can we explain lower investment in features, structures and transporting materials to a location in the Middle Paleolithic compared with the Upper Paleolithic?

In his book *The Foraging Spectrum* Robert Kelly (1995) took an ecological perspective on variation in hunter-gatherer subsistence, mobility, and trade among other areas of behavior. This work provides a starting point from which we can explore the implications of a difference in energy requirements for populations. Here, we present an example of this approach focusing on the use of space, also discussed in Verpoorte (2006).

A Central Place Foraging Model

Kelly (1990, 1991) has produced a simple model of central place foraging. This model can help us understand the implications of higher energy requirements for the effective foraging radius of Neandertals. We assume that Neandertals and AMH exploit resources and return to a central place. Another assumption of the model is that resources are homogenously distributed around the central place. As shown in Fig. 16.1, the daily net return for foraging decreases farther from camp, as the forager uses more time and energy traveling to and from the foraging area. The net return also decreases with a decrease in the environment's return rate. The effective foraging radius is the distance from the camp to the foraging patch at which the required amount of energy is equal to the net return of resources at that distance. This distance depends on the return rates of the resources as well as on the needs of the foragers, which are in turn dependent on group size, number of foragers per group, and individual energy requirements. In similar environmental conditions, if a forager requires higher daily energy returns, the distance from camp at which he or she can forage at an energetic gain becomes shorter. Thus, if Neandertals had relatively high energy requirements, we would expect them to have a shorter effective foraging radius, shown by the difference between distances d1 and d2 in Fig. 16.1. In addition, higher costs of locomotion in Neandertals (Weaver and Steudel-Numbers, 2005) would reduce net foraging returns and produce a sharper decline in returns with foraging distance, and still shorter effective foraging radius.

Kelly (1995: 132–133) highlights a strong relationship between individual foraging and camp movement. The decision to move camp is influenced by the conditions around the camp. As the effective foraging radius becomes shorter,

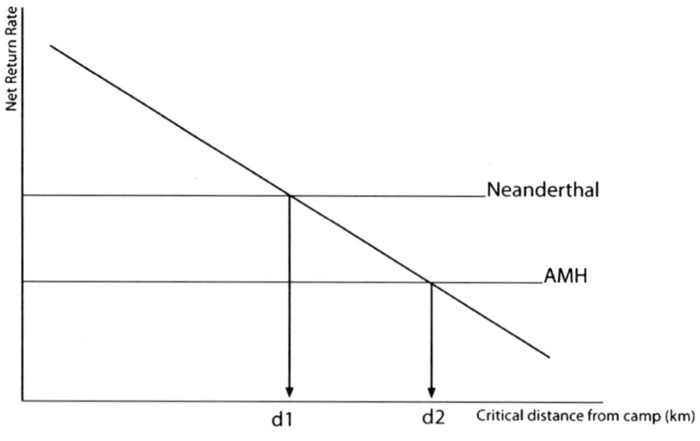

Fig. 16.1. Consequences of high energy requirements for central-place foraging under similar environmental conditions.

the forager will probably move camp after a shorter time (as resources diminish in abundance due to exploitation) and for a shorter distance. The decision to move is also likely to be influenced by expected returns from the area around the new camp, and transport costs (Kelly, 1995: 136–139). The important point here is that high Neandertal energy requirements would mean a shorter staying time at a central place.

Implications

If the central-place foraging model applies, and Neandertals used locations over a shorter time than AMH did, this has implications for investment in site structure or "furniture" (Binford, 1990). A number of authors have suggested that there should be an inverse relationship between staying time and investment in housing (Kent and Vierich, 1989; Binford, 1990; Kent, 1992). Investments in a site (for instance, constructing a shelter, building a hearth structure) have costs in time, energy, or materials, and people are more likely to invest heavily in a site if they anticipate staying a longer time. Studies of variation across cultures and also between different settlement types occupied by a particular group of people show that investment in house construction (for instance, use of durable materials and different materials for roof and walls) correlates with length of stay, among other factors (Kent and Vierich, 1989; Binford, 1990; Kent, 1992; Kelly et al., 2005). Alexander Verpoorte (2006) has modeled the costs of investing in site structure in relation to staying time. Many of the costs of investment in a site are highest to start off with. These investments pay back over the use-life of the structure, for instance in providing warmth or shelter. If Neandertals occupied locations for a shorter time than AMH, the payback for investing in a structure would have to outweigh the initial costs over a shorter time period. This implies that we should not necessarily *expect* Neandertals to have built huts, or in general terms to have invested in site "furniture" in such a way that this left clear archeological traces. Given the energetic constraints, low investment is better than high investment in locations.

A shorter staying time and smaller foraging radius has knock-on effects for mobility patterns over a longer time span, including the distance moved between camps, frequency of moves, and the total distance covered in a year. As a thought experiment, we can assume that the effective foraging radius was 3 km for Neandertals and 4 km for modern humans in a similar environment. These figures are used for ease of calculation, and do not derive from the estimates of energy expenditure discussed in the previous section. This 25% difference in foraging radius results in a foraging area 77% larger for modern humans. If we assume that Neandertals stayed in this area for 20 days compared with 35 days for humans, this results in Neandertals moving eight more times per year than modern humans. If both move just far enough to reach a camp in a completely new foraging area, this results in a single movement 2 km further for humans than Neandertals, but also in Neandertals traveling 28 km further in total camp movements in the course of a year (18 × 6 km = 108 km/year, 10 × 8 km = 80 km/year). However, if we take travel within the foraging radius into account, Neandertals move 2,190 km/year, and modern humans move 2,920 km/year (365 × 6 km and 365 × 8 km respectively). The important point as far as Neandertals are concerned is that higher total energy expenditure implies shorter stays and more frequently moving camp, but less total movement over a year.

A Neandertal mobility strategy involving fewer kilometers traveled per year may have been traded off against the locomotion costs of shorter lower limbs and higher body mass (Steudel-Numbers and Tilkens, 2004, Weaver and Steudel-Numbers, 2005). While climate has an important role in determining body proportions, Weaver and Steudel-Numbers (2005) suggest that higher mobility may have selected for energetic efficiency in mobile foraging in anatomically modern humans.

Conclusion and Discussion

We hope that this example shows that Neandertal energetics has behavioral consequences. In this example, higher total energy expenditure implies shorter stays and more frequently moving camp, and less total movement over a year. Staying for a shorter time implies lower investment in locations. The behavioral consequences of high energy requirements can help us to understand characteristics of the Middle Paleolithic record. In this case, characteristics of the current Neandertal record for the use of space – namely lack of investment in dwellings or camp structures – can be understood as a result of short staying time. This is a preliminary attempt, and other aspects of the Neandertal record reviewed in this paper could also be addressed in this way, as discussed in more detail in Verpoorte (2006). Possible areas for future research include diet breadth (Kelly, 1995, Chapter 3) and lithic technology (Ugan et al., 2003). In addition to explaining differences between the Middle and Upper Paleolithic record, this approach also yields possible explanations for variation within the Middle and Upper Paleolithic.

In addition, an energetics approach enables us to put Neandertals (and AMH) in their wider ecological setting. Species within an ecosystem are linked by transfers of energy. Energetic relations between species on different trophic levels have important implications for the energy available and for species characteristics. Ecological networks were different from the present in Pleistocene environments such as the Mammoth Steppe. As Verpoorte (2006) has suggested, a study of Middle Pleistocene ecology in energetic terms should provide new insights into Neandertal body size and the ecological constraints on Neandertal energy budgets.

The nature of the Middle Paleolithic record has been interpreted in terms of constraints on Neandertal behaviour due to limited cognitive abilities. Our paper shows that Neandertal use of space was optimal given their high energetic requirements, and was not constrained by limited cognitive abilities. However, this does not tell us anything about the cognitive abilities required for Neandertal mobility strategies. Our central-place foraging model is based on the assumption that foragers have perfect knowledge of the distribution and return rates of resources, but makes no assumptions regarding the underlying processes through which such information is acquired or processed. Questions about cognitive abilities are susceptible to study from an ecological perspective, as discussed in detail in a recent review (Dall et al., 2005), and the structures associated with cognitive ability have energetic costs (Aiello and Wheeler, 1995). Thus a similar perspective might in future provide insights into Neandertal cognition.

An energetics perspective integrates fossil and archaeological evidence in a common framework. This approach will benefit from further developments in biological anthropology, in particular, from improvements in estimates of the Neandertal energy budget. Our brief survey of research on Neandertal energetics raises a number of questions. Which of the various methods available should we use to compare Neandertal and AMH energy budgets (and those of other hominins)? Can skeletal evidence for the activity levels of Neandertals and MUP humans help us to improve our estimates of maintenance costs? What are the implications of differences in life history for Neandertal production energy? And what are the advantages of having large, energetically expensive bodies in general and in the Middle Pleistocene of Europe in particular?

This example of the use of space indicates that a difference in total energy expenditure can have important implications for behavior. We have focused on the potential of this difference for explaining contrasts between the Middle and Upper Paleolithic record. In our view, the energetics perspective opens up a major arena of integrative research on the Paleolithic. The hominin record documents a number of important changes in energy budgets, such as bipedal locomotion, reduction of sexual dimorphism, and the occupation of middle latitudes, with implications for behavior. In future research we may be able to integrate studies of the evolution of human energetics with wider issues related to the evolution of hominin behavior.

Acknowledgments. We thank the other conference participants, as well as Gerrit Dusseldorp and Raymond Corbey for helpful comments on our conference paper. This research was conducted in the framework of the 'Thoughtful hunters' project funded by the Dutch Organization for Scientific Research (N.W.O).

References

Aiello, L. C., 2007. Notes on the implications of the expensive tissue hypothesis for human biological and social evolution. In: Roebroeks, W. (Ed.), Guts and brains: an integrative approach to the hominin record. Leiden University Press, Leiden, The Netherlands.

Aiello, L. C., and Wheeler, P., 1995. The expensive-tissue hypothesis – the brain and the digestive-system in human and primate evolution. Current Anthropology 36, 199–221.

Aiello, L. C., and Wheeler, P., 2003. Neanderthal thermoregulation and the glacial climate. In: van Andel, T. H., Davies, W. (Eds.), Neanderthals and modern humans in the European landscape during the last glaciation. McDonald Institute for Archaeological Research, Cambridge, pp. 147–166.

Alley, R. B., 2000. Ice-core evidence of abrupt climate changes. Proceedings of the National Academy of Sciences of the United States of America 97, 1331–1334.

Ashton, N., 2002. Absence of humans in Britain during the last Interglacial (Oxygen isotope stage 5e). In: Tuffreau, A., Roebroeks, W. (Eds.), Le dernier interglaciaire et les occupations humaines du Paléolithique moyen. CERP, Université des Science et Technologies de Lille, Lille, pp. 93–104.

Bar-Yosef, O., Vandermeersch, B., Arensburg, B., Belfer-Cohen, A., Goldberg, P., Laville, H., Meignen, L., Rak, Y., Speth, J. D., Tchernov, E., Tillier, A.-M., Weiner, S., 1992. The excavations in Kebara Cave, Mount Carmel. Current Anthropology 33, 497–550.

Binford, L. R., 1990. Mobility, housing and environment: a comparative study. Journal of Anthropological Research 46, 119–152.

Bischoff, J. L., Williams, R. W., Rosenbauer, R. J., Aramburu, A., Arsuaga, J. L., Garcia, N., Cuenca-Bescós, G., 2007. High-resolution U-series dates from the Sima de los Huesos hominids yields 600 kyrs: implications for the evolution of the early Neanderthal lineage. Journal of Archaeological Science 34, 763–770.

Boëda, E., Connan, J., Dessort, D., Muhesen, S., Mercier, N., Valladas, H., Tisnerat, N., 1996. Bitumen as a hafting material on Middle Palaeolithic artefacts. Nature 380, 336–338.

Bratlund, B., 1999. Anthropogenic factors in the thanatocoenose of the Last Interglacial travertines at Taubach (Germany). In: Gaudzinski, S., Turner, E. (Eds.), The role of early humans in the accumulation of European Lower and Middle Palaeolithic bone assemblages. Habelt, Bonn, pp. 255–262.

Churchill, S. E., 2007. Bioenergetic perspective on Neanderthal thermoregulatory and activity budgets. In: Harvati, K., Harrison, T. (Eds.), Neanderthals revisited: new approaches and perspectives. Springer, Dordrecht.

Churchill, S. E., 2009. The evolution of the human capacity for 'killing at a distance': the human fossil evidence for the evolution of projectile weaponry. In: Hublin, J. J., Richards, M. P. (Eds.), The evolution of hominid diets: integrating approaches to the study of Palaeolithic subsistence. Springer, Dordrecht.

Dall, S. R. X., Giraldeau, L.-A., Olsson, O., McNamara, J. M., Stephens, D. W., 2005. Information and its use by animals in evolutionary ecology. Trends in ecology and evolution 20, 187–193.

Dansgaard, W., Johnsen, S. J., Clausen, H. B., Dahl-Jensen, D., Gundestrup, N. S., Hammer, C. U., Hvidberg, C. S., Steffensen, J. P., Sveinbjornsdottir, A. E., Jouzel, J., Bond, G., 1993. Evidence for general instability of past climate from a 250-kyr ice-core record. Nature 364, 218–220.

Dennell, R., Roebroeks, W., 2005. An Asian perspective on early human dispersal from Africa. Nature 438, 1099–1104.

FAO/WHO/UNU., 1985. Energy and protein requirements. Report of the Joint FAO/WHO/UNU Expert Committee. WHO, Geneva.

Féblot-Augustins, J., 1997. La circulation des matières premières au paléolithique. ERAUL (ERAUL 75), Liège.

Féblot-Augustins, J., 1999. Raw material transport patterns and settlement systems in the European Lower and Middle Palaeolithic: continuity, change and variability. In: Roebroeks, W., Gamble, C. S. (Eds.), The Middle Palaeolithic occupation of Europe. University of Leiden, Leiden, pp. 193–214.

Gamble, C. S., 1986. The Palaeolithic settlement of Europe. Cambridge University Press, Cambridge.

Gamble, C. S., 1987. Man the shoveller: alternative models for Middle Pleistocene colonization and occupation in northern latitudes. In: Soffer, O. (Ed.), The Pleistocene old world: regional perspectives. Plenum, New York, pp. 81–98.

Gaudzinski, S., 2004. A matter of high resolution? The Eemian Interglacial (OIS 5e) in north-central Europe and Middle Palaeolithic subsistence. International Journal of Osteoarchaeology 14, 201–211.

Gaudzinski, S., 2009. Middle Palaeolithic subsistence in a diachronous perspective. In: Hublin, J. J., Richards, M. P. (Eds.), The evolution of hominid diets: integrating approaches to the study of Palaeolithic subsistence. Springer, Dordrecht.

Gaudzinski, S., Roebroeks, W., 2000. Adults only. Reindeer hunting at the Middle Palaeolithic site Salzgitter Lebenstedt, Northern Germany. Journal of Human Evolution 38, 497–521.

Gehan, E. A., George, S. L., 1970. Estimation of human body surface area from height and weight. Cancer Chemotherapy Reports 54, 225–235.

Geneste, J. M., 1985. Analyse lithique d'industries Moustériennes du Périgord: une approche technologique du comportement des groupes humaines au Paléolithique moyen. University of Bordeaux, Bordeaux.

Grünberg, J. M., 2002. Middle Palaeolithic birch bark pitch. Antiquity 291, 15–16.

Guatelli-Steinberg, D., Reid, D. J., Bishop, T. A., Larsen, C. S., 2005. Anterior tooth growth periods in Neandertals were comparable to those of modern humans. Proceedings of the National Academy of Sciences of the United States of America 102, 14197–14202.

Guthrie, R. D., 1984. Mosaics, allelochemics and nutrients: an ecological theory of late Pleistocene megafaunal extinctions. In: Martin, P., Klein, R. (Eds.), Quaternary extinctions: a prehistoric revolution. University of Arizona Press, Tucson.

Guthrie, R. D., 1990. Frozen fauna of the Mammoth Steppe: the story of Blue Babe. University of Chicago Press, Chicago.

Holliday, T. W., 1997. Postcranial evidence of cold adaptation in European Neandertals. American Journal of Physical Anthropology 104, 245–258.

Holliday, T. W., 2000. Evolution at the cross-roads: modern human emergence in Western Asia. American Anthropologist 102, 54–68.

Hublin, J.-J., 1998. Climatic changes, paleogeography, and the evolution of the Neandertals. In: Akazawa, T., Aoki, K., Bar-Yosef, O. (Eds.), Neandertals and modern humans in Western Asia. Plenum Press, New York, pp. 295–309.

Jaubert, J., Lorblanchet, M., Laville, H., Slott-Moller, R., Turq, A., Brugal, J. P., 1990. Les chausseurs d'Aurochs de la Borde. Un site du Paléolithique moyen (Livernon, Lot). Editions de la maison des Sciences de l'Homme, Paris.

Kaplan, H. S., Hill, K., Lancaster, J., Hurtado, A. M., 2000. A theory of human life history evolution: diet, intelligence and longevity. Evolutionary Anthropology 9, 156–185.

Kelly, R. L., 1990. Marshes and mobility in the Western Great Basin. In: Janetski, J., Madsen, D. B. (Eds.), Wetlands adaptations in the Great Basin. Brigham Young University, Museum of Peoples and Cultures Occasional Paper No. 1. Brigham Young University, Provo, pp. 259–276.

Kelly, R. L., 1991. Sedentism, sociopolitical inequality and resource intensification. In: Gregg, S. (Ed.), Between bands and states. Southern Illinois University Press, Carbondale, Illinois, pp. 135–158.

Kelly, R. L., 1995. The foraging spectrum. Smithsonian Institution Press, Washington, DC.

Kelly, R. L., Poyer, L., Tucker, B., 2005. An ethnoarchaeological study of mobility, architectural investment, and food sharing among Madagascar's Mikea. American Anthropologist 107, 403–416.

Kent, S., 1992. Studying variability in the archaeological record: an ethnoarchaeological model for distinguishing mobility patterns. American Antiquity 57, 635–660.

Kent, S., Vierich, H., 1989. The myth of ecological determinism: anticipated mobility and site spatial organization. In: Kent, S. (Ed.), Farmers as hunters. Cambridge University Press, Cambridge, pp. 96–130.

Kleiber, M., 1961. The fire of life: an introduction to animal energetics. Wiley, New York.

Klein, R. G., 2000. Archaeology and the evolution of human behaviour. Evolutionary Anthropology 9, 17–36.

Klein, R. G., 2001. Southern Africa and modern human origins. Journal of Anthropological Research 57, 1–16.

Klein, R. G., Edgar, B., 2002. The dawn of human culture. Wiley, New York.

Kolen, J., 1999. Hominids without homes: on the nature of Middle Palaeolithic settlement in Europe. In: Roebroeks, W., Gamble, C. (Eds.), The Middle Palaeolithic occupation of Europe. University of Leiden Press, Leiden, The Netherlands, pp. 139–175.

Krause, J., Orlando, L., Serre, D., Viola, B., Pruefer, K., Richards, M. P., Hublin, J.-J., Haenni, C., Derevianko, A. P., Paabo, S., 2007. Neanderthals in central Asia and Siberia. Nature 449, 902–904.

Leonard, W. R., Ulijaszek, S. J., 2002. Energetics and evolution: an emerging research domain. American Journal of Human Biology 14, 547–550.

Leonard, W. R., Snodgrass, J. J., Sorensen, M. V., 2005. Metabolic adaptation in indigenous Siberian populations. Annual Review of Anthropology 34, 451–471.

Locht, J.-L., 2004. Le gisement paléolithique moyen de Beauvais (Oise). Thèse de Doctorat. U.F.R. de Géographie. Université des Sciences et technologies de Lille, Lille.

Locht, J.-L., 2006. Caours. La Lettre de l'Archéologie 8, 3–5.

Mazza, P. P. A., Martini, F., Sala, B., Magi, M., Colombini, M. P., Giachi, G., Landucci, F., Lemorini, C., Modugno, F., Ribechini, E., 2006. A new Palaeolithic discovery: tar-hafted stone tools in a European Mid-Pleistocene bone-bearing bed. Journal of Archaeological Science 33, 1310–1318.

Mellars, P., 2004. Neanderthals and the modern human colonization of Europe. Nature 432, 461–465.

Pérez-Pérez, A., Espurz, V., Bermúdez de Castro, J. M., de Lumley, M. A., Turbón, D., 2003. Non-occusal dental microwear variability in a sample of Middle and Late Pleistocene human populations from Europe and the Near East. Journal of Human Evolution 44, 497–513.

Ramirez Rozzi, F. V., Bermúdez de Castro, J. M., 2004. Surprisingly rapid growth in Neanderthals. Nature 428, 936–939.

Révillion, S., Tuffreau, A. (Eds.), 1994. Les industries laminaires du Paléolithique moyen. Centre National de la Recherche Scientifique, Paris.

Richards, M. P., 2009. Isotopic evidence for European Upper Palaeolithic human diets. In: Hublin, J. J., Richards, M. P. (Eds.),

The evolution of hominid diets: integrating approaches to the study of Palaeolithic subsistence. Springer, Dordrecht.

Richards, M. P., Pettitt, P. B., Trinkaus, E., Smith, F. H., Paunovic, M., Karavanic, I., 2000. Neanderthal diet at Vindija and Neanderthal predation: the evidence from stable isotopes. Proceedings of the National Academy of Sciences of the United States of America 97, 7663–7666.

Richards, M. P., Pettitt, P. B., Stiner, M. C., Trinkaus, E., 2001. Stable isotope evidence for increasing dietary breadth in the European mid-Upper Paleolithic. Proceedings of the National Academy of Sciences of the United States of America 98, 6528–6532.

Rieder, H., 2000. Die altpaläolithischen Wurfspeere von Schöningen, ihre Erprobung unf ihre Bedeutung für die Lebensumwelt des Homo erectus. Praehistorica Thuringica 5, 68–75.

Roebroeks, W., Gamble, C. S. (Eds.), 1999. The Middle Palaeolithic occupation of Europe. University of Leiden Press, Leiden.

Roebroeks, W., Speleers, B., 2002. Last Interglacial (Eemian) occupation of the North European plain and adjacent areas. In: Tuffreau, A., Roebroeks, W. (Eds.), Le Dernier Interglaciaire et les occupations humaines du Paléolithique moyen. CERP, Lille, pp. 31–40.

Roebroeks, W., Tuffreau, A., 1999. Paleoenvironment and settlement patterns of the northwest European Middle Palaeolithic. In: Gamble, C. S., Roebroeks, W. (Eds.), The Middle Palaeolithic occupation of Europe. University of Leiden Press, Leiden, pp. 121–138.

Roebroeks, W., Kolen, J., Rensink, E., 1988. Planning depth, anticipation and the organization of Middle Palaeolithic technology: the "archaic natives" meet Eve's descendants. Helinium 28, 17–34.

Roebroeks, W., Conard, N. J., van Kolfschoten, T., 1992. Dense forests, cold steppes and the Palaeolithic settlement of northern Europe. Current Anthropology 33, 551–586.

Ruff, C. B., Walker, A., Trinkaus, E., 1993. Postcranial robusticity in Homo. I: Temporal trends and mechanical interpretation. American Journal of Physical Anthropology 91, 21–53.

Ruff, C. B., Trinkaus, E., Holliday, T. W., 1997. Body mass and encephalization in Pleistocene Homo. Nature 387, 173–176.

Ruff, C. B., Niskanen, M., Junno, J.-A., Jamison, P., 2005. Body mass prediction from stature and bi-iliac breadth in two high-latitude populations, with application to earlier high-latitude humans. Journal of Human Evolution 48, 381–392.

Sandgathe, D., Dibble, H. L., McPherron, S. J. P., Turq, A., 2005. Roc de Marsal (Campagne de Bugue, Dordogne) Rapport d'Operation pour l'annee 2005.

Shea, J. J., 2006. The origins of lithic projectile point technology: evidence from Africa, the Levant, and Europe. Journal of Archaeological Science 33, 823–846.

Shea, J. J., 2009. The ecological impact of projectile weaponry in Late Pleistocene human evolution. In: Hublin, J. J., Richards, M. P. (Eds.), The evolution of hominid diets: integrating approaches to the study of Palaeolithic subsistence. Springer, Dordrecht.

Sorensen, M. V., Leonard, W. R., 2001. Neandertal energetics and foraging efficiency. Journal of Human Evolution 40, 483–495.

Speth, J. D., 2004. Hunting pressure, subsistence intensification, and demographic change in the Levantine Late Middle Palaeolithic. In: Goren-Inbar, N., Speth, J. D. (Eds.), Human paleoecology in the Levantine Corridor. Oxbow Books, Oxford, pp. 149–166.

Speth, J. D., 2006. Housekeeping, Neandertal style: hearth placement and midden formation in Kebara Cave (Israel). In: Hovers, E., Kuhn, S. L. (Eds.), Transitions before the transition: evolution and stability in the Middle Palaeolithic and Middle Stone Ages. Springer, New York, pp. 171–188.

Speth, J. D., Clark, J. L., 2006. Hunting and overhunting in the Levantine Late Middle Palaeolithic. Before Farming 2006/3 article 1.

Steegmann, A. T., Cerny, F. J., Holliday, T. W., 2002. Neandertal cold adaptation: physiological and energetic factors. American Journal of Human Biology 14, 566–583.

Steudel-Numbers, K. L., Tilkens, M. J., 2004. The effect of lower limb length on the energetic cost of locomotion: implications for fossil hominins. Journal of Human Evolution 47, 95–109.

Stewart, J. R., 2005. The ecology and adaptation of Neandertals during the non-analogue environment of oxygen isotope Stage 3. Quaternary International 137, 35–46.

Stiner, M. C., 2001. Thirty years on the "Broad Spectrum Revolution" and paleolithic demography. Proceedings of the National Academy of Sciences of the United States of America 98, 6993–6996.

Stiner, M. C., 2002. Carnivory, coevolution, and the geographic spread of the genus Homo. Journal of Archaeological Research 10, 1–63.

Thieme, H., 1997. Lower Palaeolithic hunting spears from Germany. Nature 385, 807–810.

Tomasello, M., Carpenter, M., Call, J., Behne, T., Moll, H., 2005. Understanding and sharing intentions: the origins of cultural cognition. Behavioural and Brain Sciences 28, 675–691.

Trinkaus, E., 1995. Neanderthal mortality patterns. Journal of Archaeological Science 22, 121–142.

Turq, A., 1999. Reflections on the Middle Palaeolithic of the Aquitaine Basin. In: Roebroeks, W., Gamble, C. S. (Eds.), The Middle Palaeolithic occupation of Europe. University of Leiden, Leiden, pp. 107–120.

Ugan, A., Bright, J., Rogers, A., 2003. When is technology worth the trouble? Journal of Archaeological Science 30, 1315–1329.

van Andel, T. H., Davies, W. (Eds.), 2003. Neanderthals and modern humans in the European landscape during the last glaciation. McDonald Institute for Archaeological Research, Cambridge.

Verpoorte, A., 2006. Neandertal energetics and spatial behaviour. Before Farming 2006/3 article 2.

Villa, P., 2009. Lower and Middle Paleolithic faunal exploitation in Europe. In: Hublin, J. J., Richards, M. P. (Eds.), The evolution of hominid diets: integrating approaches to the study of Palaeolithic subsistence. Springer, Dordrecht.

Weaver, T. D., Steudel-Numbers, K. L., 2005. Does climate or mobility explain the differences in body proportions between Neandertals and their Upper Palaeolithic successors? Evolutionary Anthropology 14, 218–223.

Winslow, C. E. A., Herrington, L. P., 1949. Temperature and human life. Princeton University Press, Princeton.

17. $\delta^{13}C$ Values Reflect Aspects of Primate Ecology in Addition to Diet

Margaret J. Schoeninger
Department of Anthropology
University of California at San Diego
La Jolla, CA 92093-0532, USA
mjschoen@ucsd.edu

Keywords Monkeys • apes • prosimians • hominins • diet reconstruction • stable carbon isotope ratios

Abstract Hair samples from different C_3-feeding primate species living in ecologically distinct areas of Mesoamerica, South America, Africa, and Madagascar have $\delta^{13}C$ values that vary by 6‰ in association with the amount of canopy cover. *Cebus capucinus* (capuchin) and *Ateles geoffroyi* (spider) from La Selva, Costa Rica live in an area of tropical wet forest with continuous forest canopy and have identical average $\delta^{13}C$ values, even though the former is omnivorous and the latter is frugivorous. *Alouatta palliata* (mantled howler) from La Pacifica, Costa Rica *Brachyteles arachnoides* (muriquis) from Fazenda Esmeralda, Brazil, a population of chimpanzees from East Africa, and *Galago zanzibaricus* and *Galago garnettii* from Gedi, Kenya live in areas of mixed evergreen and deciduous forest with broken forest canopy. Their $\delta^{13}C$ values are similar to each other but significantly different from capuchin and spider monkeys. The differences are of the same magnitude and in the same direction as that in leaves from open canopies compared with closed canopies, and are independent of specific primate diet. *Lepilemur leucopus* from Beza Mahafaly Special Reserve, Madagascar, and another population of chimpanzees live in dry, deciduous forests. Their $\delta^{13}C$ values are similar to each other, even though lepilemur is a folivore and chimpanzees are frugivorous. Both species are significantly less negative than the ones from closed and broken forest canopy habitats. Published data on North American cervids, East African bovids, and sifakas from Madagascar largely match the patterns observed in these primates. In reconstructing diets in archaeological human populations and in fossil hominins, this 6‰ variation must be considered rather than using the average $\delta^{13}C$ value for C_3 plants. In addition, because fossil hominin samples may be altered by 1–2‰, the range for pure C_3 diets could be more extensive than commonly appreciated.

J.-J. Hublin and M.P. Richards (eds.), *The Evolution of Hominin Diets: Integrating Approaches to the Study of Palaeolithic Subsistence*, 221–227.

Introduction

Accurate reconstructions of the diets of earlier human populations and our fossil ancestors and relatives are critical for testing various hypotheses regarding the course of human evolution. The behaviors involved in procuring, processing, and eating food provide constraints on the individual and social behavior of any organism; humans and their ancestors are no exception. One of the increasingly common methods applied in reconstructing diets is that of stable carbon isotope analysis. First applied to archaeological populations (Vogel and van der Merwe, 1977; van der Merwe and Vogel, 1978), the applications have exploded over the last few years with some of the most exciting being those of extinct populations (e.g., Bocherens, 2009; Richards, 2009; Sponheimer, and Dufour 2009). The majority of these applications are directed toward the estimation of the amounts of C_3 versus C_4 plants or plant-eating animals in the diets, since plants (and the animals who feed upon them) show a non-overlapping bimodal distributions of $\delta^{13}C$ values (O'Leary, 1988). The present chapter will summarize data from primates and other mammals and show that ecology affects animal $\delta^{13}C$ values up to 6‰ for animals eating only C_3 plants or C_3-feeding animals. This chapter will also touch lightly on the potential influence of diagenesis on the $\delta^{13}C$ values in degraded archaeological and/or fossil samples. In some regions, diagenesis can alter biogenic $\delta^{13}C$ values by several per mil. In combination, this review demonstrates that using average values for C_3 and C_4 plants, and calculating average diet compositions, can lead to significant errors when proposing diets in earlier human populations and in fossil hominins.

Variation in C_3 Foods

The distribution of $\delta^{13}C$ values across all plant species is bimodal. Herbaceous vegetation, trees, and cool season grasses (C_3 plants) display $\delta^{13}C$ values than range from −31‰ to −23‰ (O'Leary, 1988); much of this variation is explained by the specific ecological situation where these plants grow (see Heaton, 1999). Patterned differences in the variation, known for over a decade (van der Merwe and Medina, 1989; Broadmeadow et al., 1992), are due to variations in: (1) the $\delta^{13}C$ value in the carbon dioxide available to the plant during photosynthesis and (2) a set of isotope effects that occur during the process of photosynthesis (including the overall rate of photosynthesis). The $\delta^{13}C$ value of well-mixed atmospheric carbon dioxide today is around −7.8‰ (Wahlen, 1994). The value reported in deciduous tropical forests (i.e., open canopies) is similar (−8.3‰ to −7.8‰) to the atmospheric value; but the value within semievergreen tropical forests (i.e., more closed canopies) has been reported as low as −9.9‰ (Broadmeadow et al., 1992). The more negative values within forests are partially due to the addition of ^{12}C-enriched carbon dioxide respired from organic detritus in soil (van der Merwe and Medina, 1989). In open areas such as savannas or deciduous bush/woodlands, the carbon dioxide available to plants is −7.8‰, whereas that available to plants in forests varies according to the amount of soil-respired carbon dioxide and the amount of mixing with atmospheric carbon dioxide.

Yet variation in plant tissues can exceed that which can be accounted for by variation in source carbon dioxide. The additional variation is due to the second cause mentioned above, i.e., isotope effects. Although these isotope effects are not completely elucidated as yet, and much inter-annual and inter-seasonal variation in plant tissues is not yet understood (see Cerling et al., 2003), there are some generalities that apply. Isotope effects occur at particular points during photosynthesis (Farquhar et al., 1982, 1989); but of particular impact is the rate of diffusion of atmospheric carbon dioxide into the plant (Loreto et al., 1992) and its associated rate of photosynthesis (Heaton, 1999). Increased concentrations of air carbon dioxide, such as those that occur in closed canopies due to the addition of soil respired carbon dioxide and low light levels, associate with a lower rate of photosynthesis and more negative leaf values (Broadmeadow et al., 1992). High light levels, such as those that occur in more open canopies, stimulate stomatal closure and associate with higher rates of photosynthesis and less negative leaf $\delta^{13}C$ values (Yakir and Israeli, 1995). There is also variation within forests such that leaves at the top of the highest trees have the highest $\delta^{13}C$ values while those at the very base have the lowest values.

Variation in Animals Feeding on C_3 Foods

These relationships in plants have been reported as present and consistent from every continent, and data from a wide range of animal species (mostly large herbivorous mammals) demonstrate that they are recorded in the animals feeding on these plants (see Krigbaum, 2003 for a recent summary). Primates have received less attention in these studies, probably because they occur at lower densities than more-commonly reported cervid and bovid species. Yet primates are particularly relevant for studies involving the reconstruction of diets in earlier humans and our hominin ancestors because most primates feed on a wide variety of foods, in contrast to the other, more specialized mammalian species that are commonly studied. Significantly, most primates feed largely on C_3 foods (Milton, 1987), although several cercopithecine species (Jolly, 1970; Codron et al., 2005) consume significant quantities of grass seeds and grass corms as well. Many primates are arboreal, and because they feed throughout canopies they smooth out the intra-canopy variation (Heaton, 1999) within the systems in which they feed. This smoothing provides an averaging effect on that variation.

A series of studies across several species of extant primates demonstrate that their hair $\delta^{13}C$ values correlate with canopy cover, irrespective of their specific diets (Schoeninger et al., 1997, 1998, 1999; and see Table 17.1), and more recent work, summarized here, largely supports those findings. The original studies are summarized here, and compared with the more recent work (Fig. 17.1). Then, the implications of the data on extant primates are considered for reconstructing aspects of earlier human and ancestral hominin ecology and subsistence strategies. All the data on living primates come from hair, taken either during routine capture for assessing physiological status or, in the case of chimpanzees, from night nests. The data from earlier groups come from bone collagen or from tooth enamel. Published studies show the relationships between these tissues (Fig. 17.2) and are used in comparing the data on hair with the data from bony tissues.

Of the extant primates, two species (both New World monkeys) have identical $\delta^{13}C$ values (Schoeninger et al., 1997) although their diets differ. This forest (La Selva, Costa Rica) exhibits low seasonality in rainfall, with average rainfall around 4,000 mm per year (Janzen, 1983). The canopy is mostly closed, consisting of evergreen and semievergreen tree stands. One of the species, the spider monkey (*Ateles geoffroyi*), is a highly selective feeder. Fruit is its preferred food, with merely 20% of their diet consisting of young, not mature, leaves (data summarized in Strier, 1992). This adaptation is confirmed by their tooth morphology (Kay, 1975), digestive tract morphology (Chivers and Hladik, 1980), and rapid food assimilation rates (Milton, 1984). The second species, the capuchin monkey (*Cebus capucinus*), is a noted omnivore with a diet that includes a significant weight percent of insects, with ranges in different populations reported from 20% to over 50% (data cited in Fedigan et al., 1985). Tooth morphology and time spent feeding indicate that fruit is also preferred food (Kay, 1975; Fedigan et al., 1985). Even though

Table 17.1. $\delta^{13}C$ in primate hair[a]

Tropical forest type	Old World Prosimians $\delta^{13}C \pm$ s.d.(n)	New World Monkeys $\delta^{13}C \pm$ s.d.(n)	Chimpanzees $\delta^{13}C \pm$ s.d.(n)
Drought deciduous	−21.3‰ ± 0.9(9)		−22.0‰ ± 0.3(12)
Deciduous	−23.1‰ ± 0.4(18)	−23.5‰ ± 0.3(19)	−23.1‰ ± 0.3(10)
Evergreen		−24.8‰ ± 0.4(9)	

[a]data from Schoeninger et al, 1997, 1998, 1999

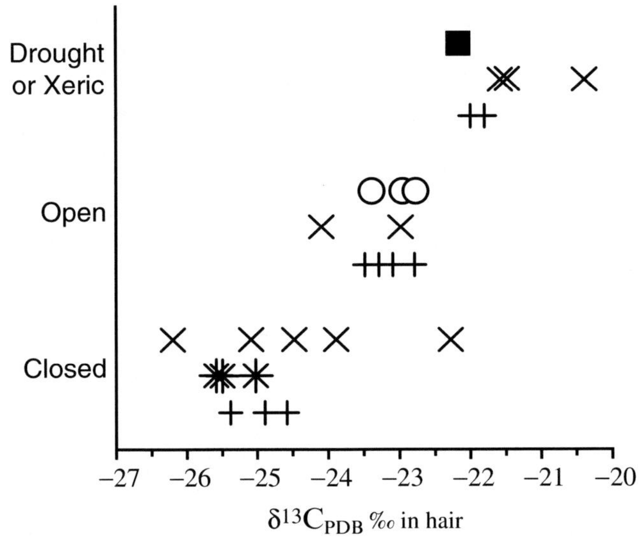

FIG. 17.1. $\delta^{13}C$ values in hair from across mammal species reflect the extent of canopy cover in the areas in which the animals lived. Plus symbols are data from Schoeninger et al., 1999; stars are from Urton and Hobson, 2005; X's are from Cerling et al., 2003, circles are from McGee and Vaughn, 2003, and the filled square is from Sponheimer et al., 2006. Two species of New World monkeys, three species of North American cervids, and three species of East African bovids from closed canopies show values around −26‰ to −24.5‰ although two other species (X's) fall within the open canopy range (see text for discussion). Two other species of New World monkeys, two species of African prosimians, a population of chimpanzees, two species of East African bovids, and three different populations of prosimians from Madagascar come from deciduous, open canopies and show values around −24‰ to −22‰. One species of prosimian from Madagascar, two different populations of chimpanzees, and three species of East African bovids show the least negative values, around −22 to −20‰. The variation is independent of the specific diet of these species.

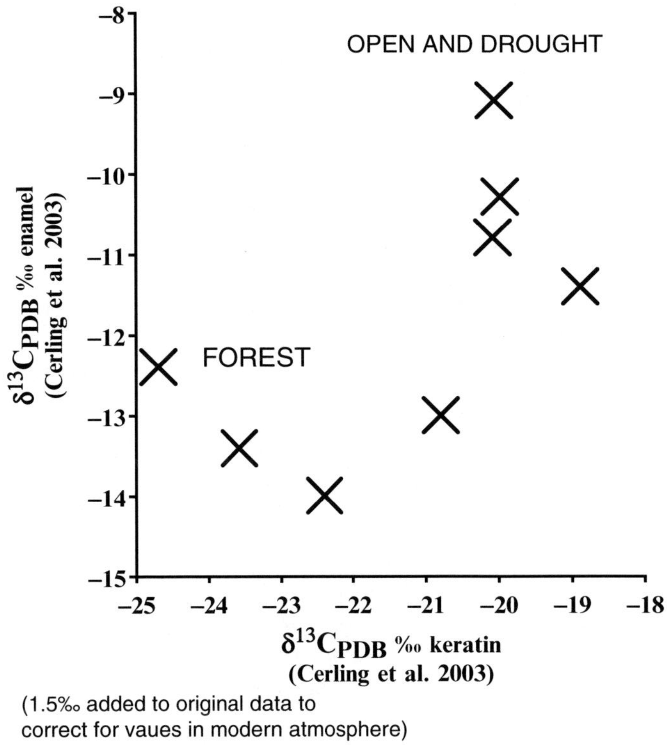

FIG. 17.2. Comparison of $\delta^{13}C$ values in keratin (hair, horn, or hoof) and tooth enamel in several species of browsing East African bovids. The closed canopy forest dwelling species have values markedly lower than those from open, dry deciduous forests. No comparison could be made for species from open deciduous forests that were not also dry forests.

fruit is eaten in greater amounts (relative to body size) by the spider monkey than by the capuchin monkey, the two species show identical average $\delta^{13}C$ values in hair (Table 17.1).

Significantly, $\delta^{13}C$ values from hair and other keratinaceous tissues in several species of North American cervids and East African bovids, which all lived in closed canopy forests, have values similar to the Costa Rican monkeys (Fig. 17.1, Cerling et al., 2003; Urton and Hobson, 2005). Moose, elk, and bison come from North American boreal forest regions within Saskatchewan, where the dominant vegetation includes a mix of deciduous (trembling aspen and balsam poplar) and evergreen (white spruce and black spruce) trees with significant representation of shrub and understory species (Urton and Hobson, 2005). All have values that are less than or equal to −25‰. African Suni antelope, bongo, and bushbuck also appear to have been collected from closed canopies. The specific location of the suni antelope is not reported, but the dietary assessment is that it "may be a pure browser" (Cerling et al., 2003: 461). The bongo is reported to come from an open forested region; the dietary assessment is that it may "closed-canopy browse" (Cerling et al., 2003: 464), and this species is noted for its closed forest preferences. The bushbuck come from three different locations, two from closed canopy forests and the others from more open canopies (Cerling et al., 2003). This variety of habitats is reflected in the large standard deviation (2.3‰) in the sample. The mean value is plotted in Fig. 17.1, although those from the closed canopy forest probably had lower values. These three African species have average $\delta^{13}C$ values that are similar to those in the Costa Rican monkeys. In contrast, two duiker species have values that are approximately 1–2‰ higher (−24‰ and −22‰), even though they come from the closed canopy Ituri forest. All duikers feed on fruits, which drop from the top of the canopy and are enriched in ^{13}C relative to other parts of the same plants within the forest (Cerling et al., 2004). Whether the apparent contradiction of these two closed canopy duiker species is due to the particular feeding habits of duikers or to the specific structure of the Ituri forest is not clear. In any case, these two duiker species clearly contrast with the monkey species from a closed canopy forest in which the more-frugivorous spider monkey does not have higher $\delta^{13}C$ values than the less-frugivorous capuchin monkey. As discussed further below, however, the tooth enamel $\delta^{13}C$ values for these two duiker species align them with closed canopy species rather than with open canopy species.

Five extant primate species, including two New World monkey species, two sympatric species of African prosimians, and a population of Tanzanian chimpanzees (Schoeninger et al., 1999), show $\delta^{13}C$ values that are similar to each other. The two New world monkey species include samples from a mantled howler monkey (*Alouatta palliata*) population in Costa Rica, and a muriqui (*Brachyteles arachnoides*) population in Brazil; both are from areas of open deciduous forest. The mantled howler monkeys showed seasonal differences in hairs collected from the rainy and dry seasons at a low level of significance even though they showed no significant differences between sexes, sampling year, or year of analysis. Howler monkeys have the tooth morphology of a folivore (Kay, 1975) with the digestive tract morphology and function of a primary leaf-eater that includes fruit in its diet (Chivers and Hladik, 1980; Milton 1984). In contrast, muriqui molar size and morphology suggest a greater emphasis on fruit-eating (Rosenberger, 1992), and observations indicate that they prefer fruit when available (Strier, 1992). The two African prosimian species are both galagos (*Galgago zanzibaricus* and *Galago garnettii*) that live in a lowland, dry forest in the Gedi Ruins National Monument in Kenya. Both species are omnivorous and focus on insects and fruits; the larger species (*G. garnettii*) eats relatively less animal prey than does the smaller species (*G. zanzibaricus*) (Harcourt and Nash, 1986). The last species is the chimpanzee. This particular population, from the Democratic Republic of the Congo, lives along a perennially flowing river lined with a gallery forest comprised mainly of deciduous species with some evergreen species. Superficially, the area looks very similar to that inhabited by the mantled howler monkey species. Tree fruits are considered the basic food for chimpanzees (Wrangham, 1977; Teleki, 1981; McGrew et al., 1988), but actual intake levels vary. When preferred fruits are less available, individuals may add hard seeds (Suzuki, 1969), stems (McGrew et al., 1988), or pith (Wrangham et al., 1991). Leaves generally provide protein; seeds, insects, or meat (McGrew, 1983; Boesch, 1994; Stanford, 1996) can also be eaten. Analysis of 113 fecal samples showed that for this particular population, the most common foods were the fleshy fruits from trees and shrubs along the forest margins (Steklis et al., 1992). In this set of five extant species, all have $\delta^{13}C$ values in hair that are around −23‰ (see Table 17.1), even though they represent different subsistence strategies (folivore, frugivore, and omnivore), different continents (Central and South America, and Africa), and taxonomic units (prosimians, monkeys, and apes). In addition, their hair $\delta^{13}C$ values are higher than those of the monkey, cervid, and bovid species from closed canopy forests.

Significantly, two extant species of East African bovids and three different populations of Madgascar prosimians from deciduous, open canopies show similar values (Fig. 17.1). Two species of fruit-eating duikers, from open canopy forests (Cerling et al., 2003), have an average of −23.5‰ and compare well with the primate species. The three populations of prosimians, sifakas from southeastern Madagascar, all have $\delta^{13}C$ values (McGee and Vaughn, 2003) that are nearly equivalent to both the other primate species living in open environments and to the two duiker species. Sifakas are fruit- and seed-eaters; these populations come from different regions within the Ranomafana National Park, which experiences high rainfall yet has an open canopy overall.

Two extant primate species, sampled from open, deciduous canopy forests that were significantly drier and more open than the previously described habitats, have values similar to each other even though their diets differed dramatically. The two species are the prosimian, *Lepilemur leucopus*, from a region of Madagascar that had been subjected to lengthy drought conditions (Schoeninger et al., 1998), and a population of chimpanzees from a dry, open woodland in the Ugalla region of the Tongwe Forest Reserve of Tanzania (Schoeninger et al., 1999). Approximately 250 h of focal follows on individual lepilemurs revealed that all feeding was on leaves, stems, or flowers; seeds were found in only one of 69 fecal pellets (Nash, 1998). In contrast, as discussed above, chimpanzees focus on tree fruits, and 194 fecal samples collected from a region similar to Ugalla from which this chimpanzee population originated also commonly included seeds from the leguminous trees that are ubiquitous in both areas (Suzuki, 1969). Both the lepilemurs and the chimpanzees have average $\delta^{13}C$ values around −22‰, even though they differ in terms of diet (folivory vs. frugivory and seed-eating) and taxonomy (prosimian vs. ape). Rather, their hair $\delta^{13}C$ values most strongly reflect the type of habitat in which they lived.

Significantly, four bovid species from east Africa (Cerling et al., 2003) and a population of west African chimpanzees (Sponheimer et al., 2006), which are from similarly dry, deciduous woodlands, have $\delta^{13}C$ values similar to the preceding two groups. The gerenuk, two species of dikdik, and eland have average values between −21.6‰ and −20.4‰. The chimpanzee population has an average value of −22.2‰, which is nearly identical to the average value for the east African chimpanzee population.

Overall, the pattern that primates and browsing bovids and cervids from similar environments show similar $\delta^{13}C$ values even though diets and geography differ is striking. There is variation in individual $\delta^{13}C$ values that remains unexplained (Cerling et al., 2003); but overall, the pattern suggests that ecologically related variation within C_3-feeding primates must be considered when attempting the reconstruction of past diets. Such variation is on the order of 6‰.

Implications for Diet Reconstruction

For reconstructing diets in extinct populations and species, the tissues analyzed most commonly are bone collagen, bone apatite, and tooth enamel. Bone collagen will match the values in hair since both are proteins, although bone collagen provides an averaging of a decade or more of diets whereas hair represents a much shorter period of time. In extant browsing bovids, enamel shows variable offsets, individually, from keratin (hair, hoof, and horn; see Fig. 17.2, data from Cerling

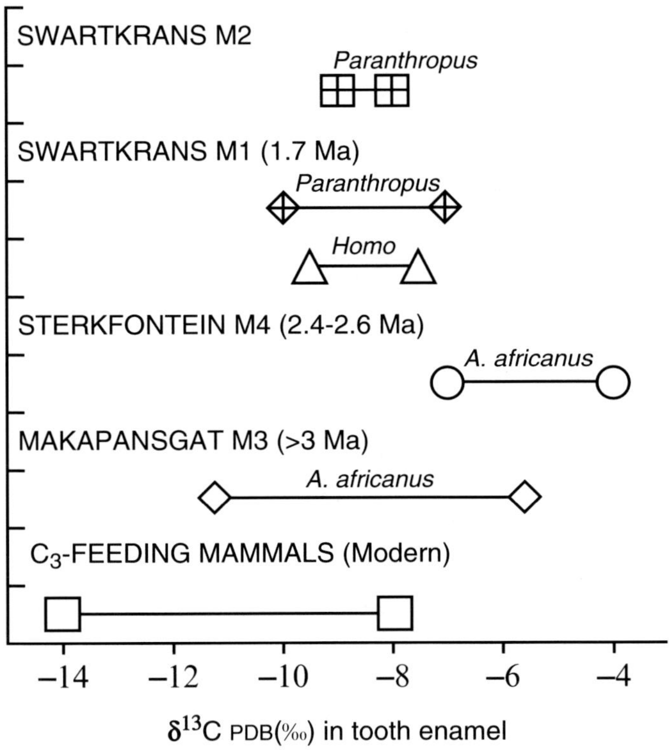

FIG. 17.3. Comparison of published $\delta^{13}C$ values from tooth enamel of several species of South African fossil hominins (data from Lee-Thorp et al., 2003) with estimated tooth enamel values for C_3-feeding mammals including primates. This comparison suggests that the majority of the hominins could have fed entirely on C_3 foods. Including an estimate of 1–2 per mil for diagenesis (with shifts to higher values due to higher values in matrix), means that only *Australopithecus africanus* falls outside the possible range for eating only C_3 foods.

et al., 2003); but the range of variation within tooth enamel shows a similar distribution as hair. Those species from closed canopy forests have lower $\delta^{13}C$ values than do those from open and dry, deciduous forests. Two of the species from closed canopies are the two duiker species, which had hair values that did not match expectations, but instead had values more similar to those from open canopy forests. Interestingly, their enamel $\delta^{13}C$ values place them with the other closed forest bovid species. This comparison, however, clearly shows that the ecological variables recorded in hair are also recorded in tooth enamel. In Fig. 17.2, 1.5‰ has been added to the data for all these extant species to offset the ^{12}C enrichment in the modern atmosphere (Friedli et al., 1986) and allow direct comparisons with extinct species as is done in Fig. 17.3.

For earlier archaeological and fossil samples, one must also consider the possible effects of diagenetic alteration. This is less of a concern with the organic fraction of bone, and there are several relatively simple means to assess preservation (e.g., C:N ratio). For bone apatite and tooth enamel, however, diagenesis can be a significant problem (Koch et al., 1997); there is no generally accepted measure of assessment (although see Shemesh, 1990). In east Africa, some 3.9 million year old tooth enamels were extensively altered mineralogically, whereas others from the same excavation site were not (Kohn et al., 1999). The altered enamels showed $\delta^{13}C$ values that were 1‰ to several per mil closer to the values in sediments, than was true of the unaltered enamels (Schoeninger et al., 2003). Because fossil sediments across many of the African sites are marine in origin, alteration toward the sediment values will make the enamel values of browsing fauna look more C_4-like. In South Africa, diagenetic alteration does not appear to be as extensive, although alteration of 1–2‰ occurs even there (Lee-Thorp, 2000).

In sum, those reconstructing past diets must consider that ecological variation (closed to dry, deciduous open canopies) can account for up to 6‰ in the organic fraction of bone even when the primate is eating only C_3 foods. Adding ≥ 1–2‰ for diagenetic alteration in fossil bone apatite or tooth enamel means that 7–8‰ range must be considered for primates eating only C_3 foods. This is especially important for early hominins because their dietary ancestry is almost assuredly that of forest and/or woodland foods. The effect can be marked. For example, a plot of published data on South African *Australopithecus*, *Paranthropus*, and *Homo* compared with the range of modern primates and other C_3-feeding mammals (based on modern data corrected by 1.5‰) shows that most individuals fall within the C_3 range (Fig. 17.3). If 1–2‰ is added to the range of C_3-feeding mammals to account for diagenetic alteration, all but the *Australopithecus africanus* fall within the C_3 range.

Interestingly, all fall within the range of mammals that live in open canopy forests and/or dry, deciduous forests. In the end, certain implications for aspects of early hominin ecology are similar however the diet is reconstructed. Whether these genera ate a significant fraction of C_4 foods (Sponheimer, 2009) combined with C_3 foods from closed canopy forests or they were eating C_3 foods in dry, deciduous forests, all of these genera dealt with the problems inherent in functioning diurnally in open regions. These include competition for food with several species of group-living monkeys and predator avoidance. While some populations of modern chimpanzees live in "savanna" regions, they rarely venture into unforested areas and then only in groups and for short periods of travel from one forested patch to another (Moore, 1992). Danger from open-country predators also implies that early hominins probably lived in groups (Cheney and Wrangham, 1987), and therefore had to find enough food for their group of large-bodied hominins. The latter has implications for the options in social arrangements which must avoid costly intragroup feeding competition (Wrangham et al., 1996).

References

Bocherens, H., 2009. Neanderthal dietary habits: review of the isotopic evidence. In: Hublin, J. J., Richards, M. P. (Eds.), The Evolution of Hominid Diets: Integrating Approaches to the Study of Palaeolithic Subsistence. Springer, Dordrecht.

Boesch, C., 1994. Chimpanzees-red colobus monkeys: a predator-prey system. Animal Behaviour 47, 1135–1148.

Broadmeadow, M. S. J., Griffiths, H., Maxwell, C., Borland, A. M., 1992. The carbon isotope ratio of plant organic material reflects temporal and spatial variations in CO_2 within tropical forest formations in Trinidad. Oecologia 89, 435–441.

Cerling, T. E., Harris, J. M., Passey, B. H., 2003. Diets of East African Bovidae based on stable isotope analysis. Journal of Mammology 84, 456–470.

Cerling, T. E., Hart, J. A., Hart, T. B., 2004. Stable isotope ecology in the Ituri Forest. Oecologia 138, 5–12.

Cheney, D. L., Wrangham, R. W., 1987. Predation. In: Cheney, D. L., Seyfarth, R. M., Smuts, B. B., Wrangham, R. W. (Eds.), Primate Societies. University of Chicago Press, Chicago, 227–239.

Chivers, D. J., Hladik, C. M., 1980. Morphology of the gastrointestinal tract in primates: comparisons with other mammals in relation to diet. Journal of Morphology 166, 337–386.

Codron, D., Luyt, J., Lee-Thorp, J. A., Sponheimer, M., de Ruiter, D., Codron, J., 2005. Utilization of savanna-based resources by Plio-Pleistocene baboons. South African Journal of Sicence 101, 245–248.

Farquhar, G. D., O'Leary, M. H., Berry, J. A., 1982. On the relationship between carbon isotope discrimination and the intercellular carbon dioxide concentration in leaves. Australian Journal of Plant Physiology 9, 121–137.

Farquhar, G. D., Ehleringer, J. R., Hubick, K. T., 1989. Carbon isotope discrimination and photosynthesis. Annual Review of Plant Physiology and Plant Molecular Biology 40, 503–537.

Fedigan, L. M., Fedigan, L., Chapman, C., 1985. Census of *Alouatta palliata* and *Cebus capucinus* monkeys in Santa Rosa National Park, Costa Rica. Brenesia 23, 309–322.

Friedli, H., Lotscher, H., Oeschger, H., Siegenthaler, U., Stauffer, B., 1986. Ice core record of the $^{13}C/^{12}C$ ratio of atmospheric CO_2 in the past two centuries. Nature 324, 237–238.

Harcourt, C. S., Nash, L. T., 1986. Species differences in substrate use and diet between sympatric Galagos in two Kenyan coastal forests. Primates 27, 41–52.

Heaton, T. H. E., 1999. Spatial, species, and temporal variations in the $^{13}C/^{12}C$ ratios of C_3 plants: implications for palaeodiet studies. Journal of Archaeological Science 26, 637–649.

Janzen, D. H., 1983. Costa Rican Natural History. University of Chicago Press, Chicago.

Jolly, C. J., 1970. The seed-eaters: a new model of hominid differentiation based on a baboon analogy. Man 5, 5–26.

Kay, R. F., 1975. The functional adaptations of primate molar teeth. Amercian Journal of Physical Anthropology 43, 195–216.

Koch, P. L., Tuross, N., Fogel, M. L., 1997. The effects of sample treatment and diagenesis on the isotopic integrity of carbonate in biogenic hydroxylapatite. Journal of Archaeological Science 24, 417–429.

Kohn, M. J., Schoeninger, M. J., Barker, W. W., 1999. Altered states: effects of diagenesis on fossil tooth chemistry. Geochimica et Cosmochimica Acta 63, 2737–2747.

Krigbaum, J., 2003. Neolithic subsistence patterns in northern Borneo reconstructed with stable carbon isotopes of enamel. Journal of Anthropological Archaeology 22, 292–304.

Lee-Thorp, J. A., 2000. Preservation of biogenic carbon isotopic signals in Plio-Pleistocene bone and tooth mineral. In: Ambrose, S. H., Katzenberg, M. A. (Eds.), Biogeochemical Approaches to Paleodietary Analysis. Kluwer Academic/Plenum Publishers, New York, 89–115.

Lee-Thorp, J. A., Sponheimer, M., van der Merwe, N. J., 2003. What do stable isotopes tell us about hominid dietary and ecological niches in the Pliocene? International Journal of Osteoarchaeology 13, 104–113.

Loreto, F., Harley, P. C., Di Marco, G., Sharkey, T. D., 1992. Estimation of mesophyll conductance to CO2 flux by three different methods. Plant Physiology 98, 1437–1443.

McGee, E. M., Vaughn, S. E., 2003. Variations in stable isotope composition in *Propithecus diadema edwardsi* from disturbed and undisturbed rainforest habitats in Ranomafana National Park, Madagascar. American Journal of Physical Anthropology 120 (S36), 149.

McGrew, W., 1983. Animal foods in the diets of wild chimpanzees (*Pan troglodytes*): why cross-cultural variation? Journal of Ethology 1, 46–61.

McGrew, W. C., Baldwin, P. J., Tutin, C. E. G., 1988. Diet of wild chimpanzees (Pan troglodytes verus) at Mt. Assirik, Senegal: I. Composition. American Journal of Primatology 16, 213–226.

Milton, K., 1984. The role of food-processing factors in primate food choice. In: Rodman, P. S., Cant, J. G. H. (Eds.), Adaptations for Foraging in Nonhuman Primates: Contributions to an Organismal Biology of Prosimians, Monkeys, and Apes. Columbia University Press, New York, 249–279.

Milton, K., 1987. Primate diets and gut morphology: implications for hominid evolution. In: Harris, M., Ross, E. B. (Eds.), Food and Evolution. Temple University Press, Philadelphia, 93–115.

Moore, J., 1992. "Savanna" Chimpanzees. In: Nishida, T., McGrew, W. C., Marler, P., Pickford, M., de Waal, F. B. M. (Eds.), Topics in Primatology, Vol. I: Human Origins. University of Tokyo Press, Tokyo, 99–118.

Nash, L. T., 1998. Vertical clingers and sleepers: seasonal influences on the activities and substrate use of *Lepilemur leucopus* at Beza Mahafaly Special Reserve, Madagascar. Folia Primatologica 69 Suppl. 1, 204–217.

O'Leary, M. H., 1988. Carbon isotopes in photosynthesis. BioScience 38, 328–336.

Richards, M. P., 2009. Stable isotope evidence for European Upper Paleolithic human diets. In: Hublin, J. J., Richards, M. P. (Eds.,) The Evolution of Hominid Diets: Integrating Approaches to the Study of Palaeolithic Subsistence. Springer, Dordrecht.

Rosenberger, A. L., 1992. Evolution of feeding niches in New World monkeys. American Journal of Physical Anthropology 88, 525–562.

Schoeninger, M. J., Iwaniec, U. T., Glander, K. E., 1997. Stable isotope ratios monitor diet and habitat use in New World Monkeys. American Journal of Physical Anthropology 103, 69–83.

Schoeninger, M. J., Iwaniec, U. T., Nash, L. T., 1998. Ecological attributes recorded in stable isotope ratios of arboreal prosimian hair. Oecologia 113, 222–230.

Schoeninger, M. J., Moore, J., Sept, J. M., 1999. Subsistence strategies of two 'savanna' chimpanzee populations: the stable isotope evidence. American Journal of Primatology 47, 297–314.

Schoeninger, M. J., Reeser, H., Hallin, K., 2003. Paleoenvironment of Australopithecus anamensis at Allia Bay, East Turkana, Kenya: evidence from mammalian herbivore enamel stable isotopes. Journal of Anthropological Archaeology 22, 200–207.

Shemesh, A., 1990. Crystallinity and diagenesis of sedimentary apatites. Geochimica et Cosmochimica Acta 54, 2433–2438.

Sponheimer, M., Dufour, D., 2009. Increased dietary breadth in Early Hominin evolution: revisiting arguments and evidence with a focus on biogeochemical contributions. In: Hublin, J. J., Richards, M. P. (Eds.), The Evolution of Hominid Diets: Integrating Approaches to the Study of Palaeolithic Subsistence. Springer, Dordrecht.

Sponheimer, M., Loudon, J., Codron, D., Howells, M., Pruetz, J., Codron, J., de Ruiter, D., Lee-Thorp, J. A., 2006. Do 'savanna' chimpanzees consume C_4 resources? Journal of Human Evolution PMID 16630647, on line 1–6.

Stanford, C. B., 1996. The hunting ecology of wild chimpanzees: implications for the evolutionary ecology of Pliocene hominids. American Anthropologist 98, 96–113.

Steklis, H. D., Sept, J. M., Harris, J. W. K., Cachel, S., 1992. Ethnoarchaeological investigations of chimpanzees in a gallery forest in eastern Zaire. Research Report submitted to National Geographic Society.

Strier, K. B., 1992. Atelinae adaptations: behavioral strategies and ecological constraints. American Journal of Physical Anthropology 88, 515–524.

Suzuki, A., 1969. An ecological study of chimpanzees in a savanna woodland. Primates 10, 103–148.

Teleki, G., 1981. The omnivorous diet and eclectic feeding habits of chimpanzees in Gombe National Park, Tanzania. In: Harding, R. S. O., Teleki, G. (Eds.), Omnivorous Primates: Gathering and Hunting in Human Evolution. Columbia University Press, New York, 303–343.

Urton, E. J. M., Hobson, K. A., 2005. Intrapopulation variation in gray wold isotope ($d^{15}N$ and $d^{13}C$) profiles: implications for the ecology of individuals. Oecologia 145, 317–326.

van der Merwe, N. J., Medina, E., 1989. Photosynthesis and $^{13}C/^{12}C$ ratios in Amazonian rain forests. Geochimica et Cosmochimica Acta 53, 1091–1094.

van der Merwe, N. J., Vogel, J. C., 1978. 13C Content of human collagen as a measure of prehistoric diet in woodland North America. Nature 276, 815–816.

Vogel, J. C., van der Merwe, N. J., 1977. Isotopic evidence for early maize cultivation in New York state. American Antiquity 42, 238–242.

Wahlen, M., 1994. Carbon dioxide, carbon monoxide and methane in the atmosphere: abundance and isotopic composition. In: Lajtha, K., Mitchener, R. H. (Eds.), Stable Isotopes in Ecology and Environmental Science. Blackwell, Oxford, 93–113.

Wrangham, R. W., 1977. Feeding behaviour of chimpanzees in Gombe National Park, Tanzania. In: Clutton-Brock, T. H. (Ed.), Primate Ecology: Studies of Feeding and Ranging Behavious in Lemurs, Monkeys and Apes. Academic Press, London, 504–538.

Wrangham, R. W., Conklin, N. L., Chapman, C. A., Hunt, K. D., 1991. The significance of fibrous foods for Kibale Forest chimpanzees. Philosophical Transactions of the Royal Society London 334, 171–178.

Wrangham, R. W., Chapman, C. A., Clark-Arcadi, A. P., Isabirye-Basuta, G., 1996. Social ecology of Kanyawara chimpanzees: implications for understanding the costs of great ape groups. In: McGrew, W. C., Marchant, L., Nishida, T. (Eds.), Great Ape Societies. Cambridge University Press, Cambridge, 45–57.

Yakir, D., Israeli, Y., 1995. Reduced solar irradiance effects on net primary productivity (NPP) and the $\delta^{13}C$ and $\delta^{18}O$ values in plantations of *Musa* sp., *Musaceae*. Geochimica et Cosmochimica Acta 59, 2149–2151.

18. Increased Dietary Breadth in Early Hominin Evolution: Revisiting Arguments and Evidence with a Focus on Biogeochemical Contributions

Matt Sponheimer
Department of Anthropology
University of Colorado at Boulder
Boulder, CO 80309, USA
msponheimer@gmail.com

and

Darna L. Dufour
Department of Anthropology
University of Colorado at Boulder
Boulder, CO 80309, USA
Darna.Dufour@colorado.edu

Keywords Diet • dietary breadth • australopith • *Homo* • biogeochemistry • trace element • carbon isotope • Sr/Ca

Abstract Increases in dietary breadth have been associated with hominin origins and the development of the genus *Homo*. For the former, researchers such as Dart noted that the environments associated with australopiths were not conducive for frugivorous diets like those of our closest living relatives the chimpanzees. Thus, many early arguments for increased dietary breadth were largely driven by perceived environmental constraints on diet. In contrast, the arguments for increased dietary breadth in early *Homo* have focused on explicating its high degree of encephalization. Biogeochemical evidence supports these ideas to varying extents. Trace element data have been used to argue for the increased utilization of animal foods among australopiths. We argue, however, that these data are equivocal and may be more consistent with the consumption of plant foods such as underground storage organs and grass seeds. Carbon isotope analysis demonstrates that australopith diets were characterized by great variability, and a tendency to consume C_4 resources such as grasses, sedges, and/or animals eating these foods. This contrasts with what has been observed in chimpanzees, which exhibit little variability and do not appear to consume significant quantities of C_4 resources even when such resources are locally abundant. Biogeochemistry has revealed little about the diets of early *Homo*, although there are data suggesting that it consumed some C_4 resources, possibly in the form of underground storage organs. At present, our best evidence for increases in dietary breadth in early *Homo* comes from the archaeological record and physiologically based models for encephalization. We surmise that increased meat consumption among early *Homo* was not in and of itself responsible for a leap in dietary quality, but rather that it served as a dietary release allowing consumption of more abundant energy-rich foods with protein of lower biological value.

J.-J. Hublin and M.P. Richards (eds.), *The Evolution of Hominin Diets: Integrating Approaches to the Study of Palaeolithic Subsistence*, 229–240.

Introduction

Understanding the role of diet in human evolution has been of interest for many years. This focus on diet is expected since an animal's diet provides an important window to its ecological niche, i.e., all of the adaptations of an organism to its environment (Pianka, 1974; Fleagle, 1999). One of the focal points of this research has been when and how the diet of our ancestors diverged from that of other hominoids. Extant apes and humans are assumed to have descended from a strongly herbivorous, forest-dwelling common ancestor, and while the diets of apes have undoubtedly changed since then, humans are assumed to have diverged furthest from the ancestral diet. In particular, there has long been a focus on hominins broadening their ancestral resource base, although there is no consensus as to the types of new foods, or first species to utilize them. Indeed, this trend can be traced back to the beginnings of African paleoanthropology (e.g., Dart, 1925, 1926), and continues to be a major research focus to this day (e.g., Leonard and Robertson, 1994; Aiello and Wheeler, 1995; Milton, 1999; Laden and Wrangham, 2005; Sponheimer et al., 2006).

Our goals herein are threefold. First, we hope to present a brief, personal, and undoubtedly incomplete history of thinking about the dietary breadth of early African hominins.

In so doing, we aim not so much to summarize over 80 years of research, but rather to briefly outline the relevant intellectual trajectories for our paleodietary work. Second, we hope to summarize and synthesize the relevant biogeochemical research of the past 15 years, with a special emphasis on what it tells us about the origin of increased dietary breadth in early hominins. Last, we hope to step back and take stock of what we have learned, and address the significance of terms such as "omnivore," "generalist," and "dietary quality" as they pertain to the evolution of our lineage.

A Brief and Not Impartial History

Dart, Australopiths, and the Desert

When Raymond Dart "discovered" the Taung child in 1924, he immediately suspected that it had broadened its resource base compared to extant apes. This was not, however, due to clues evident in the specimen's morphology that hinted at human-like dietary adaptations, but largely a function of his understanding of this new hominin's milieu (Dart, 1925, 1926). It had long been believed that our lineage arose because of fragmenting forests and subsequent adaptation to life in the savanna. Darwin (1871) had argued that our progenitors "frequented some warm, forest-clad land," and that the shrinking of these ancestral forests (*inter alia*) may have precipitated the dawn of humankind. Notably, Dart's mentor Grafton Elliot Smith (1924) echoed this sentiment, calling the earliest hominin environments "the unknown world beyond the trees." This basic idea seemed especially prescient when the Taung Child was found in 1924. Taung (Tswana "place of the lions") is situated on the edge of the Kalahari Desert, and thus provides a habitat very unlike those of great apes today. As it was believed at the time that southern African environments had not changed appreciably since the Cretaceous Period (Rogers, 1922), this presented something of an enigma. How could this southern ape survive in an environment so hostile to its forest-loving kindred (*Pan* and *Gorilla*) some thousands of miles to the north? Quite reasonably, Dart concluded that there must have been a major ecological discontinuity between the newly named *Australopithecus africanus* and the extant African apes (Dart, 1925, 1926). Given the incipient morphological similarity between *Australopithecus* and modern humans, he reasoned that this creature might also have developed ecological similarities to modern hunter-gatherers that allowed it to survive in such an arid and unforgiving environment. Specifically, he argued that the Taung child supplemented the traditional herbivorous ape diet with roots, bulbs, and a variety of animal foods including birds' eggs, insects, rodents, and young antelope (Dart, 1926).

Of course, subsequent discoveries, especially those from Makapansgat, inspired Dart to further develop his ideas about australopith ecology. His pioneering taphonomic study of the Makapansgat fossils showed that while some skeletal elements (e.g., distal humeri) were very abundant at the site, others (e.g., caudal vertebrae) were exceedingly uncommon (Dart, 1957). Dart explained these disparities as being the consequence of australopith behavior. For example, he suggested that the abundance of distal humeri reflected their status as preferred weapons, and the lack of caudal vertebrae indicated that australopiths used these bones outside the caves as whips and signals (Dart, 1957). He further described these early australopiths as mighty hunters and confirmed cannibals who "slaked their ravenous thirst on the hot blood of victims and greedily devoured livid, writhing flesh" (Dart, 1948).

Subsequent work by Brain (1967, 1969, 1981) demonstrated that the observed element frequencies at Makapansgat were most likely the result of carnivore accumulation, and that Dart's views on australopith ecology and behavior were largely incorrect. Yet, Dart's initial premise, that dietary broadening would have been necessary for early hominins, remained correct insofar as hominin environments were still considered to be fairly dry and non-forested (e.g., Brain, 1958). Yet, by the 1970s, evidence began to mount suggesting that earlier hominids of the genus *Australopithecus* inhabited more mesic and wooded habitats than the later *Paranthropus* and *Homo* (Vrba, 1975; Bonefille, 1976; Jaeger and Wesselman, 1976, 1980). Moreover, with the recent discovery of *Ardipithecus* and its association with colobus monkeys, kudu-like antelope, and canthium seeds (WoldeGabriel et al., 1994), it has become fashionable to posit heavily wooded or even forested habitats for some of the earlier australopiths.

These changes in thinking about australopiths' environments had important ramifications for our evolving understanding of australopith dietary breadth, for as woodlands and forests became increasingly associated with these creatures, ape-like frugivory became a possible dietary adaptation, and ecological analogies with human hunter-gatherers and baboons (e.g., Washburn and Devore, 1961; Laughlin, 1968) that had seemed environmentally appropriate were supplanted with the notion of australopiths *qua* chimpanzees. The emergence of chimpanzees as the dominant ecological model for australopiths found further support in a series of studies on dental morphology and microwear (e.g., Walker, 1981; Kay, 1985; Grine and Kay, 1988). For instance, Grine (1981) and Kay (1985) noted the low occlusal relief of australopith molars, which is a trait commonly found in frugivorous primates today; and Walker (1981) and Grine and Kay (1988) noted that the microscopic pits and scratches on the teeth of *P. boisei* and *A. africanus* were similar to those of modern frugivorous primates such as chimpanzees. This commonality with chimpanzees was further strengthened by molecular data demonstrating a close phylogenetic relationship between humans and chimpanzees (e.g., Goodman, 1963; Sarich and Wilson, 1967).

Homo and Its Pesky Large Brain

The above findings did not so much lead to the dismissal of the notion that early hominin diets had expanded compared to those of extant apes, as much as shift the diet change temporally. There had long been a focus on the importance of hunting in open savanna environments as the key behavior leading to the expansion of the brain in the *Homo* clade (e.g., Laughlin, 1968; Washburn and Lancaster, 1968; Tiger and Fox, 1971). The posited hunting was big game hunting, and thus fundamentally different from the hunting and animal food consumption that has been observed in chimpanzees (Goodall, 1986; Stanford, 1996). Thus, adherents of this idea postulated an increase in dietary breadth first amongst the members of our genus. This well-known man-the-hunter model posited that big game hunting was a way of life associated with intelligence, cooperation, and planning among males, technical skill, tool use, division of labor by sex, food sharing, etc. As Laughlin (1968) stated, "Hunting played a dominant role in transforming a bipedal ape to a tool-using and tool-making man who communicated by means of speech and expressed a complex culture." Ironically, the focus was more on the behavior of hunting rather than the product of the hunt, i.e., meat. The product was simply assumed to be "good," and distinct from the foods regularly consumed by apes.

More recently, related arguments have been made about the importance of increased dietary breadth for *Homo*, although these, unlike their predecessors, have a more explicit physiological grounding (e.g., Milton, 1987; Leonard and Robertson, 1991, 1994; Aiello and Wheeler, 1995). Milton (1987, 1999) has argued that the emergence of *Homo* was associated with a higher quality diet than that found in the other hominoids. She bases the argument on comparative gut anatomy, gut proportions, and digestive kinetics in hominoids. Human guts are similar to those of other hominoids at the gross structural level, but their gut proportions are different. In humans, the small intestine accounts for a greater proportion of the total gastrointestinal tract. This proportionality is the derived condition, and indicative of a higher quality diet (e.g., a diet lower in fiber and lignin), and is shared by other highly omnivorous, but distantly-related primates such as *Papio* and *Cebus* (Milton, 1987). The antiquity of this gut proportionality in our lineage is unknown, although there is some reason to believe that an incipient shift to the modern human condition may have begun over 1.78 Ma (Shipman and Walker, 1989; Aiello and Wheeler, 1995). Milton also argues that the incorporation of meat into the diets of early humans was a key element in enabling them to maintain a high quality diet while increasing body size, sociality, and levels of physical activity.

Leonard and Robertson (1991, 1994) have also argued that the emergence of *Homo* was associated with a higher quality and more omnivorous diet than that found in the other hominoids, but base their argument on the high energy cost of encephalization. They noted that the large brain – a distinguishing characteristic of hominids, and especially *Homo* – has a high metabolic cost and must have been facilitated by a change to a higher quality diet, especially one containing energy dense animal foods. Aiello and Wheeler (1995) pick up a similar line of argument in proposing the "expensive tissue hypothesis." They reasoned that since brain tissue has a high metabolic rate, a highly encephalized animal should have a higher than expected basal metabolic rate (BMR) for its body size, and since humans do not, something else has to have changed. They posit that a reduction in the size (and hence metabolic cost) of the gastrointestinal (GI) tract compensated for the increase in brain size, and further posit that size reduction was made possible by the adoption of a high quality diet. However, subsequent work by Hladik et al. (1999) demonstrated that humans do not appear to be an exception among primates in terms of the gut size and diet quality relationship, and hence do not show a trend to reduced intestinal size that could have allowed more energy for brain function. Indeed, in terms of the total absorptive capacity of the gut, they showed that humans fall on the same regression line as frugivores.

Broadhurst et al. (1998, 2002) also argue for the necessity of increased dietary breadth and a high quality diet to support the evolution of the large complex brain in *Homo*, but go further by proposing that aquatic foods would have been a necessary dietary component. Their argument rests on the fact that the brain is composed almost entirely of two polyunsaturated fatty acids, arachidonic acid (AA) and docosahexaenoic acid (DHA). AA and DHA are long chain fatty acids that are not widely available in savanna food chains, but can be synthesized endogenously, albeit slowly, from shorter chain precursors. Hence, they hypothesize that AA and DHA could have been limiting to the development of a large brain in *Homo* unless dietary sources of AA and DHA were available. Given that AA and DHA are more abundant in the marine food chain than the terrestrial, they argue that the availability of marine resources was critical to encephalization. This is an interesting argument that relies on some things we know with a good deal of certainty, as well as on some questionable assumptions. One assumption is DHA was limited in savanna environments because the tissue of local herbivores did not contain DHA. In reality, although herbivore muscle tissue is almost exclusively AA, both AA and DHA are prevalent in the neurological tissues which are highly prized by ethnographically known foragers, and may well have been consumed early hunters. Broadhurst et al. (1998, 2002) also argue that endogenous synthesis of AA and DHA from widely available PUFA precursors would not have been sufficient to meet needs of adults; however, adequate data are simply not available to support this assumption (IOM, 2005). Last, there is an implicit assumption that brain size is a function of the availability AA and DHA. Although it is clear that achieving species-typical brain size does require minimum levels of AA and DHA, it does not follow that greater availability of substrates like AA and DHA will lead to greater brain size. In other words, the availability of AA and DHA are necessary, but not sufficient to explain encephalization.

Biogeochemical Approaches

Trace Element Studies

Several chemical techniques were developed in the 1980s and 1990s in an effort provide further insight into early hominid diets. These were based upon the principle that "you are what you eat," or stated otherwise, the atoms consumed by an animal find their way into its hard tissues. The first of these techniques applied to early hominins was strontium/calcium (Sr/Ca) ratio analysis (Sillen, 1992). This technique was based upon the observation that Sr/Ca ratios decrease as one moves up the food chain, as a result of biological discrimination against strontium (Kostial et al., 1969; Spencer et al., 1973). Thus, when a zebra eats grass, the Sr/Ca ratios of its hard tissues will be lower than that of its food. When the zebra is consumed by a lion, in turn, its bones will have lower Sr/Ca ratios than the bones of its prey. Sillen (1992) demonstrated that, after application of "solubility profiling" which was designed to partition biogenic from diagenetic mineral, carnivores from Swartkrans had lower Sr/Ca ratios than many herbivores, as expected. This suggested that the biogenic chemical signatures of the bones had not been lost during fossilization and had been successfully extracted using the solubility profiling procedure. He also found that *P. robustus* had lower Sr/Ca than contemporaneous baboons and hyraxes. This led Sillen to argue that *Paranthropus* was not an exclusive herbivore, as had been posited by Robinson (1954), but rather that this taxon had begun to diversify the ancestral hominoid diet by consuming significant quantities of animal foods. Thus, the trace element data seemed to suggest that Dart was right (prior to his ideas about the osteodontokeratic at least), in that early hominins were generally characterized by an increase in dietary breadth compared to extant great apes. This made good sense intuitively, for although the fossil fauna from Swartkrans suggested that australopiths did not live in semi-desert conditions as Dart initially believed, the fossil fauna were still indicative of environments far more open and sere than those preferred by African apes today (Vrba, 1980; Reed, 1997). Thus, environment was still seen as a major constraint on diet.

Intriguingly, Sillen et al. (1995) found that two *Homo* specimens had relatively higher Sr/Ca than *Paranthropus*. This was contrary to expectations, for as was discussed above at some length, it had generally been assumed that *Homo* consumed more animal foods than the australopiths (e.g., Milton, 1987; Aiello and Wheeler, 1995). Sillen et al. (1995) argued that this apparent inconsistency might reflect increased reliance on underground storage organs (USO) by early *Homo*, which are relatively rich in Sr, an idea which continues to gain adherents (O'Connell et al., 1999; Wrangham et al., 1999). While generally unexpected, the suggestion was not unprecedented. Hatley and Kappelman (1980) had previously argued that ursids, suids, and early hominins occupied similar adaptive zones, in that highly abundant underground resources were an important dietary resource for all of them. O'Connell et al. (1999) further developed this hypothesis by noting that the evidence for hunting among early *Homo* was sparse, and that with the expedient of a crude digging stick, early *Homo* could have had access to reliable underground resources for which there is relatively little competition. Conklin-Brittain et al. (2002) argued that the utilization of these resources probably began with the australopiths. They suggested that as australopiths were forced to move out of preferred forest habitats to drier woodlands, they would have focused on resources available around water margins, and that by exploiting underground storage organs like roots, tubers, and corms they could have significantly decreased their overall dietary fiber intake (and increased dietary quality) compared to chimpanzees. Most recently, Laden and Wrangham (2005) avowed that australopith masticatory morphology, along with their co-occurrence with root eating rodents in the fossil record (demonstrating they lived in USO-rich habitats), testifies to the utilization of these resources among early members of our lineage.

Important as these trace element studies were for injecting new life into the debate on early hominin diets, they suffered from a number of limitations. For one, studies have shown that Sr/Ca ratios are quite variable in plants at the base of the food web, which seriously complicates interpretation of Sr/Ca data (e.g., Bowen and Dymond, 1955; Runia, 1987; Burton et al., 1999). For instance, leaves from trees have relatively low Sr/Ca ratios compared to grasses, and as a result browsing herbivores such as kudu have lower Sr/Ca ratios than grazing herbivores. In fact, browsers fall nicely within the range of carnivores (Sillen, 1988, 1992; Sponheimer and Lee-Thorp, 2006); thus, it is possible that the relatively low Sr/Ca ratios of *Paranthropus* observed by Sillen (1992) reflect the consumption of browse plants rather than animal foods. Furthermore, while the lack of occlusal relief on the molars of *Paranthropus* is probably inconsistent with significant leaf consumption (Kay, 1985; Teaford et al., 2002), we cannot rule out the possibility that the low Sr/Ca of *Paranthropus* results from fruit consumption, as fruit has been shown to have relatively low Sr/Ca in some instances (Haghiri, 1964).

An equally serious problem with the trace element studies is that they were carried out on bone which is particularly vulnerable to diagenetic overprinting (Sillen, 1981; Budd et al., 2000; Hoppe et al., 2003; Lee-Thorp and Sponheimer, 2003). This suggested the possibility that the hominin Sr/Ca ratios may have been altered postdepositionally. To address this possibility, renewed efforts have been undertaken to obtain trace element ratios from tooth enamel, which is known to be less susceptible to diagenesis than bone (Wang and Cerling, 1994; Budd et al., 2000; Hoppe et al., 2003; Lee-Thorp and Sponheimer, 2003). These efforts have suggested that, in contrast to previous studies, neither *Paranthropus* nor *Australopithecus* has lower Sr/Ca than contemporaneous baboons and most herbivores (Sponheimer et al., 2005a). Consequently, the Sr/Ca ratios by themselves offer little in the way of evidence for australopith omnivory. Yet, if one plots two physiologically-related trace element ratios (barium/calcium and strontium/barium) for early hominins and associated non-hominin fauna, several interesting results emerge (Fig. 18.1).

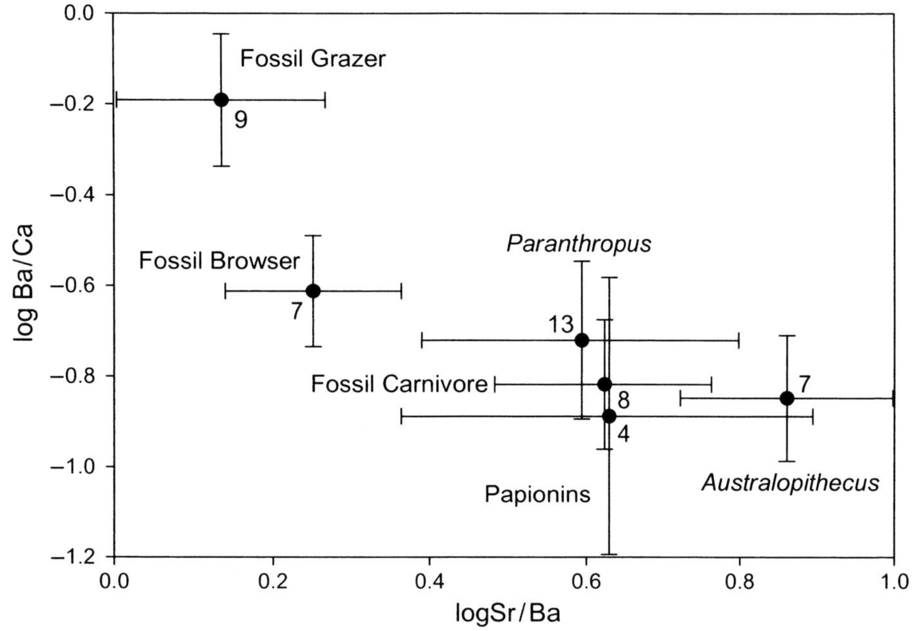

FIG. 18.1. Bivariate log(Ba/Ca) and log (Sr/Ba) plot for fossils grazers, browsers, carnivores, papionins and hominins from the Sterkfontein Valley. The dots are mean values, the whiskers represent standard deviations (1σ), and the numbers indicate sample sizes. The clear separation between grazers, browsers, and carnivores demonstrates that ecological information is retained in fossil enamel.

First, it is very clear that ecological patterning is retained in fossil enamel, as grazers, browsers, and carnivores are very nicely separated in much the same way we see with modern African fauna (Sponheimer and Lee-Thorp, 2006). Second, the hominins do look, at least superficially, similar to the carnivores; however, the same is also the case for the papionins, which are demonstrably not carnivorous – thus little should be made of this superficial resemblance between hominins and carnivores. Clearly we need to increase our knowledge of how trace element abundances are distributed throughout modern African foodwebs before we can properly interpret trace element data from early hominins and other fossil fauna.

Carbon Isotope Studies

Stable carbon isotope analysis is another biogeochemical tool that has been used to investigate early hominid diets (Lee-Thorp et al., 1994; Sponheimer and Lee-Thorp, 1999; van der Merwe et al., 2003; Sponheimer et al., 2005b). Once again, these studies rely on the principle that "you are what you eat," but unlike Sr/Ca analysis, arose out of a growing body of knowledge on plant photosynthesis. In tropical Africa, trees, bushes, shrubs, and forbs use what is known as the C_3 photosynthetic pathway. Grasses and some sedges, in contrast, use the C_4 photosynthetic pathway (Smith and Epstein, 1971; Vogel et al., 1978). As a result of anatomical and biochemical differences between these two types of plants, C_3 plants discriminate more strongly against $^{13}CO_2$ during photosynthesis than C_4 plants do, with the result that they are isotopically "lighter." Carbon isotopes are then passed into the tissues of herbivores that consume them, with the result that grass eaters (grazers) like zebras are isotopically "heavier" than C_3 tree leaf and fruit eaters (browsers) like giraffes (Vogel, 1978; Tieszen et al., 1979).

As a result, stable carbon isotope analysis is a useful method for testing hypotheses such as "Australopiths were near-exclusive frugivores," for if this were the case, they would have had carbon isotope ratios similar to those of contemporaneous C_3 vegetation consumers like giraffes and kudu. In contrast, if their carbon isotope ratios (which are expressed in parts per thousand (‰) relative to the PDB standard) were relatively enriched in ^{13}C compared to known C_3 consumers, this would falsify the hypothesis and indicate that they had broadened their diets to include savanna resources like grasses or animals eating those foods. Lee-Thorp et al. (1994, 2000) analyzed the tooth enamel of *P. robustus*, early *Homo*, and a variety of other taxa from Swartkrans, South Africa. They were able to show that fossil grazers and browsers remained isotopically distinct, and that while *Paranthropus* and *Homo* had predominantly C_3 diets, a significant portion of their diets came from C_4 "savanna" resources (see Fig. 18.2 for an updated dataset). The authors argued that savanna animal foods were the most likely utilized C_4 resource. Similar studies were carried out on older *Australopithecus* fossils from Makapansgat and Sterkfontein (Sponheimer and Lee-Thorp, 1999; van der Merwe et al., 2003) and demonstrated that this taxon consumed as much, if not more, C_4 resources than *Paranthropus*. In fact, the mean $\delta^{13}C$ value for *Australopithecus* (−6.4 ± 2.3‰; this includes all 23 specimens that have been analyzed and attributed to the taxon on morphological grounds) falls almost halfway in between the mean $\delta^{13}C$ values of penecontemporaneous browsers (−11.5 ± 1.3‰) and grazers (−0.6 ± 1.8‰) (Fig. 18.2). Moreover, six *Australopithecus* specimens have $\delta^{13}C$ values that place them within the range of the grass-eating baboon,

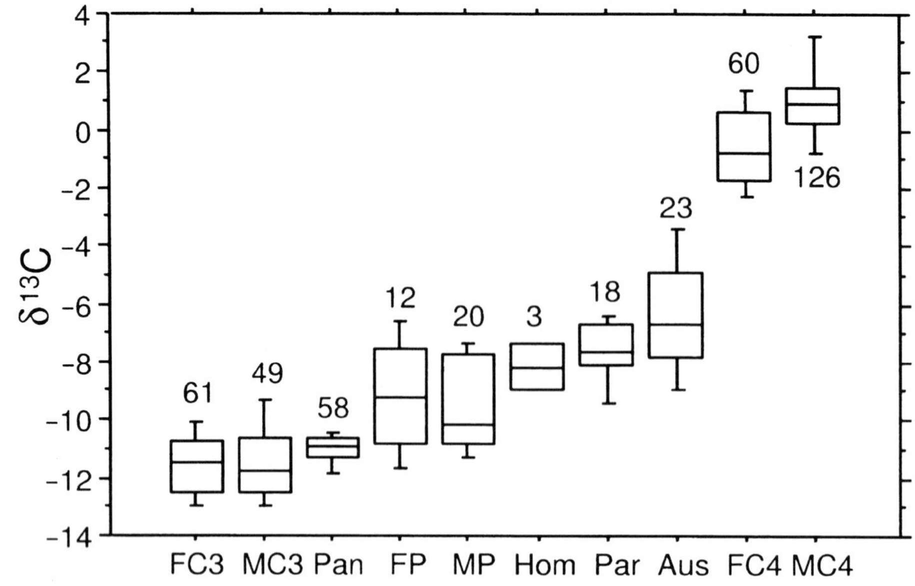

Fig. 18.2. $\delta^{13}C$ values of hominin and non-hominin fossil fauna from South Africa, as well as $\delta^{13}C$ values for modern samples. Modern hair and enamel values were converted for the fossil fuel effect and tissue differences as in Sponheimer et al. (2006). The key is as follows: FC3 is fossil C_3 consumers, MC3 is modern C_3 consumers, Pan is all "savanna" chimpanzees, FP is *Papio hamadryas robinsoni*, MP is *Papio hamadryas ursinus*, Hom is early *Homo*, Par is *Paranthropus*, Aus is *Australopithecus*, FC4 is fossil C_4 consumers, and MC4 is modern C_4 consumers. The boxes represent the 25th–75th percentiles (with the medians as horizontal lines) and the whiskers show the 10th–90th percentiles.

Theropithecus oswaldi (Lee-Thorp et al., 1989; Codron et al., 2005). This strongly suggests that a taste for C_4 resources was an important hominin trait – even if the exact nature of the C_4 resources remains unknown.

This is in striking contrast to chimpanzees, which are not known to consume significant quantities of C_4 resources, even in the most arid and open areas of their range (Schoeninger et al., 1999; Sponheimer et al., 2006). To fully appreciate this contrast, it is useful to directly compare the stable carbon isotope compositions of australopiths to "savanna" chimpanzees. Unfortunately, this cannot be accomplished directly, as there are no published stable isotope data for chimpanzee tooth enamel, but we do have $\delta^{13}C$ data for 58 chimpanzee hair samples (Schoeninger et al., 1999; Sponheimer et al., 2006). Although this is not an ideal comparison because hair and tooth enamel preserve dietary information from different periods of time and have different diet-tissue fractionations, such comparisons have been shown to be robust when making broad dietary comparisons (Sponheimer et al., 2003, 2006). The hair samples were collected from some of the most marginal, open environments inhabited by chimpanzees; yet, they betray no evidence of the consumption of C_4 foods, even though they are locally abundant. Furthermore, there is precious little variability in the chimpanzee data, even though the sample comprises data from three populations, all of which reside in different countries. This is, once again, very different from what has been observed in australopiths. As shown in Fig. 18.2, the australopiths are characterized by great isotopic variability.

Hence, australopiths are very different from chimpanzees both in terms of mean $\delta^{13}C$ values and in variability. This suggests that even if early hominins were in environments largely similar to those of chimpanzees, they utilized resources in these environments in different ways. Thus, even if early hominins retained a variety of arboreal adaptations (e.g., Vrba, 1979; Stern and Susman, 1983; Clarke and Tobias, 1995) and were dependent on trees for protection from predators and for much of their diets, they had broadened their dietary resource base in a way that would allow them to survive in habitats too open and xeric for chimpanzees.

All told, this calls into question the use of chimpanzees as referential ecological models (*sensu* Tooby and DeVore, 1987) for australopiths. In fact, australopiths are more similar to what has been observed in modern and fossil baboons, both of which tend to consume measurable amounts of C_4 resources, and which seem to be inherently variable (Fig. 18.2; see Codron et al., 2005; Sponheimer et al., 2006). This suggests that, from an ecological perspective, baboons might be a more useful analog than chimpanzees (e.g., Dunbar, 1976; Jolly, 2001), especially given the well-known association between papionins and hominins in the fossil record (Dart, 1925; Delson, 1984). This is not to suggest that australopiths and baboons were consuming the same foods, but rather that both appear to be inherently flexible and capable of utilizing forest as well as savanna resources.

Where Does This Leave Us?

So what does the biogeochemical evidence tell us about increased dietary breadth in early human evolution? The trace element data are certainly equivocal, but perhaps suggestive in some ways. For one, Sillen's data hinted that early *Homo* might have diversified its resource base, but perhaps not in the way most people anticipated. Underground storage organs are not consumed by extant apes to a significant extent (although they are consumed, albeit infrequently in some areas (McGrew et al., 1982; Doran and McNeilage, 1998)). The carbon isotope data for early *Homo* could also be consistent with this, so long as some of the USOs came from C_4 resources such as some sedges and grasses. The enamel data for *Paranthropus* and *Australopithecus* do not directly support the hypothesis that they broadened their dietary repertoires to include meat. The Sr/Ca and Ba/Ca ratios of *Australopithecus*, however, can be taken as possible evidence for the consumption of underground resources, as these are similar to those of mole rats which are known to subsist on roots and bulbs (Sponheimer et al., 2005a; Sponheimer and Lee-Thorp, 2006). Nonetheless, they are also consistent with the consumption of grass seed (Sponheimer and Lee-Thorp, 2006).

Fortunately, the carbon isotope data are less ambiguous as far as establishing dietary breadth in concerned. It is clear that australopiths and *Homo* consumed, to varying extents, C_4 savanna foods that are not consumed to a significant degree by chimpanzees or other hominoids. This in and of itself bespeaks a broadening of the ancestral ape resource base. Moreover, the high degree of variability in early hominin $\delta^{13}C$ values suggests they were highly flexible and opportunistic, which probably made them capable of utilizing a broader variety of microhabitats than extant apes. We suspect that australopiths had highly frugivorous diets, much like chimpanzees, when in heavily wooded habitats, but supplemented these foods with varying amounts of C_4 resources when in suboptimal environments (as is the case with papionins). This is consistent with Ungar's (2004) notion that australopith craniodental morphology was an adaptation for utilizing hard, brittle foods during periods of resource stress (which might be seasonal or interannual), rather than an adaptation for preferred day to day dietary resources.

Unfortunately, we are still some way from disentangling the knotty problem of what C_4 foods were consumed by australopiths and early *Homo*, for the simple reason that many different diets can result in near-identical carbon isotope compositions. At present our best evidence indicates that, for the most enriched hominins at least, there is little chance that sedges or animal foods alone could have accounted for the entire C_4 signal of australopiths (Peters and Vogel, 2005; Sponheimer et al., 2005b). More likely, these foods were consumed in tandem with grass products such as seeds and roots, both of which would have been useful seasonal resources. Other forms of paleodietary analysis (e.g., dental microwear, ecomorphology, other biogeochemical proxies) should allow us to test this hypothesis.

Discussion

So what evidence do we have for increased dietary breadth in our lineage, and when did the increase occur? The biogeochemistry compellingly suggests that dietary breadth did indeed expand well before the advent of *Homo*, as suggested by Dart. This is also supported by other lines of evidence that we have not discussed at length here (such as dental microwear) which suggest a great deal of dietary variability among some of the australopiths (Scott et al., 2005). Furthermore, carbon isotope evidence for the consumption of C_4 resources does suggest that this had to do with environmental degradation of the sort envisaged by Dart, although not to the same degree. Thus, by 3 Ma early hominins were clearly integrated into savanna ecosystems in ways we do not find with chimpanzees, which consume few if any of the novel resources in these environments. Interestingly, however, Grine et al. (2006) found that the dental microwear pattern of *Praeanthropus afarensis* (~3.9–3.0 Ma), which largely predates the hominins for which we have biogeochemical data (~3.0–1.5 Ma), overlaps significantly with the wear pattern of *Gorilla gorilla beringei*. Moreover, there is no evidence for temporal or environmental variation in its dental microwear, suggesting that, like extant apes, they utilized preferred forest resources even when in drier, more open habitats. This suggests the possibility that *Praeanthropus*, unlike the later South African australopiths, was not fully integrated into the savanna ecosystem, and that a major change in dietary breadth and flexibility occurred at about 3 Ma.

Regardless, the fact that by 3 Ma some australopiths have largely C_3 carbon isotope signatures while some have largely C_4 signatures is consistent with the idea that australopiths favored diets similar to those consumed by chimpanzees, but when in open areas with few such resources they broadened their dietary repertoire to include abundant C_4 foods. Thus, australopiths would have utilized suboptimal habitats much more efficiently and maintained higher population densities than chimpanzees (who dramatically increase their home ranges in such environments (McGrew et al., 1981; Moore, 1996)). This might lead one to speculate that chimpanzee populations that live in marginal woodland savanna habitats today (e.g., Mt. Assirik, Ishasha), would have been outcompeted by australopiths in these environments in the past, their much vaunted omnivory notwithstanding.

But, one might ask, should not omnivorous generalists like chimpanzees be capable of utilizing a variety of habitats? We think the answer to this question is yes, to a point, but that omnivorous and generalist diets in and of themselves are probably less significant in this regard than is often assumed. To be an omnivore means simply that an animal consumes both plant and animal foods, as opposed to just plants or just animals. It is a classification of animals based on their trophic level in the food chain: herbivores consume plants, carnivores consume animals, and omnivores consume both. Some terrestrial mammals fall into either the pure herbivore or pure carnivore categories, but many are considered omnivores (Allee et al., 1949; Landry, 1970). The trophic level distinctions are gross typological categories that refer to dominant and/or habitual feeding patterns, not to what an animal is capable of consuming. Most animals have the potential to consume a wider range of foods than they typically do. Pigs, for example, are primarily herbivores, but will readily consume animal foods if it is available. Reindeer, classic ruminant herbivores, can and will resort to eating fish when plant food is scarce (Allee et al., 1949). Likewise, the domestic cat, a carnivore, will, and commonly does, eat plant material in the absence of animal prey. So what an animal can eat, and does eat, in any given environment, in any given season, are different things.

To be an omnivore has ecological advantages. First, it means that the animal can meet its dietary needs with a wide range of foods and alter food preferences to match changes in the availability of specific foods. This is an important degree of flexibility. Second, by consuming a range of different foods, the omnivore minimizes the potential problems associated with the presence of toxins in plant foods. Third, omnivory is a way to maximize both energy availability and dietary quality. The first trophic level, plants, contains the most abundant source of energy on the planet. The second trophic level, herbivores, has only about 10% or less of the energy that is available in the first. However, in the second trophic level, overall dietary quality is higher because the composition of prey animals is similar to that of predators, and hence, prey animals more closely match the dietary needs of predators than do plants. In exploiting two trophic levels the omnivore has the best of both.

We also classify animals into two broad groups: generalists and specialists. Generalists are animals that consume a wide variety of other species. Specialists are animals that focus on one species, or a closely related group of species, and prosper by being good at finding and consuming their special food. Herbivores can be generalists or specialists. Cows are generalists because eat a variety of different types of leaves and other plant parts, and pandas are specialists because they only eat bamboo. Carnivores, as a group, are generalists. Omnivores are generalists by definition. There are some advantages to being a generalist. Generalists are better able to cope with seasonal variability in food resources, whereas specialists can only be active when their special food is available and are dormant the remainder of the time (Begon et al., 1990). Generalists have the advantage of not being constrained to patches of a specialized food resource, or being forced to spend large amounts of time and energy searching in mixed patches (Begon et al., 1990). Generalists have the flexibility to act as specialists in order to meet some nutritional need. A good example might be the propensity of some primates to concentrate their foraging efforts on figs when available, a specialization that has been associated with meeting calcium needs since figs are high in calcium (O'Brien et al., 1998).

So, does being an omnivore and generalist make a taxon better able to weather environmental change? Probably, but this can only take one so far as omnivores and generalists still must live with significant dietary constraints. When omnivorous generalists like chimpanzees find themselves in habitats where C_4 grass makes up the preponderance of the plant biomass, there is no evidence to suggest they can successfully utilize this resource even though they consume a vast array of species in forest environments. For instance, at Mt. Assirik, chimpanzee feeding time on USOs is less than 2%, even though they constitute much of the available nutrition in this landscape (McGrew et al., 1982). Baboons, in contrast, utilize such underground resources extensively, which might reflect some difference in their digestive capacities. Thus, while being an omnivore and generalist affords a certain buffer against environmental change, these general terms may conceal important differences in digestive adaptations. Hence, being an omnivore and generalist is probably far less meaningful in the context of human evolution than the specific capacities to efficiently utilize abundant savanna resources such as USOs, grasses, and large game, which clearly became increasingly available to hominins through time.

As for *Homo*, the biogeochemistry is also consistent with increased dietary breadth, but this is far from compelling. Our best evidence comes from scattered remains of butchered animals and stone tools which were used to process the animals (e.g., Isaac, 1983; Blumenschine, 1991). Yet, these tell us little about the relative importance of the foods in the diet of *Homo*. The physiologically flavored explanations are a bit more suggestive, yet these are not without problems. One of these is the notion of dietary quality. Unfortunately, what is meant by diet quality is often a bit fuzzy. In a general sense, quality is an essential, distinguishing attribute of something – a characteristic element. Quality can also refer to the ability of something to satisfy a stated or implied need. With regard to diet, the word quality is used in this latter sense and refers to the ability of foods to satisfy nutritional needs. Hence, a high quality food is a better source of nutrients. Similarly, a high quality diet is one that satisfies an organism's overall nutritional needs better than a low quality diet.

Nutritional needs are reasonably well known for modern humans, not very well known for extant nonhuman primates (NRC, 2003), and probably unknowable for early hominids. However, even when an organism's nutritional needs are known, assessing diet quality is tricky because foods are complex substances. They contain: Water, nutrients, non-nutrients (e.g., substances like lignin that cannot be digested), antinutrients (e.g., substances like phytate that bind minerals and make them inaccessible), and phytochemicals that can have toxic and/or pharmacological effects. Diets are even more complex because they contain a variety of different foods. In human nutrition, researchers have struggled with the problem of defining and assessing the quality of individual foods, as well as diet quality more generally (Kant, 1996). In assessing the quality of individual foods, they typically focus on single nutrients. For example, the quality of a food protein is defined in terms of its amino acid content; high quality proteins being those in which the proportions of essential amino acids most closely match human requirements. Based on this approach, animal proteins are of higher quality than plant protein, and invertebrate proteins are of intermediate value (Dufour, 1987; Robbins, 1993). Another measure of food quality is nutrient density, e.g., the concentration of a nutrient per 1,000 kcal of energy. This is a measure that makes intuitive sense, but is unduly cumbersome when extended to more than a few nutrients at a time. More recently, researchers in human nutrition have wrestled with developing indices of diet quality based on combinations of foods. For example, the Diet Quality Index provides a measure of overall diet quality that reflects, to the best of our understanding, the risk gradient for diet-related chronic diseases (Haines et al., 1999).

In paleoanthropology the concept of diet quality is important, but tends to be used rather loosely. With reference to hominins, high quality diets often seem to refer simply to those diets containing meat (Leonard and Robertson, 1991, 1994; Milton, 1999). Meat, being the flesh of mammals and similar in composition to the flesh of primates, is certainly a high quality food for primates in that it provides a nutrient package (amino acids in the proportions needed, essential fatty acids, vitamins and minerals) that is a good match for nutritional needs. Further, meat is relatively energy dense, and the nutrients in meat tend to be easily assimilated (Murphy and Allen, 2003).

In discussing non-human primates Milton (1987, 1999) uses diet quality as it is commonly used in the literature. She refers to ripe fruits as high quality foods for chimps because they are high in easily digested carbohydrates. These easily digested carbohydrates are simple sugars, and so in this case high quality refers to the concentration of simple sugars, a source of energy, but not of any other nutrients. Fruits do contain fiber, and on a dry matter basis, can be almost as high in fiber as leaves (Hart, 1985; Conklin-Brittain et al., 1998), which are typically considered a low quality food because of their high fiber content. Yet, leaves are generally higher in crude protein (dry matter basis) than fruits (NRC, 2003), and hence could be considered of higher quality, since protein is an essential dietary component.

Aiello and Wheeler (1995) refer to the fundamental importance of high quality diets to hominin evolution, but do not define what they mean by high quality except to say that these diets include high quality foods such as animal products, nuts, and underground storage organs. These three foods are nutritionally very different types of food. Animal products are a high quality food for humans and presumably earlier hominins. Nuts are primarily fat, and hence a good source of energy, but also contain plant protein, some minerals, and sometimes B-vitamins. Nuts could be considered foods of high quality given their ability to meet energy needs, but not nearly as high in quality as animal foods. Underground storage organs, i.e., roots and tubers, are predominantly starch and fiber, and hence good sources of energy, but not much else. It would be difficult to characterize them as high-quality foods.

Although not all three foods listed by Aiello and Wheeler (1995) could be considered high quality foods, a diet based on them could be high in quality. Why? Because as Milton (1999) has pointed out, animal products (in sufficient quantity) can provide the necessary amino acids, essential fatty acids, and many vitamins and minerals, leaving the plant food component of the diet primarily as a source of energy. This is an important point because it suggests that early hominins incorporating sufficient meat into their diets could have afforded to broaden their plant repertoire to include foods like roots, the nutritional value of which is primarily as an energy source. Thus, the consumption of fairly small but consistent quantities of animal foods could have been important not so much because it increased overall dietary quality in and of itself, but because it allowed utilization of a variety of energy-rich, but otherwise lower quality, resources that could only serve as fallback foods otherwise. Consequently, we might expect that what had previously been fallback foods became more consistent dietary resources after animal food consumption increased in our lineage. We might also expect that the amount of animal food consumed early on was fairly small (although more significant than has been reported for extant apes), even if the change in overall dietary quality was quite large. Moreover, such small packages of animal foods would likely have come in the form of insects, hyraxes, infant antelope, and other small prey which are well within the hunting capacities of extant chimpanzees and baboons (Goodall, 1986; Strum, 1987; Stanford, 1996). Hence, archaeological evidence for this initial increase in animal food consumption might be minimal.

Conclusions

Increased dietary breadth in one form or another has been used as a causal or enabling variable in explanatory models of hominization at two critical junctures. The first is the divergence of early hominins from the apes, and the second is in explaining the defining features of the genus *Homo*. We believe that there is considerable evidence for changes in dietary breadth at both of these transitions, although the evidence originates from very different quarters. Biogeochemical evidence suggests that a change in hominin diets occurred at least 3 Ma in South Africa (Sponheimer and Lee-Thorp, 1999), and that this was characterized by a significant increase in the utilization of C_4 savanna foods. Importantly, this suggests that baboons may be better ecological analogs for australopiths than chimpanzees, as the latter do not consume such resources to any significant extent. As for early *Homo*, the best evidence for increased dietary breadth comes from physiologically based models for hominin encephalization and the archaeological record, both of which speak to the importance of animal foods by about 2.5 Ma. We believe, however, that the notion of diet quality has not been sufficiently advanced in such studies, and that greater emphasis on animal foods as dietary releasers that allow consumption of abundant, but lower quality foods, is required.

Acknowledgments. We would like to thank Mike Richards and Jean-Jacques Hublin for giving us the chance to participate in the workshop and volume. We also thank the staff at the Max

Planck for making this such a pleasant experience and Sarah R. Taylor for helping with the manuscript. We also thank Bert Covert and Michelle Sauther for several conversations that helped shape the direction of the manuscript, as well as the reviewers.

References

Aiello, L.C., Wheeler, P., 1995. The expensive tissue hypothesis. Current Anthropology 36, 199–221.

Allee, W.C., Emerson, A.E., Park, O., Park, T., Schmidt, K.P., 1949. Principles of Animal Ecology. WB Saunders, Philadelphia.

Begon, M., Harper, J., Townsend, C.R., 1990. Ecology: Individuals, Populations and Communities. Blackwell, London.

Blumenschine, R.J., 1991. Early hominid carnivory and foraging strategies. Philosophical Transactions of the Royal Society of London B 334, 211–221.

Bonnefille, R., 1976. Palynological evidence for an important change in the vegetation of the Omo basin between 2.5 and 2.0 million years. In: Coppen, Y., Howell, F.C., Isaac, G.L.L., Leakey, R.E.F. (Eds.), Earliest Man and Environments in the Lake Rudolf Basin. University of Chicago Press, Chicago, pp. 421–430.

Bowen, H.J.M., Dymond, J.A., 1955. Strontium and barium in plants and soils. Philosophical Transactions of the Royal Society of London B 144, 355–368.

Brain, C.K., 1958. The transvaal ape-man-bearing cave deposits. Transvaal Museum Memoir 11, 1–131.

Brain, C.K., 1967. Hottentot food remains and their bearing on the interpretation of fossil bone assemblages. Scientific Papers of the Namib Desert Research Station 32, 1–11.

Brain, C.K., 1969. The contribution of Namib Desert Hottentots to an understanding of australopithecine bone accumulations. Scientific Papers of the Namib Desert Research Station 39, 13–22.

Brain, C.K., 1981. The Hunters or the Hunted? University of Chicago Press, Chicago.

Broadhurst, C.L., Wang, Y., Crawford, M.A., Cunnane, S.C., Parkington, J.E., Schmidt, W., 1998. Rift Valley lake fish and shellfish provided brain-specific nutrition for early *Homo*. British Journal of Nutrition 79, 3–21.

Broadhurst, C.L., Wang, Y., Crawford, M.A., Cunnane, S.C., Parkington, J.E., Schmidt, W.F., 2002. Brain-specific lipids from marine, lacustrine, or terrestrial food resources: potential impact on early African *Homo sapiens*. Comparative Biochemistry and Physiology Part B 131, 653–673.

Budd, P., Montgomery, J., Barreiro, B., Thomas, R.G., 2000. Differential diagenesis of strontium in archaeological human tissues. Applied Geochemistry 15, 687–694.

Burton, J.H., Price, T.D., Middleton, W.D., 1999. Correlation of Bone Ba/Ca and Sr/Ca due to Biological Purification of Calcium. Journal of Archaeological Science 26, 609–616.

Clarke, R.J., Tobias, P.V., 1995. Sterkfontein Member 2 foot bones of the oldest South African hominid. Science 269, 521–524.

Codron, D., Lee-Thorp, J.A., Sponheimer, M., De Ruiter, D., Codron, J., 2005. Utilization of Savanna-based resources by Plio-Pleistocene baboons. South African Journal of Science. 101, 245–249.

Conklin-Brittain, N.L., Wrangham, R.W, Hunt, K.D., 1998. Response of chimpanzees and Cercopithecines to seasonal variation in fruit abundance. II. Macronutrients. International Journal of Primatology 19, 971–998.

Conklin-Brittain, N.L., Wrangham, R.W., Smith, C.C., 2002. A two-stage model of increased dietary quality in early hominid evolution: the role of fiber. In: Ungar, P.S., Teaford, M.F. (Eds.), Human Diet: Its Origin and Evolution. Bergin & Garvey, Westport, pp. 61–76.

Dart, R., 1925. *Australopithecus africanus*, the man-ape of South Africa. Nature 115, 195–199.

Dart, R.A., 1926. Taungs and its significance. Natural History 26, 315–327.

Dart, R., 1948. The adolescent mandible of *Australopithecus prometheus*. American Journal of Physical Anthropology 6, 391–412.

Dart, R.A., 1957. The Osteodontokeratic culture of Australopithecus prometheus. Transvaal Museum Memoir 10, 1–105.

Darwin, C., 1871. The Descent of Man. Modern Library, New York.

Delson, E., 1984. Cercopithecid biochronology of the African Plio-Pleistocene: correlation among eastern and southern hominid-bearing localities. Courier Forschungsinstitut Senckenberg 69, 199–218.

Doran, D.M., McNeilage, A., 1998. Gorilla ecology and behavior. Evolutionary Anthropology 6, 120–131.

Dufour, D.L., 1987. Insects as food: a case study from the Northwest Amazon. American Anthropologist 89, 383–397.

Dunbar, R.I.M., 1976. Australopithecine diet based on a baboon analogy. Journal of Human Evolution 5, 161–167.

Fleagle, J.G., 1999. Primate Adaptation and Evolution, Second Edition. Academic Press, New York.

Goodall, J., 1986. The Chimpanzees of Gombe. Cambridge University Press, Cambridge.

Goodman, M., 1963. Serological analysis of the systematics of recent hominoids. Human Biology 35, 377–436.

Grine, F.E., 1981. Trophic differences between gracile and robust australopithecines. South African Journal of Science 77, 203–230.

Grine, F.E., Kay, R., 1988. Early hominid diets from quantitative image analysis of dental microwear. Nature 333, 765–768.

Grine, F.E., Ungar, P.S., Teaford, M.F., El-Zaatari, S., 2006. Molar microwear in *Praeanthropus afarensis*: evidence for dietary stasis through time and under diverse paleoecological conditions. Journal of Human Evolution 51, 297–319.

Haghiri, F., 1964. Strontium-90 accumulation in some vegetable crops. Ohio Journal of Science 64, 371–375.

Haines, P.S., Siega-Riz, A.A., Popkin, B.M., 1999. The diet quality index revisited: a measurement instrument for populations. Journal of the American Dietetic Association 99, 697–704.

Hart, J.A., 1985. Comparative dietary ecology of a community of frugivorous forest ungulates in Zaire (ruminants, feeding habits, rainforest). Ph.D. dissertation, Michigan State University, East Lansing, Michigan.

Hatley, T., Kappelman, J., 1980. Bears, pigs, and Plio-Pleistocene hominids: case for exploitation of belowground food resources. Human Ecology 8, 371–387.

Hladik, C.M., Chivers, D.J., Pasquet, P., 1999. On diet and gut size in non-human primates and humans: is there a relationship to brain size? Current Anthropology 40, 695–697.

Hoppe, K.A., Koch, P.L., Furutani, T.T., 2003. Assessing the preservation of biogenic strontium in fossil bones and tooth enamel. International Journal of Osteoarchaeology 13, 20–28.

IOM., 2005. Dietary Reference Intakes for Energy, Carbohydrates, Fiber, Fat, Fatty Acids, Cholesterol, Protein and Amino Acids. National Academies Press, Washington, DC.

Isaac, G.L., 1983. Bones in contention: competing explanations for the juxtaposition of Early Pleistocene artifacts and faunal remains. In: Clutton-Brock, J., Grigson, C. (Eds.), Animals and Archaeology 1. Hunters and their Prey. B.A.R. International Series 163, Oxford, 3–19.

Jaeger, J.J., Wesselman, H.B., 1976. Fossil remains of micromammals from the Omo group deposits. In: Coppens, Y., Howell, F.C., Isaac, G.L.L., Leakey, R.E.F. (Eds.), Earliest Man and Environments in the Lake Rudolf Basin. University of Chicago Press, Chicago, pp. 351–360.

Jolly, C.J., 2001. A proper study for mankind: analogies from the Papionin monkeys and their implications for human evolution. American Journal of Physical Anthropology 116, 177–204.

Kant, A.K., 1996. Indexes of overall dietary quality: a review. Journal of the American Dietetic Association 96, 785–791.

Kay, R., 1985. Dental evidence for the diet of *Australopithecus*. Annual Review of Anthropology 14, 315–341.

Kostial, K., Gruden, N., Durakovic, A., 1969. Intestinal absorption of calcium-47 and strontium-85 in lactating rats. Calcified Tissue International 4, 13–19.

Laden, G., Wrangham, R., 2005. The rise of the hominids as an adaptive shift in fallback foods: plant underground storage organs (USOs) and australopith origins. Journal of Human Evolution 49, 482–498.

Landry, S.O., Jr., 1970. The Rodentia as omnivores. Quarterly Review of Biology 45 (4), 351–372.

Laughlin, W.S., 1968. Hunting: An Integrating Biobehavior System and Its Evolutionary Importance. Man the Hunter. Aldine, Chicago pp. 304–320.

Lee-Thorp, J., Sponheimer, M., 2003. Three case studies used to reassess the reliability of fossil bone and enamel isotope signals for Paleodietary studies. Journal of Anthropological Archaeology 22, 208–216.

Lee-Thorp, J.A., van der Merwe, N.J., Brain, C.K., 1989. Isotopic evidence for dietary differences between two extinct baboon species from Swartkrans. Journal of Human Evolution 18, 183–190.

Lee-Thorp, J.A., van der Merwe, N.J., Brain, C.K., 1994. Diet of Australopithecus robustus at Swartkrans from stable carbon isotopic analysis. Journal of Human Evolution 27, 361–372.

Lee-Thorp, J.A., Thackeray, J.F., van der Merwe, N.J., 2000. The hunters and the hunted revisited. Journal of Human Evolution 39, 565–576.

Leonard, W.R., Robertson, M., 1991. Nutritional requirements and human evolution: a bioenergetics model. American Journal of Human Biology 4, 179–195.

Leonard, W.R., Robertson, M., 1994. Evolutionary perspectives on human nutrition: the influence of brain and body size on diet and metabolism. American Journal of Human Biology 6, 77–88.

McGrew, W.C., Baldwin, P.J., Tutin, C.E.G., 1981. Chimpanzees in a hot, dry and open habitat: Mt. Assirik, Senegal, West Africa. Journal of Human Evolution 10, 227–244.

McGrew, W.C., Sharman, M.J., Baldwin, P.J., Tutin, C.E.G., 1982. On early hominid plant-food niches. Current Anthropology 23, 213–214.

Milton, K, 1987. Primate diets and gut morphology: implications for human evolution. In: Harris, M., Ross, E.B. (Eds.), Food and Evolution: Toward a Theory of Human Food Habits. Temple Press, Philadelphia, pp. 93–116.

Milton, K., 1999. A hypothesis to explain the role of meat-eating in human evolution. Evolutionaty Anthropology 8, 11–21.

Moore, J., 1996. Savanna chimpanzees, referential models and the last common ancestor. In: McGrew, W.C, Marchant, L.F., Nishida, T. (Eds.), Great Ape Societies. Cambridge University Press, Cambridge, pp. 275–292.

Murphy, S.P., Allen, L.H., 2003. Nutritional importance of animal source foods. Journal of Nutrition 133, 3932S–3925S.

NRC., 2003. Nutrient Requirements of Nonhuman Primates. Second Revised Edition. National Academies Press, Washington, DC.

O'Brien, T.G., Kinnaird, M.F., Dierenfeld, E.S., Conklin-Brittain, N.L., Wrangham, R.W., Silver, S.C., 1998. What's special about figs? Nature 392, 668.

O'Connell, J.F., Hawkes, K., Blurton Jones, N.G., 1999. Grandmothering and the evolution of Homo erectus. Journal of Human Evolution 36, 461–485.

Peters, C.R., Vogel, J.C., 2005. Africa's wild C4 plant foods and possible early hominid diets. Journal of Human Evolution 48, 219–236.

Pianka, E.R., 1974. Evolutionary Ecology. Harper & Row, New York.

Reed, K.E., 1997. Early hominid evolution and ecological change through the African Plio-Pleistocene. Journal of Human Evolution 32, 289–322.

Robbins, C.T., 1993. Wildlife Feeding and Nutrition. Academic Press, New York.

Robinson, J.T., 1954. Prehominid dentition and hominid evolution. Evolution 8, 324–334.

Rogers, A.W., 1922. Post-cretaceous climates of South Africa. South African Journal of Science 19, 233–239.

Runia, L.J., 1987. Strontium and calcium distribution in plants: effect on paleodietary studies. Journal of Archaeological Science 14, 599–608.

Sarich, V.M., Wilson, A.C., 1967. Immunological time scale for hominid evolution. Science 158, 1200–1203.

Schoeninger, M.J., Moore, J., Sept, M., 1999. Subsistence strategies of two "savanna" chimpanzee populations: the stable isotope evidence. American Journal of Primatology 49, 297–314.

Scott, R.S., Ungar, P.S., Bergstrom, T.S., Brown, C.A., Grine, F.E., Teaford, M.F., and Walker, A., 2005. Dental microwear texture analysis shows within-species dietary variability in fossil hominins. Nature 436, 693–695.

Shipman, P., Walker, A., 1989. The costs of becoming a predator. Journal of Human Evolution 18, 373–392.

Sillen, A., 1981. Strontium and diet at Hayonim Cave, Israel: an evaluation of the strontium/calcium technique for investigating prehistoric diets. Ph.D. Dissertation, University of Pennsylvania.

Sillen, A., 1988. Elemental and isotopic analyses of mammalian fauna from Southern Africa and their implications for paleodietary research. American Journal of Physical Anthropology 76, 49–60.

Sillen, A., 1992. Strontium-Calcium ratios (Sr/Ca) of *Australopithecus robustus* and associated fauna from Swartkrans. Journal of Human Evolution 23, 495–516.

Sillen, A., Hall, G., Armstrong, R., 1995. Strontium calcium ratios (Sr/Ca) and strontium isotopic ratios (87Sr/86Sr) of Australopithecus robustus and Homo sp. from Swartkrans. Journal of Human Evolution 28, 277–285.

Smith, B.N., Epstein, S., 1971. Two categories of $^{13}C/^{12}C$ ratios for higher plants. Plant Physiology 47, 380–384.

Smith, G.E., 1924. The Evolution of Man: Essays. Oxford University Press, Oxford.

Spencer, H., Warren, J.M., Kramer, L., Samachson, J., 1973. Passage of calcium and strontium across the intestine in man. Clinical Orthopaedics and Related Research 91, 225–234.

Sponheimer, M., Lee-Thorp, J.A., 1999. Isotopic evidence for the diet of an early hominid, *Australopithecus africanus*. Science 283, 368–370.

Sponheimer, M., Lee-Thorp, J., DeRuiter, D, Smith, J., van der Merwe, N., Reed, K., Ayliffe, L., Heidelberger, C. & Marcus, W. (2003). Diets of Southern African Bovidae: The Stable Isotope Evidence. Journal of Mammalogy 84: 471–479.

Sponheimer, M., Lee-Thorp, J.A., 2006. Enamel diagenesis at South African australopith sites: implications for Paleoecological reconstruction with trace elements. Geochimica et Cosmochimica Acta 70, 1644–1654.

Sponheimer, M., Lee-Thorp, J., de Ruiter, D., Codron, D., Codron, J., Baugh, A.T., Thackeray, F., 2005a. Hominins, sedges, and termites: new carbon isotope data from the Sterkfontein valley and Kruger National Park. Journal of Human Evolution 48, 301–312.

Sponheimer, M., de Ruiter, D, Lee-Thorp, J., Späth, A., 2005b. Sr/Ca and early hominin diets revisited: new data from modern and fossil tooth enamel. Journal of Human Evolution 48, 147–156.

Sponheimer, M., Loudon, J.E., Codron, D., Howells, M.E., Pruetz, J.D., Codron, J., de Ruiter, D., Lee-Thorp, J.A., 2006. Do "Savanna" Chimpanzees Consume C4 resources? Journal of Human Evolution 51, 128–133.

Stanford, C.B., 1996. The hunting ecology of wild chimpanzees: implications for the evolutionary ecology of Pliocene hominids. American Anthropologist 98, 96–113.

Stern, J., Susman, R., 1983. The locomotor anatomy of Australopithecus afarensis. American Journal of Physical Anthropology 60, 279–317.

Strum, S.C., 1987. Almost Human: A Journey into the World of Baboons. Random House, New York.

Teaford, M.F., Ungar, P.S., Grine, F.E., 2002. Paleontological evidence for the diets of African Plio-Pleistocene hominins with special reference to early Homo. In: Ungar, P.S., Teaford, M.F. (Eds.), Human Diet: Its Origin and Evolution. Bergin and Garvey, Westport, pp. 143–166.

Tieszen, L.L., Hein, D., Qvortrup, S.A., Troughton, J.H., Imbamba, S.K., 1979. Use of d13C values to determine vegetation selectivity in East African herbivores. Oecologia 37, 351–359.

Tiger, L., Fox, R., 1971. The Imperial Animal. Transaction Pubications, New York.

Tooby, J., DeVore, I., 1987. The reconstruction of hominid behavioral evolution through strategic modeling. In: Kinzey, W.G. (Ed.), The Evolution of Human Behavior: Primate Models, State University of New York Press, Albany, pp. 183–238.

Ungar, P., 2004. Dental topography and diets of Australopithecus afarensis and early Homo. Journal of Human Evolution 46, 605–622.

van der Merwe, N.J., Thackeray, J.F., Lee-Thorp, J.A., Luyt, J., 2003. The carbon isotope ecology and diet of *Australopithecus africanus* at Sterkfontein, South Africa. Journal of Human Evolution 44, 581–597.

Vogel, J.C., 1978. Isotopic assessment of the dietary habits of ungulates. South African Journal of Science 74, 298–301.

Vogel, J.C., Fuls, A., Ellis, R.P., 1978. The geographical distribution of kranz grasses in South Africa. South African Journal of Science 74, 209–215.

Vrba, E., 1975. Some evidence of chronology and palaeoecology of Sterkfontein, Swartkrans and Kromdraai from the fossil Bovidae. Nature 254, 301–304.

Vrba, E.S., 1979. A new study of the scapula of *Australopithecus africanus* from Sterkfontein. American Journal of Physical Anthropology 51, 117–130.

Vrba, E.S., 1980. The significance of bovid remains as indicators of environment and predation patterns. In: Behrensmeyer, A.K., Hill, A.P. (Eds.), Fossils in the Making. University of Chicago Press, Chicago, pp. 247–272.

Walker, A., 1981. Diet and teeth – dietary hypotheses and human evolution. Philosophical Transactions of the Royal Society of London B 292, 57–64.

Wang, Y., Cerling, T., 1994. A model of fossil tooth and bone diagenesis: implications for paleodiet reconstruction from stable isotopes. Palaeogeography Palaeoclimatology Palaeoecology 107, 281–289.

Washburn, S.L., DeVore, I., 1961. Social behavior of baboons and early man. In: Washburn, S. (Ed.), Social Life of Early Man. Methuen, London, pp. 91–105.

Washburn, S.L., Lancaster, C.S., 1968. The evolution of hunting. In: Lee, R.B., Devore, I. (Eds.), Man the Hunter. Aldine, Chicago, pp. 293–303.

WoldeGabriel, G., White, T.D., Suwa, G., Renne, P., de Heinzelln, J., Hart, W.K., Helken, G., 1994. Ecological and temporal placement of early Pliocene hominids at Aramis, Ethiopia. Nature 371, 330–333.

Wrangham, R.W., James, J.H., Laden, G., Pilbeam, D., Conklin-Brittain, N., 1999. The raw and the stolen: cooking and the ecology of human origins. Current Anthropology 40 (5), 567.

19. Neanderthal Dietary Habits: Review of the Isotopic Evidence

Hervé Bocherens
Institute für Geowissenschaften, Biogeologie
Universität Tübingen
Sigwartstrasse 10
D-72076 Tübingen, Germany
herve.bocherens@uni-tuebingen.de

Keywords Neanderthal • diet • carbon-13 • nitrogen-15 • collagen

Abstract Carbon and nitrogen isotopic ratios of fossil bone collagen reflect those of the average diet, and can be preserved for tens of thousands of years under favorable conditions. Twelve European Neanderthal bones ranging in age from 100,000 to 32,000 years old have yielded reliable collagen. For this well-preserved collagen, isotopic signatures offer the possibility to reconstruct the dietary habits of Neanderthals. The degree of interpretation of the isotopic results depends on the paleoecological context, especially on the knowledge of the available food resources and their isotopic signatures. Animal bones associated with the studied human remains provide the most reliable source for such information. In addition, isotopic data from animal bones can be retrieved from nearby sites of similar age if they are not present in the hominid site. However, the precision of the interpretation decreases when difference in distance and age between hominids and fauna increases.

This paper illustrates how such isotopic investigations have impacted our understanding of Neanderthals' dietary habits. A critical review of the available data will be presented, with a discussion of some methodological points, such as preservation assessment and quantification of consumed protein resources. Comparisons of prey selection patterns based on isotopic results between Neanderthals and animal predators, such as hyenas, show that Neanderthals obtained much of their dietary proteins from very large herbivores in open environments by hunting. Discrepancies between prey consumption by the isotopic approach and by zooarcheology may point to individuals with special diets or transport decision that lead to the underrepresentation of very large mammal bones in archeological assemblages.

J.-J. Hublin and M.P. Richards (eds.), *The Evolution of Hominin Diets: Integrating Approaches to the Study of Palaeolithic Subsistence,* 241–250.

Introduction

Dietary subsistence is a major component of the biology of a human population. Neanderthals became extinct around 30,000 years ago, but their subsistence strategies have attracted a lot of scientific attention. An accurate estimate of their subsistence strategies is crucial to evaluate the cognitive abilities of these hominids. Numerous works, based on different approaches such as zooarcheology and tooth wear patterns have provided convincing evidence for a diet largely oriented towards the meat of large terrestrial herbivores (e.g., Lalueza Fox and Pérez-Pérez, 1993; Gaudzinski, 1996). In this context, the first isotopic study of Neanderthal bone collagen published a decade and a half ago that showed isotopic results close to those of animal predators such as wolves and hyenas did not come as a surprise (Bocherens et al., 1991). Many additional specimens have been analyzed since this first study, and this paper will evaluate their relevance in terms of questions currently asked by prehistorians about Neanderthal subsistence. For instance, it is still debated whether Neanderthals hunted actively or acquired ungulate meat through scavenging (e.g., Binford, 1988; Mellars, 1989; Patou, 1989; Marean, 1998), whether the hunting strategies of Neanderthals were different from those of early modern humans (e.g., Chase, 1987; Stiner, 1994; Marean and Assefa, 1999; Pike-Tay et al., 1999; Hardy et al., 2001; Speth and Tchernov, 2001), and whether diet and health differences can account for the "transition" between Middle and Upper Paleolithic and for the extinction of Neanderthals (e.g., Skinner, 1996; Cachel, 1997).

Material and Methods

The dietary determinations based on stable isotope ratios of fossil bones are based on the following principles (e.g., Vogel and Van der Merwe, 1977; DeNiro and Epstein, 1978, 1981; Schoeninger and DeNiro, 1984; Bocherens et al., 1999; Bocherens, 2003):

(1) "You are what you eat," therefore the isotopic composition of an organism's tissues reflect those of the diet consumed during the formation of this tissue.
(2) Some fractions of an organism's tissues can be preserved with unaltered isotopic compositions for several tens of thousands of years; bone collagen is one of those fractions.
(3) We have a reasonably good knowledge of the isotopic variations in the ecosystems.

These different points will be addressed in the next paragraphs, with emphasis on the specific conditions of Neanderthal fossil bone collagen.

Isotopic Record in Bone Collagen

The isotopic analysis of animals raised on controlled diets or monitored in the natural environments allows us to provide the general rules of isotopic record in bone collagen. Dietary reconstruction is based on the isotopic compositions of carbon ($^{13}C/^{12}C$) and nitrogen ($^{15}N/^{14}N$). Carbon is provided by the whole food consumed by an organism, while nitrogen is provided by the protein fraction of the food. The carbon isotopic signature of an animal is globally equivalent to that of its average food (DeNiro and Epstein, 1978). However, different fractions of the body present varying carbon isotopic signatures compared to the body as a whole. For example, the collagen of an organism raised on a homogenous diet exhibits $\delta^{13}C$[1] values enriched by 5‰ relative to the diet and the whole body. Flesh is enriched by around 1‰ compared to the average diet. This is why the $\delta^{13}C$ values of the collagen of a predator is enriched by around 1‰ (from 0.8‰ to 1.3‰) compared to the $\delta^{13}C$ values of the collagen of its average prey (Bocherens and Drucker, 2003b). Contrary to carbon, the isotopic composition of nitrogen is higher in a given organism than in its food, with the $\delta^{15}N$ values of an organism being 3–5‰ more positive than those of its average diet (DeNiro and Epstein, 1981; Schoeninger and DeNiro, 1984). When the $\delta^{15}N$ values of the bone collagen of a predator are compared to those of its average prey, a similar enrichment of 3–5‰ is found (Bocherens and Drucker, 2003b).

Collagen reflects the isotopic composition of the food consumed in the period during which it was synthesized. In large mammals, bone collagen reflects several years of an individual's lifetime before its death when collagen is sampled from a dead individual, as is the case for fossils. On the contrary, tooth collagen reflects a shorter period, usually during the first months or years of an individual's life. This situation has important implications in mammals, where the young individuals are raised on their mother's milk, which exhibits $\delta^{15}N$ values around 3‰ higher than those of the adult diet. For these reasons, the $\delta^{15}N$ values of tooth collagen do not always reflect the adult diet in mammals, and they should be treated cautiously if used in paleodietary reconstruction (Bocherens et al., 1994; Bocherens and Mariotti, 1997). For the same reason, bone collagen of young individuals, from birth until a few years after weaning, is affected by the consumption of ^{15}N-enriched milk, and several studies have demonstrated the consequences in the isotopic variations of the youngest individuals, up to 3 year old babies, in modern and archaeological human populations (e.g., Fogel et al., 1989; Katzenberg and Pfeiffer, 1995). The isotopic signatures of such individuals are therefore difficult to address for paleodietary reconstruction.

Criteria for Collagen Reliability

Collagen has the great advantage of exhibiting very invariable chemical conditions across taxa. Collagen from mammal bones and teeth present a very narrow range of carbon and nitrogen content, a property that is routinely used to monitor the reliability of collagen extracted from fossil bones (e.g., DeNiro, 1985; Ambrose, 1990). Different techniques can be used to extract collagen from fossil bones (e.g., Longin, 1971; Bocherens et al., 1997; Higham et al., 2006), but all of them have the same aim: eliminating any component other than collagen, including the mineral fraction containing carbonate, and any type of contaminating organic matter that could have impregnated the bone sample during burial, without losing the remaining collagen. It is widely accepted that any extract with a C/N atomic ratio lower than 2.9 or higher than 3.6 is to be discarded since these values are clearly outside the range exhibited by collagen extracted from fresh bone. However, the lower acceptable limit for carbon and nitrogen percentages in fossil bone extracts is evaluated case by case. Collagen-like extracts with carbon and nitrogen percentages lower than those measured in collagen extracted from fresh bones, i.e., 36% and 12%, respectively (Rodière et al., 1996), should be considered very cautiously, especially if they are outliers compared to other collagen extracts from the same site. Amino acid composition has also been suggested to discriminate between well-preserved and altered collagen, since collagen exhibits a very unique amino acid composition (DeNiro and Weiner, 1988; Bocherens et al., 1991). However, besides the technical complications of this type of analysis, measuring a collagen-like amino acid profile does not preclude the presence of non-amino acid contaminants in the extract that will remain invisible to the detection techniques used in amino acid biochemistry, but will be combusted together with the collagenic extract during the isotopic analysis.

Isotopic Variations in Late Pleistocene European Ecosystems

Between 150,000 and 25,000 years ago, corresponding roughly to the time range of classical Neanderthals, climatic and environmental conditions have changed dramatically in

[1] Isotopic abundances are expressed as δ (delta) values, as follows: $\delta^{E}X = (R_{sample}/R_{standard} - 1).1000$ (per mill), where X stands for C or N, E stands for 13 or 15 respectively, and R stands for the isotopic ratios $^{13}C/^{12}C$ and $^{15}N/^{14}N$ respectively. The standard, internationally defined, is a marine carbonate (VPDB) for carbon and atmospheric nitrogen (AIR) for nitrogen.

Europe (e.g., Huntley and Allen, 2003; van Andel, 2003; Jöris, 2005). Some periods were as temperate as today, for instance during the Eemian interglacial (128–115 kyr ago), while other periods were clearly colder, as during the Marine Oxygen Isotopic Stages (MOIS) 6 (around 150 kyr ago) and 4 (58–74 kyr ago). During the major part of this time range, most of Europe north of 45°N latitude was covered by a very productive terrestrial ecosystem with no modern analogue, called the "mammoth-steppe" or "steppe-tundra" (e.g., Guthrie, 1982), while in more southern latitudes, Mediterranean-like environments prevailed (e.g., Allen et al., 1999). Since all the Neanderthal specimens that yielded well-preserved collagen so far originate from northern latitudes (Bocherens et al., 2005), the following discussion will focus on the isotopic variations documented for these regions.

There is no evidence for C_4 plants in late Pleistocene Europe; therefore all $\delta^{13}C$ variations in plants correspond to different conditions of C_3 photosynthesis, differences in sources of CO_2 and to some extent the plant type (Fig. 19.1). One well-documented case is the canopy effect observed in both plants and animal tissues under dense forest closed canopies (e.g., van der Merwe et al., 1990; Schoeninger et al., 1997; Cerling and Harris, 1999; Drucker et al., 2003a). Very low $\delta^{13}C$ values due to the canopy effect are expected during the Eemian interglacial (MOIS 5e) and are possible locally during some interstadial oscillations, such as MOIS 5c and 5a. During non-forested periods, which correspond to the majority of the time range, only small differences in $\delta^{13}C$ values are expected between different types of plants, the most significant being the consistently more positive $\delta^{13}C$ values of lichens relative to vascular plants (Fizet et al., 1995; Drucker et al., 2001). Based on isotopic studies of modern plants from subarctic and boreal contexts, it is expected that the $\delta^{15}N$ values should vary widely according to the plant type (see review in Bocherens, 2003). A consistent pattern arises with $\delta^{15}N$ values lower in mycorrhized plants, especially shrubs and dwarf shrubs, than in non-mycorrhized plants, such as grass, with the whole range being as large as 12‰ (Fig. 19.1). This range of variation is significantly larger than the 3–5‰ enrichment between a herbivore and its food, but due to some degree of dietary specialization among herbivores, a distinct isotopic pattern is observed in late Pleistocene ungulates (Fig. 19.1). During the development of steppe-tundra ecosystems in northern Eurasia, consistent patterns of isotopic variations occur in fossil herbivore collagen, from western Europe to Alaska (Bocherens, 2003). Reindeer exhibit the most positive $\delta^{13}C$ values, due to their consumption of lichens, while mammoths exhibit lower $\delta^{13}C$ values and significantly higher $\delta^{15}N$ values (Fig. 19.1). Other herbivores, such as horse, woolly rhinoceros, bison, and deer exhibit less variable isotopic values but some clustering according to the species is observed and most probably reflects different food preferences for these different herbivorous species as one way to avoid excessive competition. Notably, cave bears appear to belong to the herbivorous guild, with isotopic values suggesting a diet incorporating mainly non-monocotyledon plant resources. Altogether, the total range of isotopic values is up to 7‰ for $\delta^{13}C$ values and 6‰ for $\delta^{15}N$ values (Fig. 19.1). At the next trophic level, predators also exhibit some clustering according to the species, which corresponds to the

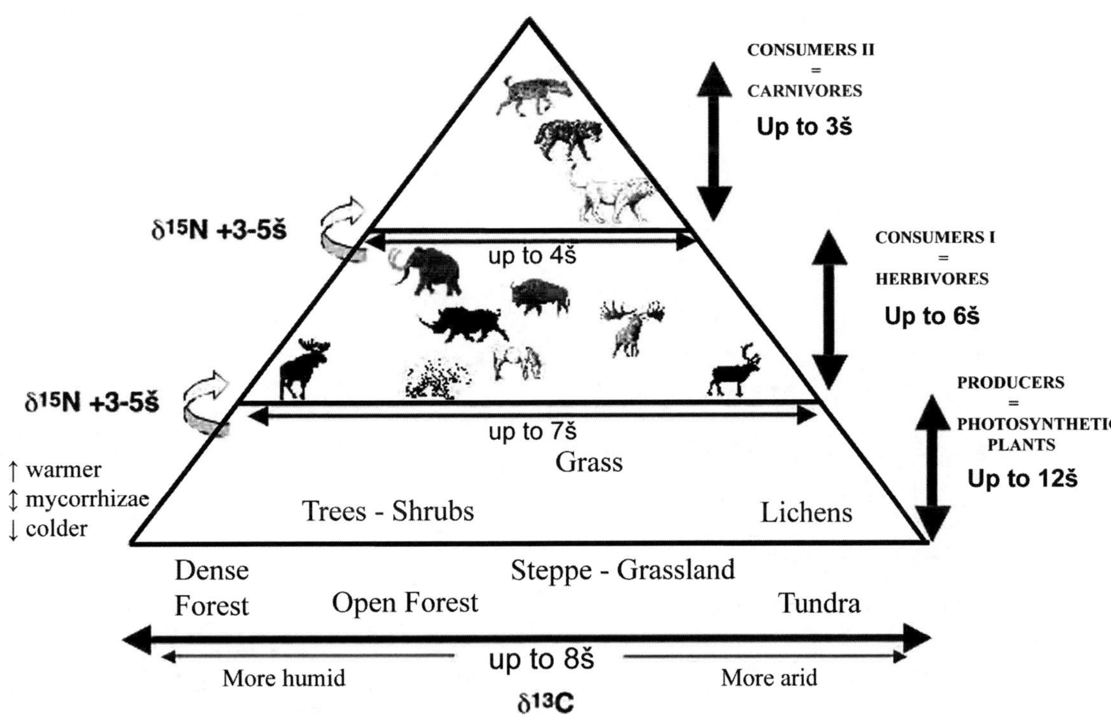

FIG. 19.1. Summary of the variations in $\delta^{13}C$ and $\delta^{15}N$ values in the terrestrial trophic webs of late Pleistocene European mammals contemporary to Neanderthals.

preferential consumption of some prey species with slightly different isotopic signatures (Fig. 19.1). This brief review of isotopic variations in late Pleistocene mammal collagen shows us that there is more than a dichotomy between herbivores and carnivores to be found in these variations. This is relevant for the interpretation of the isotopic signatures of Neanderthal bone collagen as well.

The pattern described above deals only with terrestrial resources. Aquatic resources may be relevant in the study of Neanderthal diet, and they present different isotopic signatures. Coastal Neanderthals could have had access to marine resources, as demonstrated by zooarcheological investigations in some sites from the Mediterranean area (e.g., Stiner, 1994). Such resources exhibit $\delta^{13}C$ and $\delta^{15}N$ values much higher than those of terrestrial resources (e.g., Schoeninger and DeNiro, 1984). Freshwater resources also exhibit $\delta^{15}N$ values higher than those of terrestrial resources, but their $\delta^{13}C$ values are very variable, ranging from more negative, equivalent to, to more positive than those of terrestrial resources (e.g., Dufour et al., 1999; Drucker and Bocherens, 2004).

An additional point to be considered when dealing with the possible diets of Neanderthals, as expressed in the isotopic signatures of their bone collagen, is the impact of very different nitrogen content in animal *versus* plant food. Plant food contains only 1‰ nitrogen on average while meat contains 14‰ nitrogen, though the carbon percentages are similar, averaging 44‰ and 51.5‰, respectively (Phillips and Koch, 2002). The consequence of this difference in nitrogen content is that a mixture of plant and animal food does not lead to a linear isotopic variation between pure vegetarian and pure carnivorous end-points. Hyena can be used as a representative of the pure carnivorous end-point at the time of Neanderthals in Europe, and cave bear can be used as a representative of the pure vegetarian end-point. Indeed, paleontological and isotopic data demonstrate that this species had a purely vegetarian diet (e.g., Kurten, 1976; Bocherens et al., 1994; Stiner, 1998) and that hibernation had no significant impact on the isotopic signatures of bone collagen in this species (Bocherens, 2004). Moreover, the plant food typically consumed by bears covers a very similar spectrum as the plant food edible to human beings (e.g., Rockwell, 1991), and the bears have a simple digestive tract lacking the special adaptations of most ungulates (e.g., Pritchard and Robbins, 1990; Rode et al., 2001), as human beings do. Therefore, cave bears appear to be a reasonable proxy for vegetarian humans. Fig. 19.2 presents the example of the non-linear isotopic variation between pure vegetarian (cave bear) and pure carnivorous (hyena) end-points, featuring the average collagen isotopic values of cave bears and hyena from layer 1A of Scladina cave, a Belgian cave dated around 40,000 years. A small percent of meat already increases very significantly the $\delta^{15}N$ value, and contributions of plant food as high as 50% do not yield $\delta^{15}N$ values lower than 1 standard-deviation of the average hyena collagen $\delta^{15}N$ value (Fig. 19.2). This example clearly illustrates that the collagen isotopic values of Neanderthal collagen provide data on the relative contribution of different protein resources, but it does not preclude a significant amount of plant food with low nitrogen content, as high as half the dry weight dietary intake.

After having made these methodological points, a review of the published results will allow one to retain only relevant results, i.e., (1) those that present reliable testing for the absence of diagenetical alteration, (2) those which are interpretable in trophic terms, without interference of suckling and weaning, and (3) those for which a comparison with the isotopic data of contemporary herbivores and carnivores is possible.

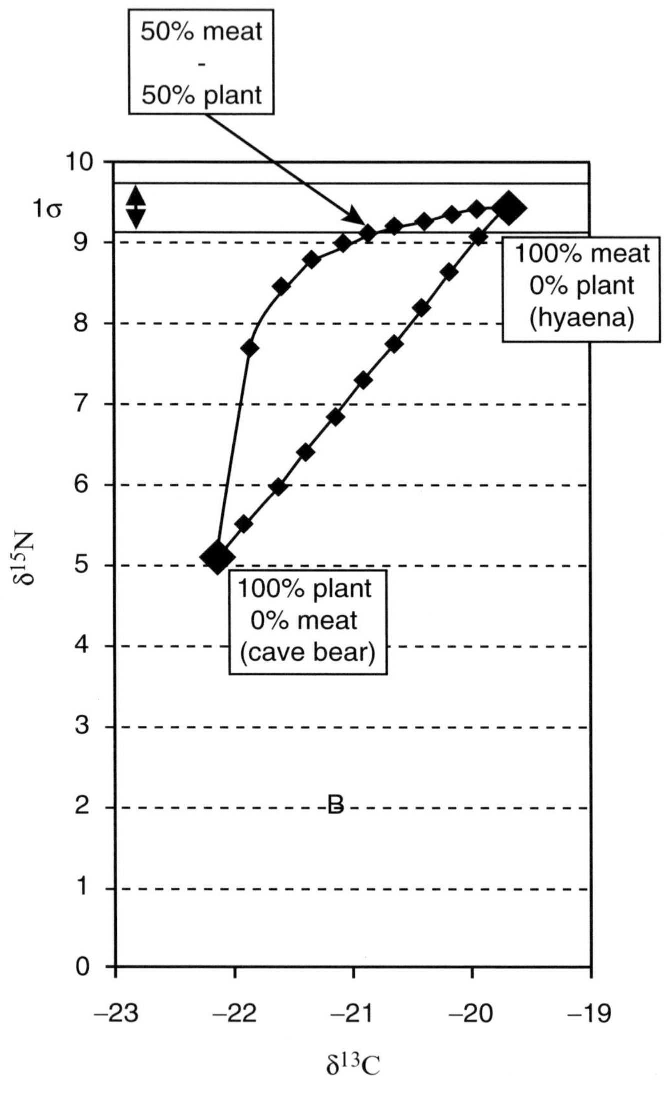

Fig. 19.2. Simulated variations in $\delta^{13}C$ and $\delta^{15}N$ values in the bone collagen of an omnivore consuming various amounts of plant food and herbivore meat. End-members are the average value for cave bear and hyena from layer 1A of Scladina Cave (from Bocherens et al., 1997). All other points are hypothetical. Carbon and nitrogen content of food resources are from Phillips and Koch (2002), namely C = 51.5% and N = 14%, and C = 44% and N = 1%, for meat and plant food, respectively.

Results

Among the 12 Neanderthal specimens for which isotopic results have been published so far (Table 19.1), only six of them fulfill the criteria for dietary interpretation. In addition to the two specimens from Les Pradelles with problematic chemical composition (see a more complete discussion in Bocherens et al., 2005), a recent paper by Higham et al. (2006) revealed that the isotopic signatures measured on the Neanderthals from Vindija previously published by Smith et al. (1999) and interpreted in paleodietary terms by Richards et al. (2000), were not reliable. Indeed, the bones from this site exhibit poor collagen preservation and routine purification techniques used by Richards et al. (2000) failed to properly purify unaltered collagen (Higham et al., 2006). Moreover, among the new analyses performed on the same specimen, only one of them yielded reliable isotopic results. However, these revised data are not yet interpretable in dietary terms since no comparison with contemporary herbivores and carnivores is possible at the moment. The same situation holds for the Neanderthal specimen from layer 3 of Scladina cave, published in Bocherens et al. (2001), which will be interpreted once isotopic analysis of contemporary herbivores and carnivores currently underway is completed. In addition, the child specimen from Engis 2 may be affected by suckling and is therefore not discussed in paleodietary terms here (Bocherens et al., 2005).

The six remaining isotopic results correspond to four sites, with three specimens coming from the same site, Les Pradelles (Fig. 19.3). All these Neanderthals have isotopic signatures in the same range as the animal predators, such as hyena, wolf and lion. These results lead to the conclusion that Neanderthals were possibly as carnivorous as large predators, although as seen in a previous paragraph, these results mean instead that the source of protein was mostly the meat of terrestrial ungulates, without excluding a complement of plant food in the diet.

For one site, Saint-Césaire, the good quality of the isotopic data on faunal elements allowed the use of a mathematical mixing model for calculating the range of possible proportions of different prey consumed by Neanderthals and hyenas. This was possible due to the variety of potential prey species that could be analyzed in the vicinity of the human specimen, as well as the clear isotopic differences observed in the major prey items.

Discussion

The isotopic results on all the Neanderthal specimens studied so far point to a similar trophic ecology. In each case, the $\delta^{15}N$ values of Neanderthals are 1–2 ‰ more positive than those of contemporaneous predators, such as hyenas. Such results indicate that the proteins were mainly provided by

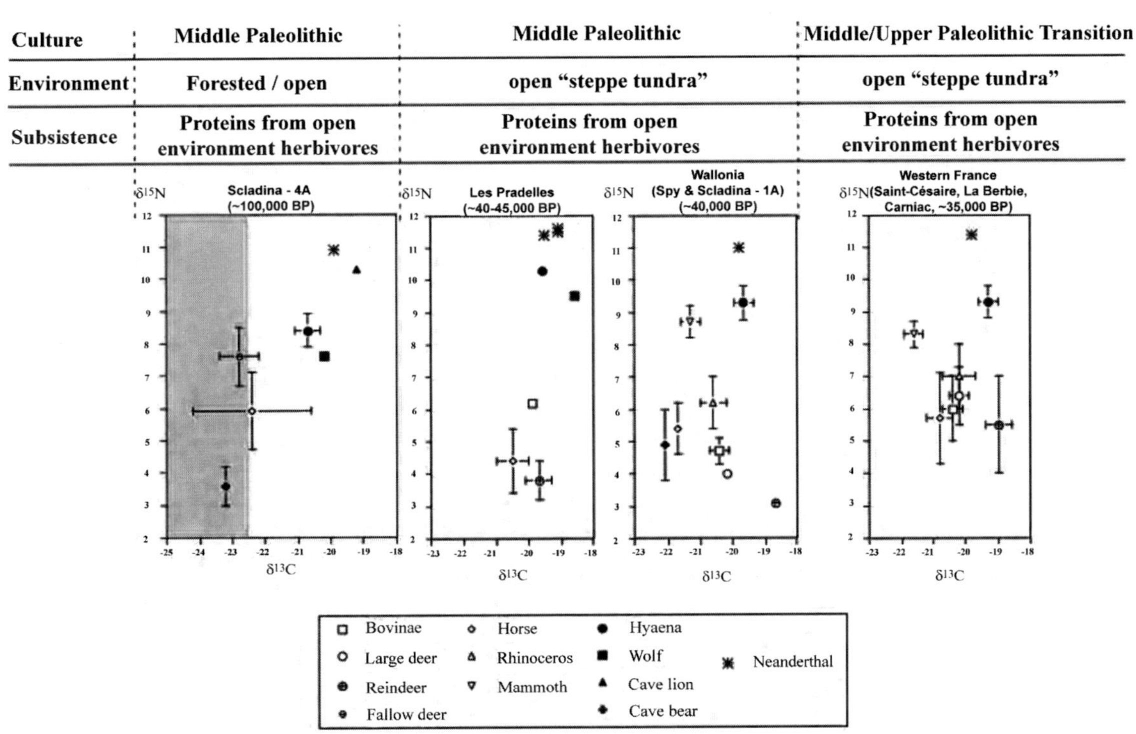

FIG. 19.3. $\delta^{13}C$ and $\delta^{15}N$ values of Neanderthal collagen from southwestern France (Saint-Césaire, Camiac, La Berbie), Les Pradelles, Wallonia (Spy and Scladina-1A), compared to those of their respective contemporary mammal fauna (average ± standard-deviation for different herbivorous and carnivorous species).

TABLE 19.1. List of published isotopic results for Neanderthal collagen. $\delta^{13}C$ and $\delta^{15}N$ results that present good chemical integrity are in bold, those that cannot be interpreted readily in dietary terms due to ontogenic stage or lack of trophic context are in italics. In the case of Vindija samples 8295 and 8296, two $\delta^{15}N$ values are indicated since those reported by Smith et al. (1999) and Higham et al. (2006) differ by 0.5‰.

No. analysis (No. excavation) ell Site, layer	Age	Piece	Chemical integrity	Ontogeny	Trophic context	$\delta^{13}C$	$\delta^{15}N$	Reference	
drl SC18800 (SCLA-4A 2)	Scladina Cave (Sclayn, Belgium), layer 4A	OIS5c	skull	+	+	+	−19.9	10.9	Bocherens *et al.* (1999)
djustright MT500 (SCLA-1B 4)	Scladina Cave (Sclayn, Belgium), layer 3	OIS5b?	phalanx	+	+	-	−21.2	11.8	Bocherens *et al* (2001)
27801	Les Pradelles (Marillac-le-Franc, Charente, France)	OIS3	skull	?	+	+	ar-20.2	9.3	Bocherens *et al.* (1991)
64801	Les Pradelles (Marillac-le-Franc, Charente, France)	OIS3	skull	+	+	+	−19.1	11.6	Fizet *et al.* (1995)
M300	Les Pradelles (Marillac-le-Franc, Charente, France)	OIS3	skull	+	+	+	−19.1	11.5	Bocherens *et al.* (2005)
M400	Les Pradelles (Marillac-le-Franc, Charente, France)	OIS3	skull	+	+	+	−19.5	11.4	Bocherens *et al.* (2005)
M100	Les Pradelles (Marillac-le-Franc, Charente, France)	OIS3	skull	-	+	+	−21.8	8.4	Bocherens *et al.* (2005)
MT100 (Engis 2)	Awirs Cave (Belgium)	OIS3	skull	32+	-	+	−19.6	12.6	Bocherens *et al.* (2001)
MT200 (SPY OMO 1)	Spy (Betche-al-Roche) Cave (Belgium)	OIS3	scapula	+	+	+	−19.8	11.0	Bocherens *et al.* (2001)
stright RPB7000	La Roche-à-Pierrot (Saint-Césaire, Charentes-Maritimes, France)	OIS3 (~36,000 BP, TL)	fibula	+	+	+	−19.8	11.4	Bocherens & Drucker (2003a)
8296 (VI-207)	Vindija Cave (Croatia)	OIS3 (29,080 ± 400)	mandible	(+) -	+	-	−20.5	10.8 (11.3?)	Smith *et al.* (1999), Richards et al. (2000) Higham *et al.* (2006)
2089-07 (VI-207)	Vindija Cave (Croatia)	OIS3 (32,400 ± 1,800)	mandible	-	+	-	−24.6	11.1	Higham *et al.* (2006)
8295 (VI-208)	Vindija Cave (Croatia)	OIS3 (28,020 ± 400)	parietal	(+) -	+	-	−19.5	10.1 (10.6?)	Smith *et al.* (1999) Richards et al. (2000) Higham *et al.* (2006)
10 2089-06 (VI-208)	Vindija Cave (Croatia)	OIS3 (**32,400 ± 800**)	parietal	+	+	-	−20.2	10.3	Higham *et al.* (2006)

herbivores from open environments with high $\delta^{15}N$ values, such as woolly rhinoceros, mammoth, and large bovids. This result is particularly informative in the case of the specimen from layer 4A from Scladina cave, where paleoenvironmental proxy indicates that forest was developed at that time around the site, as confirmed by the carbon isotopic signatures of large herbivores (Bocherens et al., 1999). It seems, therefore, that the trophic ecology of Neanderthals was rather stable through time, depending largely on the meat of large herbivores from open environments, even during the milder episodes allowing forest development. Also, the Châtelperronian Neanderthal from Saint-Césaire did not deviate from this pattern at a time when anatomically modern humans were already present in some parts of Europe (Trinkaus et al., 2003). If this pattern holds when more isotopic data become available, this could reflect a trophic "rigidity" in Neanderthals that may have put them at a disadvantage in the competition with modern humans at the "transition" between middle and upper Paleolithic in Europe (Bocherens and Drucker, 2006).

The case of Saint-Césaire allowed further interpretation of the subsistence strategy. The fact that the isotopic signatures of contemporary hyena indicate much less consumption of mammoth and woolly rhinoceros meat for the Neanderthal individual strongly suggests that the meat of these mega-herbivores was not readily available through scavenging (Bocherens et al., 2005). The recent publication of zooarcheological results for this same site by Patou-Mathis (2006) allows an interesting comparison with those of the isotopic approach (Fig. 19.4). It seems that the relative contribution of very large herbivores, such as mammoth and woolly rhinoceros is lower when reconstructed by the zooarcheological approach than when reconstructed by the isotopic approach, while the relative contribution of large bovids is higher (Fig. 19.4). This could reflect a special diet for the studied human individual in comparison with the average group that left the bone remains behind in the Châtelperronian layer. Another explanation could be linked to transport decisions: filleted meat of very large herbivores could have been transported to the camp and therefore did not leave as many bone remnants as those of less bulky prey species (Bocherens et al., 2005).

So far, only Neanderthals from the northern latitudes could be studied with the isotopic approach. Problems of collagen preservation occur in warmer contexts, such as the Middle East (Ambrose, 1998). New technologies aimed at purifying molecules that are sometimes more stable than collagen but appear in very small quantities may be a way to overcome these difficulties (Nielsen-Marsh et al., 2005). However, extensive modern reference datasets for such molecules based on monitored wild animals and captive ones under controlled diets will be necessary to interpret the isotopic results in paleodietary terms.

Conclusions

Since the first use of stable isotopes in Neanderthal collagen to reconstruct the dietary habits of this extinct hominid, the addition of new data and the improvement of the methodology yielded valuable information about the subsistence strategies of these extinct hominids. The implications of this field of research are now widely incorporated into more general investigations about Neanderthal adaptation and extinction

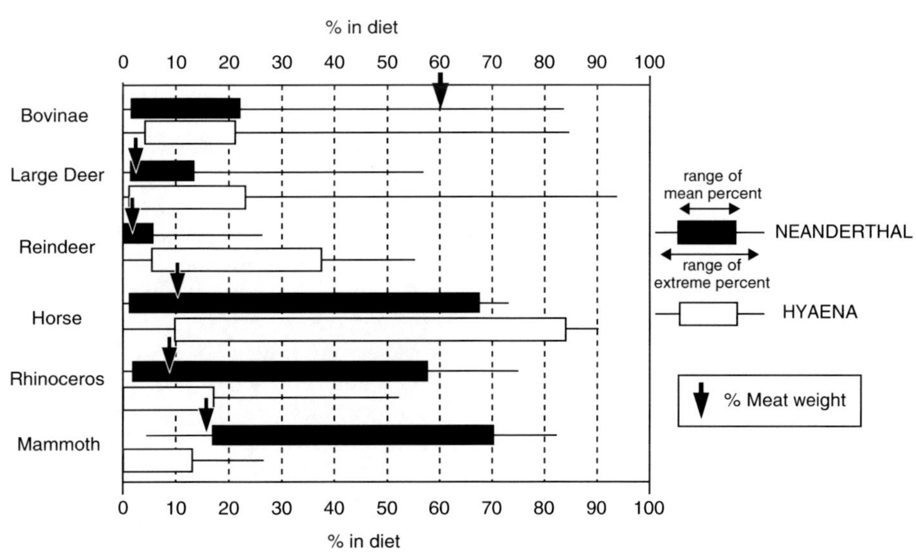

FIG. 19.4. Paleodiet of Saint-Césaire I: Summary of the mean and extreme possible ranges of prey percentage for Neanderthal and hyena in southwestern France around 36,000 years BP. The solid black bar takes the means from the distributions of solutions using the mixing model for Neanderthal and hyena (Bocherens et al., 2005) and shows the range of those means. The vertical arrows indicate the average meat weight for each prey species as deduced by zooarchaeological studies, calculated by multiplying the minimum number of individuals by the average meat weight for this species (Patou-Mathis, 2006).

(e.g., Klein, 2000; Langbroek, 2001; Roebroeks, 2001; Hockett and Haws, 2003; Finlayson, 2004; Jankovic, 2004; Hockett and Haws, 2005; Strauss, 2005). It is anticipated that more data will be generated during the coming years, especially as side products of AMS radiocarbon dating of Neanderthal bones. However, although these new data on Neanderthals are very welcome, it is advocated here that this work should be performed with the requirements of paleodietary interpretations, i.e. with the appropriate additional isotopic data on contemporary herbivores and predators that will allow the trophic interpretation of these Neanderthal isotopic data. Otherwise, the only possible interpretation will be to compare these isolated isotopic data with those obtained on other Neanderthal specimens, assuming that no significant changes occurred in the baseline of nitrogen isotopic signatures at the basis of the foodweb. Such an assumption is likely to be wrong during Marine Isotopic Stage 3, since large temperature excursions have been identified, for instance between 60 and 70 kyr ago (e.g., Jouzel et al., 2006), and $\delta^{15}N$ in plants has been shown to track temperature changes in the late Pleistocene (e.g., Drucker et al., 2003b; Richards and Hedges, 2003; Drucker and Bocherens, 2004). Whenever possible, the use of mathematical mixing models will allow for quantifying the possible ranges of consumption of different prey species in Neanderthals' diet (Bocherens et al., 2005), and the comparison with the conclusions of zooarcheological investigations will be particularly instructive. Another promising line of research is to investigate sites with several Neanderthal individuals to document the extent of inter-individual isotopic variation and to compare this range to that of Upper Paleolithic, Epipaleolithic and Mesolithic modern humans. With all these remarks in mind, I am convinced that progress will continue to be made on Neanderthal paleobiology as tracked by bone stable isotopic composition.

Acknowledgments. I thank very much J.-J. Hublin and M. Richards for their kind invitation to participate to this highly interesting meeting in Leipzig. Also thanks are due to the CNRS, Naturalia et Biologia, and the Foundation Alexander von Humboldt, who provided me with research and travel grants during the last years.

References

Allen, J.R.M., Brandt, U., Brauer, A., Hubberten, H.-W., Huntley, B., Keller, J., Kraml, M., Mackensen, A., Mingram, J., Negendank, J.F.W., Nowaczyk, N.R., Oberhänsli, H., Watts, W.A., Wulf, S., Zolltschka, B., 1999. Rapid environmental changes in southern Europe during the last glacial period. Nature 400, 740–743.

Ambrose, S.H., 1990. Preparation and characterization of bone and tooth collagen for isotopic analysis. Journal of Archaeological Science 17, 431–451.

Ambrose, S.H., 1998. Prospects for stable isotopic analysis of later Pleistocene hominid diets in West Asia and Europe. In: Akazawa, T., Aoki, K., Bar-Yosef, O. (Eds.), Origin of Neandertals and humans in West Asia. Plenum Press, New York, pp. 277–289.

Binford, L.R., 1988. Etude taphonomique des restes de la grotte Vaufrey, couche VIII. In: Rigaud, J.-P. (Ed.), La grotte Vaufrey à Cénac et St Julien (Dordogne). Paléoenvironnements, chronologie et activités humaines. Paris. Mémoires de la Société préhistorique Française 19, pp. 535–563.

Bocherens, H., 2003. Isotopic biogeochemistry and the paleoecology of the mammoth steppe fauna. In: Reumer, W.F., Braber, F., Mol, D., Vos, J. de (Eds.), Advances in mammoth research. Deinsea, Rotterdam 9, pp. 57–76.

Bocherens, H., 2004. Cave bear palaeoecology and stable isotopes: checking the rules of the game. In: Philippe, M., Argant, A., Argant, J. (Eds.), Proceedings of the 9th International Cave Bear Conference, Cahiers scientifiques du Centre de Conservation et d'Etude des Collections (Muséum d'Histoire naturelle de Lyon) Hors Série No 2, pp. 183–188.

Bocherens, H., Drucker, D., 2003a. Reconstructing Neandertal diet from 120,000 to 30,000 BP using carbon and nitrogen isotopic abundances. In: Patou-Mathis, M., Bocherens, H. (Eds.), Le rôle de l'environnement dans les comportements des chasseurs-cueilleurs préhistoriques. British Archaeological Reports International Series 1105, pp. 1–7.

Bocherens, H., Drucker, D., 2003b. Trophic level isotopic enrichments for carbon and nitrogen in collagen: case studies from recent and ancient terrestrial ecosystems. International Journal of Osteoarchaeology 13, 46–53.

Bocherens, H., Drucker, D., 2006. Dietary competition between Neanderthals and modern humans: insights from stable isotopes. In: Conard, N.J. (Ed.), When Neanderthals and modern humans met. Tübingen Publications in Prehistory, Kerns, Tübingen, pp. 129–143.

Bocherens, H., Mariotti, A., 1997. Comments on: diet, physiology and ecology of fossil mammals as inferred from stable carbon and nitrogen isotope biochemistry: implications for Pleistocene bears by Bocherens et al. – Reply. Palaeogeography, Palaeoclimatology, Palaeoecology 128, 362–364.

Bocherens, H., Fizet, M., Mariotti, A., Lange-Badré, B., Vandermeersch, B., Borel, J.-P., Bellon, G., 1991. Isotopic biogeochemistry (^{13}C, ^{15}N) of fossil vertebrate collagen: implications for the study of fossil food web including Neandertal man. Journal of Human Evolution 20, 481–492.

Bocherens, H., Fizet, M., Mariotti, A., 1994. Diet, physiology and ecology of fossil mammals as inferred by stable carbon and nitrogen isotopes biogeochemistry: implications for Pleistocene bears. Palaeogeography, Palaeoclimatology, Palaeoecology 107, 213–225.

Bocherens, H., Billiou, D., Patou-Mathis, M., Bonjean, D., Otte, M., Mariotti, A., 1997. Isotopic biogeochemistry (^{13}C, ^{15}N) of fossil mammal collagen from Scladina cave (Sclayn, Belgium). Quaternary Research 48, 370–380.

Bocherens, H., Billiou, D., Patou-Mathis, M., Otte, M., Bonjean, D., Toussaint, M., Mariotti, A., 1999. Palaeoenvironmental and palaeodietary implications of isotopic biogeochemistry of late interglacial Neandertal and mammal bones in Scladina Cave (Belgium). Journal of Archaeological Science 26, 599–607.

Bocherens, H., Toussaint, M., Billiou, D., Patou-Mathis, P., Bonjean, D., Otte, M., Mariotti, A., 2001. New isotopic evidence for dietary habits of Neandertals from Belgium. Journal of Human Evolution 40, 497–505.

Bocherens, H., Drucker, D.G., Billiou, D., Patou-Mathis, M., Vandermeersch, B., 2005. Isotopic evidence for diet and subsistence pattern of the Saint-Césaire I Neanderthal: review and use of a multi-source mixing model. Journal of Human Evolution 49, 71–87.

Cachel, S., 1997. Dietary shifts and the European Upper Palaeolithic transition. Current Anthropology 38, 579–603.

Cerling, T.E., Harris, J.M., 1999. Carbon isotope fractionation between diet and bioapatite in ungulate mammals and implications for ecological and palaeocological studies. Oecologia 120, 347–363.

Chase, P.G., 1987. Spécialisation de la chasse et transition vers le Paléolithique supérieur. L'Anthropologie (Paris) 91, 175–188.

DeNiro, M.J., 1985. Postmortem preservation and alteration of in vivo bone collagen isotope ratios in relation to palaeodietary reconstruction. Nature 317, 806–809.

DeNiro, M.J., Epstein, S., 1978. Influence of diet on the distribution of carbon isotopes in animals. Geochimica et Cosmochimica Acta 42, 495–506.

DeNiro, M.J., Epstein, S., 1981. Influence of diet on the distribution of nitrogen isotopes in animals. Geochimica et Cosmochimica Acta 45, 341–351.

DeNiro, M.J., Weiner, S., 1988. Use of collagenase to purify collagen from prehistoric bones for stable isotopic analysis. Geochimica et Cosmochimica Acta 52, 2425–2431.

Drucker, D., Bocherens, H., Pike-Tay, A., Mariotti, A., 2001. Isotopic tracking of seasonal dietary change in dentine collagen: preliminary data from modern caribou. Comptes Rendus de l'Académie des Sciences, Série II, Paris 333, 303–309.

Drucker, D., Bocherens, H., Bridault, A., Billiou, D., 2003a. Carbon and nitrogen isotopic composition of Red Deer (*Cervus elaphus*) collagen as a tool for tracking palaeoenvironmental change during Lateglacial and Early Holocene in northern Jura (France). Palaeogeography, Palaeoclimatology, Palaeoecology 195, 375–388.

Drucker, D., Bocherens, H., Billiou, D., 2003b. Evidence for shifting environmental conditions in Southwestern France from 33,000 to 15,000 years ago derived from carbon-13 and nitrogen-15 natural abundances in collagen of large herbivores. Earth and Planetary Science Letters 216, 163–173.

Drucker, D.G., Bockerens, H., 2004. Carbon and nitrogen stable isotopes as tracers of diet breadth evolution during middle and upper Palaeolithie in Europe. International Journal of Osteoarchaeology 14,162–177.

Dufour, E., Bocherens, H., Mariotti, A., 1999. Palaeodietary implications of isotopic variability in Eurasian lacustrine fish. Journal of Archaeological Science 26(6), 627–637.

Finlayson, C., 2004. Neanderthals and modern humans, an ecological and evolutionary perspective. Cambridge University Press, Cambridge.

Fizet, M., Mariotti, A., Bocherens, H., Lange-Badré, B., Vandermeersch, B., Borel, J.P., Bellon, G., 1995. Effect of diet, physiology and climate on carbon and nitrogen isotopes of collagen in a late Pleistocene anthropic paleoecosystem (France, Charente, Marillac). Journal of Archaeological Science 22, 67–79.

Fogel, M.L., Tuross, N., Owsley, D.W., 1989. Nitrogen isotope traces of human lactation in modern and archeological populations. Annual Report of the Director of the Geophysical Laboratory, Carnegie Institution 1988–1989, 111–117.

Gaudzinski, S., 1996. On bovid assemblages and their consequences for the knowledge of subsistence patterns in the Middle Palaeolithic. Proceedings of the Prehistoric Society 62, 19–39.

Guthrie, R.D., 1982. Mammals of the mammoth steppe as paleoenvironmental indicators. In: Hopkins, D.M., Matthews, J.V., Schweger, C.E., Young, S.B. (Eds.), Paleoecology of Beringia, Academic Press, New York, pp. 307–326.

Hardy, B.L., Kay, M., Marks, A.E., Monigal, K., 2001. Stone tool function at the paleolithic sites of Starosele and Buran Kaya III, Crimea: behavioral implications. Proceedings of the National Academy of Sciences 98, 10972–10977.

Higham, T., Bronk Ramsey, C., Karavanic, I., Smith, F.H., Trinkaus, E., 2006. Revised direct radiocarbon dating of the Vindija G1 Upper Paleolithic Neandertals. Proceedings of the National Academy of Sciences 103 (3), 553–557.

Hockett, B., Haws, J., 2003. Nutritional ecology and diachronic trends in Paleolithic diet and health. Evolutionary Anthropology 12(5), 211–216.

Hockett, B., Haws, J.A., 2005. Nutritional ecology and the human demography of Neandertal extinction. Quaternary International 137, 21–34.

Huntley, B., Allen, J.R.M., 2003. Glacial environments III: Palaeo-vegetation patterns in Last Glacial Europe. In: van Andel, T.H., Davis, W. (Eds.), Neanderthals and modern humans in the European landscape during the last glaciation. McDonald Institute Monographs, Oxbow Books, Oxford, pp. 79–102.

Jankovic, I., 2004. Neandertals ... 150 years later. Collegium Antropologicum 28(suppl. 2), 379–401.

Jöris, O., 2005. Aus einer anderen Welt – Europa zur Zeit des Neandertalers. In: Conard, N.J., Kölbl, S., Schürle, W. (Eds.), Vom Neandertaler zum modernen Menschen. Jan Thorbecke, Ostfildern, pp. 47–79.

Jouzel, J., Lorius, C., Raynaud, D., 2006. Climat et atmosphère au Quaternaire: de nouveaux carottages glaciaires. Comptes Rendus Palevol 5, 45–55.

Katzenberg, M.A., Pfeiffer, S., 1995. Nitrogen isotope evidence for weaning age in a nineteenth century canadian skeletal sample. In: Grauer, A.L. (Ed.), Body of evidence. Wiley-Liss, New York, pp. 221–235.

Klein, R.G., 2000. Archeology and the evolution of Human behavior. Evolutionary Anthropology 9(1), 17–36.

Kurtén, B., 1976. The cave bear story, life and death of a vanished animal. Columbia University Press, New York.

Lalueza Fox, C., Pérez-Pérez, A., 1993. The diet of the Neanderthal child Gibraltar 2 (Devil's Tower) through the study of the vestibular striation pattern. Journal of Human Evolution 24, 29–41.

Langbroek, M., 2001. Debating neandertals and modern humans in Late Pleistocene Europe. Archaeological Dialogues 8(2), 123–151.

Longin, R., 1971. New method of collagen extraction for radiocarbon dating. Nature 230, 241–242.

Marean, C.W., 1998. A critique of the evidence for scavenging by Neandertals and early modern humans: new data from Kobeh Cave (Zagros Mountains, Iran) and Die Kelders Cave 1 layer 10 (South Africa). Journal of Human Evolution, 35, 111–136.

Marean, C.W., Assefa, Z., 1999. Zooarcheological evidence for the faunal exploitation behavior of Neandertals and early modern humans. Evolutionary Anthropology 8, 22–37.

Mellars, P., 1989. Major issues in the emergence of Modern Humans. Current Anthropology 30, 349–385.

Nielsen-Marsh, C.M., Richards, M.P., Hauschka, P.V., Thomas-Oates, J.E., Trinkaus, E., Pettitt, P.B., Karavanic, I., Poinar, H., Collins, M.J., 2005. Osteocalcin protein sequences of Neanderthals and modern primates. Proceedings of the National Academy of Sciences 102(12), 4409–4413.

Patou, M., 1989. Subsistance et approvisionnement au Paléolithique moyen. In: Freeman, L., Patou, M. (Eds.), L'Homme de Neandertal, vol.6. La subsistance, Etudes et Rercherches Archéologiques de l'Université de Liège (E.R.A.U.L.), 33, Liège, pp. 11–18.

Patou-Mathis, M., 2006. Comportements de subsistance des Néandertaliens du niveau châtelperronien de Saint-Césaire (Charente-Maritime). Munibe (Antropologia-Arkeologia) 57, 197–204.

Phillips, D.L., Koch, P.L, 2002. Incorporating concentration dependence in stable isotope mixing models. Oecologia 130, 114–125.

Pike-Tay, A., Valdés, V.C., de Quiros, F.B., 1999. Seasonal variations of the Middle-Upper Paleolithic transition at El Castillo, Cueva Morin and El Pendo (Cantabria, Spain). Journal of Human Evolution 36, 283–317.

Pritchard, G.T., Robbins, C.T., 1990. Digestive and metabolic efficiencies of grizzly and black bears. Canadian Journal of Zoology 68, 1645–1651.

Richards, M.P., Hedges, R.E.M., 2003. Variations in bone collagen $\delta^{13}C$ and $\delta^{15}N$ values of fauna from Northwest Europe over the last 40 000 years. Palaeogeography, Palaeoclimatology, Palaeoecology 193, 261–267.

Richards, M.P., Pettitt, P.B., Trinkaus, E., Smith, F.H., Paunovic, M., Karavanic, I., 2000. Neandertal diet at Vindija and Neandertal predation: the evidence from stable isotopes. Proceedings of the National Academy of Sciences 97, 7663–7666.

Rockwell, D., 1991. Giving voice to bear. North American Indian myths, rituals, and images of the bear. Roberts Rinehart, Lanham.

Rode, K.D., Robbins, C.T., Shipley, L.A., 2001. Constraints on herbivory by grizzly bears. Oeocologia 128, 62–71.

Rodière, É., Bocherens, H., Angibault, J.-M., Mariotti, A., 1996. Particularités isotopiques chez le chevreuil (*Capreolus capreolus* L.): implications pour les reconstitutions paléoenvironnementales. Comptes Rendus de l'Académie des Sciences, Série II, Paris 323, 179–185.

Roebroeks, W., 2001. Hominid behaviour and the earliest occupation of Europe. Journal of Human Evolution 41(5), 437–461.

Schoeninger, M.J., DeNiro, M.J., 1984. Nitrogen and carbon isotopic composition of bone collagen from marine and terrestrial animals. Geochimica et Cosmochimica Acta 48, 625–639.

Schoeninger, M.J., Iwaniec, U.T., Glander, K.E., 1997. Stable isotope ratios indicate diet and habitat use in New World monkeys. American Journal of physical Anthropology 103, 69–83.

Skinner, M., 1996. Development stress in immature Hominines from Late Pleistocene Eurasia: evidence from enamel hypoplasia. Journal of Archaeological Science 23, 833–852.

Smith, F.H., Trinkaus, E., Pettitt, P.B., Paunovic, M., 1999. Direct radiocarbon dates for Vindija G1 and Velika Pecina Late Pleistocene hominid remains. Proceedings of the National Academy of Sciences 96, 12281–12286.

Speth, J.D., Tchernov, E., 2001. Neandertal hunting and meat-processing in the Near East, evidence from Kebara Cave. In: Stanford, C.B., Bunn, H.T. (Eds.), Meat-eating and human evolution. Oxford University Press, Oxford, pp. 52–72.

Stiner, M.C., 1994. Honor among thieves, a zooarchaeological study of Neandertal ecology. Princeton University Press, Princeton.

Stiner, M.C., 1998. Mortality analysis of Pleistocene bears and its paleoanthropological relevance. Journal of Human Evolution 34, 303–326.

Strauss, L.G., 2005. A mosaic of change: the Middle-Upper Paleolithic transition as viewed from New Mexico and Iberia. Quaternary International 137, 47–67.

Trinkaus, E., Moldvan, O., Milota, S., Bilgar, A., Sarcina, L., Athreya, S., Bailey, S.E., Rodrigo, R., Mircea, G., Higham, T., Bronk Ramsey, C., van der Plicht, J., 2003. An early modern human from the Pestera cu Oase, Romania. Proceedings of the National Academy of Sciences 100 (20), 11231–11236.

van Andel, T.H., 2003. Glacial environments I: the Weichselian climate in Europe between the end of OIS-5 interglacial and the Last Glacial Maximum. In: van Andel, T.H., Davis, W. (Eds.), Neanderthals and modern humans in the European landscape during the last glaciation. McDonald Institute Monographs, Oxbow Books, Oxford, pp. 9–19.

van der Merwe, N.J., Lee-Thorp, J.A., Thackeray, J.F., Hall-Martin, A., Kruger, F.J., Coetzee, H., Bell, R.H.V., Lindeque, M., 1990. Source-area determination of elephant ivory by isotopic analysis. Nature 346, 744–746.

Vogel, J.C., Van der Merwe, N.J., 1977. Isotopic evidence for early maize cultivation in New York State. American Antiquity 42(2), 238–242.

20. Stable Isotope Evidence for European Upper Paleolithic Human Diets

Michael P. Richards
Department of Human Evolution
Max Planck Institute for Evolutionary Anthropology
Deutscher Platz 6
D-04103 Leipzig, Germany
Richards@eva.mpg.de

Keywords Stable isotopes • carbon • nitrogen • diet • aquatic foods • marine foods • Upper Palaeolithic

Abstract This paper presents the published and unpublished stable carbon and nitrogen isotope values for 36 European Upper Paleolithic humans from 20 sites. The isotope data were measured to determine the sources of dietary protein in Upper Paleolithic diets; the evidence indicates that animal, not plant, protein was the dominant protein source for all of the humans measured. Interestingly, the isotope evidence shows that aquatic (marine and freshwater) foods are important in the diets of a number of individuals throughout this period.

Introduction

The use of the stable isotope ratios of carbon and nitrogen to reconstruct human diets has been widely applied in archaeology since the late 1970s (see Katzenberg, 2000; Sealy, 2001; Lee-Thorp and Sponheimer, 2006 for general reviews of the method). Bone collagen carbon and nitrogen stable isotope ratios are related to the carbon and nitrogen isotope values of protein consumed over the lifetime of a human or animal (Ambrose and Norr, 1993). The isotope ratios can differentiate between some general categories, indicating whether the main dietary protein was from marine or terrestrial sources ($\delta^{13}C$ value) (Chisholm et al., 1982; Schoeninger et al., 1983) and also whether the main protein sources were plant or animal ($\delta^{15}N$ value, especially when there is also comparative data from contemporary fauna from the site) (Schoeninger and DeNiro, 1984). It can also indicate the consumption of freshwater foods ($\delta^{15}N$), but only when consumed in significant quantities. Isotope analysis may appear to provide limited information, however, in the right circumstance and addressing the right questions, it can be a powerful tool for reconstructing past subsistence adaptations.

The majority of the applications of isotope analysis have been to Holocene humans and animals, where there are often large numbers of skeletal samples available for analysis, and where the results of isotope analysis can provide often dramatic evidence of changing diets through time, such as at the adoption of agriculture in Europe (Tauber, 1981; Richards et al., 2003) and North America (Vogel and van der Merwe, 1977). For the Paleolithic, carbon and nitrogen isotope analysis has so far only been applied to Neanderthals and modern humans – not earlier hominids. This is largely due to the fact that bone collagen, the material extracted and analysed, usually only survives in bone and teeth from samples which are less than a hundred thousand years old, and often only when the sample has been exceptionally preserved in a relatively temperate or cold environment. This precludes the measurements of earlier *Homo* in Europe, Africa, and Asia, as well as australopithecines from Africa. However, the application of this method to Neanderthals has provided key information on Neanderthal subsistence adaptations, showing that in each case, they were likely top-level carnivores likely consuming large herbivores, but did not consume significant quantities of aquatic (freshwater or marine) foods (Bocherens et al., 1991; Richards et al., 2000b; Bocherens et al., 2001; Richards et al., 2001; Bocherens and Drucker, 2003). Due to the better preservation of material, and the survival of more specimens, there have been many more applications of the method to Upper Paleolithic humans in Europe, often linked with radiocarbon dating projects, as the extraction method for collagen for ^{14}C dating and isotope analysis is often identical. Increasingly, isotope measurements and other checks on the collagen quality are undertaken to ensure the extracted collagen is well enough preserved for dating.

In this paper I present the carbon and nitrogen isotope values for Upper Paleolithic humans from Europe, including both unpublished and previously published data. I then consider the changes in isotope value (and therefore diets) chronologically throughout the Upper Paleolithic to look for evidence of general dietary patterns, as well as dietary diversity, throughout the Upper Paleolithic.

J.-J. Hublin and M.P. Richards (eds.), *The Evolution of Hominin Diets: Integrating Approaches to the Study of Palaeolithic Subsistence*, 251–257.

Upper Paleolithic Human Isotope Values

To date, carbon and nitrogen isotope measurements have been produced for 36 Upper Paleolithic humans from 20 European sites. This can further be subdivided into 16 individuals dating to the Early Upper Paleolithic (defined here as pre-Late Glacial Maximum period, before 20 ka), including two dating to before 30,000 BP, 14 later Gravettian humans, and 20 individuals from the Late Upper Paleolithic (post-LGM) period. The sites, isotope values, and sources of the data are given in Table 20.1. The sites are from a wide range of European countries and environments, including inland and coastal sites as well as sites close to major freshwater river sources.

To better understand the data, and to look for changes in diets through time, the individual results are plotted chronologically below. First, the Early Upper Paleolithic humans are discussed, followed by the Late Upper Paleolithic. Finally, all of the data are considered together, to illustrate the changing dietary adaptations of modern humans throughout the entirety of the Upper Paleolithic.

Early Upper Paleolithic (ca. 35–20 ka)

The Early Upper Paleolithic data have been plotted chronologically in Fig. 20.1a, b. There are fewer individuals plotted on the second graph (Fig. 20.1b), as $\delta^{15}N$ values were not measured for all of the humans.

In Europe a human $\delta^{13}C$ value of approximately −12 ± 1‰ would indicate that almost all of the dietary protein in their diet came from marine foods, whereas a value of approximately −20 ± 1‰ would indicate a diet of mainly terrestrial foods. Using a simple mixing model (i.e. Schwarcz, 1991) then a value of −16‰ would indicate a diet of approximately 50% marine and 50% terrestrial protein. The carbon isotope data for the Early Upper Paleolithic humans (Fig. 20.1a) show that most of the humans had a diet in which all of the protein came from mainly terrestrial sources (i.e. ca. −20‰). Two individuals from the only two coastal sites from this time period, Arene Candide and La Rochette, have $\delta^{13}C$ values that are shifted away from the terrestrial end-point and towards the marine end-point, indicating some

TABLE. 20.1. Table of bone collagen carbon and nitrogen isotope values for European Upper Paleolithic humans.

Site/individual	Country	$\delta^{13}C$	$\delta^{15}N$	Age	Source
Duruthy	France	−19.4	n/a	ca. 11,150 BP	Hayden et al. 1987
Sandalja	Croatia	−20.8	13.0	11,025 ± 60 (KIA 23,489)	Richards et al. unpublished data
Cap Blanc	France	−18.8	n/a	ca. 12,000 BP	Hayden et al. 1987
Kendrick's	UK	−18.1	13.7	11,760 ± 90	Richards et al. 2005
Kendrick's	UK	−17.7	13.9	12,090 ± 90	Richards et al. 2005
Kendrick's	UK	−18.0	13.4	11,930 ± 90	Richards et al. 2005
Kendrick's	UK	−17.9	13.8	11,880 ± 90	Richards et al. 2005
Gough's Cave	UK	−18.6	8.0	11,820 ± 120	Richards et al. 2000
Gough's Cave	UK	−18.5	7.1	ca. 12 ka	Richards et al. 2000
Gough's Cave	UK	−18.6	6.5	12,300 ± 100	Richards et al. 2000
Gough's Cave	UK	−19.1	5.4	11,700 ± 100	Richards et al. 2000
Gough's Cave	UK	−19.2	6.2	11,480 ± 90	Richards et al. 2000
Sun Hole Cave	UK	−19.8	7.2	12,210 ± 160	Richards et al. 2000
La Madelaine	France	−20.0	n/a	ca. 13,000 BP	Hayden et al. 1987
Saint-Germain-la-Riviere	France	−19.2	10.2	15,780 ± 200 (GifA95456)	Drucker and Gambier 2005
Neuessing	Germany	−19.7	11.6	16,200 cal BC	Grupe et al. 2003
Abri Pataud	France	−20.4	n/a	ca. 18,000 BP	Hayden et al. 1987
Abri Pataud	France	−19.9	n/a	ca. 18,000 BP	Hayden et al. 1987
Abri Pataud	France	−20.3	n/a	ca. 18,000 BP	Hayden et al. 1987
Abri Pataud	France	−20.0	n/a	ca. 18,000 BP	Hayden et al. 1987
Kostenki 18	Russia	−19.1	13.1	21,020 ± 180 (OxA-7128)	Richards et al. 2001
Dolni Věstonice 35	Czech Rep.	−18.8	12.3	22,840 ± 200 (OxA-8292)	Richards et al. 2001
Arene Candide	Italy	−17.6	12.4	23,440 ± 190 BP (OxA-10700)	Pettitt et al. 2003
Brno-Francouzská 2	Czech Rep.	−19.0	12.3	23,680 ± 200 (OxA-8293)	Richards et al. 2001
Sunghir 3	Russia	−18.9	11.3	24,100 ± 240 (OxA-9038)	Richards et al. 2001
Sunghir 2	Russia	−19	11.2	23,830 ± 220 (OxA-9037)	Richards et al. 2001
Sunghir 1	Russia	−19.2	11.3	22,930 ± 200 (OxA-9036)	Richards et al. 2001
Eel Point	UK	−19.7	11.4	24,470 ± 110 BP (OxA-14164)	Schulting et al. 2005
La Rochette	France	−17.1	11.2	ca. 25 ka	Orschiedt 2002 and unpublished
Paviland 1	UK	−18.4	9.3	25,840 ± 280 (OxA-8025)	Richards et al. 2001
Cro-Magnon	France	−19.4	n/a	ca. 25–30,000 BP	Hayden et al. 1987
Cro-Magnon	France	−19.5	n/a	ca. 25–30,000 BP	Hayden et al. 1987
Cro-Magnon	France	−19.6	n/a	ca. 25–30,000 BP	Hayden et al. 1987
Cro-Magnon	France	−19.9	n/a	ca. 25–30,000 BP	Hayden et al. 1987
Kostenki 1	Russia	−18.2	15.3	32,600 ± 1100 (OxA-7073)	Richards et al. 2001
Oase	Romania	−18.8	14.2	34,950 + 990, −890 BP	Trinkaus et al. 2003 and unpublished

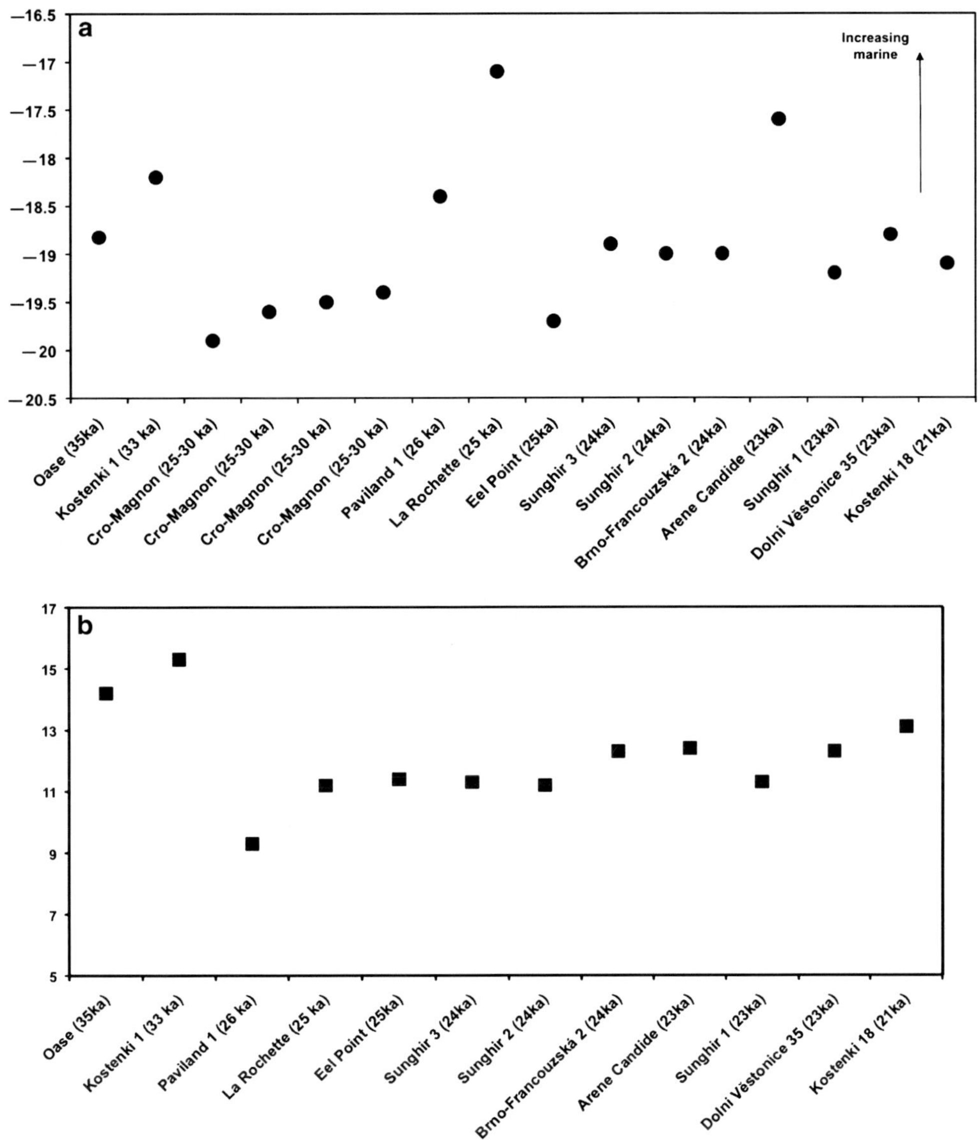

FIG. 20.1. Chronological plot of carbon (**a**) and nitrogen (**b**) isotope values of European Early Upper Paleolithic humans.

consumption of marine foods. Using a simple mixing model we can determine that approximately 20–25% of the protein in the diets of the Arene Candide ($\delta^{13}C = -17.6‰$) and La Rochette ($\delta^{13}C = -17.1‰$) humans was from marine foods. This could be indicative of a seasonal intensive use of marine foods in the course of a year or of regular consumption of some marine foods.

The nitrogen isotope data (Fig. 20.1b) from this time period is particularly interesting. Most of the humans have values that are best interpreted as deriving from the consumption of animal protein, namely terrestrial herbivores, as the main source of dietary protein. This is based on the comparison between the $\delta^{15}N$ values of these humans, and the $\delta^{15}N$ values of herbivores and carnivores from a number of Eurasian sites dating to this time period. However, as most of the human values presented here are from sites where there have been no isotopic measurements of associated fauna, the interpretation of the $\delta^{15}N$ values as indicating a diet of mainly terrestrial protein must be considered preliminary.

A number of individuals have higher $\delta^{15}N$ values than would be expected from a diet in which the protein derived purely from terrestrial herbivores. Of particular interest are the two earliest individuals from the sites of Oase and Kostenki 1, which both have very high $\delta^{15}N$ values. These elevated values are indicative of the consumption of large amounts (proportionally) of freshwater foods, likely fish, which is a reasonable assumption as both of these sites are close to freshwater sources. These nitrogen isotope values are similar to those reported for Mesolithic humans found along the Danube River at the Iron Gates region (Bonsall et al., 1997), where the $\delta^{15}N$ values were also very high and the sites contained large numbers of fish remains from species such as

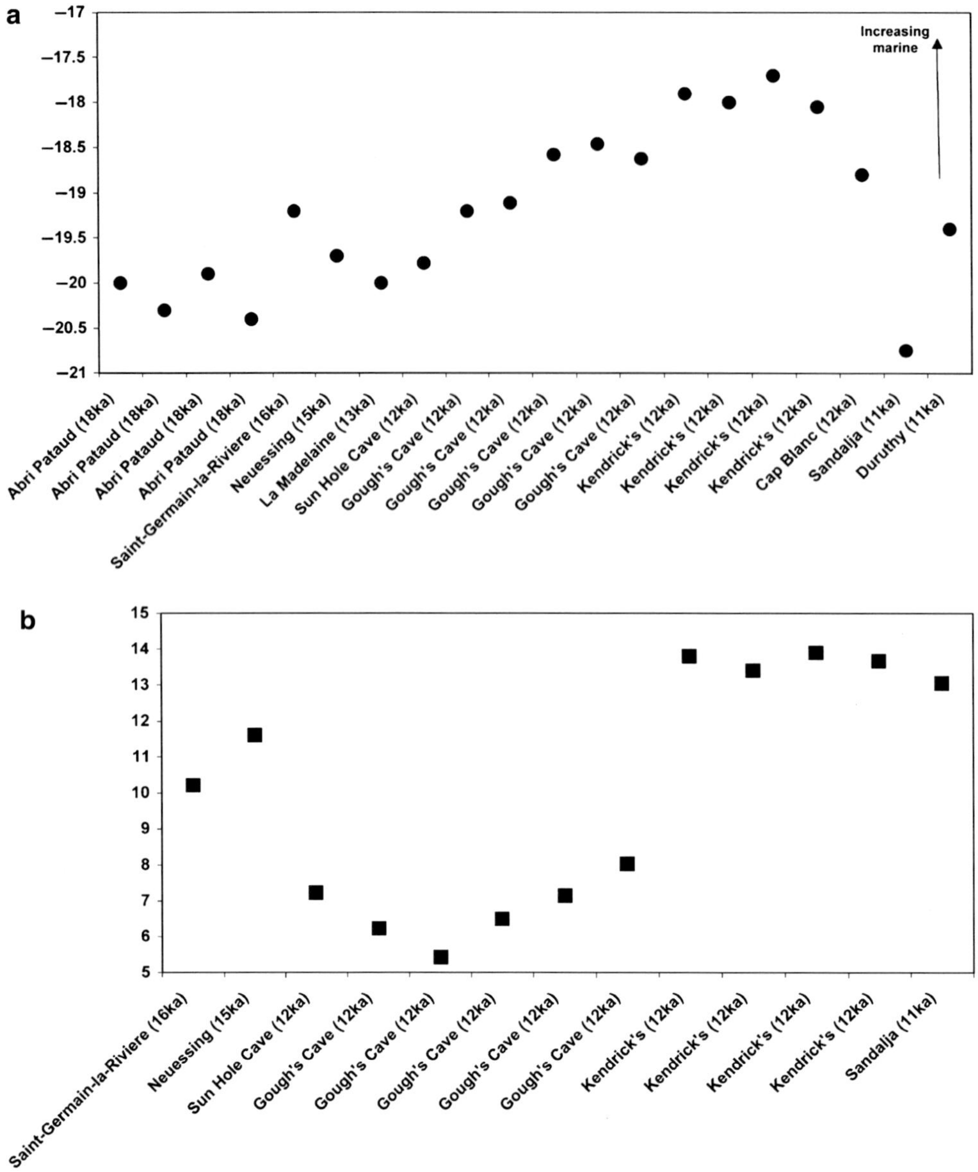

FIG. 20.2. Chronological plot of carbon (**a**) and nitrogen (**b**) isotope values of European Late Upper Paleolithic humans.

sturgeon. There is no evidence that there was an unusual phenomenon where the underlying terrestrial fauna had elevated $\delta^{15}N$ values (Richards and Hedges, 2003) at these sites and at the time periods when these humans were alive, so the most parsimonious interpretation of this data is the consumption of large amounts (i.e. over 50% of dietary protein) of freshwater fish. The humans from Arene Candide and La Rochette also have elevated $\delta^{15}N$ values, which, as discussed above, are also from the consumption of aquatic protein, in this case marine fish, as indicated by the associated $\delta^{13}C$ values.

Late Upper Paleolithic (ca. 20–10 ka)

The late Upper Paleolithic data have been plotted chronologically in Fig. 20.2a, b. There are more data available from this time period, reflecting the larger number of human remains that have been recovered from this later period.

The carbon isotope data (Fig. 20.2a) largely indicate that most of the humans were consuming terrestrial protein, as was also seen in the Early Upper Paleolithic. However, as also observed for the earlier period, a number of individuals have carbon isotope values indicating significant consumption of marine foods. In some cases these individuals have values that are more shifted towards the theoretical marine endpoint, indicating a higher proportion of marine protein in the diet (at the site of Kendrick's Cave, for example) than in the Early Upper Paleolithic.

The nitrogen isotope data (Fig. 20.2b) are spread over a wide range of values, although there are no individuals that have nitrogen isotope values as high as observed in the two earliest Upper Paleolithic humans (Oase and Kostenki 1).

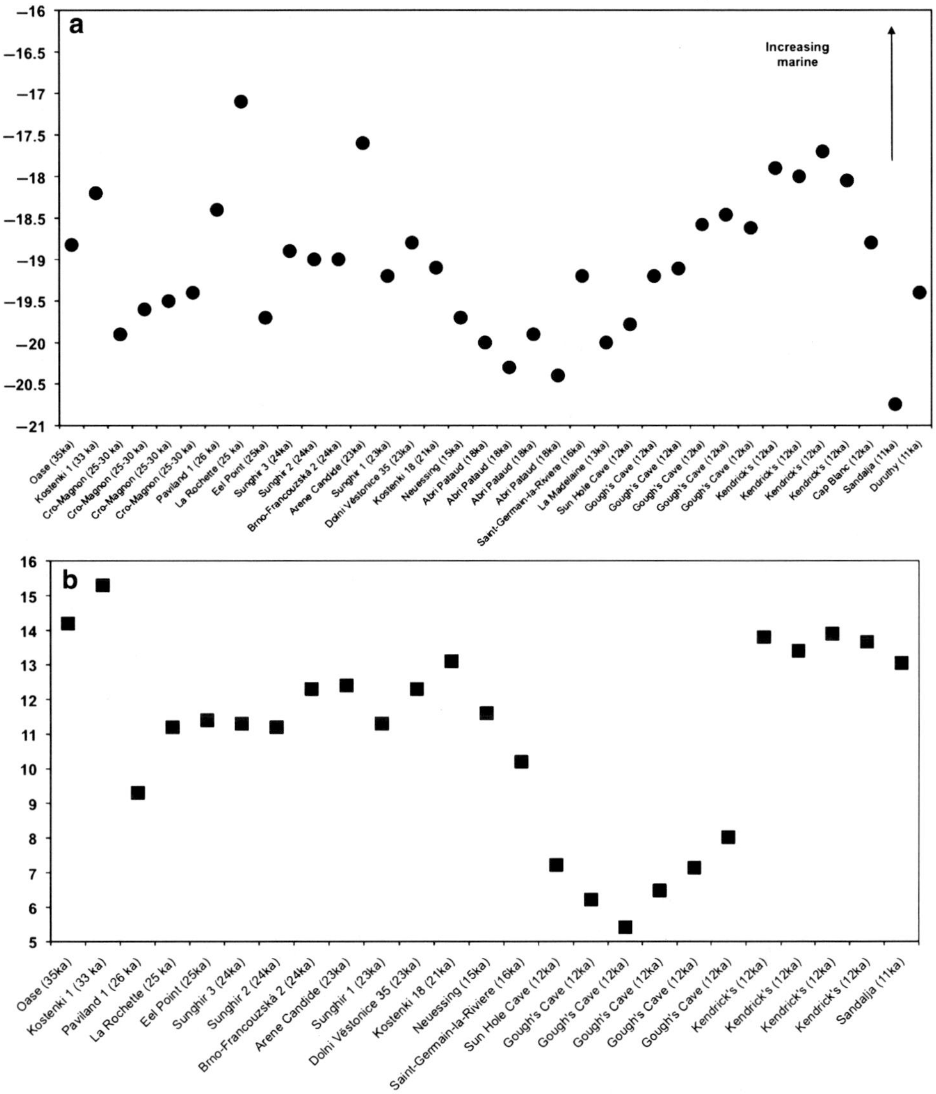

FIG. 20.3. Chronological plot of carbon (**a**) and nitrogen (**b**) isotope values of all Upper Palaeolithic humans in Europe.

The humans with high $\delta^{15}N$ values from Kendrick's Cave also have shifted $\delta^{13}C$ values; their elevated $\delta^{15}N$ values are most likely due to marine protein consumption. The human from Sandalja has a high $\delta^{15}N$ value, but not a shifted $\delta^{13}C$, so the elevated $\delta^{15}N$ value is likely due to the consumption of a significant amount of freshwater fish.

The individuals from the sites of Sun Hole Cave and Gough's Cave dating to approximately 13–12 ka have relatively very low nitrogen values, which could be misinterpreted as indicating a diet of mainly terrestrial plant protein. However, this is due to a phenomenon observed in Northern Europe during this period in which all of the mammal $\delta^{15}N$ values are lower than in the previous and subsequent periods, likely linked to climate changes (Richards and Hedges, 2003). Analysis of fauna associated with the humans from Gough's and Sun Hole caves (Richards et al., 2000a) showed that they were lower than other herbivores, averaging approximately 3‰ therefore, these human values were interpreted as indicating a diet where dietary protein mainly came from terrestrial herbivores.

What is especially interesting is the diversity of carbon and nitrogen isotopes values from roughly contemporary sites in similar geographic areas. This is especially evident when we compare the isotope data from Kendricks Cave and Gough's Cave. These sites, and humans, are located relatively close to each other geographically, and date to a similar time period (although Gough's is slightly older). However, their diets are very different; the Gough's Cave humans derived most of their dietary protein from terrestrial herbivores, likely deer and/or cattle (Richards et al., 2000a), while the Kendrick's humans had up to 25% of their dietary protein from marine foods, likely marine mammals (Richards et al., 2005). This shows that there was a range of dietary and subsistence adaptations by Late Upper Paleolithic humans, and not just a single type of adaptation. Interestingly, the Kendrick's Cave data also marks the start of specialization in marine foods in the UK that intensifies in the subsequent Mesolithic period, in which humans obtained up to 50% (Schulting and Richards, 2002), and even up to 100% (Richards and Mellars, 1998) of their dietary protein from marine sources.

Discussion

All of the human isotope data are plotted together in Fig. 20.3a, b to highlight the trends through time. There are some clear patterns in the data. There are only two humans dating to the earliest period of modern human presence in Europe. These humans from Oase and Kostenki 1 have the highest $\delta^{15}N$ value seen in any of the Upper Paleolithic humans; these values indicate a significant consumption of freshwater fish. This use of aquatic foods is also clearly seen in the subsequent Gravettian period, where a number of individuals have elevated $\delta^{15}N$ values that likely indicate some consumption of aquatic foods – this is particularly the case for the humans from Arene Candide and La Rochette, who likely derived up to 20% of their dietary protein from marine foods. In the Late Upper Paleolithic we see humans with the clearest evidence of marine food consumption at the site of Kendrick's Cave. Yet, we also see the persistence of the consumption of herbivore protein, likely from hunted animals as the main form of subsistence. This is the main form of subsistence throughout the Upper Paleolithic, and there are only a few cases in which this differs due to the consumption of aquatic foods.

Summary and Conclusions

This paper presented published and unpublished bone collagen carbon and nitrogen stable isotope data for European Upper Paleolithic humans. The isotope data shows that animal protein, mainly from terrestrial herbivores, was the dominant source of dietary protein for most humans in this period. However, in some circumstances, aquatic protein was an important source of dietary protein. This was the case for the earliest modern humans in Europe, where freshwater resources were important, in the Gravettian period when marine foods contributed up to 20% of dietary protein at two coastal sites, and in the Late Upper Paleolithic, where there is evidence for higher amounts of marine protein consumption at a coastal site. Clearly, though, animal protein was the main source of dietary protein for all humans, and must have been obtained through active hunting (and fishing). Plant foods were not an important part of diet, at least in terms of dietary protein, which means that gathering protein-rich plants like hazelnuts was unlikely to have been an important subsistence activity in the Upper Paleolithic, at least for the humans measured here.

There is a great need, however, for more results to see if this picture of an animal-protein rich diet holds for all regions in the Upper Paleolithic of Europe. What are especially needed are more results from warmer Mediterranean regions, where plant foods would have been more readily available than in Northern Europe. As the work progresses, it will also be important to compare these results to those from the preceding Neanderthals, and also to hunter-gatherers in Holocene Europe, to see if, and how, Upper Paleolithic peoples may have adapted differently in the same geographic regions.

Acknowledgments. I would like to thank my colleagues Erik Trinkaus, Paul Pettitt, Ivor Karavaniç and Jörg Orscheidt for permission to use unpublished data arising from collaborative projects. I would especially like to thank Jean-Jacques Hublin for providing the opportunity to organise this conference.

References

Ambrose, S.H., Norr, L., 1993. Experimental evidence for the relationship of the carbon isotope ratios of whole diet and dietary protein to those of bone collagen and carbonate. In: Lambert, J., Grupe, G. (Eds.), Prehistoric Human Bone: Archaeology at the Molecular Level. Springer, New York, pp. 1–37.

Bocherens, H., Drucker, D., 2003. Reconstructing Neandertal diet from 120,000 to 30,000 BP using carbon and nitrogen isotopic abundances. In: Patou-Mathis, M., Bocherens, H. (Eds.), Le rôle de l'environnement dans les comportements des chasseurs-cueilleurs préhistoriques. British Archaeological Reports International Series, 1105, pp. 1–7.

Bocherens, H., Fizet, M., Mariotti, A., Lange-Badre, B., Vandermeersch, B., Borel, J.P., Bellon, G., 1991. Isotopic biogeochemistry (13C, 15N) of fossil vertebrate collagen: application to the study of a past food web including Neanderthal man. Journal of Human Evolution 20, 481–492.

Bocherens, H., Billiou, D., Mariotti, A., Toussaint, M., Patou-Mathis, M., Bonjean, D., Otte, M., 2001. New isotopic evidence for dietary habits of Neandertals from Belgium. Journal of Human Evolution 40(6):497–505.

Bonsall, C, Lennon, R., McSweeney, K, Stewart, C., Harkness, D., Boroneant, V., Bartosiewicz, V., Payton, R., Chapman, J., 1997. Mesolithic and early Neolithic in the Iron Gates: A palaeodietary perspective. Journal of European Archaeology 5.1, 50–92.

Chisholm, B.S., Nelson, D.E., Schwarcz, H.P., 1982. Stable carbon ratios as a measure of marine versus terrestrial protein in ancient diets. Science 216, 1131–1132.

Drucker, D., Gambier, D., 2005. Determination of the dietary habits of a Magdalenian woman from Saint-Germain-la-Rivie're in south-western France using stable isotopes. Journal of Human Evolution 49, 19–35.

Grupe, G., Mikić, Z., Peters, J., Manhart, H., 2003. Vertebrate food webs and subsistence strategies of Meso- and Neolithic populations of central Europe. In: Grupe, G., Peters, J. (Eds.), Decyphering Ancient Bones: The Research Potential of Bioarchaeological Collections. Verlag Marie Leidorf, Rahden, Germany, pp. 193–213.

Hayden, B., Chisholm, B., Schwarcz, H.P., 1987. Fishing and foraging: marine resources in the Upper Palaeolithic of France. In: Soffer, O. (Ed.), The Pleistocene Old World: Regional Perspectives, Plenum Press, New York, pp. 279–291.

Katzenberg, M.A., 2000. Stable isotope analysis: a tool for studying past diet, demography and life history. In: Katzenberg, M.A., Saunders, S.R. (Eds.), The Biological Anthropology of the Human Skeleton. Wiley, New York, pp. 305–327.

Lee-Thorp, J.A., Sponheimer, M., 2006. Contributions of biogeochemistry to understanding hominin dietary ecology. Yearbook of Physical Anthropology 49, 131–148.

Orschiedt, J., 2002. Datation d'un vestige humain provenant de la Rochette (Saint-Leon-Sur-Vezere) par la methode du carbone 14 en spectrometrie de masse. Paléo 14, 15.

Pettitt, P.B., Richards, M.P., Maggi, R. and V. Formicola (2003). The Gravettian burial known as the Prince ('Il Principe'): new evidence for his age and diet. Antiquity 295, 15–19.

Richards, M.P., Hedges, R.E.M., 2003. Bone collagen δ^{13}Cand δ^{15}N values of fauna from Northwest Europe reflect palaeoclimatic variation over the last 40,000 years. Palaeogeography, Palaeoclimatology, Palaeoecology 193, 261–267.

Richards, M.P., Mellars, P., 1998. Stable isotopes and the seasonality of the Oronsay middens. Antiquity 72 (275), 178–184.

Richards, M.P., Jacobi, R., Currant, A., Stringer, C., Hedges, R.E.M., 2000a. Gough's Cave and Sun Hole Cave human stable isotope values indicate a high animal protein diet in the British Upper Palaeolithic. Journal of Archaeological Science 27, 1–3.

Richards, M.P., Pettitt, P.B., Trinkaus, E., Smith, F.H., Karavanić, I., Paunović, M., 2000b. Neanderthal diet at Vindija and Neandertal predation: the evidence from stable isotopes. The Proceedings of the National Academy of Sciences, USA 97, 7663–7666.

Richards, M.P., Pettitt, P.B., Stiner, M.C., Trinkaus, E., 2001. Stable isotope evidence for increasing dietary breadth in the European mid-Upper Paleolithic, The Proceedings of the National Academy of Sciences, USA 98, 6528–6532.

Richards, M.P., Schulting, R.J., Hedges, R.E.M., 2003. Sharp shift in diet at onset of Neolithic. Nature 425, 366.

Richards, M.P., Jacobi, R., Cook, J., Pettitt, P.B., Stringer, C.B., 2005. Isotope evidence for the intensive use of marine foods by Late Upper Palaeolithic humans. Journal of Human Evolution 49, 390–394.

Schoeninger, M., DeNiro, M., 1984. Nitrogen and carbon isotopic composition of bone collagen from marine and terrestrial animals. Geochimica et Cosmochimica Acta 48, 625–639.

Schoeninger, M., DeNiro, M., Tauber, H., 1983. Stable nitrogen isotope ratios of bone collagen reflect marine and terrestrial components of prehistoric human diet. Science 220, 1381–1383.

Schulting, R.J., Richards, M.P., 2002. Finding the coastal Mesolithic in Southwest Britain: AMS dates and stable isotope results on human remains from Caldey Island, South Wales. Antiquity 76, 1011–1025.

Schulting, R.J., Trinkaus, E., Higham, T., Hedges, R., Richards, M.P., Cardy, B., 2005. A Mid-Upper Palaeolithic human humerus from Eel Point, South Wales, UK. Journal of Human Evolution 48, 493–505.

Schwarcz, H.P., 1991. Some theoretical aspects of isotope paleodiet studies. Journal of Archaeological Science 18, 261–275.

Sealy, J., 2001. Body tissue chemistry and Palaeodiet. In: Brothwell, D.R., Pollard, A.M. (Eds.), Handbook of Archaeological Sciences. Wiley, Chichester, pp. 269–279.

Tauber, H., 1981. 13C evidence for dietary habits of prehistoric man in Denmark. Nature 292, 332–333.

Trinkaus, E., Moldovan, O., Milota, S., Bîlgar, A., Sarcina, L., Athreya, S., Bailey, S., Rodrigo, R., Mircea, G., Higham, T., Bronk-Ramsey, C., and J. van der Plicht 2003. An early modern human from the Pestera cu Oase, Romania PNAS 100, 11231–11236.

Vogel, J.C., van der Merwe, N.J., 1977. Isotopic evidence for eary maize cultivation in New York State. American Antiquity 42, 238–242.

Index